The Spatial Analysis of Radiocarbon Databases

The spread of the first farmers in Europe and of the fat-tailed sheep in Southern Africa

Thembi M. Russell

BAR International Series 1294
2004

This title published by

Archaeopress
Publishers of British Archaeological Reports
Gordon House
276 Banbury Road
Oxford OX2 7ED
England
bar@archaeopress.com
www.archaeopress.com

BAR S1294

The Spatial Analysis of Radiocarbon Databases: The spread of the first farmers in Europe and of the fat-tailed sheep in Southern Africa

ISBN 1 84171 644 8

Printed in England by The Basingstoke Press

All BAR titles are available from:

Hadrian Books Ltd
122 Banbury Road
Oxford
OX2 7BP
England
bar@hadrianbooks.co.uk

The current BAR catalogue with details of all titles in print, prices and means of payment is available free from Hadrian Books or may be downloaded from www.archaeopress.com

Contents

List of figures

List of Tables

Preface

Two spatially referenced radiocarbon databases, for the spread of Europe's first farmers and for the spread of the first sheep in southern Africa respectively, are analysed. Methods for visualising large-scale diffusion processes are compared using the European database. These include chronology maps (after Clark 1965), isochron maps (after Ammerman & Cavalli-Sforza 1984) and a new simulation using the probability distribution of calibrated dates. The patterns in the radiocarbon data are then quantified by using linear regression of both calibrated and uncalibrated dates to calculate rates of spread and possible departure points for the European data at two scales of analysis, continent-wide and by demic and cultural region (after Zvelebil and Lillie 2000). The new analytical technique, using the whole of a date's calibrated range, is then applied to the database for the spread of the first sheep in present day South Africa. Two competing hypotheses for the route of the spread, a western coastal route from present day Namibia southwards (Stow (1905), Cooke (1965)), and an interior route (Elphick 1977), are tested by analysing the spatial and temporal patterns in the radiocarbon data.

This work was submitted for the degree of Doctor of Philosophy in the University of Southampton in 2002. I thank James Steele for his help throughout my time there. I thank and acknowledge Tien Tran who helped me tremendously with all things computational. I thank Margo Russell, who spent much time reading my work, and through her Martin Russell .

Chapter 1: Introduction

Nine thousand years ago the first farming appeared in Greece, within four thousand years it had spread throughout Europe. Two thousand years ago the first sheep farming appeared in southern Africa. These processes are the setting for this research. There is a distinction that must be made for clarity. In southern Africa farming involving crop cultivation, livestock herding and iron working is restricted to the eastern half of the country. The spread of stock herding without plant cultivation or iron working technology is restricted to the Western half of southern Africa. Only this western spread of stock herding is studied here.

Though geographically and temporally distant and distinct, these processes raise questions that are strikingly similar. On both continents there is yet to be resolution about how these innovations spread, why they spread and by whom. Was farming carried with an exotic migrating people or was it adopted by and transmitted between indigenous hunter-gathering people? Numerous questions naturally lead on from this. Under what circumstances do hunter-gatherers become farmers? What impact would an immigrant population have had on a local population? At what rate did the spread take place? What drove the spread of farming? Can the spread be traced in the archaeological record at a resolution that is fine enough to understand its underlying mechanisms? Are the two mechanisms, the spread of people and the spread of ideas, distinguishable archaeologically?

It is within this framework of questions that this research arose. Presently, research into the European Neolithic has a regional focus often determined by modern day political boundaries. Since the early work of Clark (1965a/b) and Ammerman and Cavalli-Sforza (1971, 1984) there has been little synthesis at the continent-wide scale. To address this, a database of Neolithic dates was compiled with which to look at the spread of farming as a large-scale diffusion process. The natural starting point was to re-examine past work and methodologies that had worked at a similar scale of analysis, to find out whether the new data would challenge or confirm old theories. It became apparent, however, that before any answers could be found, the method by which radiocarbon dates are used in spatial analysis required scrutiny. The consequence is that the research has evolved into an examination of how to use radiocarbon dates in spatial analyses, a problem that needed to be addressed before there could be any progress towards tackling problems that are common to the Neolithic of Europe and the introduction of domestic stock to southern Africa.

In southern Africa, although the questions are the same, the limited nature of previous research into the spread of the domestic stock and the kind of data available make it different to the European case. The spread of domestic stock has not been previously examined using spatial mapping techniques nor had a spatially referenced database of its spread been assembled. This meant that there was not the body of previous research upon which to build and to use in a comparative analysis. There are other differences too. The number of dates associated with the spread of domestic stock is small. This has both advantages and disadvantages. Because there are few dates, the process of compiling the database gave a far more intimate knowledge of the process at the site level (this experience gave me a greater appreciation of the potential problems that might be associated with the data collected for the Neolithic database over which I had no control). The disadvantage with the southern African database is that for statistical purposes a smaller database is likely to be far more sensitive to errors and outliers in the data. So in many ways these studies are complementary: in both cases the traces of the spread of a new process are being followed. For the European database the sample is large with little knowledge of the fine detail of the archaeology at site level. For the southern African database the opposite is true: the sample is small but the archaeology at the level of site stratigraphy has been thoroughly examined.

The questions posed in the research can be summarised as follows:

1. Are there better ways of examining spatially and temporally referenced data than those that past and current research continue to use? Can answers be found by improving the methods of analysis or are the answers simply beyond the reach of large-scale spatial-temporal analysis?
2. Does the recent larger body of data throw new light on the spread of farming across Europe?
3. Does the collection and manipulation of spatially and temporally referenced data throw new light on the spread of domestic stock to southern Africa?
4. Finally, can key questions about the mechanisms and rate of spread of farming and domestic animals be resolved by looking at this aspect of the archaeological record?

The remainder of the introduction outlines the chapters. Chapter Two describes the European and the southern African databases, the ingredients of the research. Chapter Three gives an overview of radiocarbon dating, the nature of calibrated dates and the problems of using them in statistical analyses. In Chapter Four the previous methods that have been used to visualise the spatial distribution of dates associated with the large scale spread of the first farming into Europe are reproduced and the results using the new database are compared to those from previous methods. Apart from providing direct evidence of how similar the patterns remain despite the increased sample size, this comparison provides tangible information on the composition of the new database (i.e. comparison of our database with that of Pinhasi *et al.* (2000) showed us where there might be gaps in our database). Chapter Five, Chapter Six and Chapter Seven present methods to quantify the patterns seen in the radiocarbon determinations and calibrated dates. In Chapter Five an approach that aims to locate the most likely origin point for a set of spatially referenced dated sites is given. Linear regression to calculate the rate of spread of the first farming sites across Europe is also discussed. Rate measurements at the continent wide scale have been commonly criticized for masking an underlying complex process. To examine this the European database is broken down in to smaller regional groups in Chapter Six. The demarcation of these groups is based on recent work in which Europe was divided into acculturation zones, where farming spread by cultural diffusion, and demic diffusion zones, where farming was spread by an immigrant population. In Chapter Seven the issue of calibration is addressed. A visualisation technique to display the entire probability distribution of a calibrated radiocarbon date is shown. A technique that enables the whole of the calibrated probability range of dates to be used in statistical analyses is also presented in this chapter. In Chapter Eight the fundamentals of the debate about the introduction of the first domestic stock to southern Africa are introduced. Then in Chapter Nine the analytical methods that were developed and tested on the European database are applied to the southern African data. Chapter Ten contains an appraisal of the methods and the findings of this research and presents ideas for work that would continue and improve on this research.

Chapter 2: The Databases

2.1 The European Mesolithic/Neolithic database:

Marina Gkiasta, from University College London, compiled the European database used in this research[1] (for clarity it will be referred to as the Gkiasta database in this research to distinguish it from other databases with European content). The aim was to obtain a geographically referenced database of radiocarbon determinations of European late Mesolithic and early Neolithic sites with the accompanying detailed information on sample provenance and site stratigraphy, using publications, online radiocarbon laboratory databanks and datasets held by universities and archaeological institutions. It was soon apparent that the detailed information on date provenance was available only in site reports. Time and funding constraints ruled out its retrieval. Indeed the project highlighted how inaccessible this kind of information is to the researcher. A similar problem exists in the USA. In his discussion of radiocarbon dating in the USA, Taylor (2000) notes that there are no comprehensive databases for the estimated 150 000 radiocarbon measurements to date for the US, a number which is rapidly increasing due to the heightened rate of processing made possible by the accelerator mass spectrometry (AMS) technique and also the large number of determinations that are being produced by commercial labs for environmental impact assessment reports (often not reported in conventional scientific literature). Archaeologists in Europe face a similar predicament. Few laboratories regularly publish date-lists in the journal *Radiocarbon* (the traditional publication point for radiocarbon data) so that, "...results are unpublished or are hidden away in obscure places, making it difficult for archaeologists to compile data of interest" (Bowman & Balaam 1990:317). There are some online databases available[2].

The Gkiasta database is confined to Europe here defined as the area to the west of an imaginary line running from the western edge of the Black sea to the eastern Baltic (Figure 2.1). Two cultural periods are encompassed, the Mesolithic (599 determinations[3]) and the Neolithic (1716 determinations) (Table 2.1).

Gkiasta aimed to collect the **latest** Mesolithic determinations and the **earliest** Neolithic determinations in each area of Europe. This becomes apparent in the figures in Chapter 4 where maps depicting the most recent time periods have the Neolithic confined to the British Isles. Of course the Neolithic was continuing elsewhere in Europe, but only the British Isle sites are shown in these figures as this part of Europe was still in the initial stages of Neolithisation (e.g. Figure 4.5, pg 28).

The assignment of the cultural period simply follows that given by the archaeologist in charge. The final database, available online, consists of five tables, summarised in Table 2.2[4].

Initially Gkiasta collected 4000 determinations, many of which she subsequently discarded. All determinations with a standard deviation greater than 150 years were excluded. Also excluded were determinations that had poor cultural association (following Gob 1990). Radiocarbon determinations obtained prior to 1980 are not included in the database. The resulting database has 2543 determinations. The online database and the work contained in this research use these 2543 determinations (a total which includes both Mesolithic and Neolithic determinations). These 2543 radiocarbon determinations span a 6500 year time period from nearly 10 000 years before present to 3370 years before present (Figure 2.2). Gaps occur in many fields (based on the difficulties in obtaining this information, as mentioned above). Many sites in the database do not yet have latitude and longitude coordinates, so they had to be excluded from the analyses.

---

[1] This data base compilation was undertaken in 2000 with funding from the UK Arts and Humanities Board to Stephen Shennan from University College London and James Steele from the University of Southampton.

[2] For example, a database of 2000 Neolithic and Bronze Age dates for Switzerland, Netherlands and Germany is available for download at Http://www.jungsteinsite.de/radon/radon.htm

[3] Uncalibrated dates are referred to as determinations to distinguish them from calibrated dates.

[4] The database is accessible online through the UK Archaeology Data Service (http://www.ads.ahds.ac.uk).

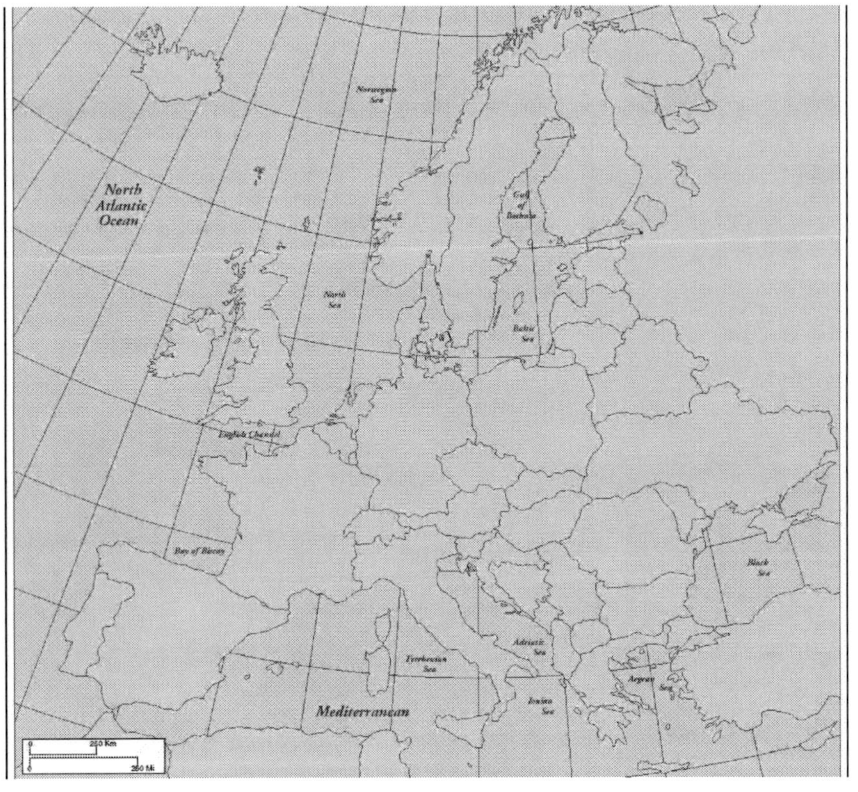

Figure 2.1 Europe as defined for the Gkiasta database: the area to the west of an imaginary line from the western shores of the Black sea to the eastern shores of the Baltic Sea.

Period Classification	Number of determinations
Early Mesolithic	24
Mesolithic	465
Middle Mesolithic	3
Late Mesolithic	107
Late Mesolithic to Early Neolithic transition	3
Meso/Neo	101
Epipalaeolithic	47
Middle Epipalaeolithic	1
Late Epipalaeolithic	1
Early Neolithic	641
Early Neolithic/Middle Neolithic	3
Middle Neolithic	350
Neolithic	593
Middle/Late Neolithic	8
Late Neolithic	16
Final Neolithic	1
Stone Age	2
Unclassified	177
Total	2543

Table 2.1 The Gkiasta database structure listed according to cultural period. The cultural period is the classification as assigned by the archaeologist in charge of the particular site.

Table	Table Fields
Sample	Sample ID, determination BP, standard deviation, calibrated dates ranges, treatment, delta ^{13}C correction, material, comments, date of process, source, lab code, submitters, lab type (AMS or conventional)[5],
References	References used to compile this dataset
Site	Sample ID, site name, site ID, area, country, cultural ID, site type, latitude, longitude, comments, period (this refers to the archaeological temporal characterisation of the site made by the archaeologist in charge), references, association
Context	Sample ID, context, context type
Site Phase[6]	Sample ID, site phase, absence or presence of pottery, domesticated animals and plants and human remains.

Table 2.2 The Gkiasta database is made up of five tables as summarized here.

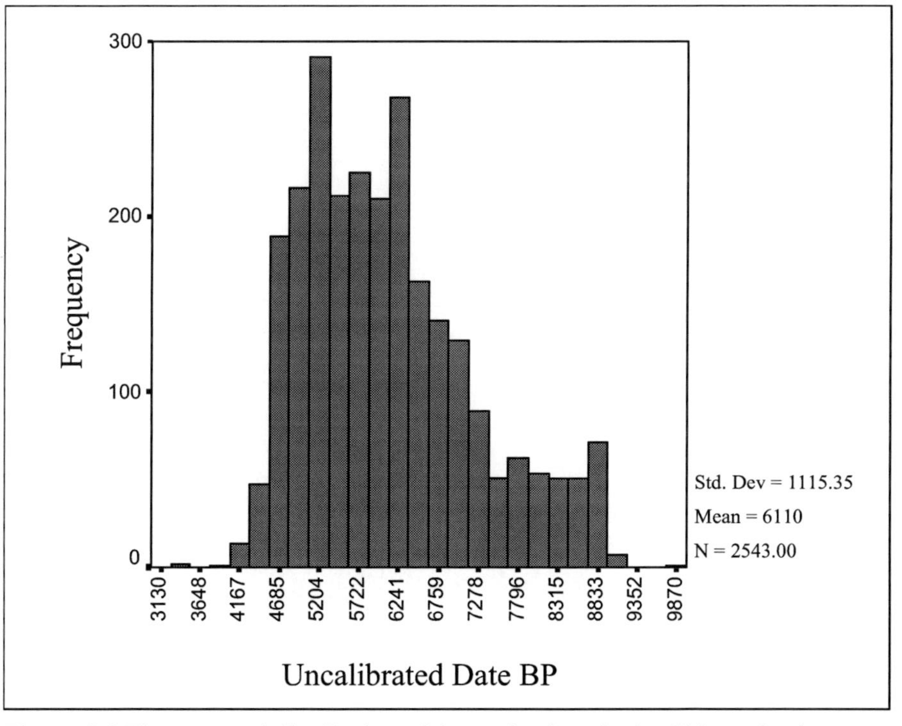

Figure 2.2 The temporal distribution of determinations in the Gkiasta database.

---

[5] Subsequent to the database compilation an error was discovered in the classification of the dating technique used; some AMS dates are incorrectly classified as conventional dates in the 'Sample' table.
[6] The 'Site Phase' and 'Context' tables must be used together as there is no consistency in the way these two terms have been used in the data sources.

2.1.1 *Selecting the Neolithic subsets to use in analyses*

There are 697 Neolithic sites (1705 determinations) in the Gkiasta database[7]. The maximum number of determinations at any site varies from 1 to 28 determinations per site (Figure 2.3).

Spatial analyses require a single point value (namely, a single date) at a particular set of longitude and latitude coordinates. For the work contained in this research, the interest is in the first occurrence of farming at each site. The date should ideally be the oldest one in a sequence of dates associated with farming. This becomes the date and place of the first farmers for that particular point on the map for all subsequent analyses. Of the 697 Neolithic sites in the database 367 have only one determination (Figure 2.3). The other 330 sites have two or more determinations. It is at the 330 sites with multiple determinations that the question of determination selection must be carefully considered.

The Gkiasta database does not contain sufficient information on the stratigraphic sequence of the determinations at a particular site. The only way of giving a temporal order at a site is by following the cultural period label, middle Neolithic, late Neolithic or Early Neolithic for example, as assigned by the archaeologist in charge. If there was stratigraphic information and if the relative order of the dates in the excavated stratigraphy were in agreement with their radiocarbon age, this would greatly strengthen the date's validity[8].

The correct way to select dates for the first occurrence of archaeological phenomena is unresolved. The main discussion of this problem is amongst those working on the first appearance of modern humans in Europe (Bocquet-Appel and Yves Demars 2000). Two questions summarize the problem; is it correct to take the earliest date from the batch of dates associated with each site, especially when there is no certainty on their stratigraphic relationship to one another? Is it more precise to take an average of the dates, even though this might lose the very information that is of particular interest, namely, the *first* occurrence of farming?[9].

Analyses concerned with the first occurrence of an archaeological phenomenon, be it the first appearance of modern humans on the European continent or the first appearance of domesticated plants in the Near East, are open to criticism when the oldest date is used without due consideration and justification. For those involved in this type of analysis the temptation is to take the oldest date, as this seems a rational choice when looking for the 'first'. A review of the literature shows that authors working on the spatial and temporal analysis of the first appearance of the Neolithic in Europe have all yielded to the temptation by taking the date from the lowest (or oldest) stratigraphic layer associated with the Neolithic.

In the absence of consensus on how to proceed, the decision was taken to run the analyses with a number of different datasets in order to assess whether and how the date selection process affects the results. If the results were to show very little difference then it could be argued that the exact selection method was unimportant.

Each dataset is selected from the 697 Neolithic sites using slightly different criteria. The reason for the shortfall in site numbers (e.g. 697 to 509 sites for Datasets A to D below) is that some sites do not yet have geographic coordinates.

The combinations are as follows:

Dataset A: Weighted means or oldest determination (509 sites)(Dataset A, Appendix 1). This list contains the weighted means of determinations (see discussion below). In cases where the determinations could not be averaged then the *oldest* determination at the site was selected.

Dataset B: Weighted means or most recent determination (509 sites)(Dataset B, Appendix 1). As above in Dataset A but for determinations that could not be averaged the *most recent* determination was selected.

---

[7] Summing the Neolithic dates in Table 1 gives a total of 1716 Neolithic dates, but 11 dates have been omitted as they come from an unnamed site, giving 1705 dates.
[8] This information must be added to the database, a task that could not be carried out within the time constraints of this project.

[9] Leaving aside, for the moment, the problems of averaging uncalibrated radiocarbon dates (Chapter 3).

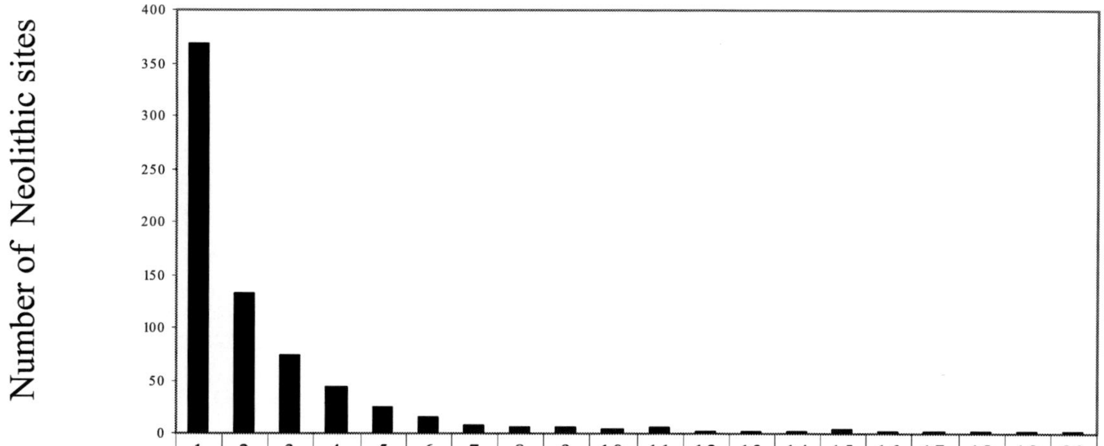

Number of dates per Neolithic site	1	2	3	4	5	6	7	8	9	10	11	12	13	14	15	16	17	18	19	20
■ Number of sites	367	132	74	43	25	15	7	5	6	4	5	2	2	1	3	2	1	1	1	1

Number of dates per Neolithic site

Figure 2.3 The number of determinations per site for the 697 Neolithic sites.

Dataset C: Oldest determinations only (509 sites)(Dataset C, Appendix 1). At sites with more than one determination, the *oldest* determination was selected.

Dataset D: Most recent determinations only (509 sites)(Dataset D, Appendix 1). At sites with more than one determination, the *most recent* determination was selected.

Dataset E: Most recent determinations, multiple dated sites only (264 sites)(Dataset E, Appendix 1). At sites with more than one determination, the *most recent* determination was selected. Sites with only one determination were excluded.

Dataset F: Oldest determinations, multiple dated sites only (264 sites)(Dataset F, Appendix 1). At sites with more than one determination, the *oldest* determination was selected. Sites with only one determination were excluded.

Dataset G: Weighted means or median determination (229 sites[10]) (Dataset G, Appendix 1). This list contains the weighted means of determinations. In cases where the determinations could not be averaged the *median* determination at the site was selected. Sites with only one determination were excluded.

2.1.2 The Chi square test and its results:

The Chi square test was applied at sites with multiple determinations to see whether the determinations should be averaged. Sites were sorted by the total number of determinations at the site and according to the type of material dated (Table 2.3). Following the advice of Bowman & Balaam (1990) and Buck *et al.* (1994) the Chi square test was only applied to sets of determinations that were made on samples of the same organic material, with the exception of the bone and charcoal category[11]. Only determinations made on the same type of sample material are compared. This reduces the error of

[10] There are only 229 sites (rather than 264 sites) because in addition to excluding all one-date sites, all two-date sites that failed the Chi square test were excluded (Table 2.4). This is because the central date or nearest to central date was used as the median value to avoid adding and averaging two dates.

[11] Bone and charcoal frequently occurred together in groups of dates at a site. Archaeologists are familiar with the 'old wood' effect where wood at archaeological sites may have ceased carbon exchange many years before the prehistoric people of interest use it. To explore this effect the Chi square test was applied to the radiocarbon results as reported and then repeated with 200 years subtracted from the radiocarbon determinations on the wood samples. There was a marginal improvement in the Chi square test results.

assuming that they all ceased carbon exchange at the same time.

The categories shown in Table 2.3 need some explanation. The column 'seed' includes all short-lived plant material. For example hazelnuts, grape pips and grasses come under this category heading. The category 'shell' includes both marine and freshwater shell species. However these two were not compared with each other in the Chi square test due to the marine reservoir effect (Chapter 3). The category 'miscellaneous' includes materials and combinations of materials that occurred too infrequently to warrant their separation in to a unique category. Examples are sod, soil and pottery. It also includes groups of determinations on mixed materials. There are cases where the dated material is not listed in the database. These are grouped as 'unknown'.

The results of the Chi square test are listed in Table 2.4. Of the 330 sites Chi square tested, 49% have determinations that are statistically comparable at the 0.05 level. These are listed in column three 'Grouped' in Table 2.4. Determinations at these sites were averaged to use in datasets A, B and G. Half the sites have Neolithic determinations that at the 0.05 level are not likely to be dating the same event. These are shown under the column heading 'can not be grouped' (Table 2.4). For those determinations that could not be grouped, the oldest, the most recent or the median determinations were selected to use in the datasets A, B and G. A fifth column shows the percentage of sites that could not be grouped. Unsurprisingly, the more determinations per site the greater the apparent spread of determinations. For sites with greater than 9 determinations per site (n = 23) only 3 had determinations that could be grouped together (note the result for one site with 22 determinations which were found to be significantly similar, Table 2.4).

The Chi square test results show that in approximately half of the cases the determinations at multiple sites are so similar that they can be averaged. In these cases the determination that is selected from a site's sequence should make very little difference to the final result. When the Chi square test shows a difference between determinations in a sequence the reasons for this needs to be considered. The determinations could show the early first occurrence of the Neolithic and its continuation over a long time period. In this case the early determination in a sequence is the determination that should be selected, but there are other possibilities. An early determination in a sequence may be an anomalous outlier, particularly in cases where there is a large time difference between the earliest determination and the next in the sequence.
The only way to satisfactorily resolve this is by detailed analysis of a site's stratigraphy and further radiocarbon dating. Indeed, ^{14}C results in instances like these may provide the stimulus for archaeologists to revisit sites to re-examine apparently anomalous determinations or to collect new samples with a better knowledge of the issues that need to be addressed. This analysis suggests 167 sites deserve such re-examination.

2.2 The southern African herder/stock keeper database

A parallel analysis of the introduction of the first domestic sheep into the western half of South Africa was undertaken, replicating the methods used in Europe for the Neolithic. In both cases domesticated animals were being introduced into an area already occupied by hunter-gathers. In both cases the mode of diffusion of domesticated stock (whether by the spread of people or by the spread of ideas) is disputed. The databases differ in the amount of data available and in the detail of data collected. The instances of first domestic stock in South Africa are limited to 10s rather than 1000s of sites. This meant that it was possible to collect detailed information on site stratigraphy, evidence that was beyond the scope of the European data collection. So the southern African data provides the opportunity to look at movement over a large area in minute detail, whilst the European data provides the advantages of a large sample size, statistically more likely to be representative of the phenomena under question.

The southern African database was assembled during fieldwork in South Africa in 1999 and represents a comprehensive list of all the non-agriculturalist sites in the western half of South Africa that contain evidence for the first appearance of domesticated sheep in the archaeological record (Figure 2.4). The database was compiled from published site reports, date lists from the journal *Radiocarbon* and from site lists provided by the National Monuments Council of South Africa[12]. There are a total of 27 sites and 116 determinations (Appendix 2). Detailed information is given for each site (Table 2.5).

2.3 Summary:

The chapter has introduced the data that will be used in this research. Chapters 4 to 7 analyse the Gkiasta database and Chapters 8 and 9 analyse the southern African database.

---

[12] My thanks to Dr Jeanette Deacon of the National Monuments Council (now replaced by SAHRA, South African Heritage Resource Agency) for providing me with information on sites and their geographical references.

Determinations per site	No of sites	Charcoal	Bone	Seed	Bone & charcoal	Shell	Miscellaneous.	Unknown
1	367	228	71	14	0	4	22	28
2	132	81	18	7	6	0	14	6
3	74	39	18	4	2	0	8	3
4	43	19	7	2	4	0	9	2
5	25	11	1	0	5	0	5	3
6	15	4	3	0	1	2	5	0
7	7	4	0	0	1	0	2	0
8	5	2	1	0	0	1	1	0
9	6	3	1	0	0	0	2	0
10	4	2	0	0	2	0	0	0
11	5	4	0	0	0	0	1	0
12	2	0	2	0	0	0	0	0
13	2	1	0	0	0	0	1	0
14	1	0	0	0	0	0	1	0
15	3	1	1	0	0	0	1	0
17	2	2	0	0	0	0	0	0
18	1	1	0	0	0	0	0	0
21	1	0	1	0	0	0	0	0
22	1	0	1	0	0	0	0	0
23	1	1	0	0	0	0	0	0
TOTAL	697	403	126	27	21	7	72	42

Table 2.3 Neolithic sites subdivided by number of determinations per site and type of material dated.

Number of Neolithic determinations at site	Number of sites containing this number of determinations.	Grouped (insignificant difference at the 0.05 level)	Can not be grouped (significant difference at 0.05 level)	Cannot be grouped (%)
2	132	81	51	38.6
3	74	38	36	48.6
4	43	23	20	46.5
5	25	9	16	64
6	15	4	11	73
7	7	1	6	86
8	5	2	3	60
9	6	2	4	67
10	4	0	4	100
11	5	0	5	100
12	2	1	1	50
13	2	1	1	50
14	1	0	1	100
15	3	0	3	100
17	2	0	2	100
18	1	0	1	100
21	1	0	1	100
22	1	1	0	0
23	1	0	1	100
Total	330	163	167	

Table 2.4 The results of the Chi square-square test for Neolithic sites with multiple determinations.

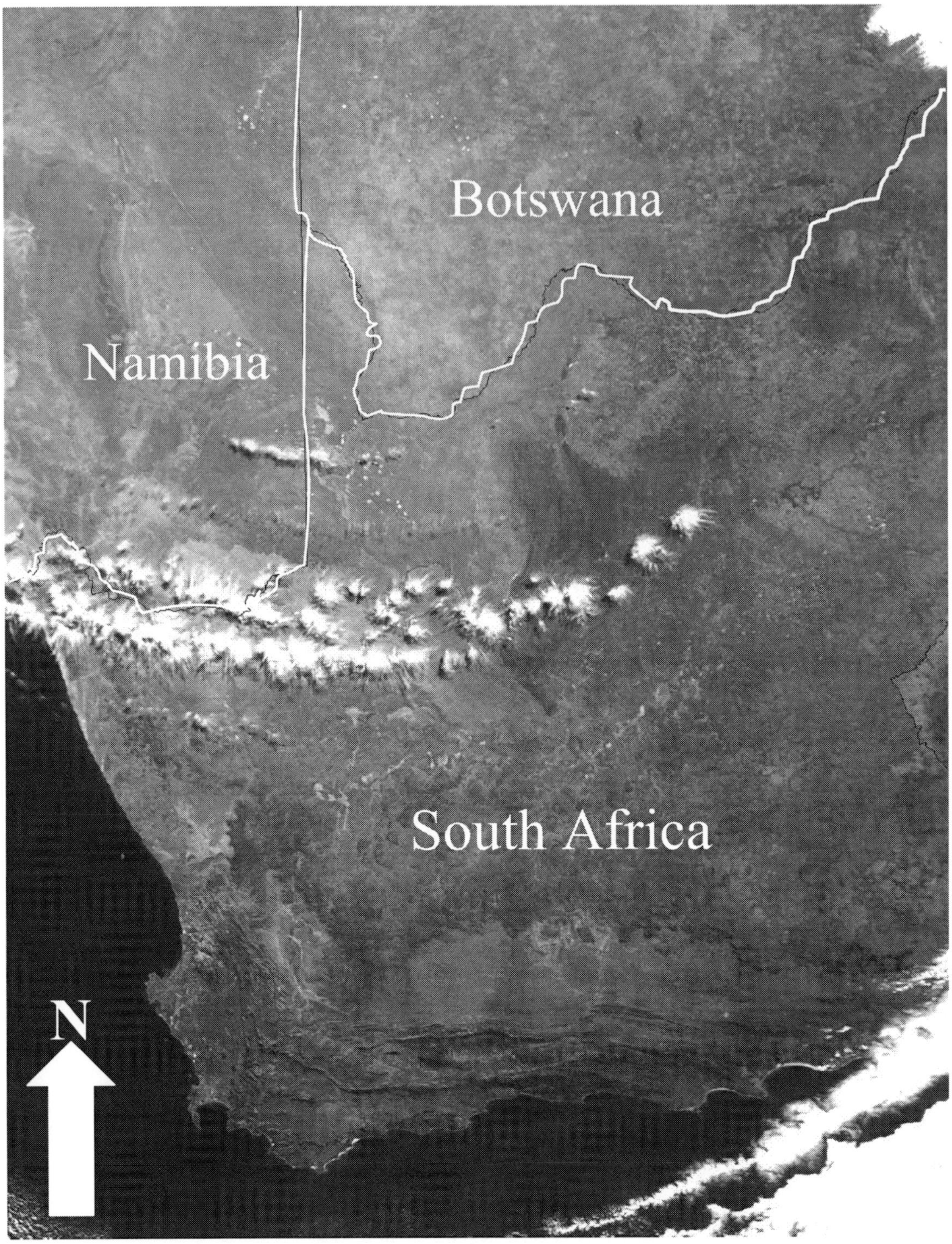

Figure 2.4 The sites in the southern African database fall within the area of present day South Africa as shown above. The majority of the sites lie along the west and east coastlines.

Fields	Information
Latitude and longitude coordinates	Geo-reference for the location of each site
Site stratigraphy	To show the provenance of the date sample and its association to the archaeological event.
Lab number	Sample Identification
Date in uncalibrated years bp	The radiocarbon determination
Material dated	Organic sample submitted to the radiocarbon lab.
Dating technique	AMS or conventional radiocarbon dating
Context and properties of site	This describes the type of site, e.g. cave or midden.
Total domestic stock at site	The total number by bone count using MNI (minimum number of individuals)
Determination at which sheep first occur and the MNI	Uncalibrated radiocarbon determination at the first occurrence of sheep and the number of sheep present (bone count presented as MNI).
Determination at which sheep peak and their MNI	Uncalibrated radiocarbon determination at the point at which the number of sheep (MNI bone count) reaches its peak at the site.
Presence/absence of stock in final occupation layer	This notes whether or not sheep bones are found in the final stratigraphic layer at the site.
References	This lists the sources used to collect the data (these are published site reports in most cases authored by the excavator).
Site type	A description of this classification system is given in Chapter 9.
Determination Grading of site	A description of this classification system is given in Chapter 9.
Dating method	AMS or conventional radiocarbon dates

Table 2.5 Information recorded for each of the sites in the database for the first occurrence of domestic stock in the western half of southern Africa.

Chapter 3 Radiocarbon dates – the backbone of the research

3.1 Introduction

Radiocarbon dates are the temporal variable used in all the analyses in this research. This chapter has two sections. The first describes how a radiocarbon date is measured and calibrated and the second deals with the use of these dates in analysis.
In this research uncalibrated radiocarbon ages are simply given the suffix, 'BP'. Dates that have been calibrated to calendar ages are given the suffix 'cal AD' or 'cal BC'. Calibrated dates reported as years before 1950 are given the suffix 'cal BP'[1].

3.2 Getting a date: radiocarbon activity of sample to calibrated date.

3.2.1 The principles of radiocarbon dating

There are numerous sources that summarise these principles (Van der Plicht & Mook 1987, Bowman 1995, Taylor & Aitken 1997, Taylor 2000, 2001). A basic outline is given here.

There are three isotopes of carbon. Their chemical properties are identical although they differ in weight (^{14}C is the heaviest, followed by ^{13}C and ^{12}C). ^{14}C is produced in the upper atmosphere; it combines with oxygen to form carbon dioxide, which is then distributed around the atmosphere by winds, dissolved into the ocean and assimilated into living systems. There is a balance between the amount of ^{14}C produced in the upper atmosphere and its decay in the biosphere. The overall level of ^{14}C in living organisms is in approximate equilibrium with its levels in the atmosphere.

Upon death, living organisms no longer exchange carbon with the atmosphere (by breathing, photosynthesising and eating for example). From this time on the ratio of the three carbon isotopes in the body of the dead organism changes. As radioactive decay begins the amount of ^{14}C diminishes relative to the other two isotopes: ^{12}C and ^{13}C are stable and so their amount at death remains constant.

Radioactive decay of ^{14}C occurs at a known rate, governed by the law of radioactive decay[2]. It takes 5730 years for half of the atoms in ^{14}C to decay. The radiocarbon age of a sample is the measure of its residual ^{14}C content. The less ^{14}C an organism contains relative to ^{12}C and ^{13}C the older it is. A radiocarbon date is in fact a measurement of the time of death of an organism i.e. the time at which it ceased to exchange carbon with the atmosphere and at which radioactive decay of the ^{14}C began. The ^{14}C content of an internationally recognized modern reference standard is used to define a zero age point with which to compare the ^{14}C activity in the archaeological sample (Aitken 1990).

If this were the whole story it would be correct to equate measured radiocarbon age with calendar age. This is not the case. There is a global carbon cycle (Bowman 1995:13). During this cycle carbon is passed between the carbon reservoirs (for example, the atmosphere, the oceans and the biosphere). The speed with which exchange and mixing of ^{14}C takes place between these reservoirs affects radiocarbon dating. If reservoirs hold on to 'old' radiocarbon and then release it at some later stage it will produce misleadingly old results for young samples.

The following reasons for radiocarbon age and calendar age discrepancies have been identified.

1. The ^{14}C concentration in each carbon reservoir has not remained constant over time (Watkins1975, Bowman 1995, Taylor 2000). Changes in the earth's magnetic field, sun spot

---

[1] This follows the convention of the journals *Radiocarbon* and *Antiquity* (Mook 1986). Variations of this are B.P. (see *American Antiquity*) or the use of small letters to refer to uncalibrated dates (e.g. bp) and capitals to denote calibrated dates (e.g. BP)(see old issues of the journal *Antiquity*) (Taylor 2001).

[2] Radioactive decay is exponential. Radiocarbon age (T) is calculated using the equation:

$$T = -\frac{1}{\lambda}\ln\frac{A}{A_0}$$

Where λ = decay constant
A = measured ^{14}C activity
A_0 = original ^{14}C activity

(Van der Plicht & Mook 1987)

activity, the burning of fossil fuels and nuclear weapons testing have all altered ^{14}C levels. The natural causes of variation in ^{14}C activity over time have a temporal trend. There are long-term oscillations due to reversals in the earth's magnetic field, medium-term variations ('wiggles') due to solar fluctuations[3], and lastly, short-term variations caused by eleven-yearly sunspot cycles (Van der Plicht & Mook 1987:174).

2. Complete and rapid mixing of ^{14}C does not occur in all the carbon reservoirs (Watkins 1975, Bowman 1995, Taylor 2000). ^{14}C in the oceans mixes slowly. Deep oceans are huge reservoirs for ^{14}C. They contain carbon with varying levels of ^{14}C due to the fact that there is not rapid mixing of old and new ^{14}C in their waters. Deep water contains ^{14}C-depleted carbon. Its up-welling and incorporation into marine and terrestrial organisms leads to misleadingly old dates for what may, in reality, be very young samples. Temperature also affects the rate at which CO_2 dissolves into water. Glacial and interglacial periods have altered the degree of mixing in the water bodies of the world. Glaciation periods will store what is effectively old carbon as ice, which, upon melting is re-released into the carbon cycle.

3. Both before and after the death of an organism its carbon isotopes ratios are changed by factors other than the radioactive decay of ^{14}C (Watkins 1975, Bowman 1995, Taylor 2000).

Before death:

In the biological pathways of living organisms the lighter carbon isotopes (^{12}C and ^{13}C) are taken up by the organism in preference to the heavier ^{14}C (a process known as fractionation). This preferential uptake differs in degree from one species to another. The result is that contemporaneous species will have different measured radiocarbon ages even though their real ages are equivalent. It is convention that a correction factor for fractionation is applied to radiocarbon results to account for this effect. The effect is quantified by the measurement of the ^{13}C concentration ($\delta^{13}C$) in the sample relative to a standard (a Cretaceous belemnite from the Peedee formation, known as PDB) and the value is then normalised to another standard (the standard is wood with a $\delta^{13}C$ value of –25 ‰) (Bowman 1995).

After death:

Other changes to the carbon isotope ratios can occur due to contamination of the sample by other carbon-containing substances after the death of the organism. Water soluble substances such a limestone (calcium carbonate) and humic acids may dissolve in ground water and contaminate an archaeological sample. Prior to dating, samples are treated in the laboratory to try to remove these contaminants (Bowman 1995).

3.2.2 Time limits of radiocarbon dating

Radiocarbon measurements can currently be obtained only on samples that fall between the age ranges of 300 years and 60 000 years old. Constraints on dating carbon samples younger than 300 years old are a consequence of (1) the seventeenth-century variability in ^{14}C production rates caused by changes in the sun's magnetic fields, (2) the effect of the combustion of fossil fuels at the end of the nineteenth century and (3) the production of artificial ^{14}C as a result of the detonation of nuclear devices in the mid-twentieth century (Taylor 2000). The main constraint on the dating of samples older than 60 000 years is the limitation caused by laboratory equipment. This range will eventually be pushed back as techniques are refined.

3.2.3 Measurement of the radiocarbon age of a sample

There are two techniques used to measure the ^{14}C depletion of a sample. These are accelerator mass spectrometry (AMS) and conventional radiocarbon dating. Conventional radiocarbon dating (or radiometric dating) measures the amount of beta decay[4] emitted by the sample over a certain time period. The older the sample, the less beta decay it will emit. In AMS dating the proportions of ^{14}C relative to ^{12}C and ^{13}C are measured directly (Bowman 1995, Taylor & Aitken 1997, http://www.rlaha.ox.ac.uk/orau/index.htm). Both of these methods produce radiocarbon results that can be interpreted in the same way (Bowman 1995). The AMS technique requires much smaller samples of carbon. AMS requires milligrams of carbon not grams of carbon. This allows archaeologists to select samples at the site without the size constraints that came with conventional radiocarbon dating. This also gives greater precision when, for example, one seed can be dated. It also allows specific fractions of the sample to be selected, namely those that experience shows produce the most reliable

[3] Sun spots (Aitken 1990:69).

[4] The unstable ^{14}C atom decays to a nitrogen atom and on so doing emits an electron. Beta decay is the name given to this process of electron emission during radioactive decay (Bowman 1995).

results. Another advantage of the AMS technique is the ease with which measurements can be repeated: the run time for AMS is hours rather than days (Aitken 1990, Bowman 1995).

3.2.4 Radiocarbon age as reported by the laboratory:

In order to standardize the measurement of radiocarbon dates and their reporting to facilitate inter-laboratory comparisons, radiocarbon dates are reported following the convention that Stuiver and Polach outlined in their 1977 paper. The parameters for reporting a conventional radiocarbon age are as follows (*ibid.* 1977):

1. Radiocarbon dating standard: the use of an internationally recognized standard (US National Bureau of Standards (NBS) oxalic acid) from which to calculate ^{14}C concentration at time 'zero'.
2. Isotopic fractionation correction: The normalisation of all samples to a common $\delta^{13}C$ value (the one used is –25‰, the value for wood) relative to the ratio of ^{13}C to ^{12}C in a common reference sample (PDB).
3. Radiocarbon age calculation: the radiocarbon age of a sample is calculated under the assumption that atmospheric ^{14}C has been constant over time. The Libby half-life (5568 years[5]) is used as the half-life for the radioactive decay of ^{14}C. Time zero from which all ^{14}C time is counted is taken as AD 1950. The age is expressed in years BP (years before 1950 AD).
4. Statistical uncertainty of age determination: The conventional ^{14}C age is accompanied by a standard error (based on ^{14}C activity in the sample, the modern reference standard and the background activity). Other errors such as laboratory removal of contamination and the measurement of the $\delta^{13}C$ value of the sample should also be included in this error term by the dating laboratory. The standard error of the conventional radiocarbon age is reported as ± one standard deviation.

3.2.5 The statistical uncertainty of a radiocarbon determination

There are approximately 140 radiocarbon dating laboratories worldwide[6] (Taylor 2000) and yet there is no set convention on how they should calculate/estimate the error term. The international radiocarbon laboratory intercomparison exercise is an initiative to control the quality of radiocarbon measurements in laboratories worldwide (Aitchison *et al.* 1990, Scott *et al.* 1990). Laboratories are required to measure the radiocarbon age of a series of samples of unknown age and the results are publicly available[7], though details of the performance of individual laboratories is not given.

A radiocarbon date is a probability distribution with a normal (Gaussian) distribution[8] (Figure 3.1). The sigma ranges are a statement of the error associated with the radiocarbon determination. The error term accompanying any radiocarbon date is important because the size of the error term of the date (combined with the error associated with the calibration curve) determines the range of the calibrated calendar date (Bowman 1994).

It is normal scientific practice that experimental error is measured as the amount by which replications of the same experiment differ from each other. For radiocarbon dating the cost of the technique, the time it takes and the scarcity of archaeological samples all prohibit the replication of measurements on the same sample. Instead an error term is estimated (Bowman 1994,1995, Buck *et al.* 1994). It is most commonly based on counting error (Bowman 1995). For conventional radiocarbon dating this is a measurement of the rate of beta decay (Spaulding 1958). Beta decay is a random process so there is no escape from the uncertainty of the estimation (Spaulding 1958). The number of decays counted in any given period will vary. The uncertainty associated with each date is a statistical statement of the uncertainty of counting random events (Aitken 1990). The Poisson distribution[9] is used to quantify the error associated with counting discrete random events (Spaulding 1958: 309, Long & Rippeteau 1974). This error can be reduced by either increasing the length of the counting period or by

[5] The true half-life of radiocarbon is 5730 ± 40 years, but by convention the Libby value is used to avoid confusion with dates reported in the past which were based on calculations made using the Libby half-life value (Van der Plicht & Mook 1987)

[6] Lists of radiocarbon laboratories are published in the journal Radiocarbon. Lists are all accessible via its website at http://www.radiocarbon.org/Info/

[7] An online report of the 4th International radiocarbon intercomparison exercise (FIRI) in 1998 is at http://www.stats.gla.ac.uk/~marian/intercomp.html

[8] Since radioactive decay occurs at random about a mean, the Poisson distribution best approximates it. However, when considering large samplings of radioactivity, as in the case of radiocarbon dates, the Gaussian distribution curve is an acceptable approximation of the Poisson (Long & Rippeteau 1974)

[9] Poisson statistics states that the expected number of events (e.g. beta decay) in a given time (i.e. the counting period) can be estimated by the measured number of events (n) with a standard deviation (error) of $\sqrt{n}$ (Bowman 1995, Spaulding 1958).

increasing the size of the sample. However, counting errors are not the only source of error and laboratories should include an estimation of the other errors (chemical fractionation, quenching and environmental conditions within the laboratory) in the final error value (Bowman 1994,1995, Buck *et al.* 1994). It is possible that in estimating this error term, the laboratories do not include all possible sources of error (Bowman 1994). To overcome the problem of underestimation of error it has been suggested that all standard deviations accompanying radiocarbon ages should be automatically doubled. However this can also mislead; for example, increasingly the standard deviation may give erroneous contemporaneity to dates when they are Chi square tested (Bowman 1994).

3.2.6 Calibration of Radiocarbon ages to Calendar Ages

Brief history:
Libby (1952) first suggested a relationship that equated radiocarbon age with calendar age (radiocarbon age is equivalent to years before present when the present is taken as AD 1950). This was subsequently shown to be invalid. The concentration of ^{14}C has not been constant over time. Figure 3.3 shows how the calibration curve varies from the relationship that Libby first proposed (Van der Plicht & Mook 1987:175, Pearson 1987:99, Aitken 1990).

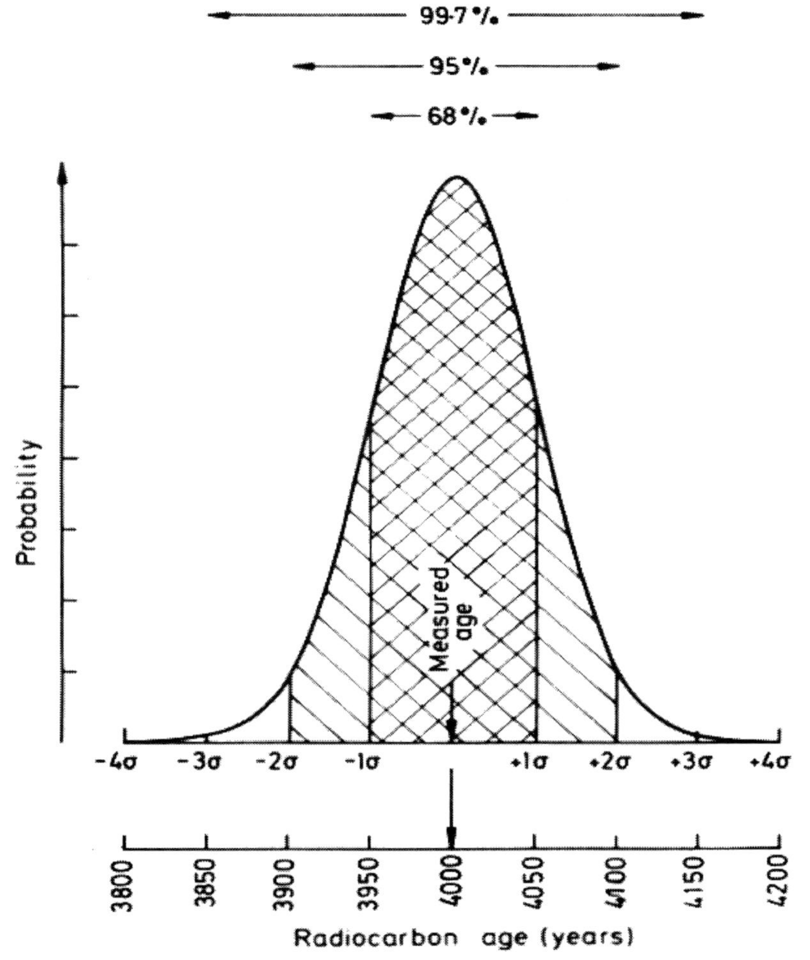

Figure 3.1 The probability curve of the uncalibrated radiocarbon date follows the Gaussian distribution. The probabilities of the true age of the sample lying within the 1 sigma, 2 sigma and 3 sigma ranges are shown in the diagram (Aitken 1990: 80).

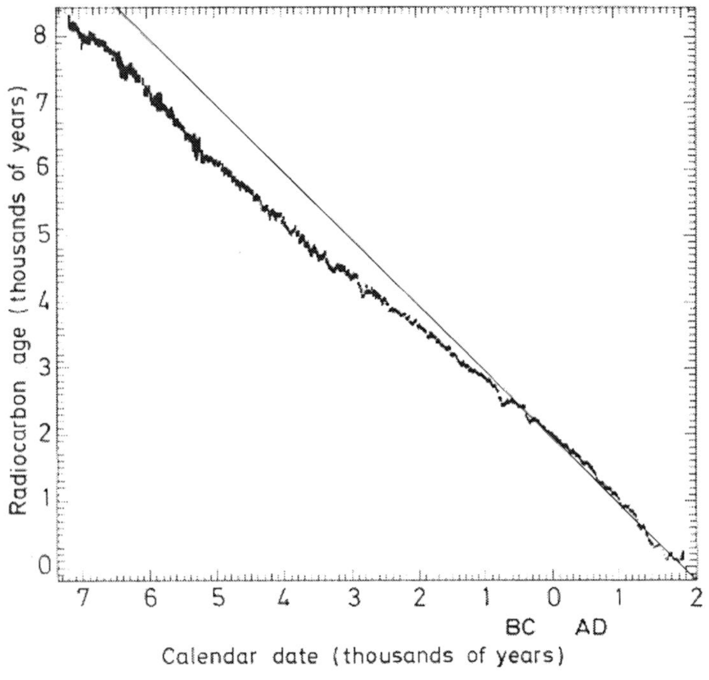

Figure 3.2 Libby's one to one relationship between radiocarbon determination and calendar age of a sample is shown by the straight line. The irregular line shows the true relationship caused by the fluctuating levels of ^{14}C through time (Aitken 1990).

In a 1958 paper De Vries first noticed discrepancies between the ages of the trees calculated by yearly tree ring growth (dendrochronology) and their measured radiocarbon age. Stuiver (1961, 1965), Suess (1965) and Stuiver & Suess (1966) confirmed that the similar variation they were observing was due to changes in solar activity. The need for calibration of radiocarbon ages to give calendar ages was becoming apparent.

In 1966 Stuiver and Suess published a table for the conversion of radiocarbon age to calendar age for the last 1000 years, based on measurements on samples of wood of known age. In the years that followed the first calibration curve (Suess 1967,1970) and further calibration tables and curves appeared (e.g. Clark 1975, Klein *et al.* 1982). It was now clearly established that dendrochronology was the perfect tool with which to calculate the fluctuations in ^{14}C activity with time. This information was used to construct calibration curves that could be used to change radiocarbon age measurements in to calendar dates. The calendar age, established from the tree ring growth, is plotted on the x-axis. Radiocarbon measurements (y-axis) on each tree ring correlates known calendar ages with radiocarbon measurements and so the calibration curve is formed (Figure 3.3).

Measurements also established that the variation of ^{14}C was a largely a worldwide phenomena (Stuiver & Suess 1966). There is a difference between the northern and southern hemispheres equivalent to a forty-year[10] radiocarbon age difference (the southern hemisphere radiocarbon ages appear 40 years older)(Vogel 1965, Vogel *et al.* 1993).

A large part of the work on developing a calibration curve was based on data obtained from the measurement of ^{14}C activity of the Bristlecone pine (*Pinus aristata*). Bristlecone pines grow at an altitude of 3000 metres above sea level in the White mountains of east-central California. They have a long life span of 4500 years and more. Both living and dead trees have been used to establish a calibration curve stretching back thousands of years. The high resin content of these pines coupled with the dry environment has preserved dead trees *in situ* (Bowman 1995). Their longevity and excellent preservation allowed for the construction of a continuous tree-ring sequence stretching back 11 000 years.

[10] There is ongoing research as to the exact number of years by which the two hemispheres differ (Vogel 2000). The figure varies between 20 to 40 years (Vogel *et al.* 1986, Vogel *et al.* 1993, Stuiver *et al.* 1998,Vogel 2000, Bronk Ramsey 2000). This value should be subtracted from the radiocarbon age prior to calibration (Bowman 1995). In the work in this thesis the southern African dates were calibrated by subtracting 40 years from the radiocarbon determination.

Other species of wood have also been used to create long chronologies and calibration curves (e.g. tree rings from waterlogged Oak (*Quercus sp.*) from Ireland and Germany)(Pearson *et al.* 1993).

There is growing evidence that there is more location-dependent variation in ^{14}C than previously realised (Bowman 1994, McCormac *et al.* 1995, Levin *et al.* 1985). For example, a comparison of contemporary samples of Bristlecone pine and Irish oak shows that Bristlecone pine appears older than the oak (McCormac *et al.* 1995). These differences are unlikely to give noticeably different calibrated results, although it makes good sense to use calibration curves that are based on wood samples from the archaeologist's area of research (i.e. to use calibration curves based on Irish oak for radiocarbon dates from the British Isles) (Bowman 1994).

Many calibration curves have been produced (Figure 3.4). The first calibration curves to achieve international consensus on their use were those developed by Pearson & Stuiver (Pearson and Stuiver 1986, Stuiver & Pearson 1986) when, for the first time two laboratories (Seattle and Belfast) working on different tree species, produced near identical calibration curves (Bowman 1994,1995).

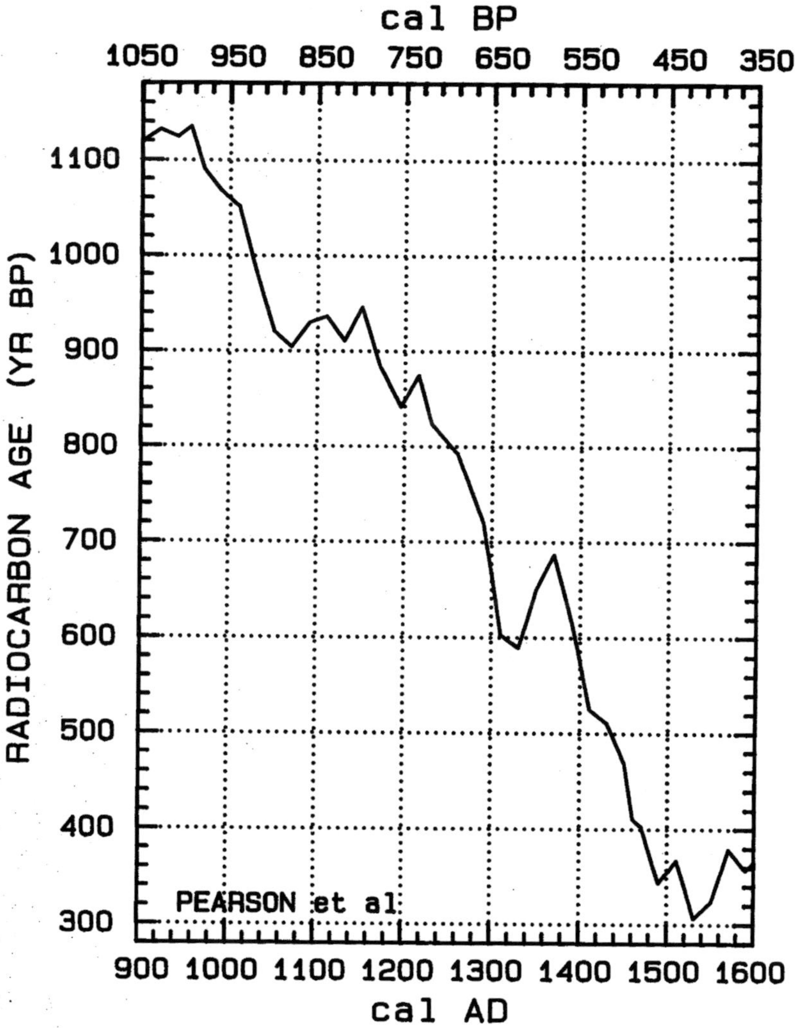

Figure 3.3 Calibration curve made on the 14C measurements of bidecadal sample of Irish Oak (Pearson, Becker & Qua 1993:107 Figure1B).

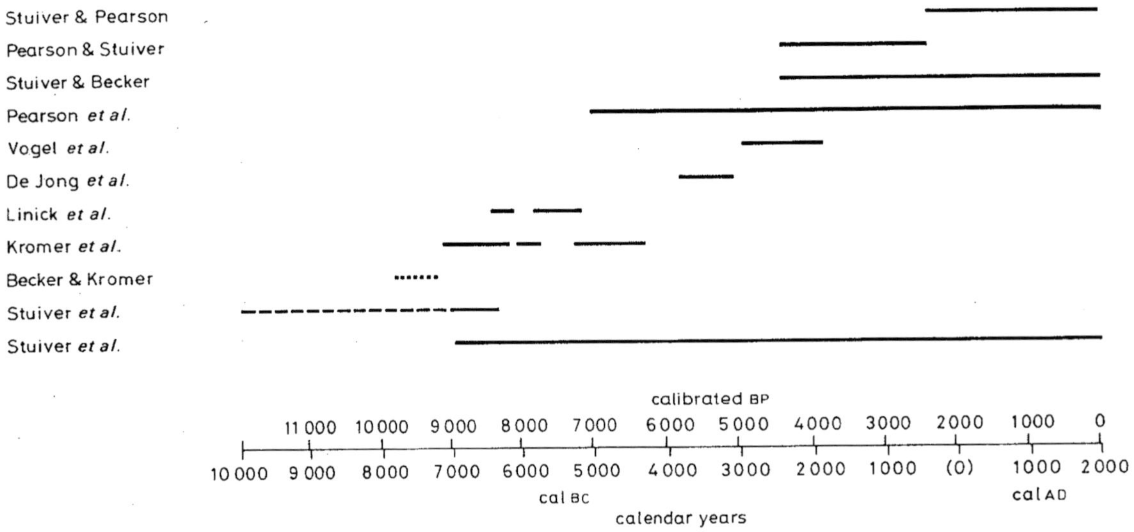

Figure 3.4 Comparison of the time periods covered by various calibration curves (Pearson 1987:100 Figure 2).

Updated versions of their calibration curves were released in 1993 (Pearson & Stuiver 1993, Stuiver & Pearson 1993). These have been superseded by the calibration curve Intcal98 produced by Stuiver *et al.* (1998) for the period 0 to 24 000 years BP (Figure 3.5).

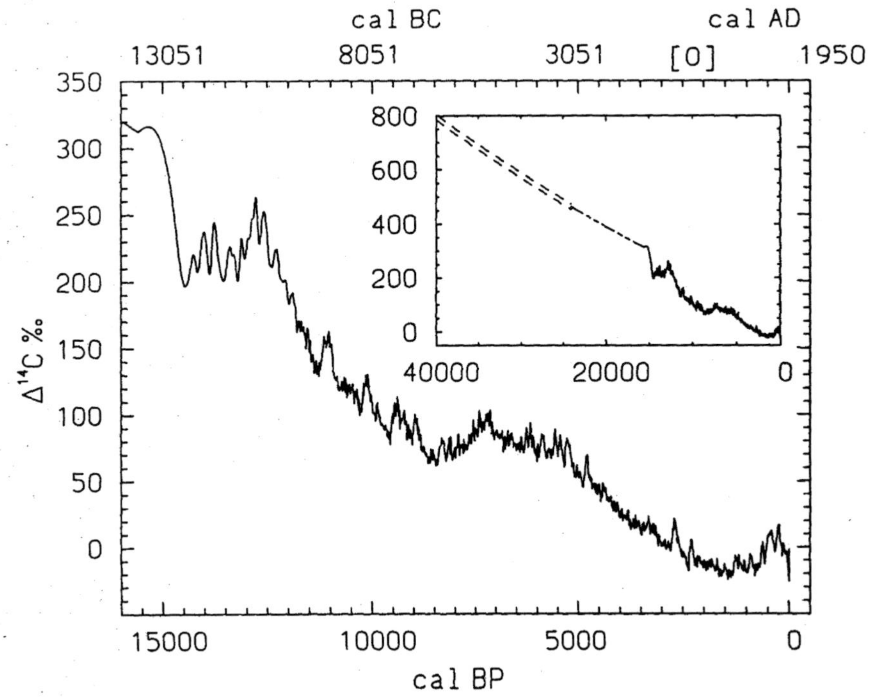

Figure 3.5 Calibration curve for the period 15 500 – 0 cal BP (tree ring data was used to construct the curve to 11 854 cal BP, the portion 11 854 to 15 500 cal BP used marine coral and varve data). The inset shows the calibration curve from 40 000 to 0 Cal BP, the solid portion of this is the INTCAL98 calibration curve (24 000 – 0 cal BP) (Stuiver *et al.*1998: 1059 Figure 12).

3.2.7 The nature of the calibrated date

There is no mathematically definable relationship between a radiocarbon measurement and the calendar age of the sample (Bowman 1995). The relationship is governed by the shape of the calibration curve at the point at which the radiocarbon age intercepts with it. This means that the transformation of radiocarbon years to calendar years is not straightforward (Dehling & Van der Plicht 1993, Van der Plicht & Mook 1987). The resultant calendar age may be older or younger than the radiocarbon age depending on the 'wiggles' in the calibration curve. A single radiocarbon measurement can correspond to more than one calendar age (Aitken 1990, Bowman 1995, Buck *et al.* 1994, Van der Plicht & Mook 1987, Aitken 1990)(Figure 3.6 and Figure 3.7).

Flat areas on the calibration curve give very long time ranges in calendar years (Figure 3.8). So even very high precision measurements with a small error term calibrate to give dates that span a long time period.

Whilst the radiocarbon age is a normal distribution with a central measure and an error term, the calendar age it produces is simply a probability range, or a series of ranges.

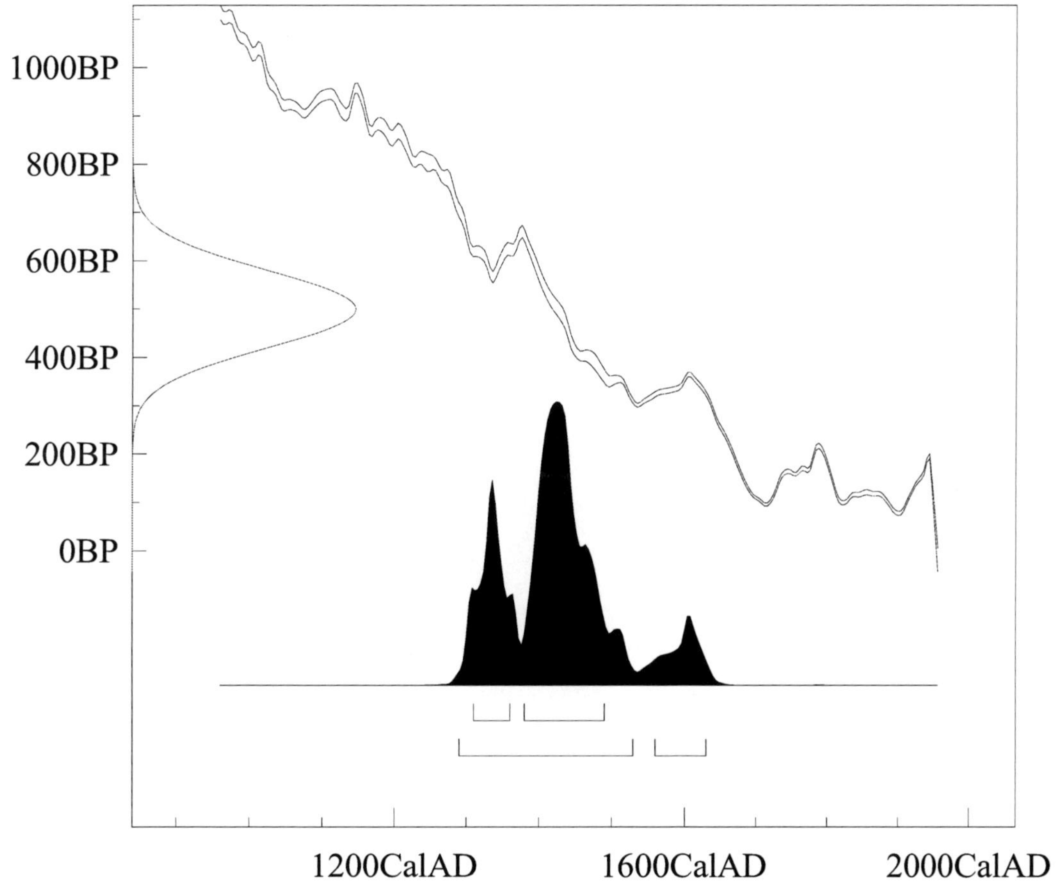

Figure 3.6 The radiocarbon determination has a Gaussian distribution (left) (y-axis). Due to the irregular nature of the calibration curve (squiggly line) the calendar date is a non-Gaussian probability distribution (x-axis). This was created using OxCal ver. 3.5.

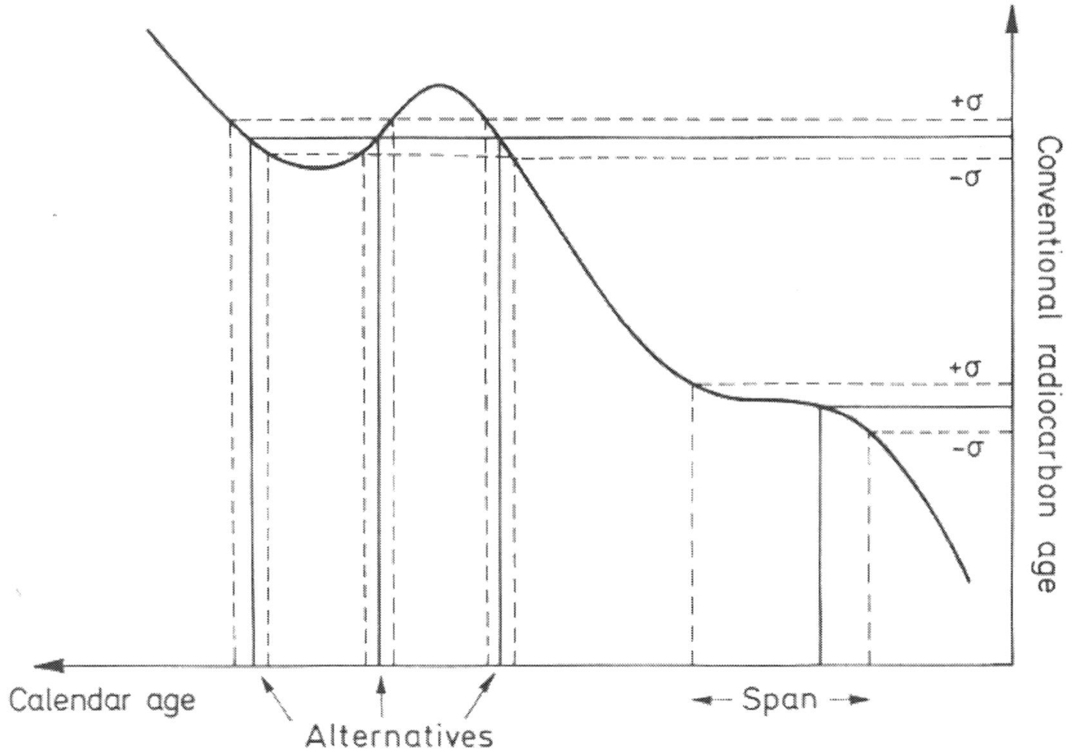

Figure 3.7 The effect of the shape of the calibration curve on the conversion of a radiocarbon determination to a calendar age (Aitken 1990:100 Figure 4.7).

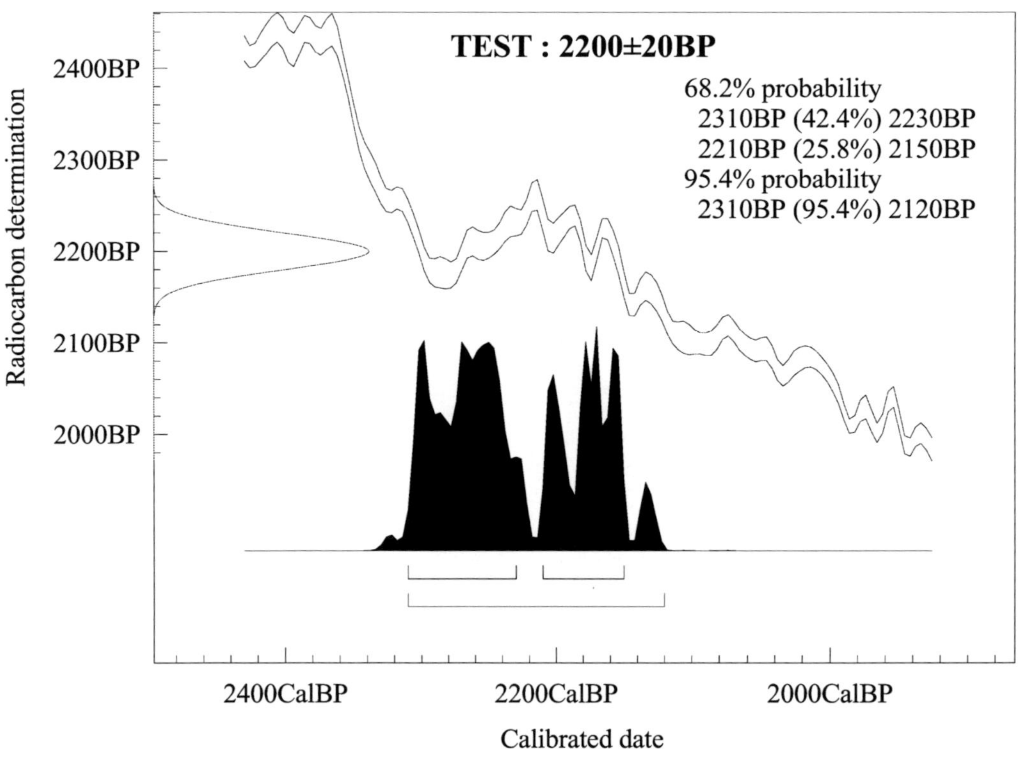

Figure 3.8 A high precision date (2200 ± 20 BP) has a 1-sigma range of 40 years prior to calibration. Due to the wiggly nature of the calibration curve the calibrated calendar date has a 1-sigma range of 160 years. (This was produced using OxCal 3.5)

The problem is succinctly illustrated and described by Van der Plicht and Mook (1987) and summarised here. Figure 3.9 shows the portion of the calibration curve used to calibrate a date, y. The date, y, is from the measurement of the ^{14}C content of an archaeological sample and comes with laboratory's estimated error (σ). The date has a normal distribution, y ± σ (there is 68% probability that the date lies in this range). Calibration of y gives three possible calibrated calendar dates (x_1, x_2, x_3) (Figure 3.9a). In Figure 3.9b, the sigma ranges are included in the calibration process; this extends the ranges of the calibrated dates to Δx_1, Δx_2, and Δx_3. The error in the calibration curve must also be considered, so the ranges of Δx_1, Δx_2, and Δx_3 are further extended (Figure 3.9c) (Van der Plicht & Mook 1987: 174).

The problem is how this conversion to calibrated dates should be achieved. There are two methods used to calibrate dates: the intercept method and the probability method.

3.2.7.1 The intercept method

In this method the radiocarbon age and an error term are used to find the intercept points on the calibration curve (Pearson 1987, Van der Plicht & Mook 1987). The error term is calculated using both the error term on the radiocarbon date and the error term on the calibration curve[11]. The points at which the boundaries of the error term (radiocarbon date ± error term) intercept with the calibration curve are used as the outer limits of the 68% probability range of the calibrated date. This is frequently not one range but a series of ranges, depending on the shape of the calibration curve at this particular point. The calibration programme (Calib)[13] gives users the option of calibrating dates with this method. There are problems with this method. It is not mathematically correct (Van der Plicht & Mook 1987) and it takes no cognisance of the shape of the normal distribution of the uncalibrated date; it simply uses endpoints of the error distribution (Bowman 1995). With this method there is an equal probability that any of the calendar dates in this range(s) is the correct one (Pearson 1987, Van der Plicht & Mook 1987, Bowman & Balaam 1990, Bowman 1995). The assumption that all calibrated dates have an equal probability is not true (Van der Plicht & Mook 1987).

3.2.7.2 The probability method

This method uses the shape of the uncalibrated radiocarbon age distribution to produce a calibrated probability distribution (Van der Plicht & Mook 1987, Bowman 1994, 1995, Bowman & Balaam 1994). The normal curve for the uncalibrated date is broken down into segments. Each of these is calibrated by its interception with the curve and each of these calibrated segments is attached to its associated Gaussian probability (as calculated from its situation in the normal distribution). The calibrated segments are then grouped together in intervals (for example, ten-yearly groups) and the probabilities of the group are summed. The heights of the resulting histogram shows those date ranges that are most probable (Bowman 1995). There is no consensus on how the probability method should be applied (Bowman & Balaam 1990)[12]. There is, however, a growing consensus that calibration must produce probability distributions that use both the radiocarbon error term and the wiggles in the calibration curve. This rules out the use of the intercept method (Bowman 1994).

The mathematics behind the calibrations/transformations is beyond the reach of the most non-mathematicians so their validity must be taken in good faith. Dehling and Van der Plicht (1993) in their investigation of the mathematics used in the CALIB calibration programme (Stuiver & Reimer 1993) and the Groningen calibration programme (Van der Plicht & Mook 1989) draw the conclusion that, "...the computer programmes used for radiocarbon calibration are correct" (Dehling & Van der Plicht 1993:244). Bronk Ramsey (2000) states that the mathematical methods used to convert radiocarbon ages to calendar ages in the OxCal calibration programme are similar to those used by Stuiver and Reimer (1993) and Van der Plicht (1993) and directs the reader to Dehling and Van der Plicht (1993). The procedure may be complicated but there is general consensus amongst the practitioners that it is sound.

---

[11] Equation for total error, σ

$$\sigma = \sqrt{\sigma_d^2 + \sigma_c^2}$$

Where,
σ_d = error on radiocarbon date
σ_c = error on calibration curve
(Pearson 1987:102, Van der Plicht & Mook 1987:174).

[12] Van der Plicht & Mook (1987) give an extremely comprehensible account of the calibration of dates using this method in the Groningen calibration programme.

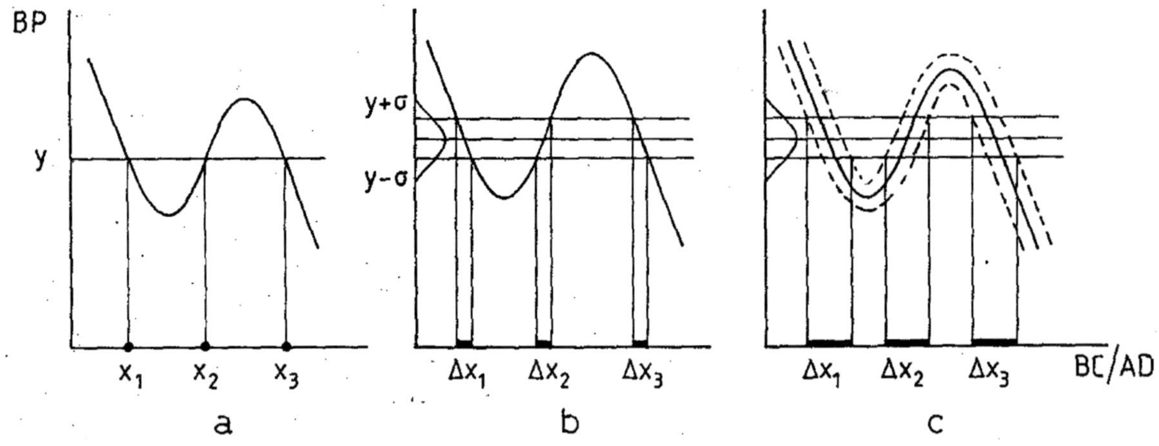

Figure 3.9 Effect of the shape of the calibration curve on the calibration of a radiocarbon age, y (Van der Plicht & Mook 1987:175 Figure 3).

3.2.8 The effect of time-width on calibration:

The ^{14}C activity measured in a sample may reflect a small period of time (i.e. a year in the growth of a hazelnut or one tree ring) or a long period (the whole trunk of a 3000 year old Bristlecone pine). This determines the choice of calibration curve (bidecadal, decadal, yearly) used. The ^{14}C sample and the calibration data must have the same time width (growth period) in order to avoid ambiguities (Van der Plicht & Mook 1987, Mook 1983). For example, it is inaccurate to calibrate a short-lived species on a bidecadal calibration curve (Bowman 1995:47)

3.2.9 The calibration programmes:

In this research three calibration programmes have been used. These are: OxCal 3.5 developed by Bronk Ramsey from the Oxford Radiocarbon Accelerator Unit[13] (Bronk Ramsey 1995), Calib 4.1 from the Universities of Washington and Belfast [14] (Stuiver & Reimer 1993) and the Groningen calibration programme (Van der Plicht 1993) from the University of Groningen[15]. Other calibration programmes are available on the Internet[16].

Different calibration programmes give slightly different results even though they use the same calibration data (Van der Plicht & Mook 1987). The variation is caused by the different mathematical methods that each calibration programme uses to draw the calibration curve, to calculate the uncertainty in the radiocarbon age (using either the 1-sigma, 2-sigma or 3-sigma range), and to calibrate the probability distribution of the calibrated date (*ibid.*). Aitchison *et al.* 1989 consider the differences in the calibrated results produced by the various programmes to be negligible.

3.3 Using the date: from calibrated date to archaeological interpretation.

Bowman (1994) defines two aspects of the radiocarbon date that must be evaluated before there is any attempt to interpret the date in archaeological terms (also see Nash & Dean 2000). They are (1) accuracy (how close the result is to the true value) and (2) precision (a description of the error inherent in any experimental procedure: the higher the precision, the more replicable is the experiment).

[13] This is available from the web pages of the Oxford Radiocarbon Accelerator Unit, ORAU (http://www.rlaha.ox.ac.uk/orau/index.htm).

[14] This is available from the web pages of the University of Washington (http://depts.washington.edu/qil/calib/) or the University of Belfast (http://radiocarbon.pa.qub.ac.uk/calib/calib.html).

[15] This is available from the web pages of the Groningen University centre for isotope research (http://www.cio.phys.rug.nl/HTML-docs/carbon14/cal25.html).

[16] The journal Radiocarbon's web page has links to the calibration programmes (http://www.radiocarbon.org/Info/).

Precision has already been discussed. The focus here is on accuracy.

Buck *et al.* 1994 identify three main sources of error affecting the accuracy of a date. (1) At the archaeological site poor sampling technique may contaminate the sample with extraneous carbon. (2) In the radiocarbon laboratory poor handling and pre-treatment of the sample may contaminate it. (3) At the archaeological site poor stratigraphy, poor excavation or poor interpretation of the stratigraphy may lead to the poor association between the sample to be dated and the event to which it is being linked (*ibid*: 252).

The first two are self-explanatory and are beyond the control of anyone working on data subsequent to its collection and dating. However, site reports, site stratigraphy and a knowledge of radiocarbon dating makes it possible to make some judgement about the degree to which the sample and the archaeological event are correlated.

3.3.1 Correlation between radiocarbon date and archaeological event:

This problem is familiar to any archaeologist working with radiocarbon dates. It is best illustrated with an example. Radiocarbon dates are often made on pieces of charcoal recovered during excavation. If the consideration of post-depositional mixing of stratigraphy, poor excavation techniques and laboratory contamination are put to one side, then the radiocarbon date from a piece of charcoal is actually the date of death of the wood from which the charcoal was made. But since a radiocarbon date is a measure of the time that has passed since a sample stopped exchanging carbon with its environment, the value of its radiocarbon date to archaeologists is the degree to which this moment in time and the archaeological event in which they are interested coincide.

Dean (1978) distinguishes between targeted events and dated events. The former describes what archaeologists are actually interested in, for example, the time that a site was first occupied. Dated events are the event that is actually dated (e.g. the time at which the piece of wood from which the sampled charcoal is made ceased carbon exchange). Whether the targeted and dated event correlate is down to the skill of the excavator and the amount of post-depositional mixing that has occurred at a site. Careful consideration of the strength of this correlation must be made by anyone about to undertake analysis using radiocarbon dates.

3.3.2 Using radiocarbon date in statistical analyses:

Radiocarbon results are not always carefully used.

> "There is a lamentable tendency to ignore or play down the importance of context in publication, especially secondary publication. Results are frequently combined or eliminated to produce a more "archaeologically acceptable" story. Standard statistical tests can be used for combining radiocarbon results only in circumstances where it can be legitimately thought that the samples represent the same event in radiocarbon time and where the results are statistically consistent. However, the same standard techniques are too frequently used in ignorance to combine the results of determinations of dissimilar events or materials (e.g. heartwood charcoal and animal bone). Such combinations give an unjustified air of precision to a date, and they disguise real uncertainty."
>
> Bowman & Balaam (1990:317)

3.3.2.1 Should one be content with uncalibrated dates?

Ideally all analyses would use the calibrated calendar age of the sample. However, the problem is that common statistical tests cannot be used on calibrated dates because, unlike the radiocarbon determination, the calendar equivalent cannot be expressed as a mean plus or minus a standard deviation. Due to the shape of the calibration curve calendar dates can only be reported as probability ranges.

3.3.2.2 Statistical techniques[17] to apply prior to calibration

Ideally statistical analysis should be carried out only on radiocarbon dates that have been derived by comparable procedures. This can assumed to be true of dates that come from the same laboratory. However its validity may be questionable for dates that come from different laboratories (Ward & Wilson 1978:20).

[17] The statistical techniques discussed all make the assumption that the radiocarbon age distribution follows a normal curve.

3.3.3 The Chi-square test and weighted averaging of radiocarbon dates

A weighted average of dates can be used when several dates are from the same context and it is believed that they represent a short-lived episode (Bowman 1995:60). Averaging is unwise in cases where one does not know if all the determinations are estimating the age of the same event or phenomenon, or when samples have been derived from different objects or from different archaeological contexts within a site (Levine 1990).

Dates from different types of organic material cannot be combined because of the possible difference between the time of the use of the material in its archaeological context (the date of interest) and the time at which the living organism ceased its carbon exchange with the biosphere (i.e. the 'old wood' problem; the dates from an archaeological site reflect the time of death of a tree rather than the use of the wood by the people in whom the interest lies) (Levine 1990, Bowman 1995). The assumption that all samples for dating the event ceased carbon exchange at the same time is not a sound one in cases where the samples are of mixed material (bone, wood, shell, seed etc.). What are obtained here are radiocarbon results from several different events each with close temporal proximity to the event of interest (Buck *et al.* 1994:254). Similarly dates from marine shell will be skewed by the marine reservoir effect[18]. They cannot be directly compared to those from terrestrial sources.
There is also the possibility that a set of different calendar years gives the same radiocarbon result due to the fluctuation of the ratio of the carbon isotopes over time (as represented in the calibration curve) or that dates once calibrated show a different order to that which they showed prior to calibration (Figure 3.10) (Bowman 1995: 58).

The test to use to see whether dates can in fact be combined is the chi-square test (Ward & Wilson 1978, Bowman 1995). Ward & Wilson (1978) recommend this test for comparing radiocarbon determinations that have been made on the same object or different parts of the same object. Under these circumstances it can be assumed that all the dates have the same true mean and that differences that occur are a result of changes in the conditions under which the measurements were made (*ibid.*). The formula for the chi-square is as follows (Ward & Wilson 1978, Bowman 1995):

$$\chi^2 = \sum \frac{(d_i - d)^2}{\sigma_i^2}$$

Where:

d_i are radiocarbon dates (d_1, d_2, $d_i \ldots d_n$)
d is the pooled mean, given by the formula

$$d = \frac{\sum d_i / \sigma_i^2}{\sum 1/\sigma_i^2}$$

σ_i are the standard deviations (σ_1, σ_2, $\sigma_i \ldots \sigma_n$)

This test compares the variation of the dates amongst themselves with the size of the standard deviations quoted with each date. If the variability between dates is larger than the standard deviations allowed then the dates are not comparable since it is assumed that they do not date the same archaeological event. Only if the dates can be combined can a weighted average of the group be made.

Dates with different standard deviations can be averaged by weighting them according to the size of their standard deviations. In this case weighting is inversely proportional to the variances (standard deviation squared) (Long & Rippeteau 1974, Ward & Wilson 1978). The formula for calculating the weighted average (D_w) is:

$$D_w = \frac{(d_1 \times w_1) + (d_2 \times w_2) + \mathrm{K}\ (d_n \times w_n)}{w_1 + w_2 + \mathrm{K}\ w_n}$$

Where the weight, $w = \dfrac{1}{\sigma^2}$

d = radiocarbon date (numbering 1,2,3….*n*).
w = the radiocarbon date's corresponding weight factor.
σ = the standard deviation of the radiocarbon date

The standard deviation for the weighted average, σ_w, is

$$\sigma_w = \frac{1}{\sqrt{\dfrac{1}{\sigma_1^2} + \dfrac{1}{\sigma_2^2} + \mathrm{K} + \dfrac{1}{\sigma_n^2}}}$$

(Long & Rippeteau 1974:208-209, Ward & Wilson 1978, Bowman 1995)

[18] Marine shells from near coastal waters have radiocarbon ages about 350 years greater than contemporary land organisms because the sea has a lower ^{14}C content than the atmosphere (Bowman 1995).

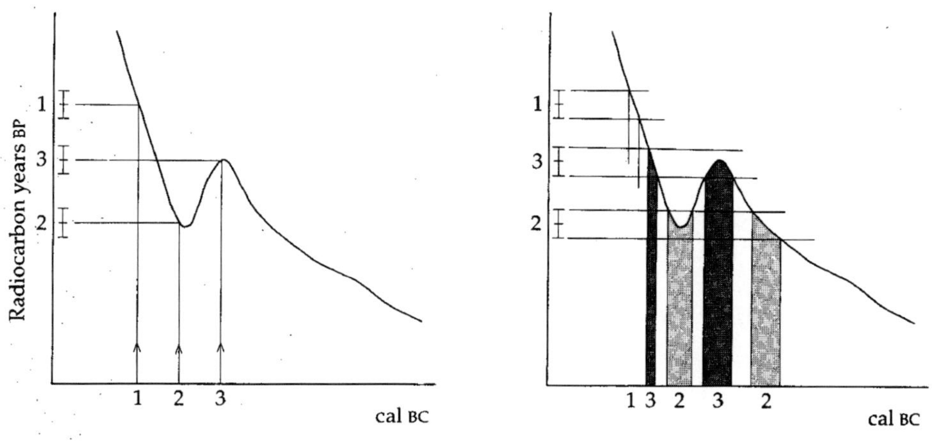

Figure 3.10 Three calendar dates (horizontal axis) appear in a different order on the radiocarbon timescale (Vertical axis). The figure on the right shows how calibration of the dates can explain the apparent discrepancy (Bowman 1995:60 Figure 25)

	OxCal	Calib	Bowman's formula
Chi-square χ^2	47.866	46.56	48.05
Value that χ^2 should not exceed P = 0.05	6.0	5.99	5.99

Table 3.1 The results of the chi-square at the site of Bavans, France. The group of radiocarbon determinations are statistically different.

The chi-square test can be performed in the calibration programmes OxCal and Calib. The three Neolithic dates from the site of Bavans, France[19] were used to test the comparability of the chi-square test result carried out in OxCal, Calib and using the formula given above (Table 3.1).

It is important to note that averaging has the effect of reducing the standard error of the mean; this may give a misleading impression of the accuracy of the radiocarbon age of a sample.

3.4 Summary

This rather pedantic account of radiocarbon dating has been considered necessary in view of the centrality of radiocarbon dates in this research. In Chapter 7 the problem of how to use the probability distribution of an irregular calibrated date in statistical analysis is addressed in more detail.

19

Sample ID	Radiocarbon determination (years BP)	Standard deviation
Lv-1415	7130	70
Lv-1588	6500	100
Lv-1590	6410	95

Table 2: Neolithic dates from the sites of Bavans, France.

Chapter 4: Patterns in the Neolithic of Europe: Part One

4.1 Introduction:

This chapter reviews the research on techniques to visualise the spatio-temporal patterns in radiocarbon record for the European Neolithic and Mesolithic as summarized by Clark (1965a/b), Ammerman and Cavalli-Sforza (1971, 1984) and Pinhasi *et al.* (2000). They are presented in chronological order of publication. Each method is replicated using the appropriate dataset from the Gkiasta database.

4.2 Clark (1965a, 1965b):

In 1965 Clark published the first radiocarbon date distribution map for the spread of the Neolithic to Europe, North Africa and the Near East (Figure 4.1).

For the area of Europe as defined in this research Clark uses 51 dates (excluding his African and Near Eastern sites). The dates are uncalibrated and they are grouped into three temporal categories. His map confirms what Childe had argued on archaeological grounds, namely that the Neolithic way of life penetrated Europe from the south-east, spreading from Greece and the south Balkans and then, with side branches northwards towards and across the Lower Danube and into Bessarabia and westward across the Adriatic to middle Italy, pushing to the middle Danube and so into middle Europe (Clark 1965a/b). Clark's method was reproduced using the oldest Neolithic radiocarbon determination at each site (Dataset C) (Figure 4.2 to Figure 4.6). Comparison of the >5200 BC range in Figure 4.1 and Figure 4.2 shows little difference between the current radiocarbon determinations and those from 1965. In the next category, 4000-5200 BC that the difference is great (Figure 4.1 and Figure 4.3). In Figure 4.1 the 4000-5200 BC group of radiocarbon determinations fall into a narrow corridor lying in a north-westerly direction. In Figure 4.3 determinations are spread out to the east, west and north of this corridor.

The 2800 – 4000 BC group is once again remarkably similar to the 1965 group, except for the greater sample size (Figure 4.1 and Figure 4.4). Clark does not have the category shown in Figure 4.5. These are radiocarbon determinations more recent than 2800 BC. Figure 4.6 shows all the temporal groups on one map, following Clark (1965a/b).

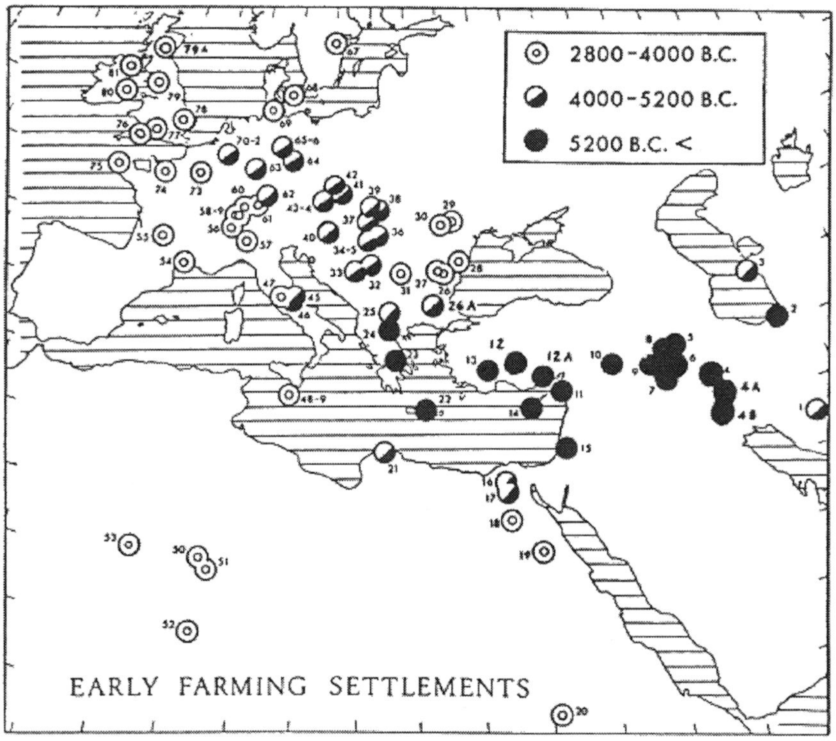

Figure 4.1 The distribution of early farming sites (Clark 1965:46).

Figure 4.2 Neolithic sites, radiocarbon determinations greater than 5200 BC.

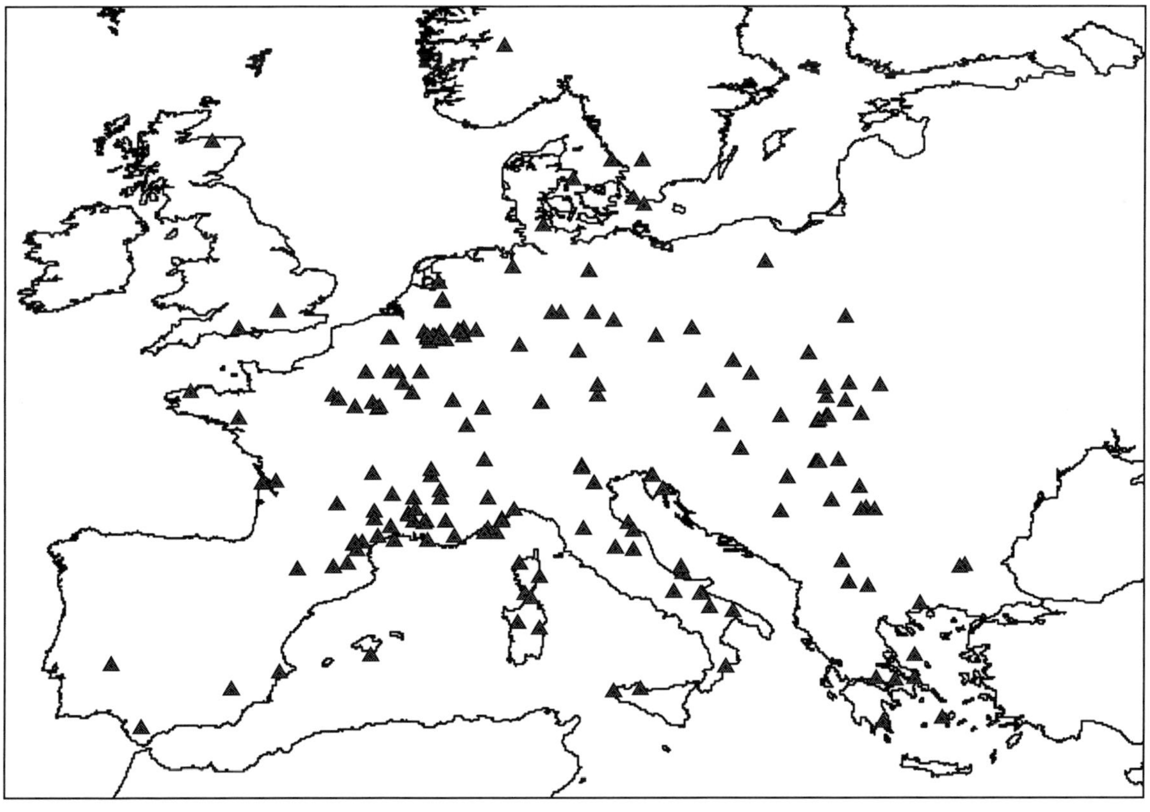

Figure 4.3 Neolithic sites, radiocarbon determinations for the period 4000 to 5200 BC.

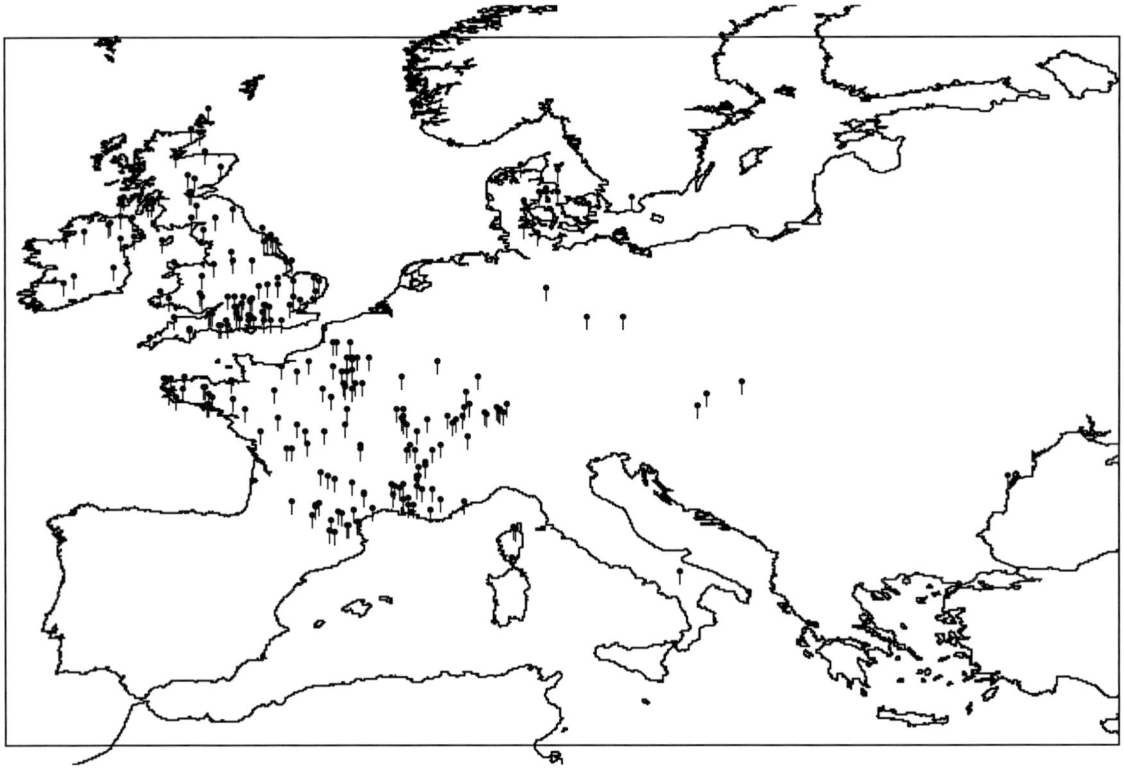

Figure 4.4 Neolithic sites, radiocarbon determinations within the time period 2800 to 4000 BC.

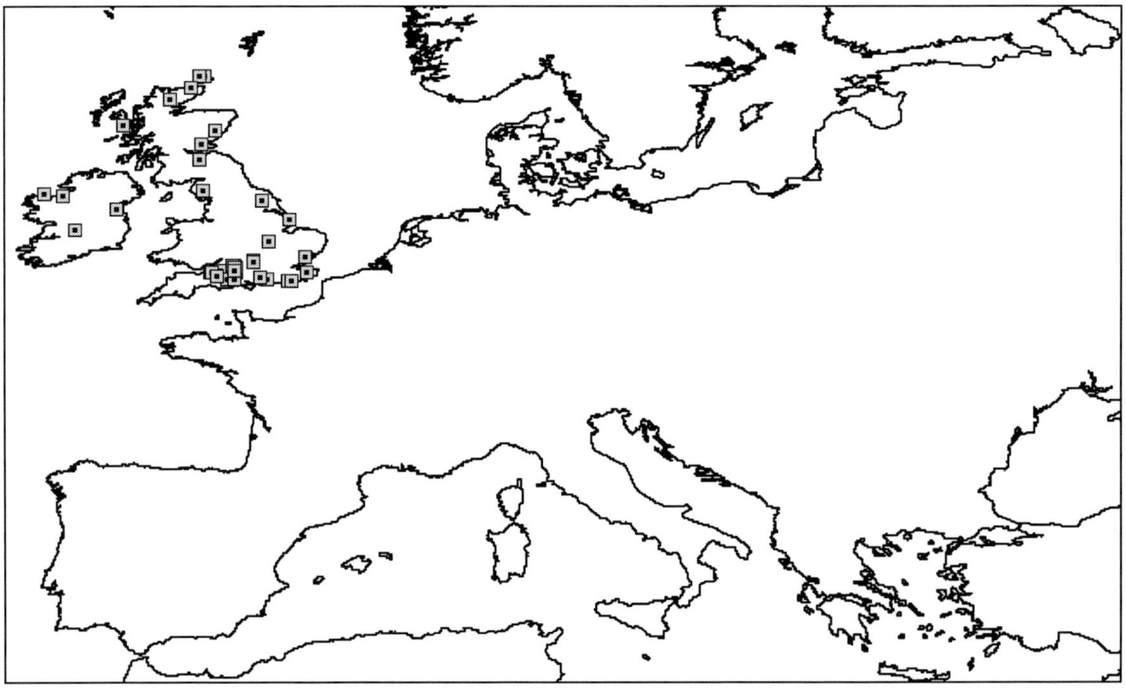

Figure 4.5 Neolithic sites, radiocarbon determinations that are more recent than 2800 BC.

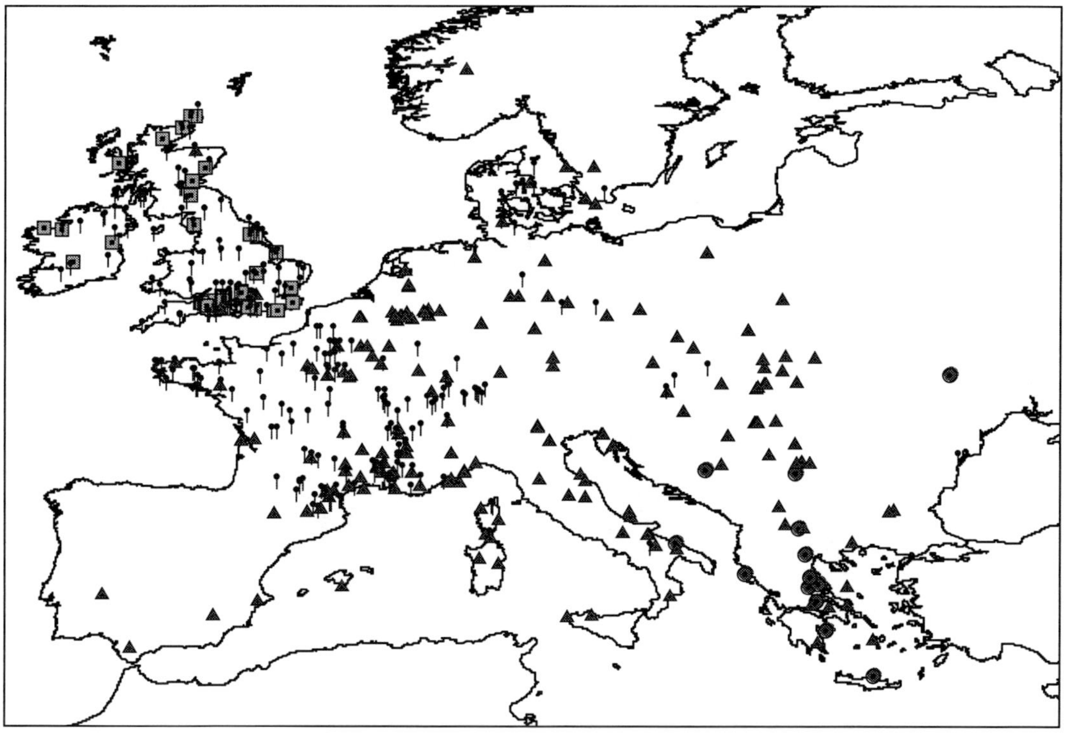

Figure 4.6 All the sites grouped within the 1200-year time interval (after Clark 1965a/b).

4.3 Ammerman and Cavalli-Sforza 1984: Isochron maps

4.3.1 The Neolithic:

In 1984 an isochron map was generated for the Neolithic of Europe using 106 radiocarbon determinations (Ammerman & Cavalli-Sforza 1984:59) (Figure 4.7). In this method contours are fitted to dates representing the same time interval. This method has advantages. Firstly, as with Clark's map, there is no need to specify an origin. Secondly, the map shows variations in the rate of spread of farming in different areas of Europe. Lines that are widely spaced show rapid spread, whereas delays in spread are shown by closely packed contour lines. For example, in Figure 4.7, the time delay for the spread of farming to the Alpine region is clearly shown by the contour line lag in this region.

To test that the method proposed to replicate Ammerman and Cavalli-Sforza's work was sound, their 1984 data list was used to replicate the 1984 isochron map (Figure 4.7). The geographic information systems (GIS) package, GRASS[1] was used to do this analysis. To create the continuous surface to which the contours could be fitted the method of inverse distance weighting (using the GRASS command: *s.surf.idw*) was used. This fills the spaces between the archaeological sites on the map by using inverse distance weighting to the 12 nearest neighbour sites for each point in the surface. This means that values on the interpolated surface are determined by the values of 12 sites closest to each point on the surface. Contours were fitted to the interpolated surface using an interval of 500 years, after Ammerman and Cavalli-Sforza. The isochron map produced is shown in Figure 4.8. Comparison of this figure and the isochron map in Figure 4.7 show that the surface interpolation and contour fitting method used here is indeed similar to that of Ammerman and Cavalli-Sforza.

This method was then applied to each of the Datasets A to G from the Gkiasta Neolithic database (Figure 4.9 to Figure 4.15).

The contour maps (Figure 4.9 to Figure 4.15) are all very similar to each other. They all show the overall trend seen in Ammerman and Cavalli-Sforza's 1984 contour map: the spread from old to more recent sites in a north westerly direction and the delay in the alpine region is still striking. However the pattern is not as smooth as that seen in Ammerman and Cavalli-Sforza's map. This is due to the fact that they filtered their data by only choosing the earliest date in each area.

[1] A detailed description of surface interpolation technique and the software used are found at http://www.geog.uni-hannover.de/grass/index2.html

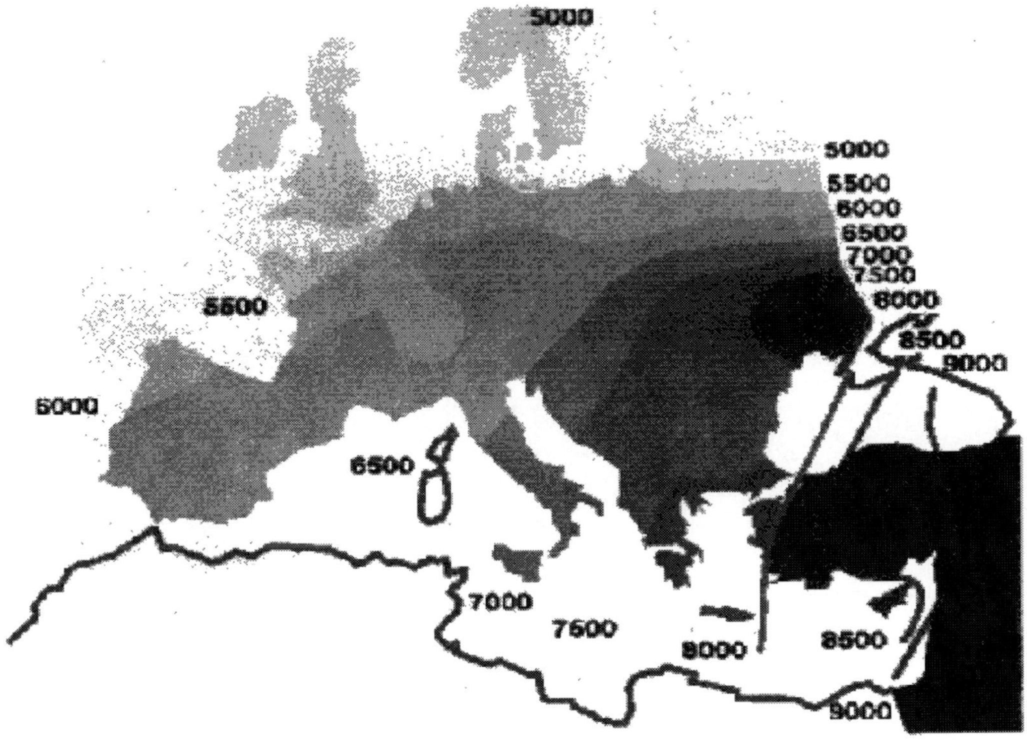

Figure 4.7 Isochron map for the spread of early farming to Europe (Ammerman & Cavalli-Sforza 1984:59, Figure 4.5).

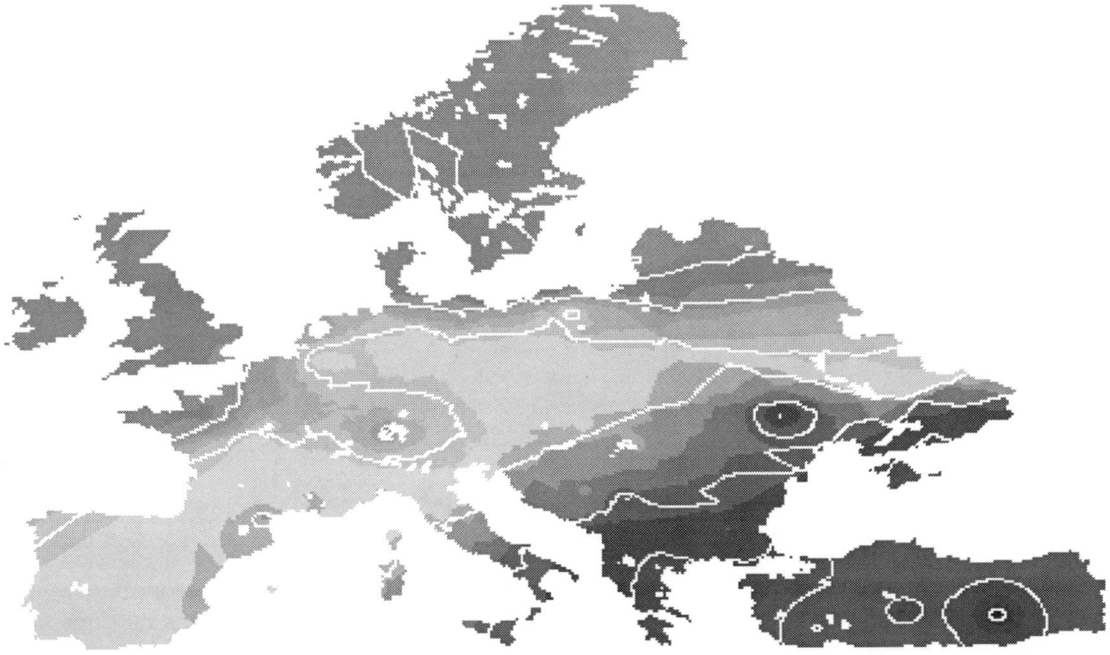

Figure 4.8 Contour map produced using Ammerman & Cavalli-Sforza's (1984) site list and the surface interpolation and contour fitting using the GIS software, GRASS.

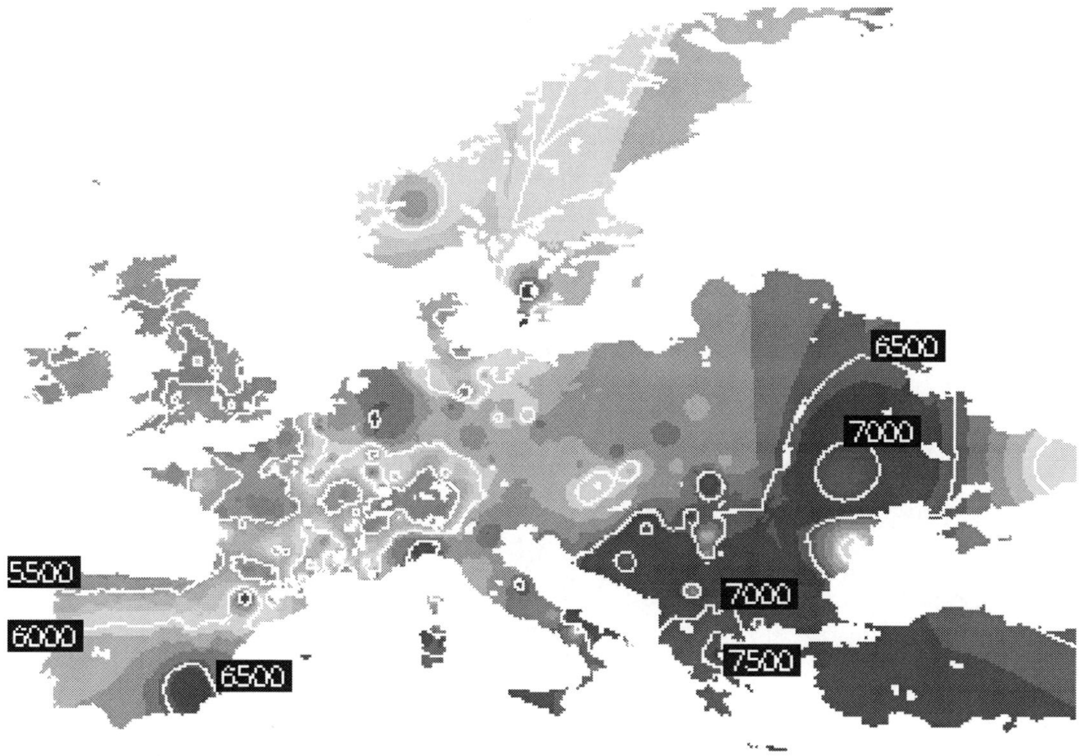

Figure 4.9 Contours fitted at 500-year intervals to dataset A (oldest or grouped radiocarbon determinations). This used 516 radiocarbon determinations.

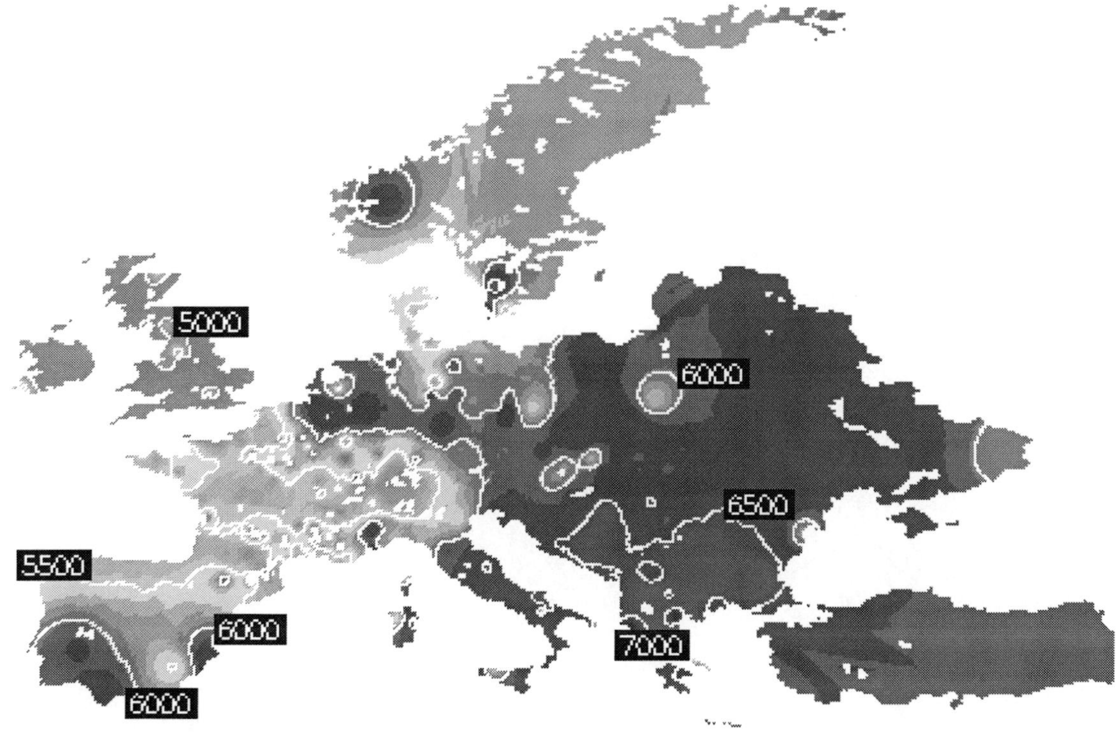

Figure 4.10 Contours fitted at 500-year intervals to Dataset B (most recent or grouped radiocarbon determination). This used 516 radiocarbon determinations.

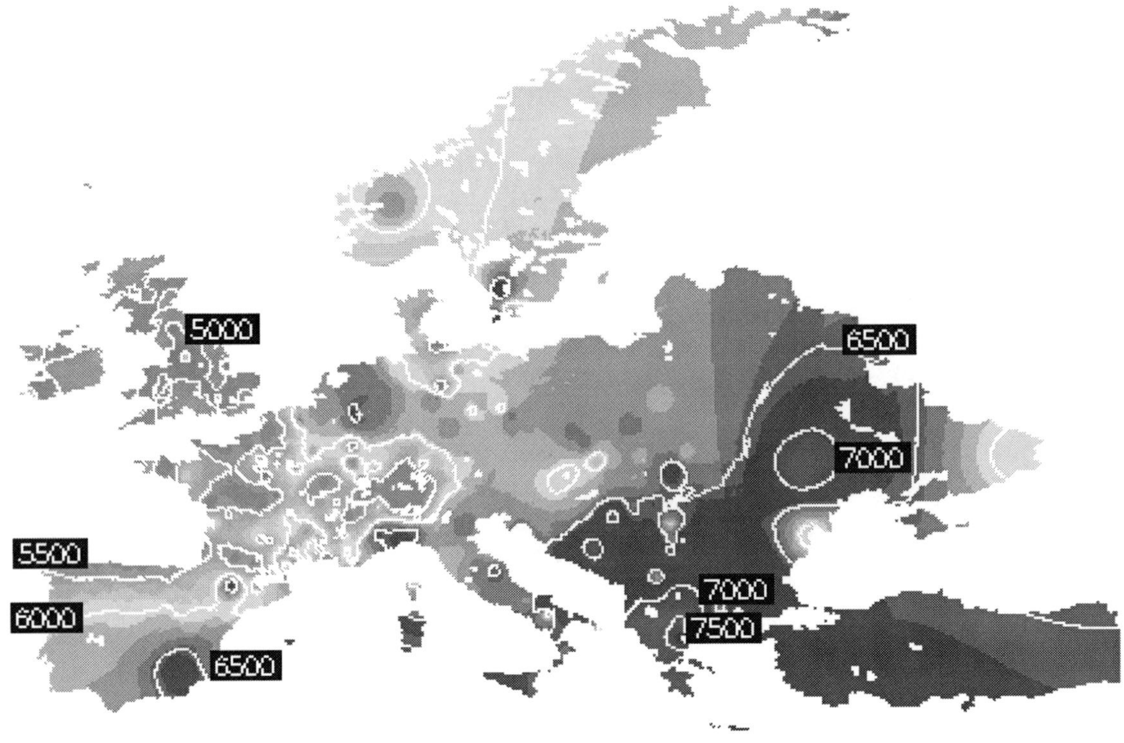

Figure 4.11 Contours fitted at 500-year intervals to Dataset C (oldest radiocarbon determination). This used 516 radiocarbon determinations.

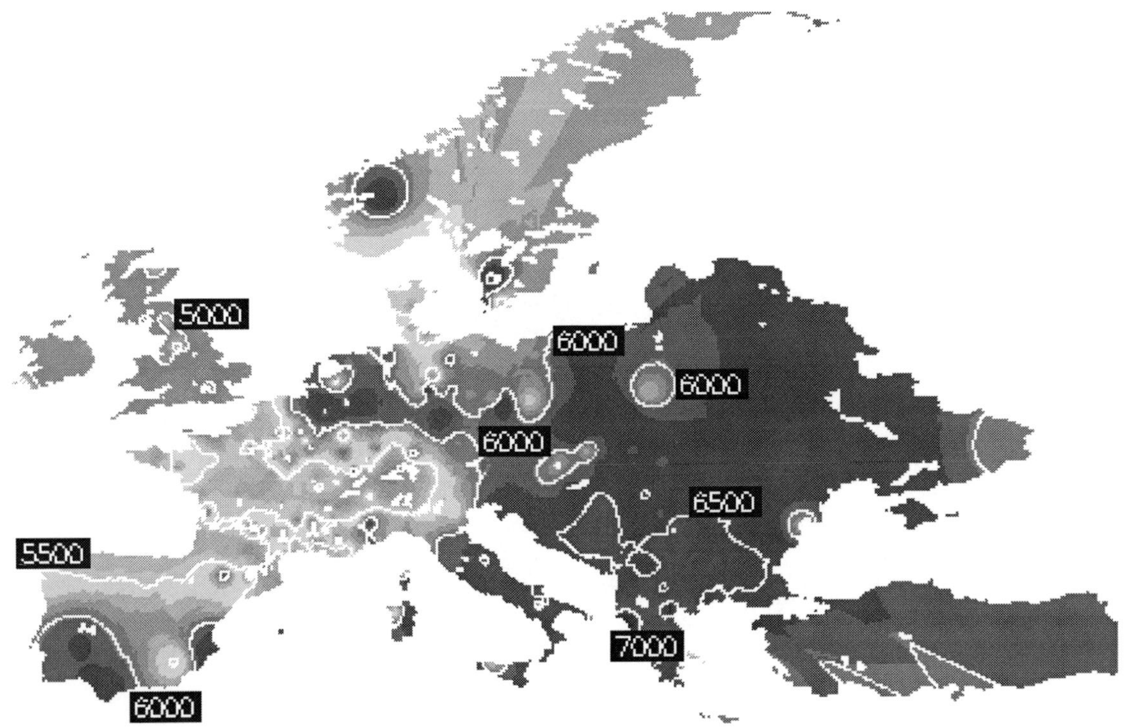

Figure 4.12 Contours fitted at 500-year intervals to Dataset D (most recent radiocarbon determination). This used 516 radiocarbon determinations.

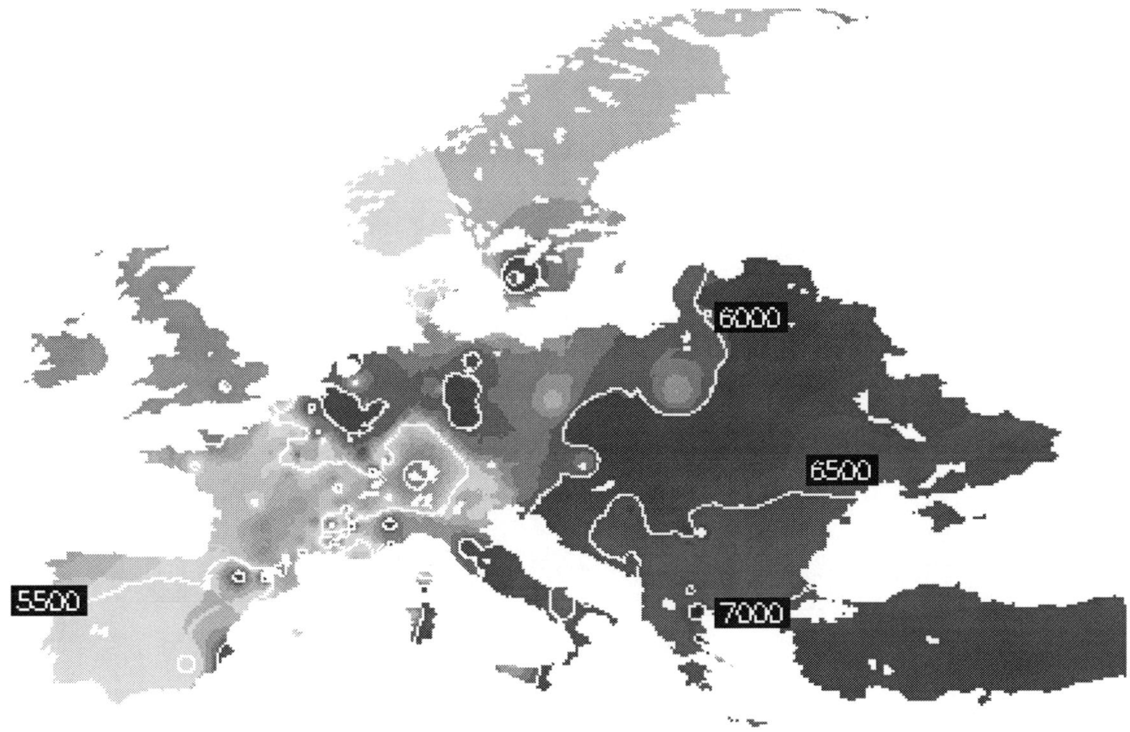

Figure 4.13 Contours fitted at 500-year intervals to Dataset E (most recent radiocarbon determination, multiple dated sites only). This used 268 radiocarbon determinations.

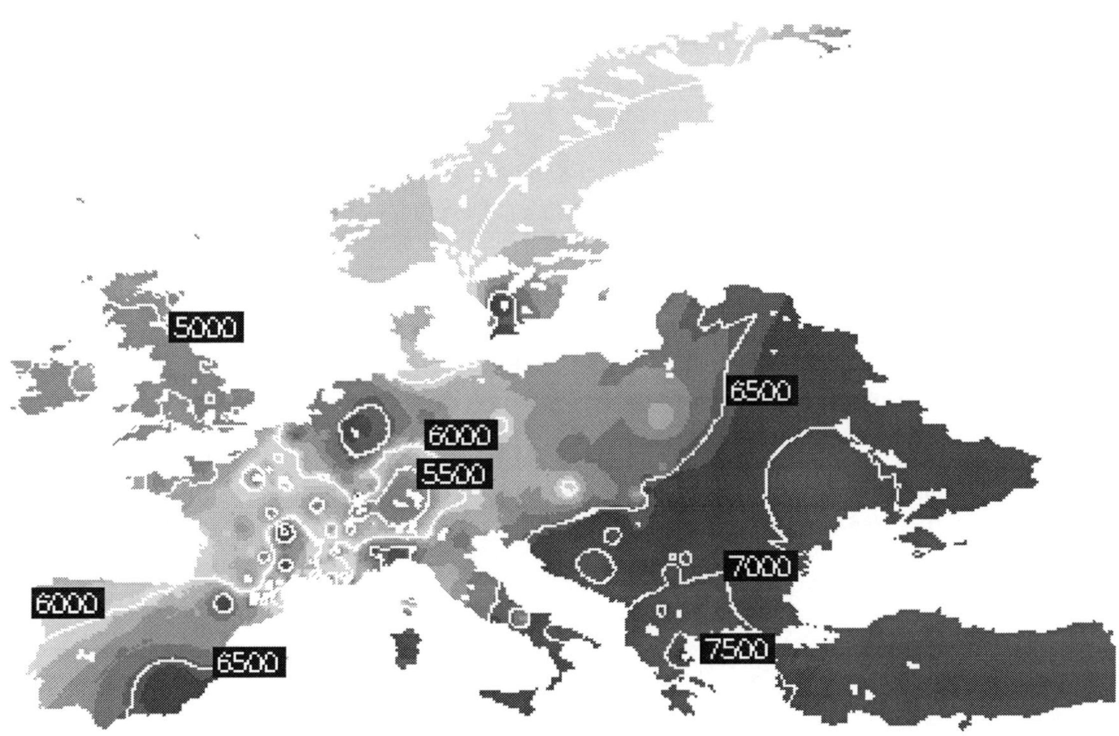

Figure 4.14 Contours fitted at 500-year intervals to Dataset F (oldest radiocarbon determinations, multiple dated sites only). This used 268 radiocarbon determinations.

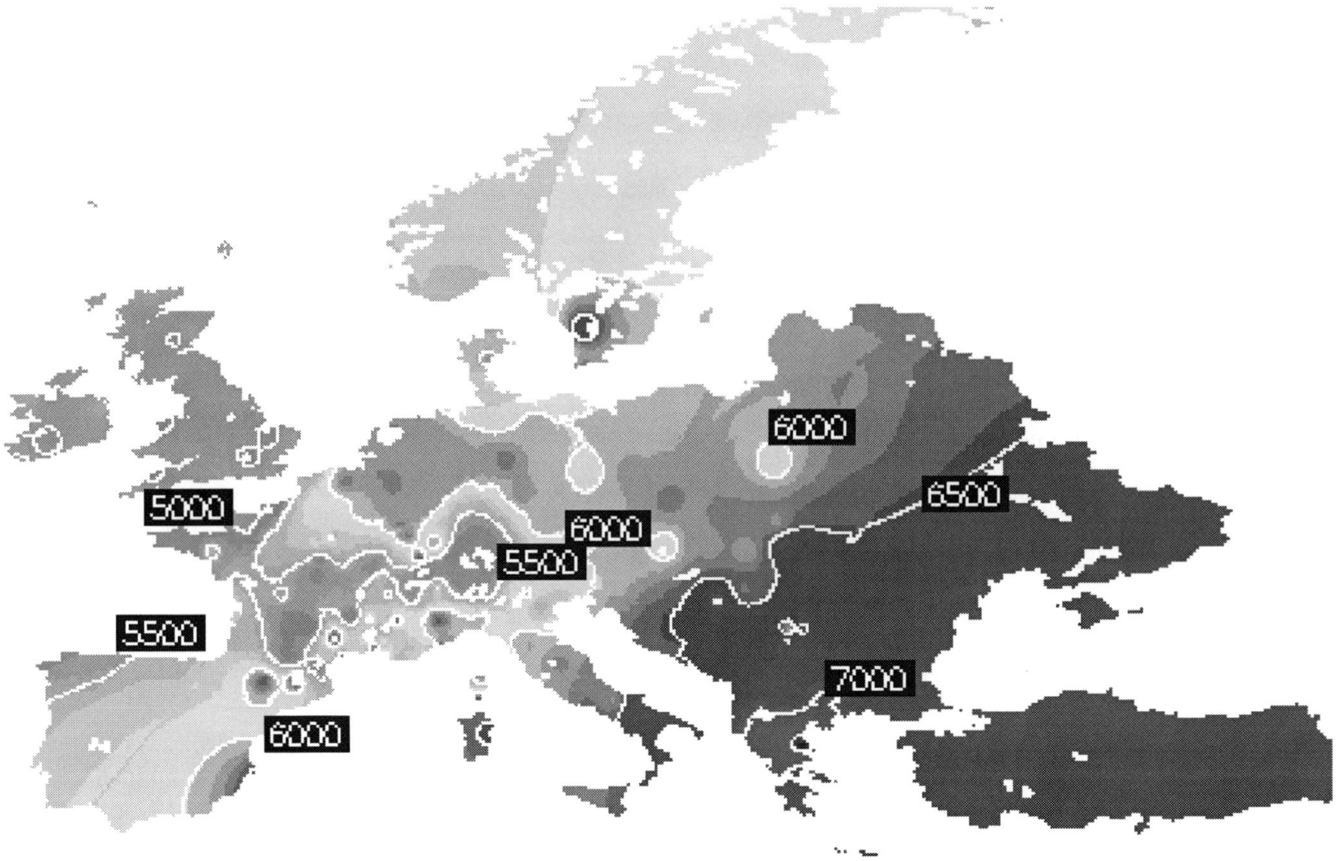

Figure 4.15 Contours fitted at 500-year intervals to Dataset G (grouped or median radiocarbon determination). This used 233 radiocarbon determinations.

4.3.2 The Mesolithic

Ammerman and Cavalli-Sforza applied the same technique to the analysis of 62 radiocarbon determinations for the 'latest' Mesolithic occupation of Europe (Figure 4.16). They then compare the isochron maps of the Neolithic and the Mesolithic to estimate the degree of overlap between the final Mesolithic and the earliest Neolithic, and interpret the good correspondence between the two as an indication of the lack of their chronological overlap (*ibid.* 1984), and therefore the lack of overlap between the two populations on the landscape.

An isochron map was generated using 107 Late Mesolithic radiocarbon determinations from the Gkiasta database (Figure 4.17). A site list is in Appendix 3. The Mesolithic isochron map (Figure 4.17) and those generated for the Neolithic (Figure 4.9 to Figure 4.15) show less correspondence than those of Ammerman and Cavalli-Sforza (therefore, more chronological overlap between the two groups).

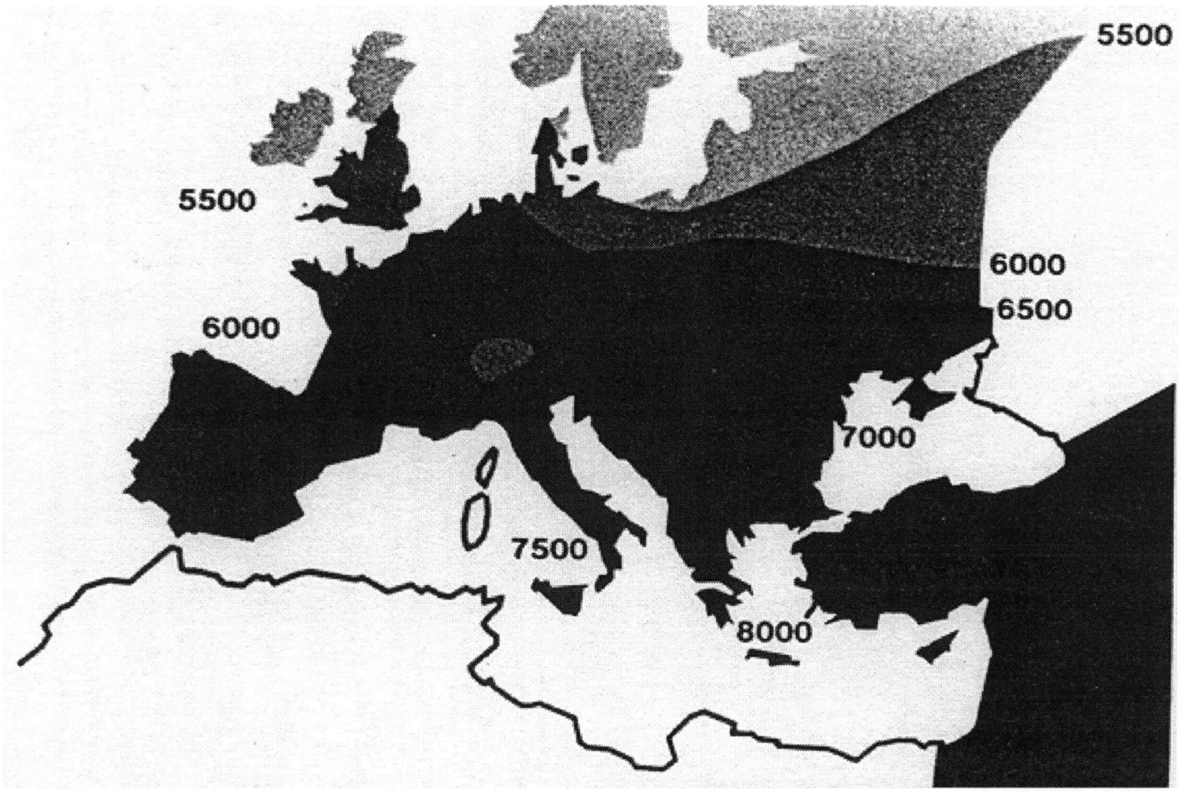

Figure 4.16 Ammerman and Cavalli-Sforza's isochron map of 62 of the 'latest' Mesolithic sites in Europe. Contours were fitted with an interval of 500 years (Ammerman and Cavalli-Sforza 1984:60, Figure 4.6)

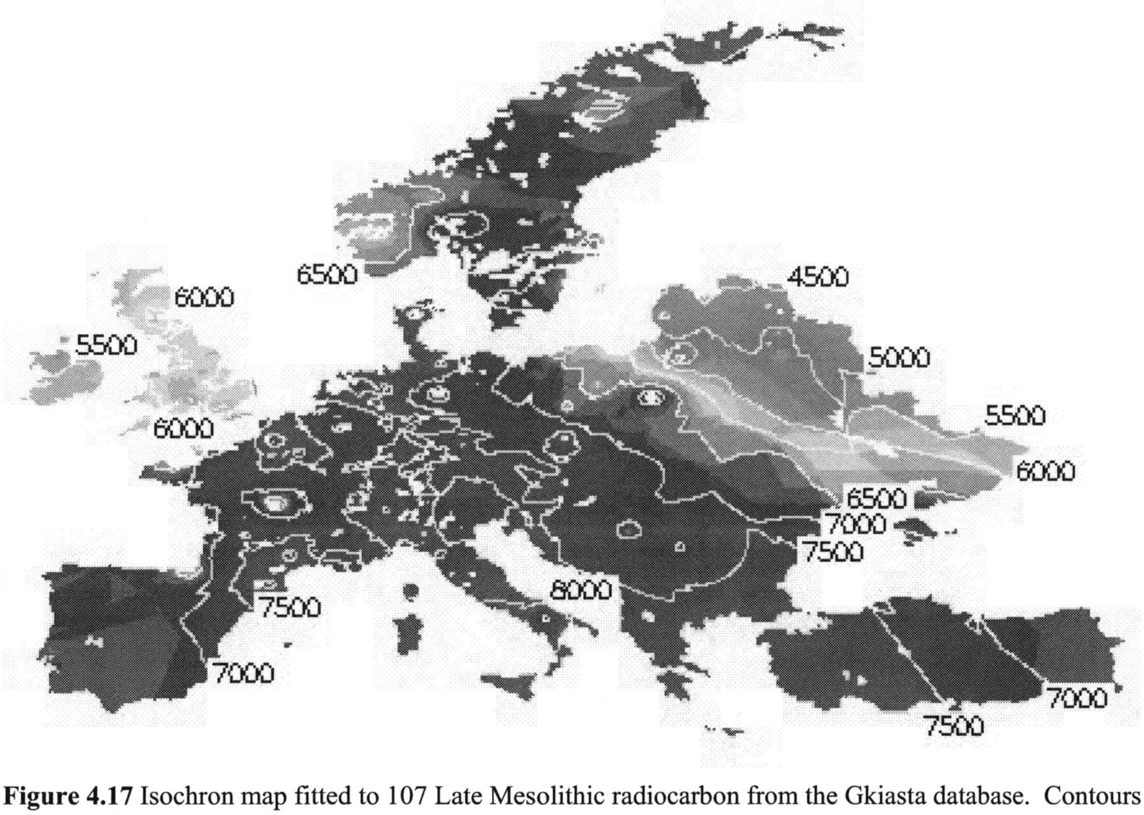

Figure 4.17 Isochron map fitted to 107 Late Mesolithic radiocarbon from the Gkiasta database. Contours have an interval of 500 years.

4.4 Pinhasi, Foley and Lahr (2000): Chronologically divided maps of archaeological data.

This is the most recent work on the spatial and temporal patterns in the archaeological record of the Palaeolithic, Mesolithic and Neolithic of Europe. The authors use 502 Mesolithic sites and 625 Neolithic sites in the analysis. Radiocarbon determinations are grouped into time bins with an interval of 2000 years for the periods 14000 to 12000 and 12000 to 10000 years BP. For 10 000 to 5000 BP an interval of 1000 years was used. Their results are shown in Figure 4.18 to Figure 4.22 (for the time period 10 000 to 5000 BP only). Note that Pinhasi *et al.*'s maps in Figure 4.18 and Figure 4.19 are identical: confirmation is needed that this is in fact a true reflection of the archaeological record in the period 10 000 BP to 8000 BP.

Pinhasi *et al.*'s method is identical to that of Clark (1965) except that the time bins they use cover a one thousand year period rather than a 1200-year period. Just as Clark did, they also take the earliest radiocarbon determination at each Neolithic site to use in the analysis. Pinhasi *et al.*'s method was reproduced using the oldest radiocarbon determination at each Neolithic site (Dataset C) and the Mesolithic sites (Appendix 3) from the Gkiasta database (Figure 4.23 to Figure 4.27). The reason for reproducing Pinhasi *et al.*'s method is that it provides a direct means of checking the new Gkiasta database. This is particularly useful as a check for gaps in the database caused by compilation oversight rather than real gaps in the archaeological record.

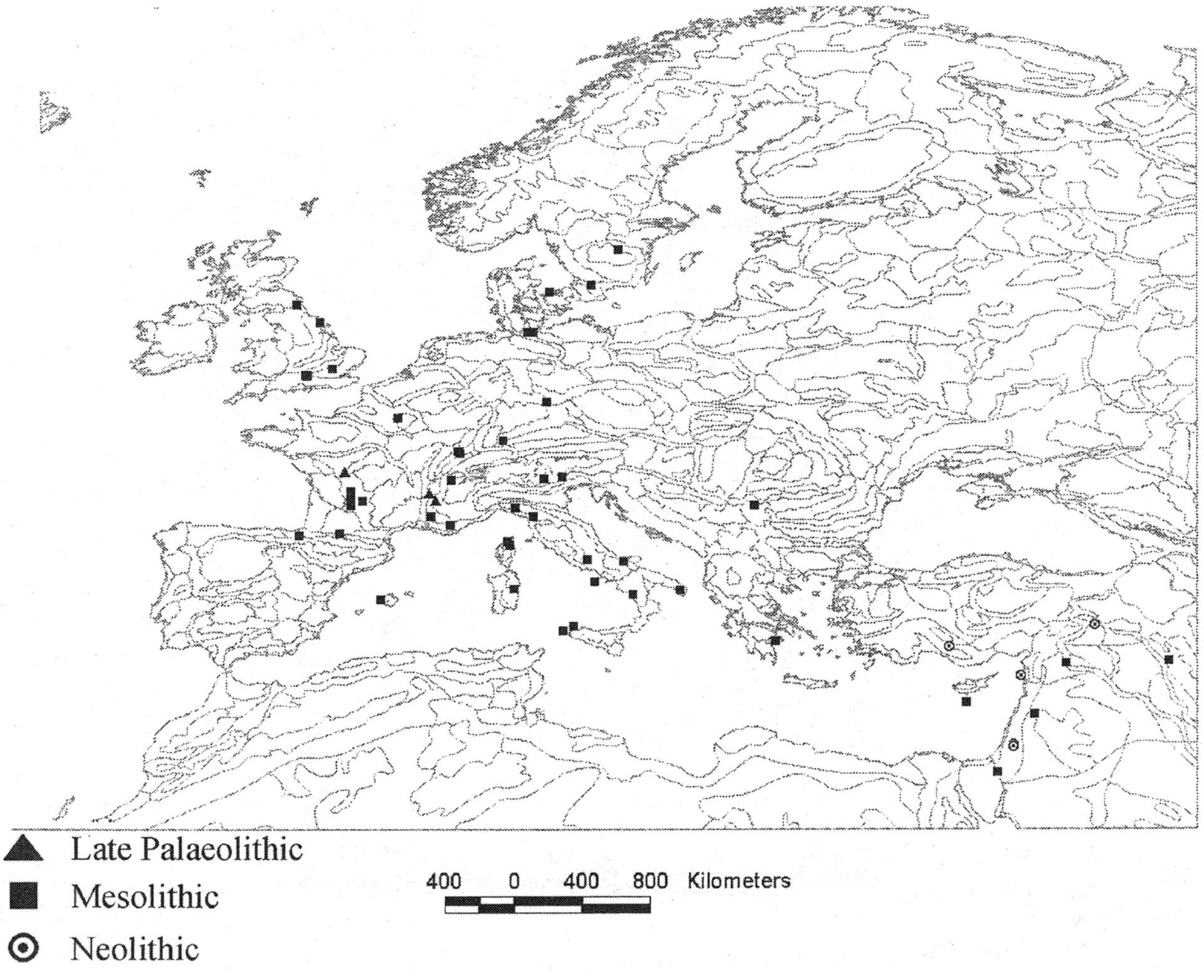

Figure 4.18 Archaeological sites, 10 000 to 9000 BP (uncal.)(Pinhasi *et al.* 2000:48, Fig. 6.3)

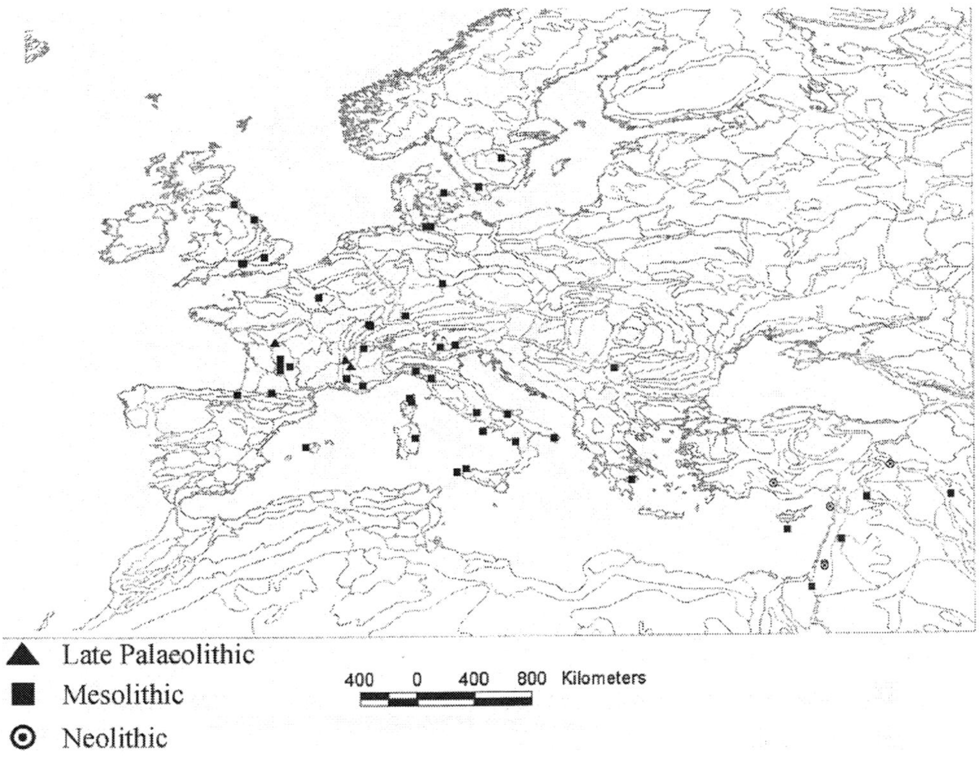

Figure 4.19 Archaeological sites, 9000 to 8000 BP (uncal.)(Pinhasi *et al.* 2000:49, Figure 6.4)

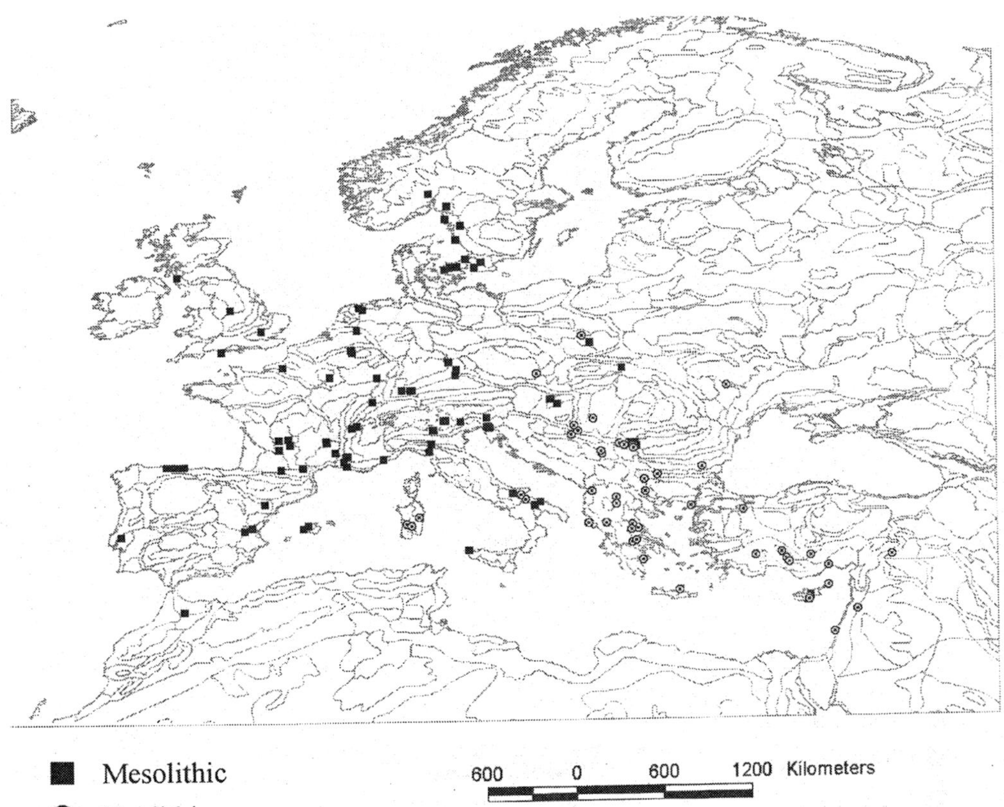

Figure 4.20 Archaeological sites, 8000 to 7000 BP (uncal.)(Pinhasi *et al.* 2000:50, Figure 6.5)

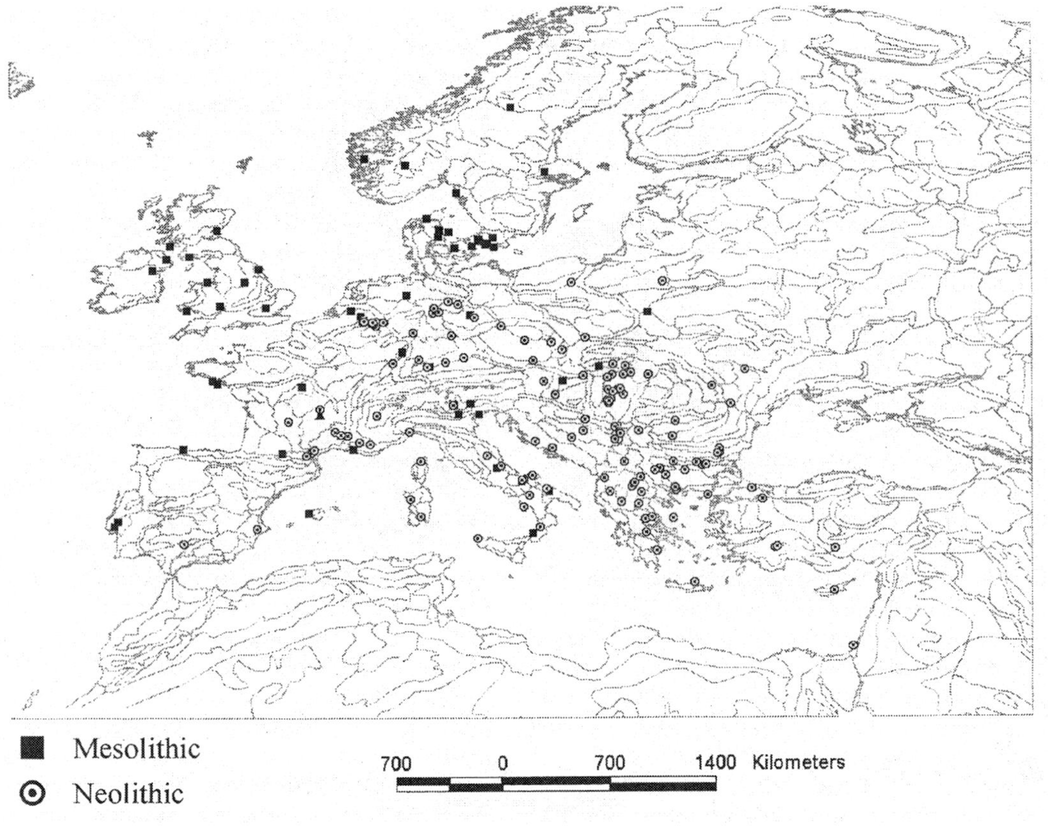

Figure 4.21 Archaeological sites, 7000 to 6000 BP (uncal.)(Pinhasi *et al.* 2000:51, Figure 6.6)

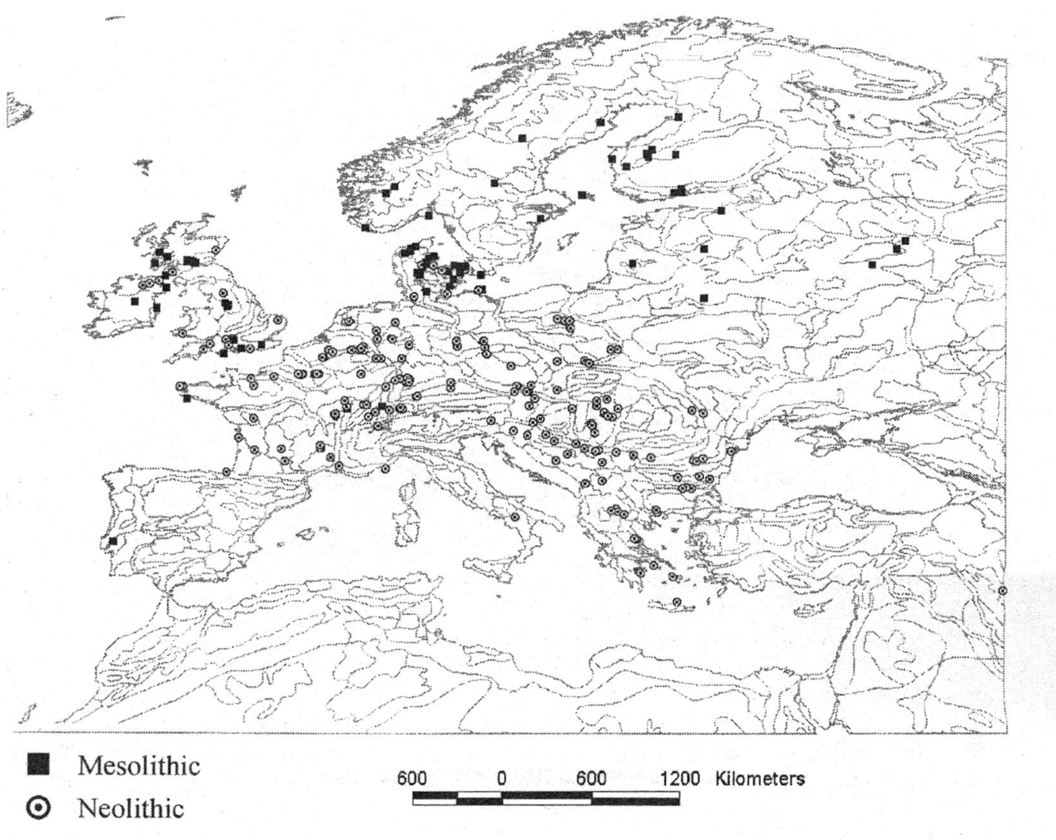

Figure 4.22 Archaeological sites, 6000 to 5000 BP (uncal.)(Pinhasi *et al.* 2000:52, Figure 6.7)

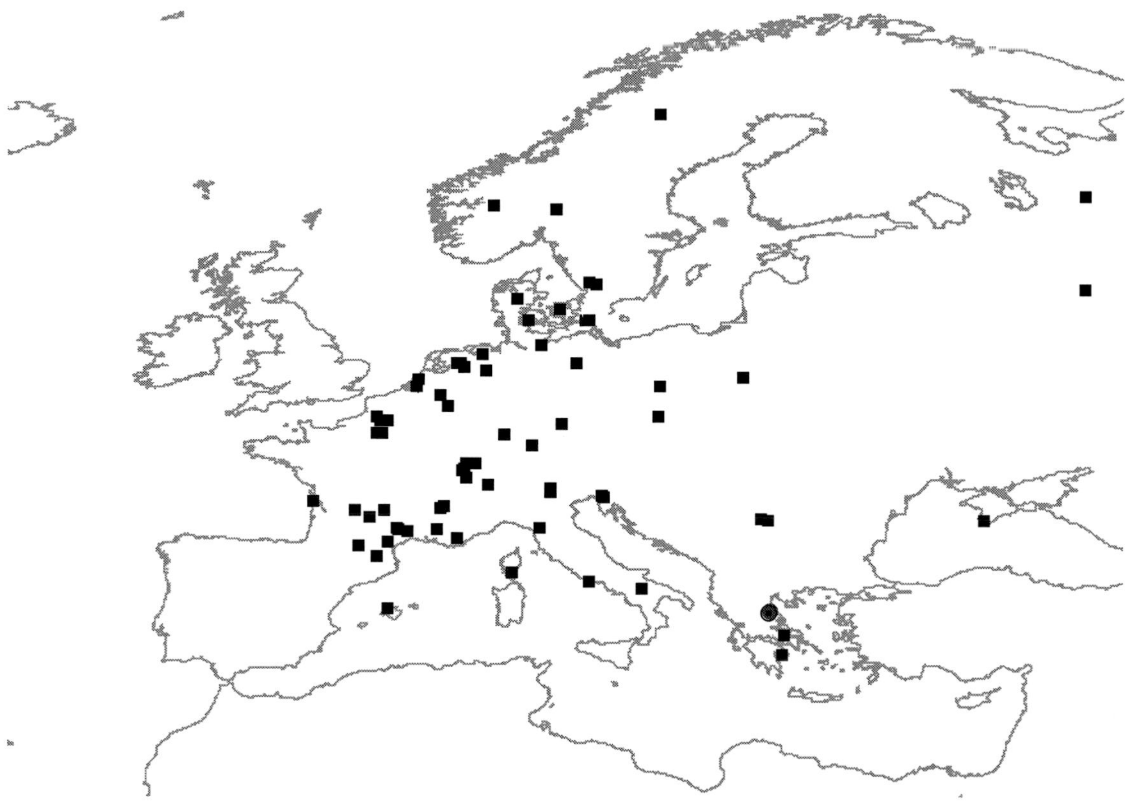

Figure 4.23 Mesolithic sites (square) and Neolithic sites (circle) for the period 9000 – 8000 years BP.

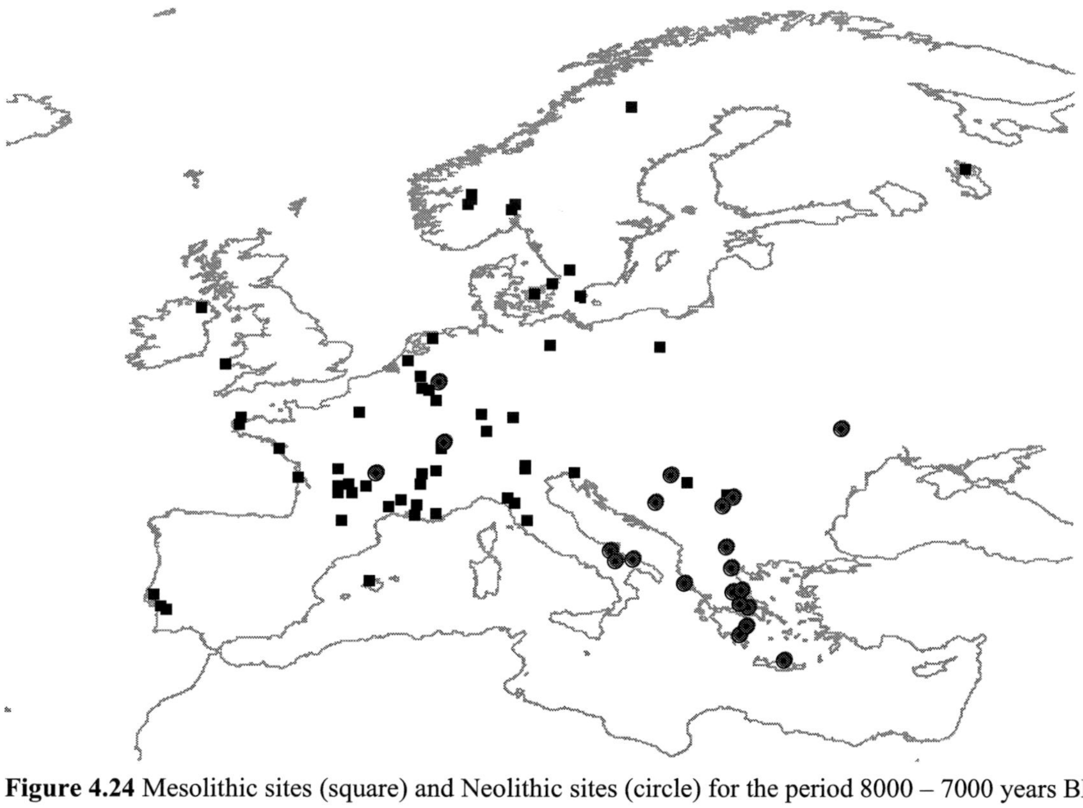

Figure 4.24 Mesolithic sites (square) and Neolithic sites (circle) for the period 8000 – 7000 years BP.

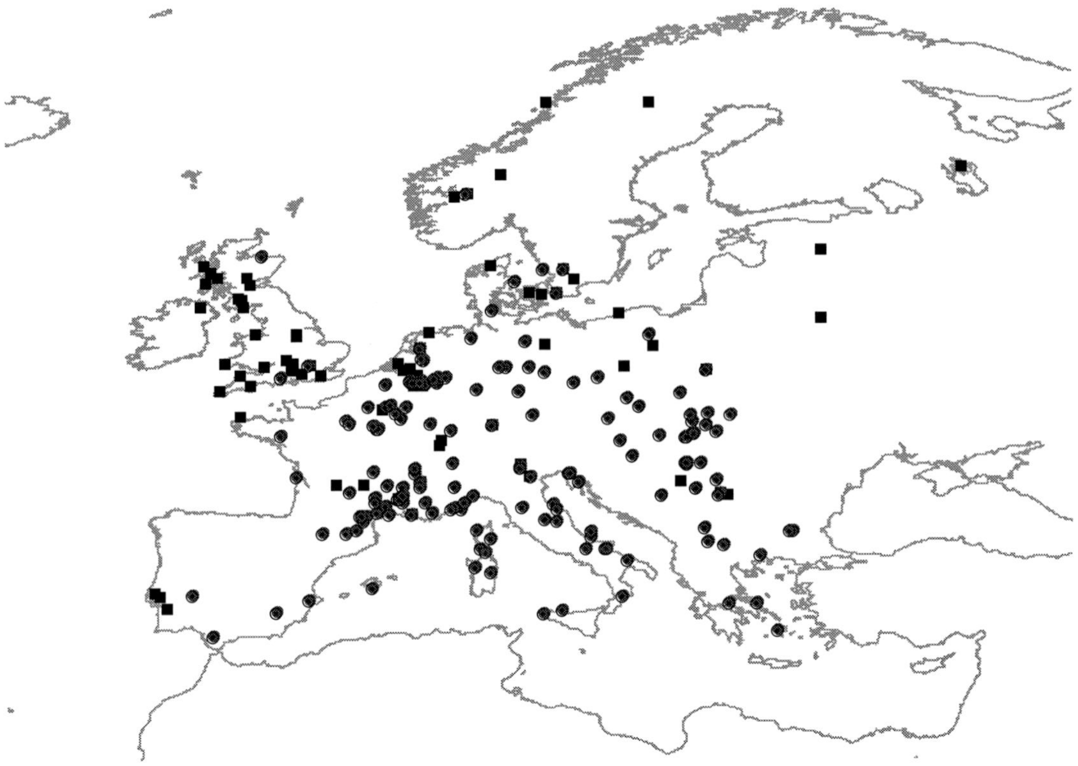

Figure 4.25 Mesolithic sites (square) and Neolithic sites (circle) for the period 7000 – 6000 years BP.

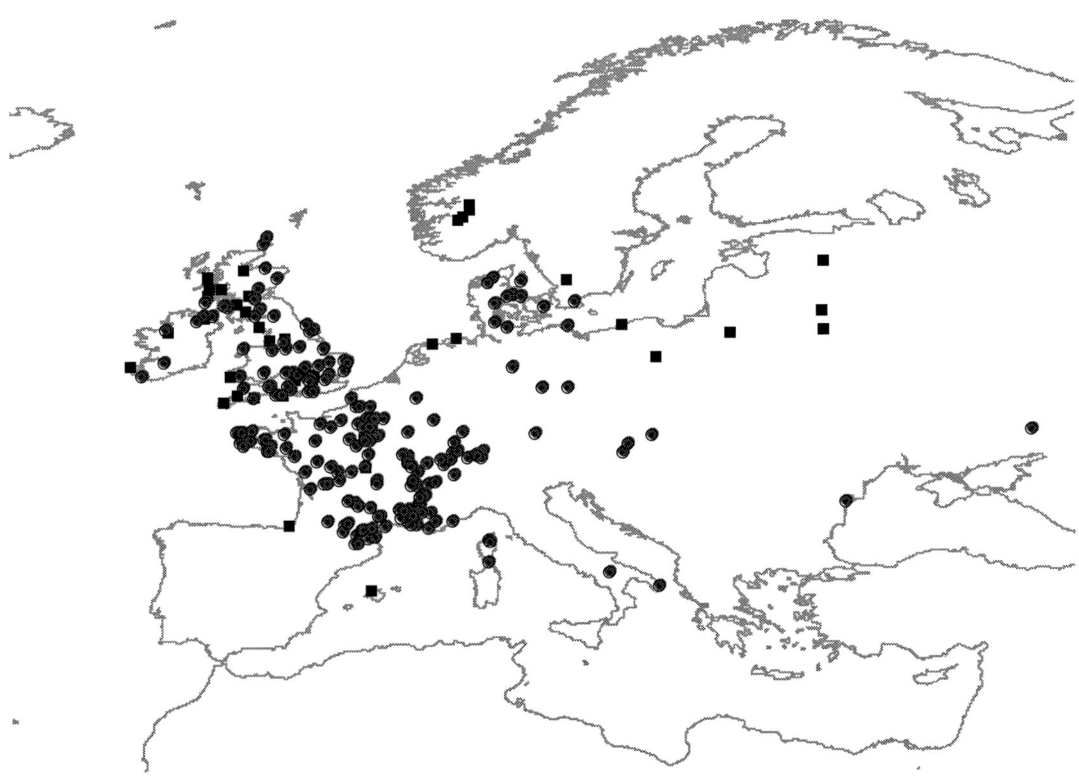

Figure 4.26 Mesolithic sites (square) and Neolithic sites (circle) for the period 6000 – 5000 years BP.

Figure 4.27 Neolithic sites (circle) for the period 5000 – 4000 years BP.

The Gkiasta database does not allow comparison with Pinhasi *et al.*'s 10 000 – 9000 BP map (Figure 4.18), because the Neolithic radiocarbon determinations that occur in this time period fall outside the geographic boundaries within which the Gkiasta database was collected and the identical distributions given for 10 000 - 9000 BP and 9000 – 8000 BP lead us to distrust the data for these periods. Comparison of our data in the 7000 – 6000 and 6000 – 5000 year periods with those of Pinhasi *et al.* shows that our database has more data for the France and the British Isles whilst Pinhasi *et al.* have more data for Greece and the south eastern European countries. An additional time period 5000 – 4000 years BP, absent in Pinhasi *et al.*'s work is shown for our data (Figure 4.27). There are no Mesolithic sites in this time period. This comparison highlights the need and scope for more comprehensive data collection in assembling databases.

4.5 Summary

Two different methods of visualising geographically referenced radiocarbon determinations: isochron maps and chronological maps have been presented and replicated using the Gkiasta database. For the Neolithic, they confirm the overall trend in the data: the first farming sites occur in the south east and become progressively more recent in the north westerly direction. In the next chapter methods to quantify trends in the data are presented.

Chapter 5 Patterns in the Neolithic of Europe: Part Two

5.1 Introduction:

Visualisation is a useful tool for initial exploratory data analysis but methods are required that quantify the trends that the eye might see in order that they can be tested and explained. In this vein, Ammerman and Cavalli-Sforza (1971) used linear regression from an assumed origin to measure the rate of spread of farming across Europe. A problem with this technique was the need to assume an origin. The spread of the European Neolithic is not the best process to illustrate how problematic this assumption is, since an origin in the Near East is generally undisputed. There are other cases where the origin of spread is unclear, for example the peopling of the Americas, the spread of the Bantu farmers to southern Africa, and the spread of domestic stock to the western half of southern Africa. To overcome this, a method is presented in this chapter to search for the most likely origin point for a set of radiocarbon determinations.

5.2 Locating the origin of diffusion:

In a paper published in 1971, Ammerman and Cavalli-Sforza calculated the rate of spread of farming into Europe by regression of radiocarbon determinations from the archaeological sites with their distance from an origin. To do this they needed to identify an origin point for this spread. To locate such an origin they tested four possible centres of spread. These were the Near Eastern sites of Jericho, Cayonu, Ali Kosh and Jarmo (Figure 5.1). This choice was based on archaeological evidence for the early domestication of cereals in this area. Additionally, they used eight early farming sites in the Near East to locate a 'centre of gravity' of spread (34°04'N latitude, 41°09'E longitude) to use in similar analyses (Figure 5.1).

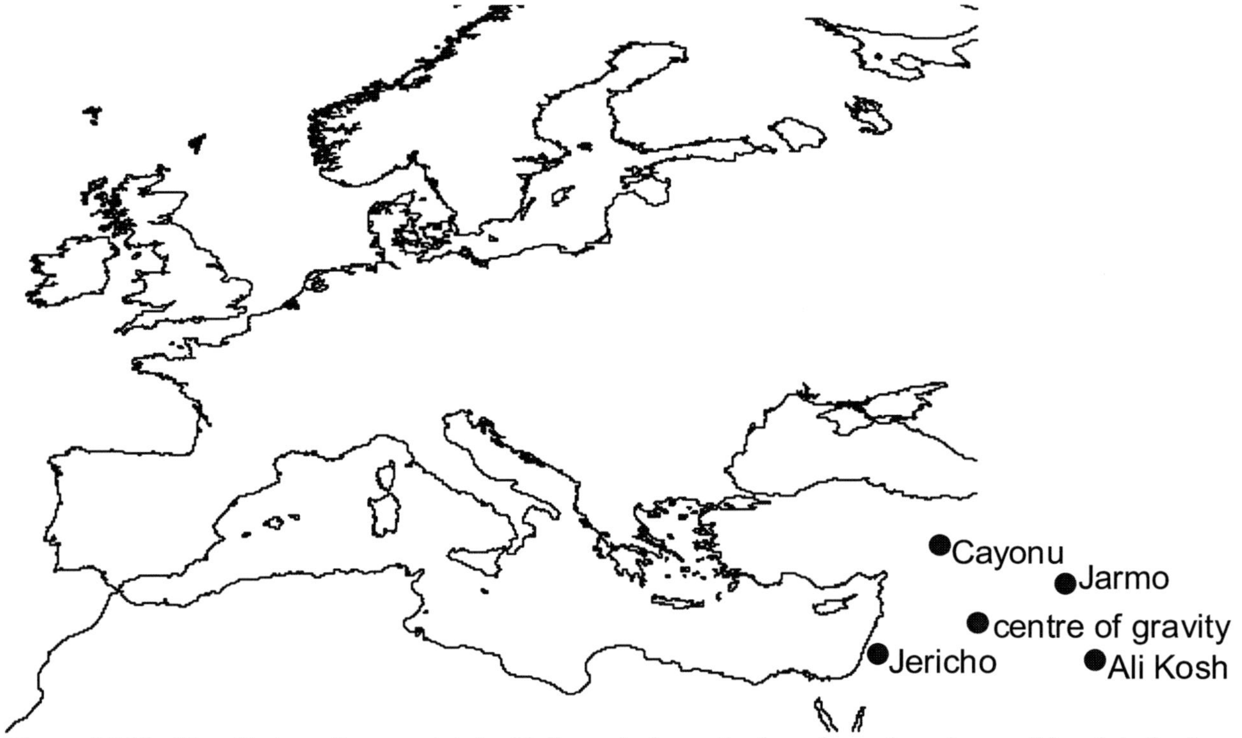

Figure 5.1 The Near Eastern sites associated with the early domestication of cereals and a possible origin for the European Neolithic spread as suggested by Ammerman and Cavalli-Sforza (1971).

5.2.1 A search for an origin of spread:

A paragraph in Ammerman and Cavalli-Sforza's 1971 paper suggested the procedure that will be outlined in this section.

> "We should also consider the case where the centre of diffusion is not known. In theory if the data are abundant, it is possible to search for centres of diffusion. We can compute correlation coefficients for a large number of different potential centres and the centre giving the highest correlation coefficient can be taken as that most likely for the set of data".
> (Ammerman & Cavalli-Sforza 1971:676)

Rather than assuming an origin in the Near East the exercise outlined here searched for the origin across Europe. The numeric computation and visualisation software, MATLAB was used to write a script (Appendix 6) that was designed to systematically move through latitude and longitude coordinates in increments of 2 decimal degrees (equivalent to ± 160 km). The latitude range is 30° N to 70° N and the longitude range is from –10° W to 36° E, giving a total of 504 possible origins across Europe (Figure 5.2 and Figure 5.3).

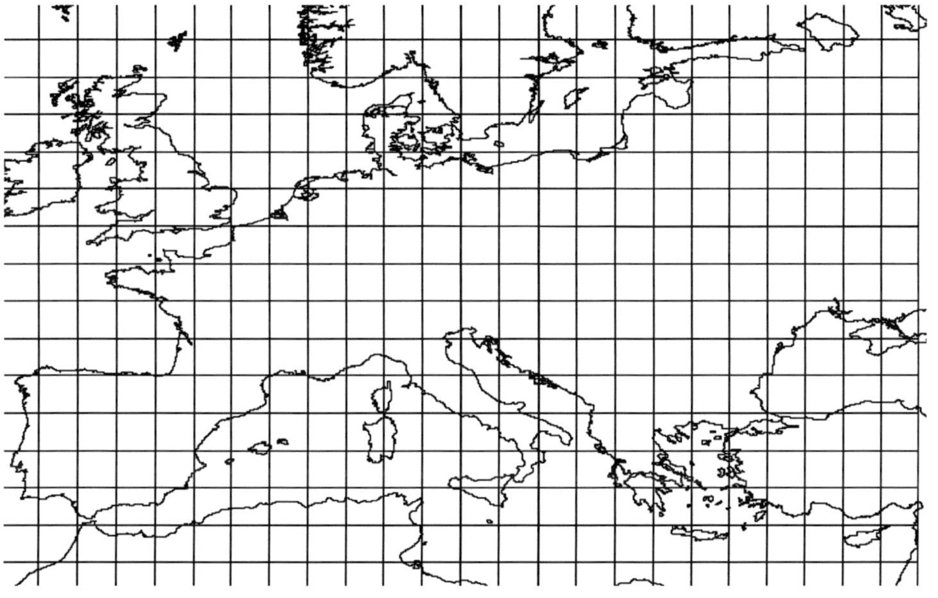

Figure 5.2 Grid to mark potential origin points.

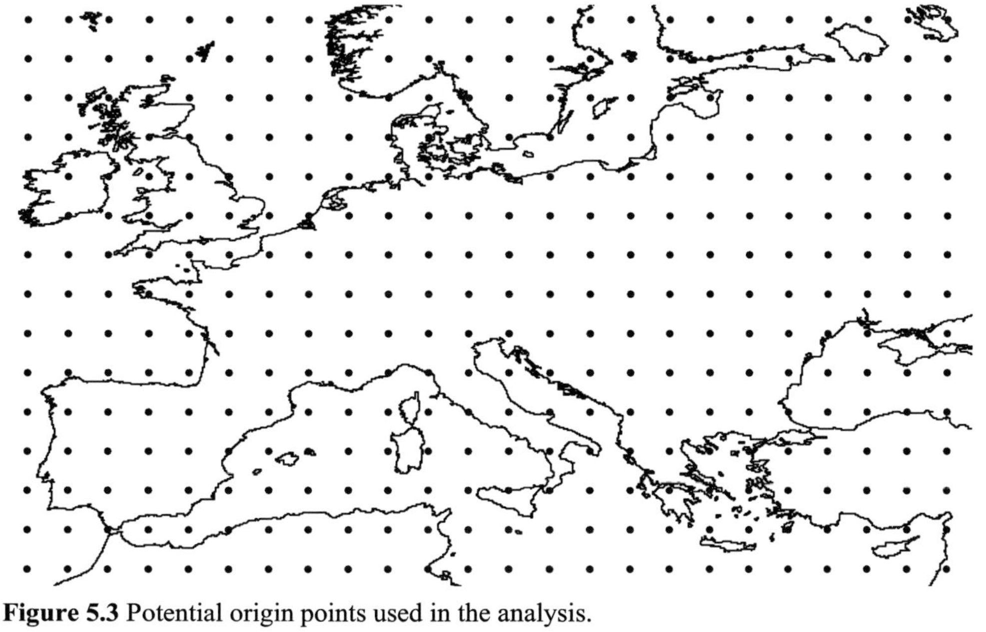

Figure 5.3 Potential origin points used in the analysis.

At each of these 504 potential origins the geodesic distance was calculated to each of the sites (as listed in each of the seven datasets). The equation used to calculate the geodesic distance, d, between two points (1 and 2) on a sphere was:

$$d = a\cos^{-1}\left[\cos\delta_1 \cos\delta_2 \cos(\lambda_1 - \lambda_2) + \sin\delta_1 \sin\delta_2\right]$$

a = 6378 km (equatorial radius of the earth).
δ = longitude (radians).
λ = latitude (radians).
(http://mathworld.wolfram.com/GreatCircle.html).

These results were then used to find the correlation between radiocarbon determination and distance from each origin point to all Neolithic sites. Pearson's product moment of correlation, r, was calculated in each case[1].

5.2.2 *Results: most probable origin areas calculated for Datasets A to G.*

The results for all of the datasets A to G show similarities. The mapped result of all the correlation coefficients calculated for each database shows two areas of strong correlation in all the datasets. There is a strong positive correlation between radiocarbon determination and distance that peaks in north western Europe (the red peak in Figure 5.4, Figure 5.7, Figure 5.10, Figure 5.13, Figure 5.16, Figure 5.19, Figure 5.22). There is an equally strong negative correlation in south eastern Europe and the Near East (the dip in the same figures). The interest is in the location of an origin. Potential origins are at points with negative correlation coefficients. These reflect the strong and inverse trend between large distances and large determinations. For a full listing of all the results for each dataset see Appendix 5.

5.2.2.1 *Dataset A (grouped radiocarbon determinations or oldest radiocarbon determinations):*

The origin with the best fit to Dataset A gives a correlation coefficient of –0.74 (the dip in Figure 5.4). This corresponds to an origin in south-western Jordan. Contour lines were fitted to the correlation coefficients to show the trend across Europe (Figure 5.5). Widely spaced lines in the area of the strong negative correlation correspond to a wide region that has large negative correlations (and therefore, potential as an area of origin).

In Figure 5.6 only correlation coefficients equal to or smaller[2] than -0.7 are mapped, with the size of the symbols decreasing with decreasing correlation coefficient size. The strongest correlation coefficient, shown as a paler circle, occurs in south western Jordan. High correlation coefficients in Egypt must be treated with some caution. They are calculated by taking geodesic distance to sites in Europe by treating the land as a homogenous surface. This ignores the sea that lies between Egypt and the most direct geodesic line to Europe.

[1] Pearson's product moment correlation coefficient, r, is:

$$r_{xy} = \frac{W_T \sum_k w_k x_k y_k - \left(\sum_K w_K x_k\right)\left(\sum_k w_k y_k\right)}{\sqrt{\left[W_T \sum_k w_k x_k^2 - \left(\sum_k w_k x_k\right)^2\right]\left[W_T \sum_k w_k y_k^2 - \left(\sum_k w_k y_k\right)^2\right]}}$$

x, y = values of variables (i.e. geodesic distance and radiocarbon determination).
w = value of weight (i.e. 1/ standard deviation of radiocarbon determination).
k = subscript for case.
W_T = total sum of the weights.
(http://www.unesco.org/webworld/idams/Doc/ManualHtml/E2pearso.htm)

[2] These are negative numbers so the smaller the correlation coefficient the stronger the correlation.

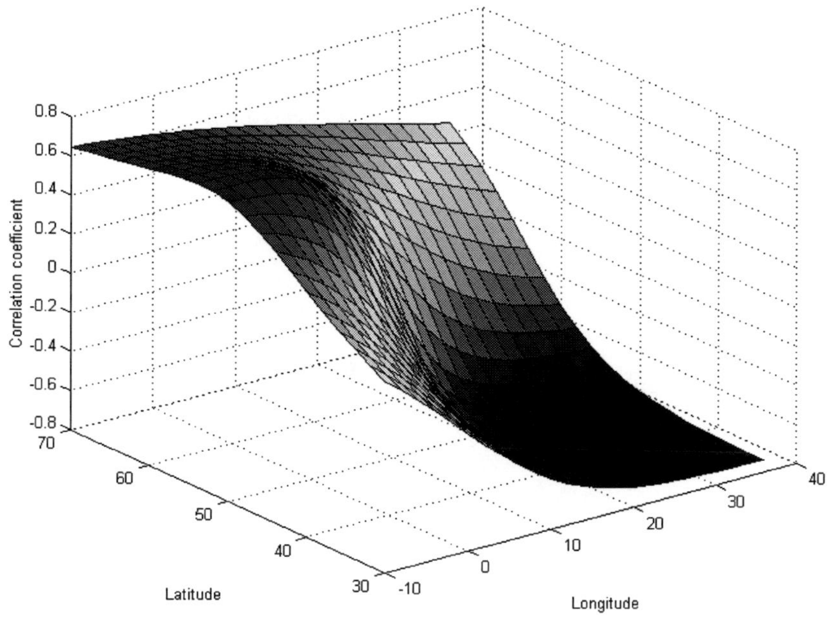

Figure 5.4 Correlation coefficients for Dataset A.

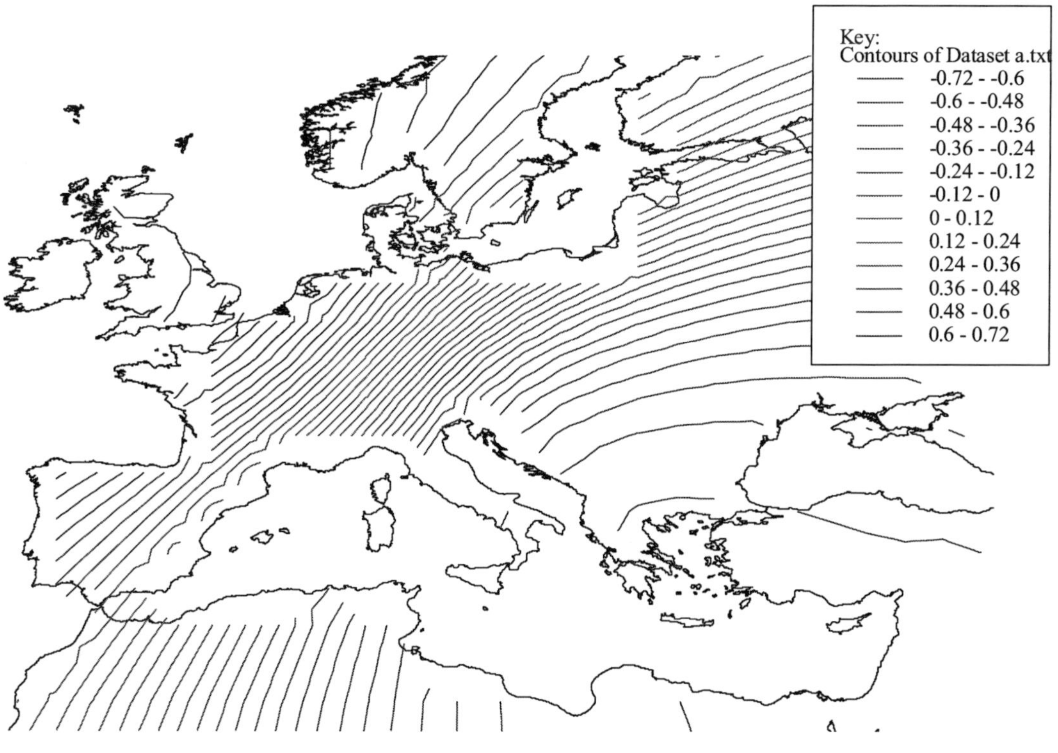

Figure 5.5 Contour lines fitted to the correlation coefficients calculated for Dataset A.

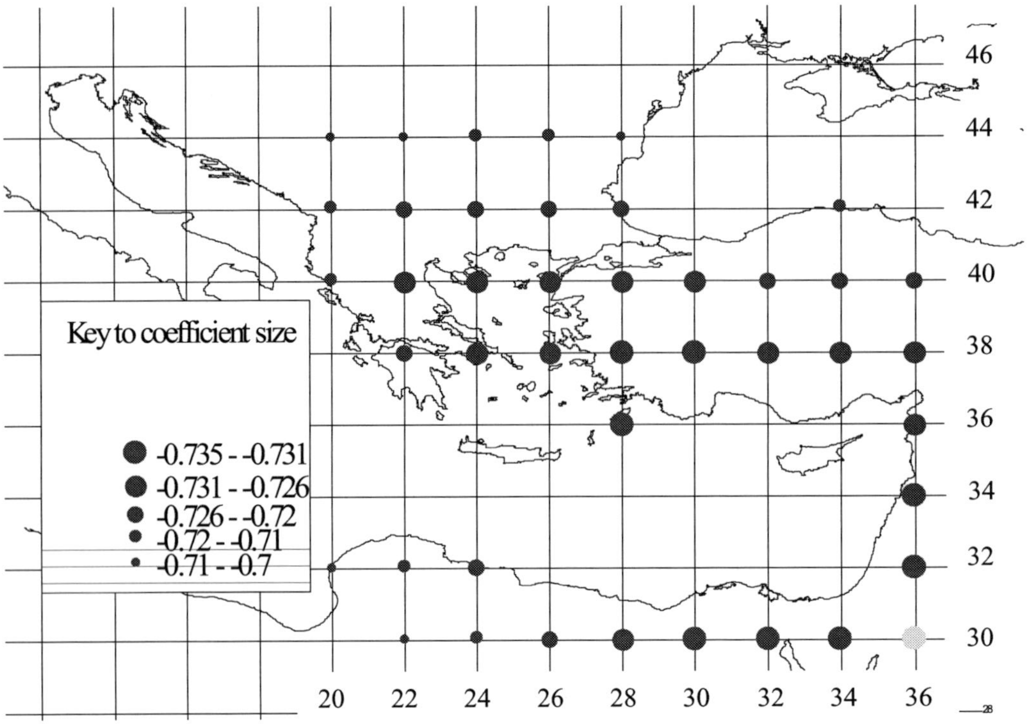

Figure 5.6 The origin points that gave the best fit to Dataset A (-0.7 or smaller). The marker with the strongest correlation coefficient (-0.7352, at 36°E, 30°N) is shown as a paler circle.

5.2.2.2 Dataset B (grouped radiocarbon determination or most radiocarbon determination):

The origin point that gives the strongest fit to Dataset B has a correlation coefficient of –0.73 (the dip in Figure 5.7). As with Dataset A the contour lines fitted to the correlation coefficients are widely spaced over a wide region where the correlation coefficient is large and negative (Figure 5.8).

The potential origin centre has moved in this case. It is located in the Rhodope Mountain range in southern Bulgaria (Figure 5.9).

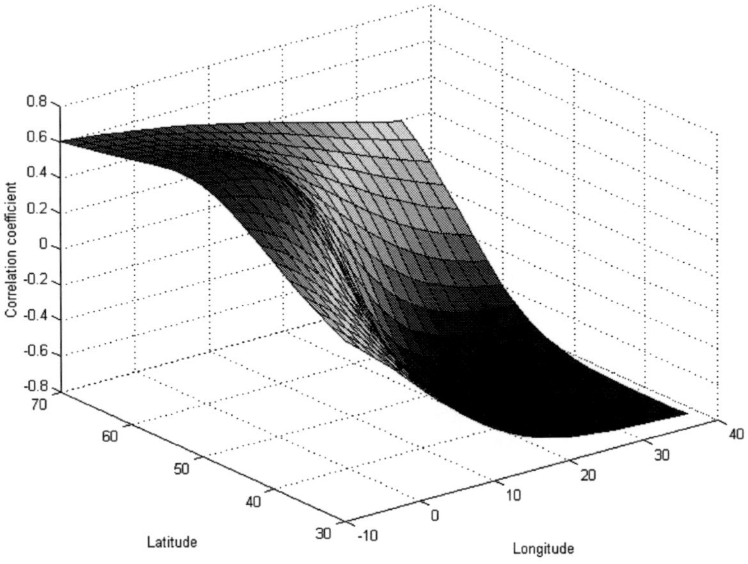

Figure 5.7 Correlation coefficients for Dataset B.

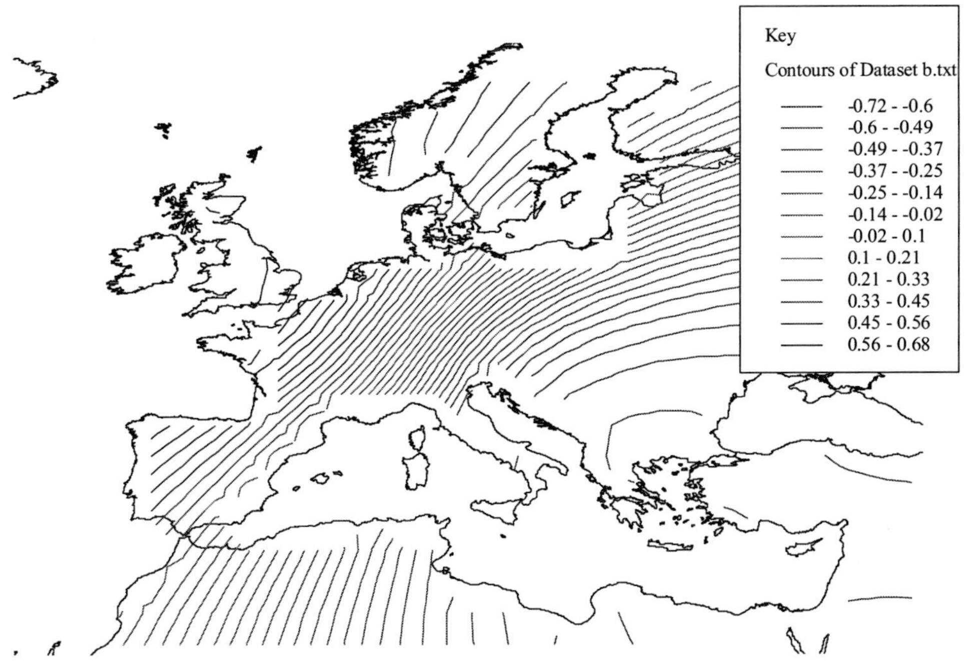

Figure 5.8 Contour lines fitted to the correlation coefficients calculated for Dataset B.

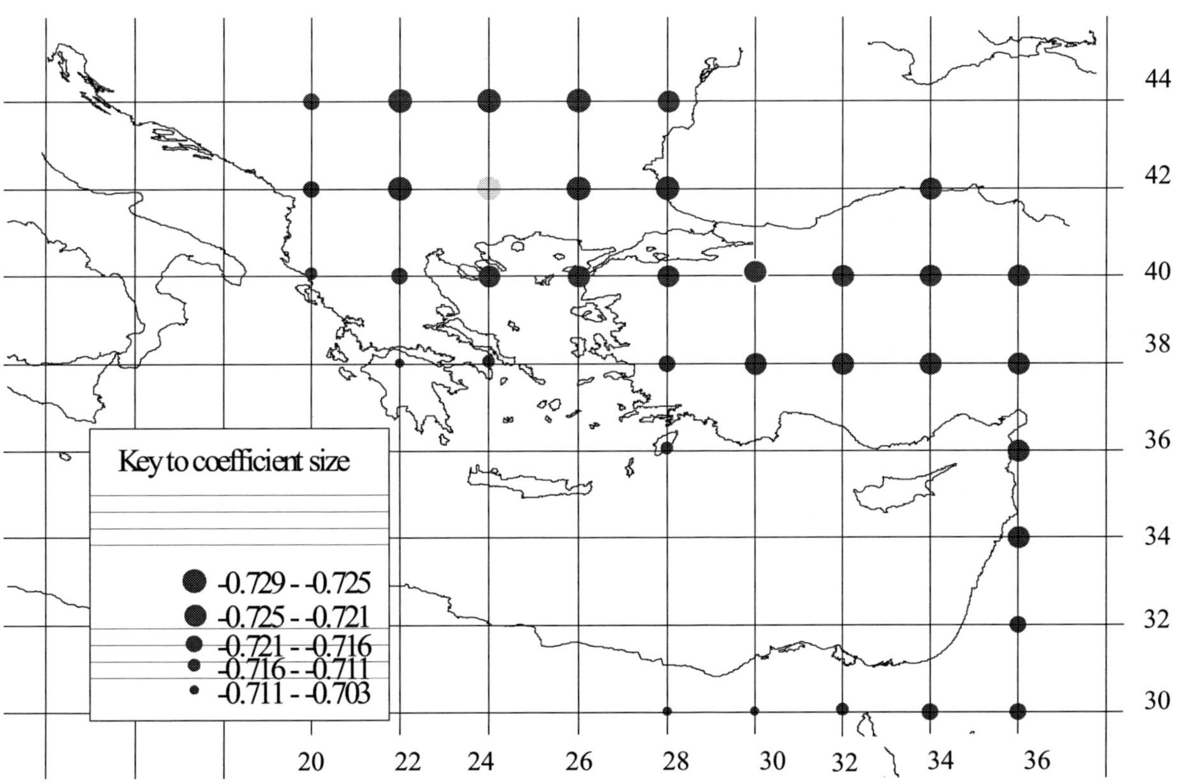

Figure 5.9 Origin points that gave the best fit to Dataset B (-0.7 and smaller). The strongest correlation coefficient, -0.7287, is shown as a paler circle at 24°E, 42°N.

5.2.2.3 Dataset C (oldest radiocarbon determinations):

The origin point with the best fit to Dataset C has a correlation coefficient of –0.72 and is found at exactly the same coordinates as the best fit origin point for Dataset A, namely at 36°E, 30°N in south western Jordan (Figure 5.10, Figure 5.11 and Figure 5.12).

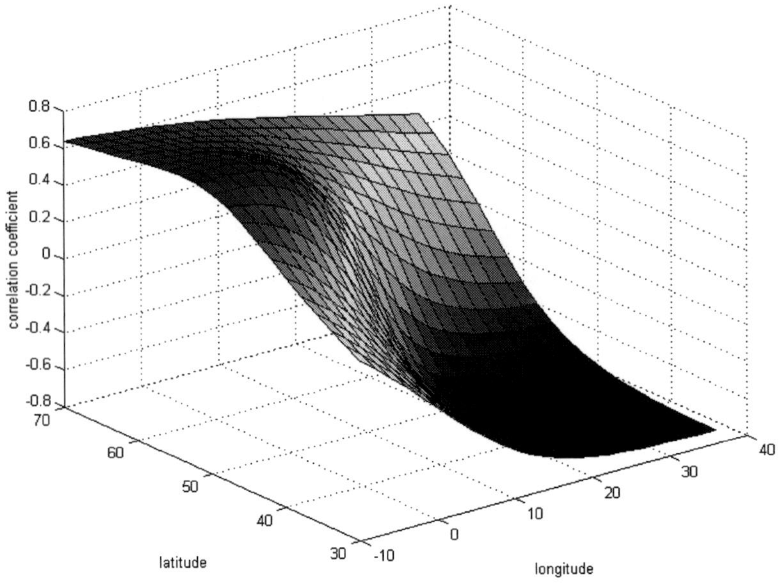

Figure 5.10 Correlation coefficients for Dataset C.

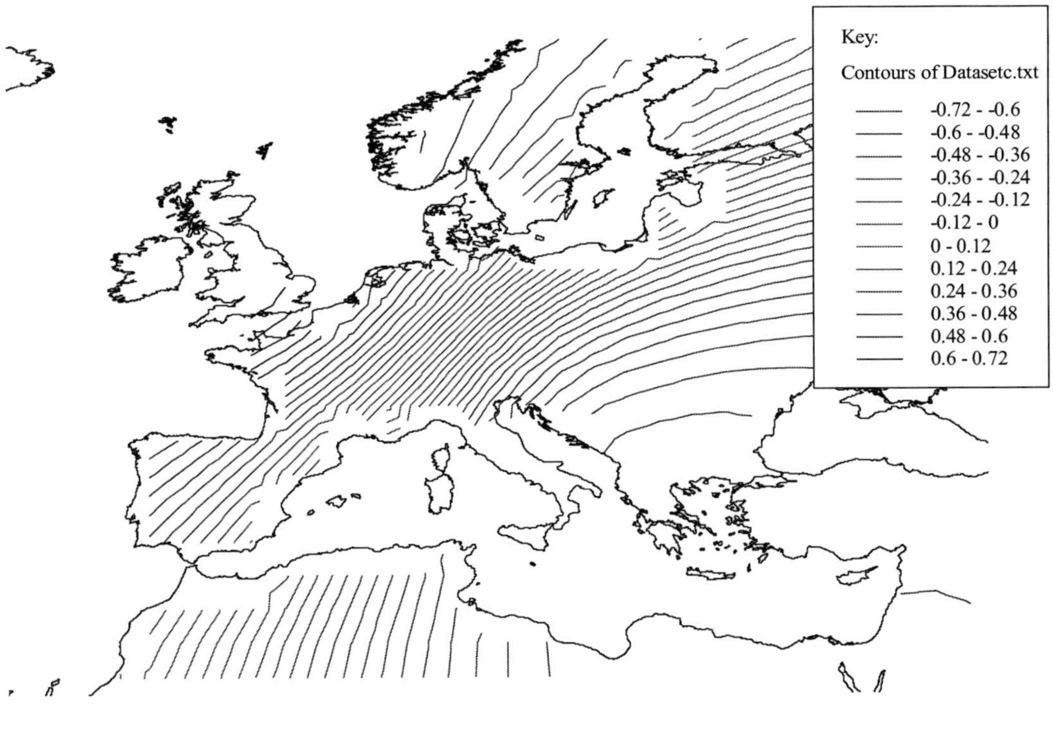

Figure 5.11 Contour lines fitted to the correlation coefficients calculated for Dataset C.

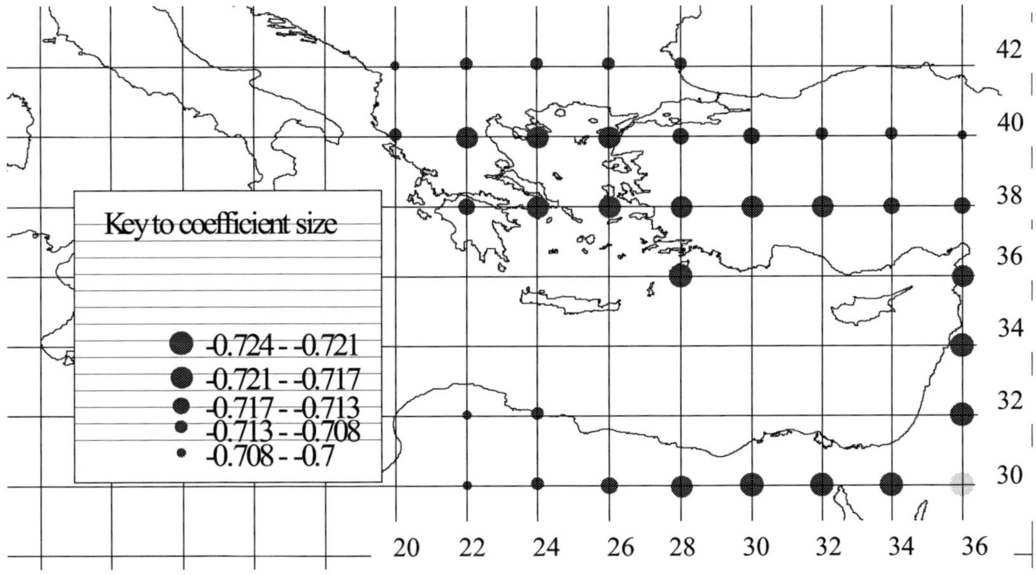

Figure 5.12 The origin points that gave the best fit to Dataset C (-0.7 and smaller). The strongest correlation coefficient, -0.7236, is shown as a paler circle at 36°E, 30°N.

5.2.2.4 Dataset D (most recent radiocarbon determination):

The origin that gives the best fit to Dataset D is the same as that for Dataset B (namely in the Rhodope mountain range in southern Bulgaria). The correlation coefficient is –0.71 (Figure 5.13, Figure 5.14 and Figure 5.15).

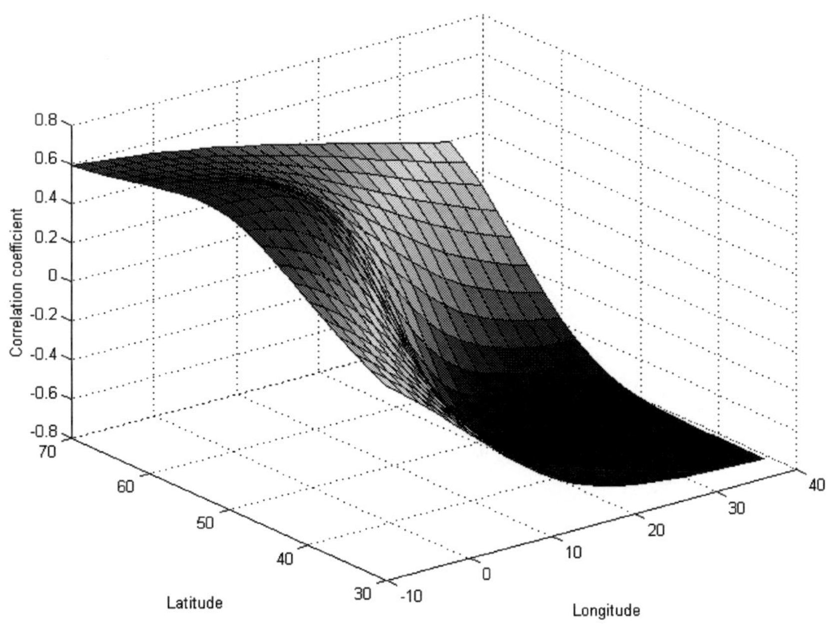

Figure 5.13 Correlation coefficients for Dataset D.

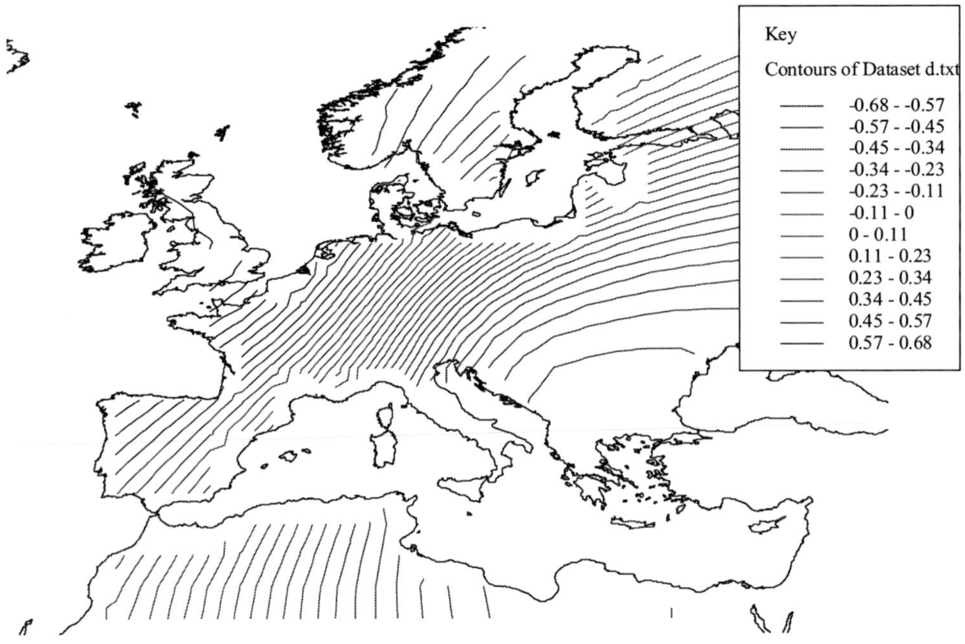

Figure 5.14 Contour lines fitted to the correlation coefficients calculated for Dataset D.

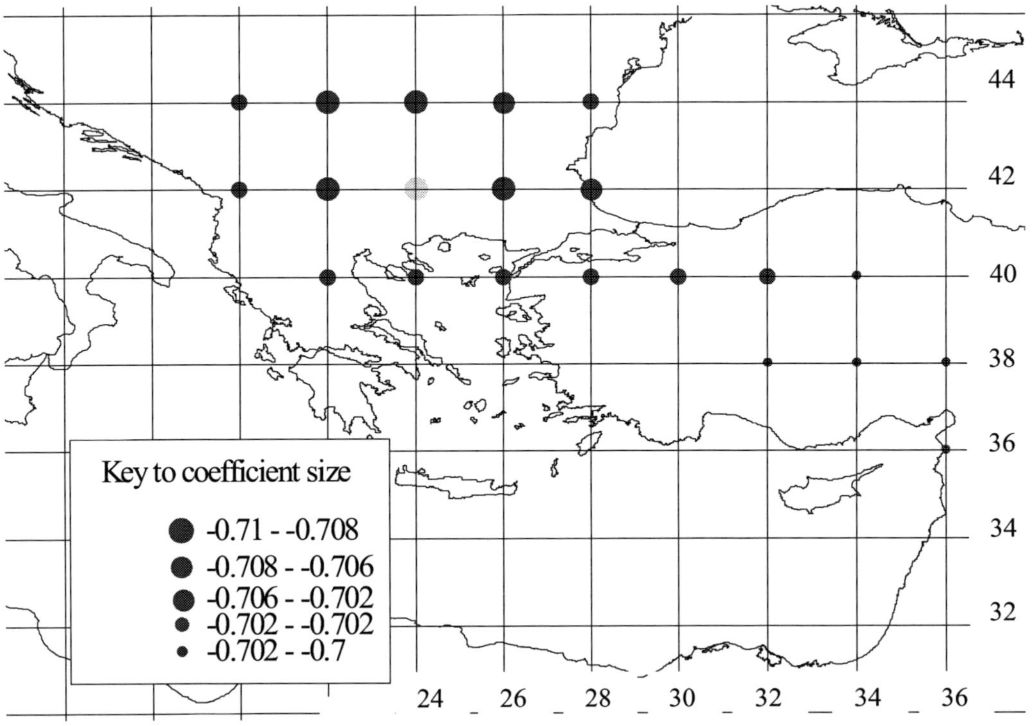

Figure 5.15 The origin points that gave the best fit to Dataset D (-0.7 and smaller). The strongest correlation coefficient, -0.7099, is shown as a paler circle at 24°E, 42°N.

5.2.2.5 Dataset E (most recent radiocarbon determination, multiple dated sites only):

The origin that gives the best fit to Dataset E has a correlation coefficient of –0.76 and is found just south of the Carpathian Mountains in southern Romania (Figure 5.16, Figure 5.17 and Figure 5.18).

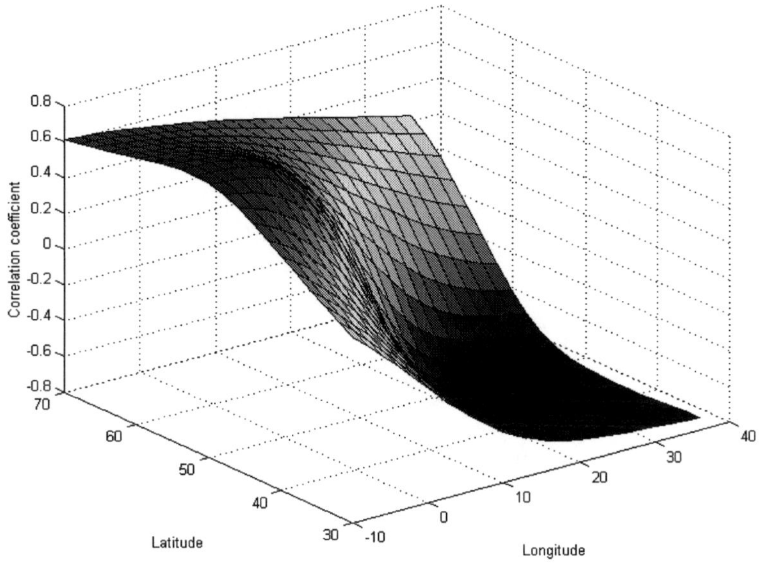

Figure 5.16 Correlation coefficients for Dataset E

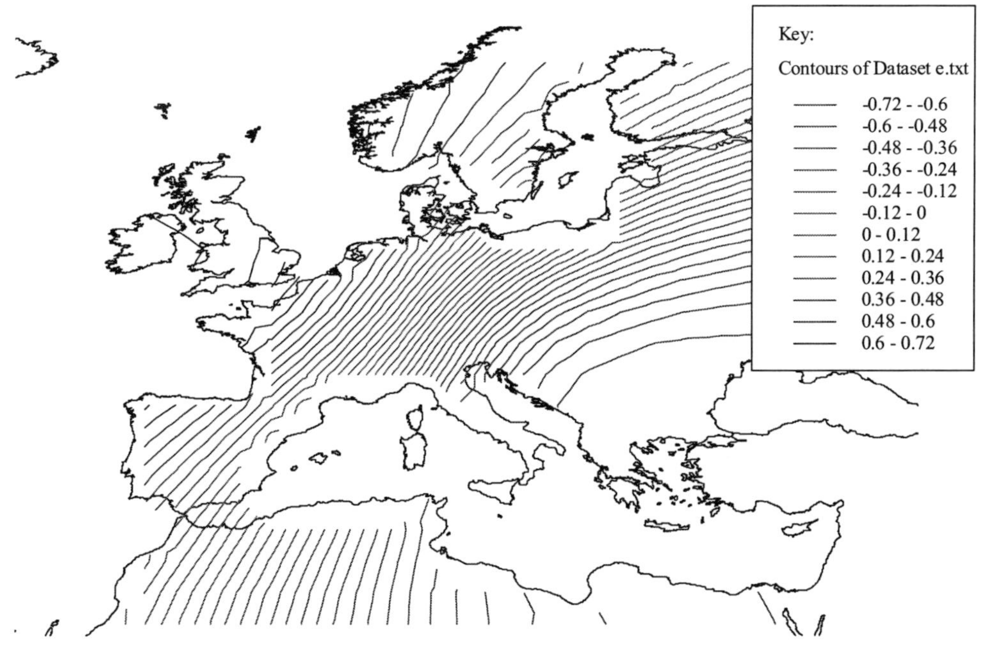

Figure 5.17 Contour lines fitted to the correlation coefficients calculated for Dataset E.

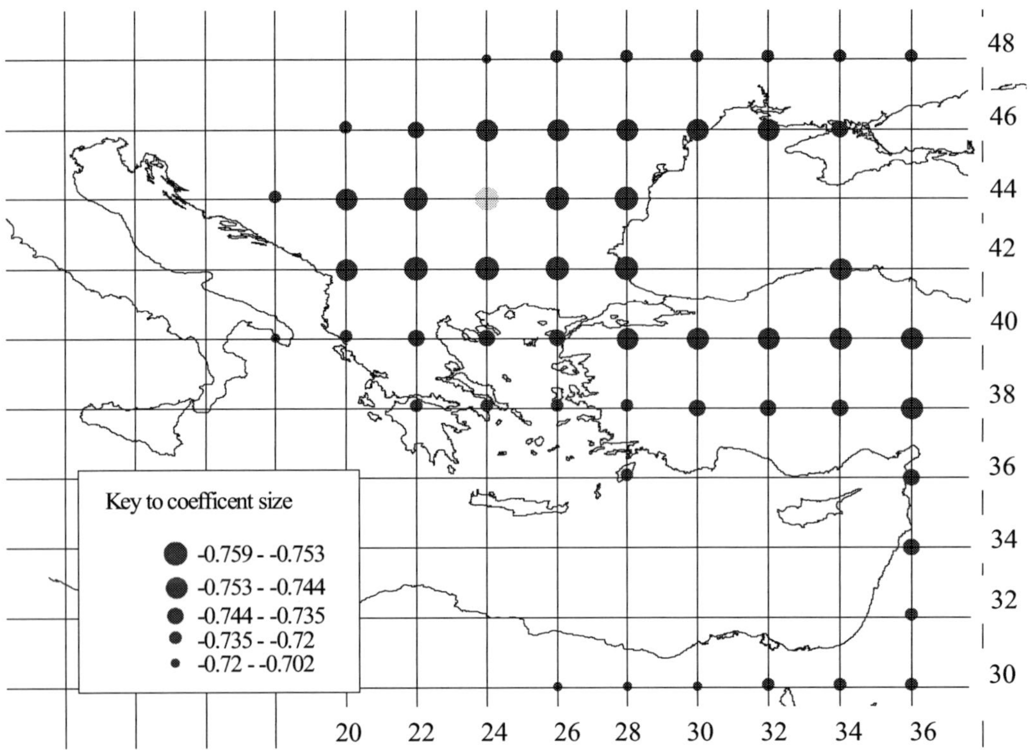

Figure 5.18 The origin points that gave the best fit to Dataset E (-0.7 and smaller). The strongest correlation coefficient, -0.7593, is shown as a paler circle at 24ºE, 44ºN.

5.2.2.6 Dataset F (oldest radiocarbon determination, multiple dated sites only):

The origin point that gives the best fit to Dataset F is identical to that for Datasets A and C, namely southwestern Jordan and it has a correlation coefficient of –0.75 (Figure 5.19, Figure 5.20 and Figure 5.21).

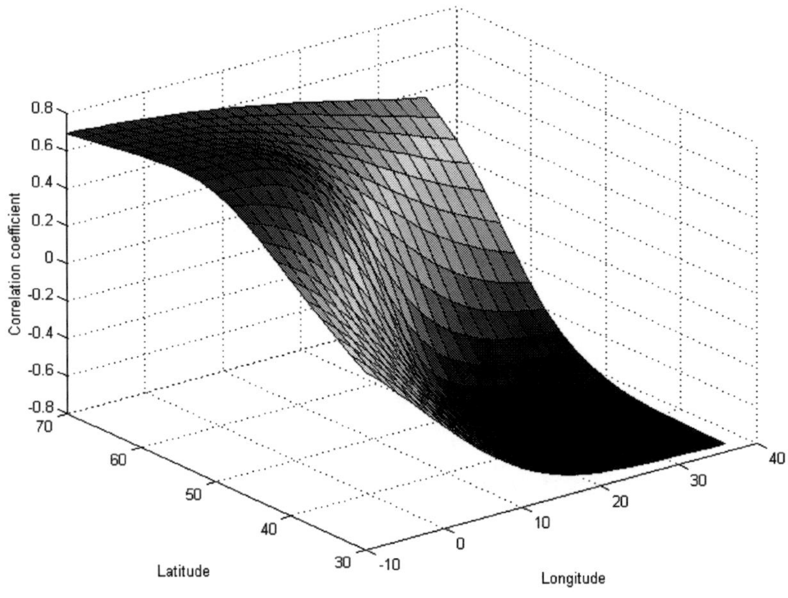

Figure 5.19 Correlation coefficients calculated for Dataset F.

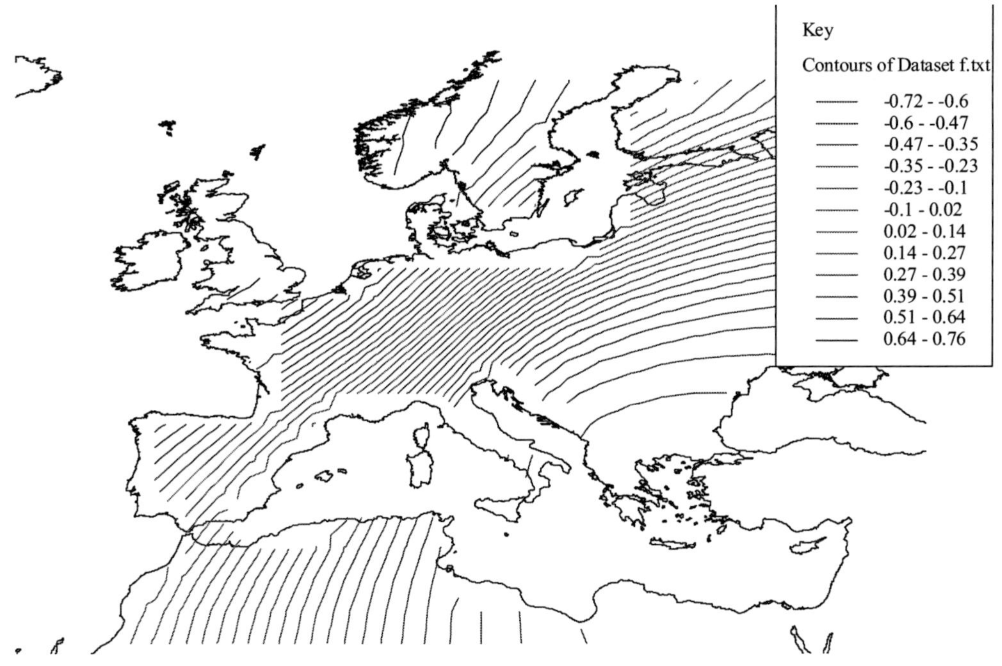

Figure 5.20 Contour lines fitted to the correlation coefficients calculated for Dataset F.

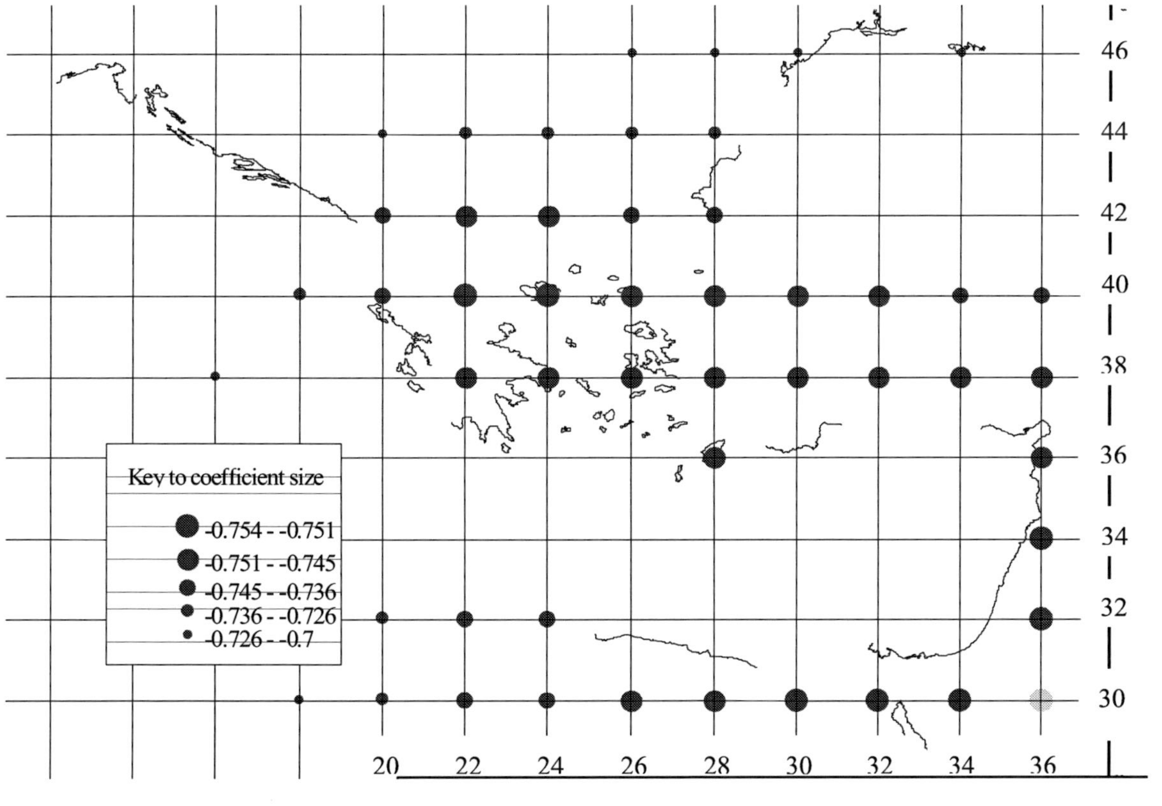

Figure 5.21 The origin points that gave the best fit to Dataset F (-0.7 and smaller). The strongest correlation coefficient, -0.7544, is shown as a paler circle at 36°E, 30°N.

5.2.2.7 Dataset G (weighted means or median radiocarbon determination):

The origin that gives the best fit with Dataset G is found in Turkey on the Mediterranean coast just north of the border with Syria (Figure 5.22, Figure 5.23, Figure 5.24). The correlation coefficient is –0.79 at this origin.

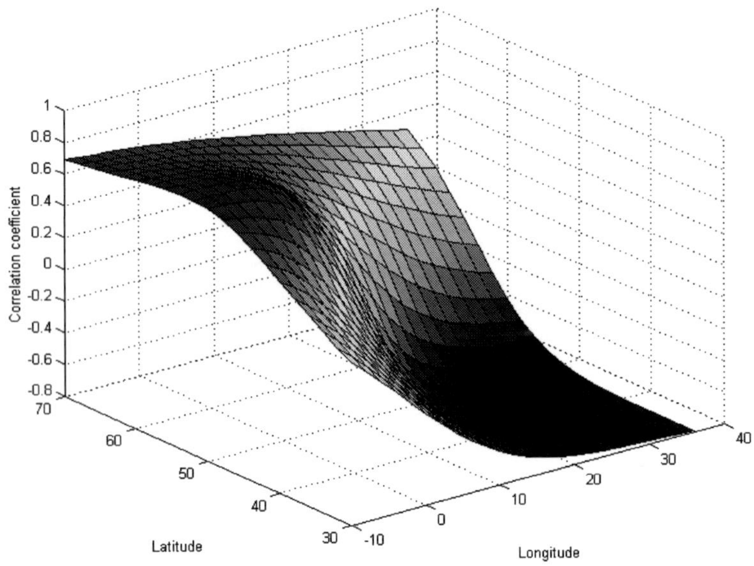

Figure 5.22 Correlation coefficients calculated for Dataset G.

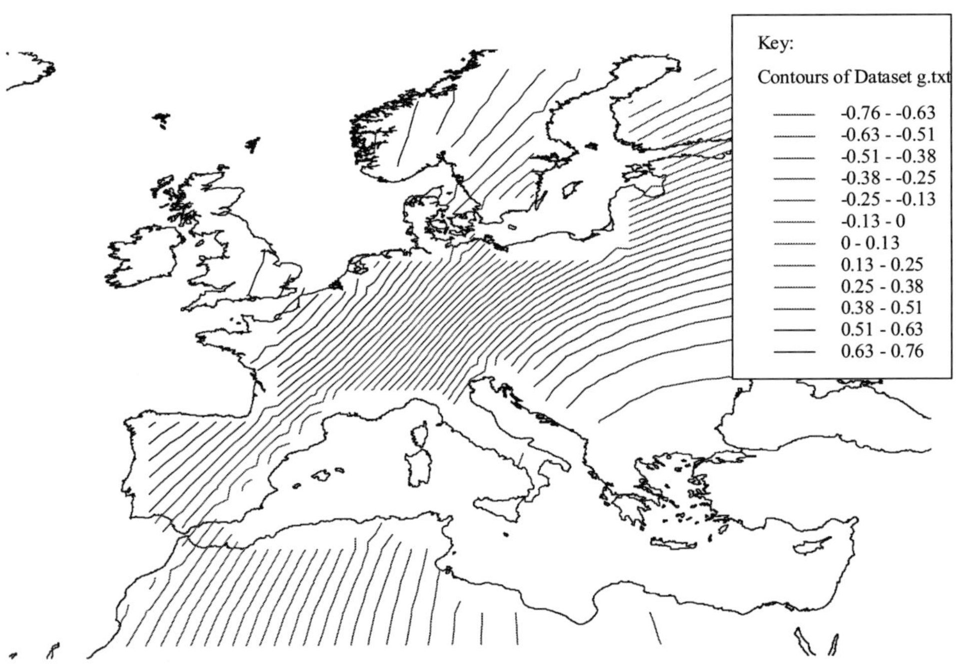

Figure 5.23 Contour lines fitted to the correlation coefficients calculated for Dataset G.

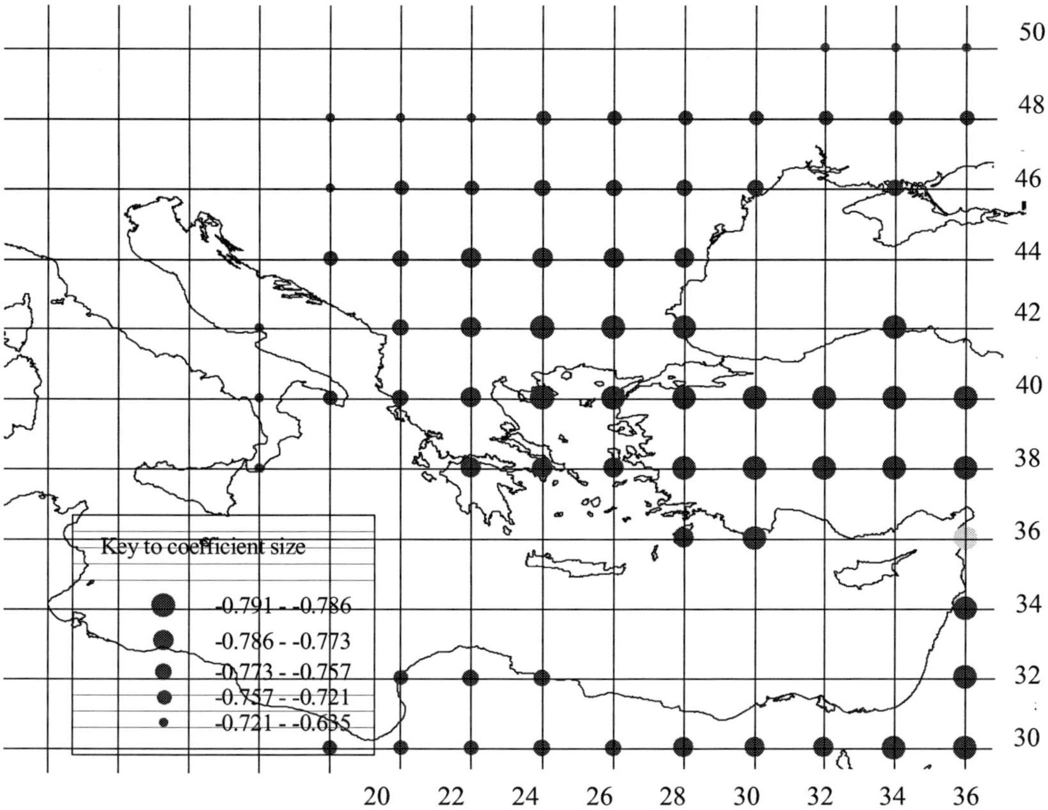

Figure 5.24 The origin points that gave the best fit to Dataset G (-0.7 and smaller). The strongest correlation coefficient, -0.7914, is shown as a paler circle at 36°E, 36°N.

5.2.3 Potential origin points for the European Neolithic database:

Most research to find the spread of the first Neolithic to Europe has used the single oldest radiocarbon determination at each site in analyses[3]. This approach has been criticized (Buck *et al.* (1994), Blockley *et al.* (2000)). It is argued that the oldest determination is not necessarily the correct one: a single radiocarbon determination is notoriously error prone, and statistically the use of a single value is unwise. There might be more confidence in the use of the single earliest determination if it is followed by a series of coherent more recent determinations in the site's stratigraphy. Unless one has detailed stratigraphic knowledge of the of the intra-site coherence between the relative order of the determinations in the ground and their radiocarbon age, it is hard to withstand such criticisms. In an attempt to overcome this the search for an origin for the Neolithic spread was conducted using the seven subsets of the Neolithic dataset (Datasets A to G).
Unsurprisingly the results show that data selection protocols do influence results. Reassuringly they do not directly contradict each other: they all point to an origin somewhere to the south east of central Europe. Additionally, they do not contradict what is already accepted about the spread of the first Neolithic, namely, that the Early Neolithic originated in the Near East and spread through Turkey and the Caucasus to Bulgaria and Romania and westwards across the Mediterranean. This is in agreement with the generally accepted position, that colonising groups introduced farming to the Balkans from Anatolia (Childe 1957, Ammerman & Cavalli-Sforza 1984, Van Andel and Runnels 1995, Zvelebil & Lillie 2000).

The different datasets A to G produced four different potential origin points for the Neolithic database (Table 5.1).

These cover a land area of thousands of kilometres from the Near East to Turkey to the Balkans (Figure 5.25). The results can be summarised as follows:
Datasets A, C and F all place the origin in south-western Jordan (with slightly different correlation coefficients). Dataset A contains grouped results or the oldest determination at a site. Dataset F contains the oldest determinations at each site but excludes sites with only one determination. Dataset C contains the oldest

[3] Clark (1965a/b), Ammerman and Cavalli-Sforza (1971, 1984), Dolukhanov *et al.* (1998), Pinhasi *et al.* (2000) and Russell and Steele (forthcoming).

Correlation coefficient	Origin	Dataset
-0.79	Turkey	Dataset G
-0.76	Southern Romania	Dataset E
-0.75, -0.74, -0.72	South western Jordan	Dataset A, C and F
-0.73, -0.71	Rhodope Mts. Southern Bulgaria	Dataset B and D

Table 5.1 The strongest correlation coefficients, their location and the dataset used.

determination at each site. This dataset is thus the best dataset to compare to that used by Ammerman and Cavalli-Sforza (1971, 1984) as they chose the oldest determination at each site for use in the analyses. They found that Jericho had the strongest correlation coefficient (0.89). Figure 5.25 shows its location in relation to the results of this analysis. With datasets A to G, the correlation coefficient with the origin at Jericho lies in the range –0.7 to –0.79. Datasets B and D place the origin in southern Bulgaria. Dataset B contains grouped determinations or the most recent determination at a site. Dataset D contains the most recent determination at each site.

Unexpectedly dataset E placed the origin in southern Romania rather than southern Bulgaria. Like datasets B and D, this dataset contains the most recent determinations at each site but used only sites with more than one determination. An identical pattern of selection, but taking the oldest determinations (Datasets A, C and F) had each yielded the same origin point.
Dataset G gives the strongest correlation coefficient of all the results and an origin in Turkey. The dataset includes grouped determinations and median determinations.

The effects of dataset selection on results is most clear if datasets that are identical in sample size and site location but opposite in determination selection (oldest or youngest determination selected) are compared:

1. Most recent determination at each site, multiple dated sites only (Dataset E, 264 sites) compared to oldest determination at each site, multiple dated sites only (Dataset F, 264). The results are an origin in southern Romania and correlation coefficient of 0.7593 for recent determinations and an origin in south-western Jordan and a correlation coefficient of 0.7236 for oldest determinations.

2. Most recent determination at each site, multiple and single determination sites (Dataset D, 509 sites) compared to oldest determination at each site, multiple and single determination sites (Dataset C, 509 sites). The results are an origin in southern Bulgaria and a correlation coefficient of 0.7099 for recent determinations and an origin in south-western Jordan and a correlation coefficient of 0.7352 for oldest determinations.

3. Most recent determination or weighted mean determination (Dataset B, 509 determinations) compared to oldest determination or weighted mean determination (Dataset A, 509 determinations). The results are an origin in southern Bulgaria and a correlation coefficient of -0.7287 for recent determinations and an origin in south-western Jordan and a correlation coefficient of -0.7544 for oldest determinations.

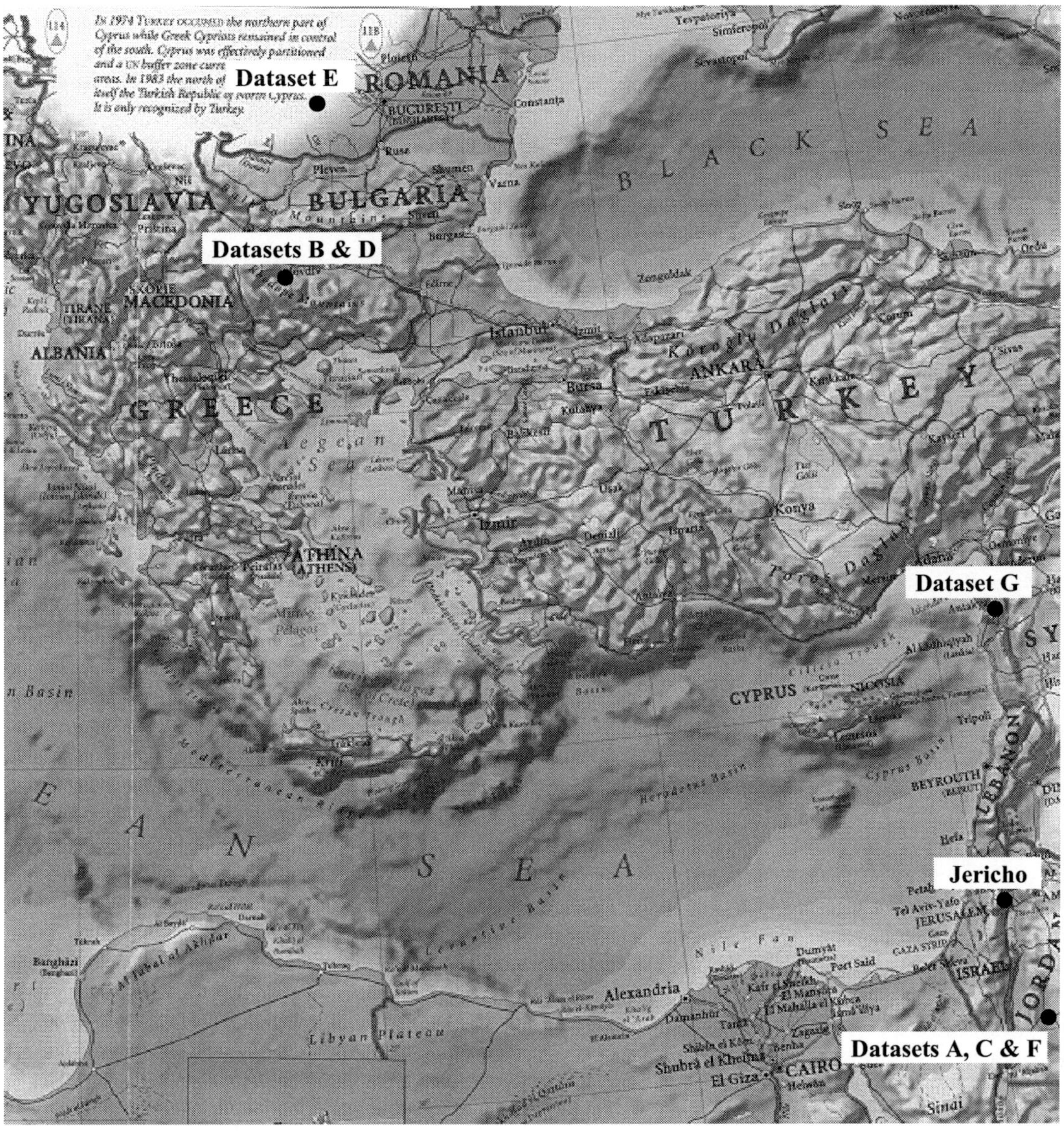

Figure 5.25 The location of the origin points that gave the best correlation coefficient with each dataset and their location relative to Jericho.

The results show that the data is fairly robust. Whether the earliest or latest determination is taken in a sequence of Neolithic determinations at a site, the linear relationship between the determination and the distance is strong. They show that when the oldest determinations at each site are used then the highest correlation coefficient is found with an origin in the Near East. When the most recent determination at each site is used in the dataset selection then the origin is found to be in the Balkans. When a median value for oldest and youngest determination is used (Dataset G) then the origin point with the highest correlation coefficient is found to be in between these two areas (Turkey and the Caucasus). The shift in origin point, further away from Europe when the oldest determinations are used, and closer to Europe when the more recent determinations are used is explained by Figure 7.11 (Chapter 7).

The isochron maps presented in Chapter Four, Figures 4.9 to 4.15 provide another method to find the most likely origin to a group of radiocarbon determinations. The isochron maps clearly show the antiquity of farming sites in the area of Greece and the Aegean.

5.3 Summary

Early work by Ammerman and Cavalli-Sforza with an assumed origin at Jericho is confirmed by the routine to search for the most likely origin point for the spread of the European Neolithic. Unsurprisingly the chapter has shown how the results will change according to which radiocarbon determination is selected. But what is also shown is that for this set of determinations, at least, the effect does not alter the general picture. For work at this scale of analysis generalities are the essence of the results. It is also a rough check on the accuracy of determinations at each site. If intra-site chronology were poor then the impact of determination selection in the sequence would have a greater effect on the results.

The linear correlation coefficients for determination and distance are large at the optimal origin point for each of the seven datasets. This justifies the use of linear regression: the results show that there is a strong linear relationship between distance and determination from an origin. This means that measures of rate from the slope of the regression line and time at departure from the origin are also valid parameters for summarising the trend in the data.

5.4 Measuring the rate of spread of farming in Europe.

Ammerman and Cavalli-Sforza (1971) tested the relationship between the age of farming sites and their distance from an origin using principal axis linear regression of uncalibrated date and distance. The measure of the relationship is the correlation coefficient, r.

This was calculated using the variables date (uncalibrated years BP) of 53 Neolithic sites and their great circle distance[4] (in kilometres) from an assumed origin at Jericho. The correlation coefficient was found to be 0.89 for an origin at Jericho (Figure 5.26).

This high positive value for the correlation coefficient lends support to their hypothesis that farming spread from an origin in the Near East to the rest of Europe. Figure 5.26 shows that sites close to Jericho are old. With increasing distance from Jericho the sites become gradually younger. Using the equation of the line they calculated the rate of spread (the slope in the line) and the time at the origin point (the intercept of the line with the axis at distance zero). This gave a rate of spread of 1.08 km/year and a time of departure from the origin at Jericho of 8845 uncalibrated years BP.

They noted that it was incorrect simply to use the regression of determination on distance or distance on determination because of the errors inherent in both types of data. Without knowing the relative importance of each type of error, they chose to use the principal axis of the regression to calculate the rate of spread (slope of the principal axis) and the starting date of the spread (the intercept of the principal axis with the plot axis at distance zero) (Figure 5.27). They did not use the error on the determination (the standard deviation) in the regression calculation.

The high correlation coefficients measured for the relationship between radiocarbon determination and distance from potential origins across Europe to sites in the Gkiasta database have shown that there is a strong linear relationship between these two variables. Having established the linearity of the trend, linear regression was applied to the Datasets A to G to find the rate of the spread and the starting determination at the origin. Before carrying out the analysis the different linear regression methods were compared in order to understand the effect that they would have on the analysis (Peltzer 2000). The numerical computation and visualisation software, Matlab was used to carry out the different regressions. The scripts for running each of the Matlab regression methods are available online (see Peltzer 2000). This section investigates the different methods of linear regression to find the one that is most suitable to the regression of geodesic distance and radiocarbon determinations. Regression is then applied to the Neolithic datasets to find measurements of rate of spread and time of departure from potential origin points.

[4] Great circle distance is equivalent to geodesic distance; it is the shortest distance between two points on a curved surface.

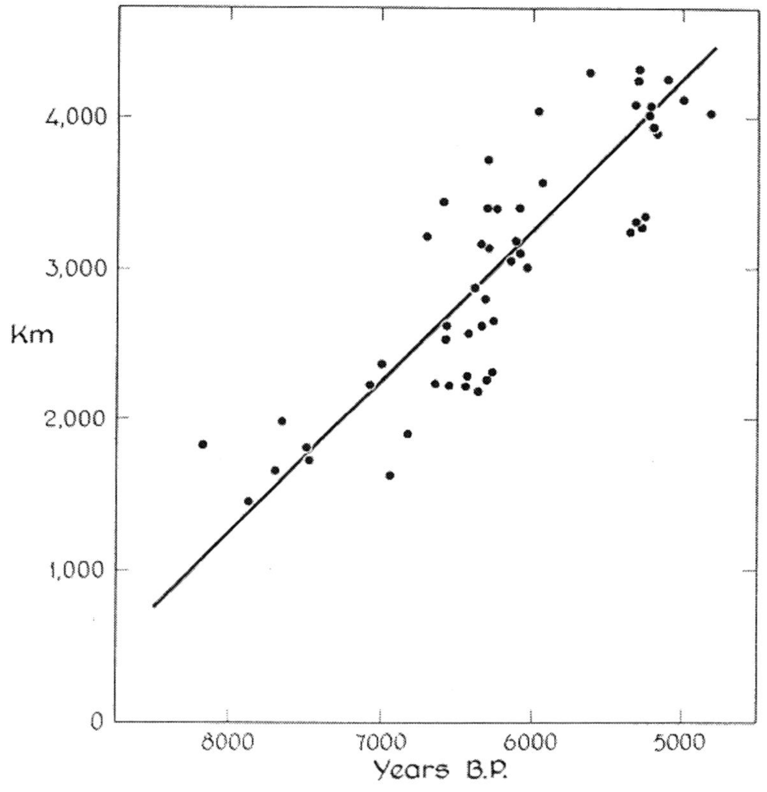

Figure 5.26 Regression line for the age of Neolithic sites with their distance from an origin at Jericho. The correlation coefficient is 0.89 (Reproduced from Ammerman and Cavalli-Sforza (1984: 53)).

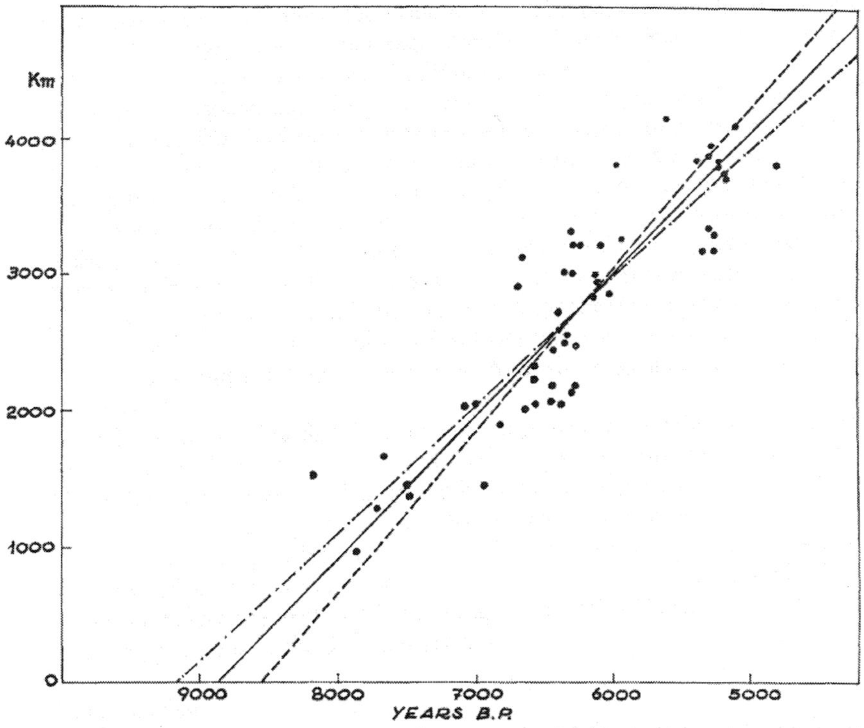

Figure 5.27 Linear regression of distance from Jericho and radiocarbon determination. The dashed lines are the two regressions (determination on distance, and distance on determination). The solid line is the principal axis, which is an average of the two regression lines (Ammerman and Cavalli-Sforza 1971:680, Figure 2).

5.4.1 *Model I regression (Peltzer 2000).*

This is the regression of Y on X or X on Y. The 'best fit' line is determined by adjusting the slope of the line through the data such that the sum of the squares of the offsets between the line and the data is a minimum (thus the method is sometimes referred to as 'least squares' regression). In model I regression these offsets are measured parallel to either the X-axis or the Y-axis, depending on which are defined as the independent and dependent variables. For example, for the regression of Y on X, offsets are measured parallel to the Y-axis. In this example Y is the dependent variable and X is the independent variable. One assumes that X has no error and that all the error is associated with Y.

Model I regression assumes that only one of the variables, either Y or X have an associated error. If both the variables have any variability or measured error then the slope of the line is underestimated and this method should not be used. This method can also be used with weights on one of the variables. The variations of the Model I type regressions are summarised in Table 5.2. Note that weighted model I regression A and weighted model I regression B are almost identical except that the uncertainties in the slope and the intercept are much smaller with weighted model I regression A.

5.4.2 *Model II regression (Peltzer 2000).*

There are several variations on the Model II regressions. They are summarised in Table 5.3. This type of regression is used to fit a straight line to the data when both the X and the Y variables are subject to error. Both Y and X are treated as independent variables. As is often the case with statistical techniques there is no definitive method and the method used depends on the situation to hand.

1. 'Major axis' regression: this is used when both the variables, Y and X, have the same units of measurement.

2. 'Reduced major axis' regression or 'geometric mean' regression. This is used when the X and Y variables have different units of measurement. This is the method that Ammerman and Cavalli-Sforza (1971, 1984) used.

3. 'Least square bisector' regression. This is a slight variation on method above, 'reduced major axis' regression.

4. 'Least squares cubic' regression. This is used when the X and Y variables have different units of measurement and different associated weights/errors.

	Standard model I linear regression	Inverted model I linear regression	Weighted model I regression A	Weighted model I regression B
Variables	Y on X	X on Y	wY on X	wY on X
Line fitting procedure	Y-offsets are minimized	X-offsets are minimized	Y-offsets are minimized	Y-offsets are minimized
Weighting Procedure	Equal weights	Equal weight	Y data weighted	Y data weighted
Application scenario	X controlled and Y is measured	Y controlled and X is measured	X controlled and Y is measured with varying error.	X controlled and Y is measured with varying error.
Matlab shell script	Lsqfity	Lsqfitxi	Lsqfityw	Lsqfityz

Table 5.2 Model I regressions

	Major Axis regression	Geometric mean regression	Least squares bisector regression	Least squares cubic regression
Also known as	First principal component regression	'Reduced major axis' or 'standard major axis' regression	None	None
Variables	X and Y	X on Y Y on X	X on Y Y on X	wY and wX
Line fitting procedure	X and Y offsets minimized simultaneously	Line is the geometric mean of the line from the regression of X on Y and Y on X	Line is the bisect of the minor angle from the regression of Y on X and X on Y	X and Y offsets are minimized simultaneously
Weighting procedure	Equal weights	Equal weights	Equal weights	Weighted X and weighted Y
Application scenario	Units and ranges of X and Y are the same.	Units and ranges of X and Y differ.	Units and ranges of X and Y differ	Measurement error of X and Y varies.
Matlab shell script	Lsqfitma	Lsqfitgm	Lsqbisec	Lsqcubic

Table 5.3 Model II regressions.

5.4.2.1 Comparison of the linear regression techniques using Dataset G:

To quantify the effects that the different techniques have on the data, they were applied to Dataset G. The results of Model I regressions on Dataset G are summarised in Table 5.4 and those for Model II regressions on Dataset G are summarised in Table 5.5. Figure 5.28 shows the graph of the results for all methods.

Model 2w requires that a weight be given to the distance variable as well as the determination variable (see discussion below). Note that the results for weighted model I regression A are identical to those for weighted model I regression B, the only difference is in the calculation of the error terms (see section 5.1.1).

	Standard model I linear regression (Model 1A)	Inverted model I regression (Model 1B)	Weighted model I regression A (Model 1Aw (i) and (ii))		Weighted model I regression B (Model 1Bw (i) and (ii))	
Weight	None	None	Weighted determination, $1/\sigma$	Weighted determination, $1/\sigma^2$	Weighted determination, $1/\sigma$	Weighted determination, $1/\sigma^2$
Slope (m)	-0.8553	-1.3512	-0.8498	-0.8426	-0.8498	-0.8426
Intercept (c)	8051.5	9471.8	8001.6	7952.5	8001.6	7952.5
Correlation coefficient (r)	-0.7956	-0.7956	-0.7976	-0.7942	-0.7976	-0.7942
Error on slope (e_m)	±0.0432	±0.0683	$\pm 7.002 \times 10^{-4}$	±0.0049	±0.0427	±0.0428
Error on intercept (e_c)	±127.496	±285.6472	±2.0941	±14.8557	±127.5778	±130.5863
Equation of line	y = -0.86x + 8052	y = -1.35x + 9472	y= -0.85x + 8002	y = -0.84x + 7953	y= -0.85x + 8002	y = -0.84x + 7953

Table 5.4 Comparison of Model I regressions using dataset G.

	Geometric mean regression (Model 2A)	Least squares cubic regression (model 2w)				Least squares bisector regression (model 2b)
Weight on X (distance)	None	$1/\sigma^2$ (1/10^2)	$1/\sigma^2$ (1/1^2)	$1/\sigma^2$ (1/50^2)	$1/\sigma^2$ (1/5^2)	None
Weight on Y (determination)	None	$1/\sigma^2$	$1/\sigma^2$	$1/\sigma^2$	$1/\sigma^2$	None
Slope (m)	-1.075	-0.866	-0.8429	-1.064	-0.8488	-1.0731
Intercept (c)	8680.8	8024.1	7953.2	8620.2	7971.6	8675.1
Correlation coefficient (r)	-0.7956	-0.7997	-0.8008	-0.7959	-0.8005	-0.7956
Error on slope (e_m)	±0.0456	±0.0428	±0.0428	±0.0447	±0.0428	±0.0456
Error on intercept (e_c)	±134.5578	±130.1412	±130.5809	±133.1986	±130.4569	±134.4341
Equation of line	Y = -1.08x + 8681	Y = -0.87x + 8024	Y= -0.82 x + 7953	Y = -1.06x + 8620	Y = -0.85x + 7972	Y=-1.07x + 8675

Table 5.5 Comparison of Model II regressions using Dataset G.

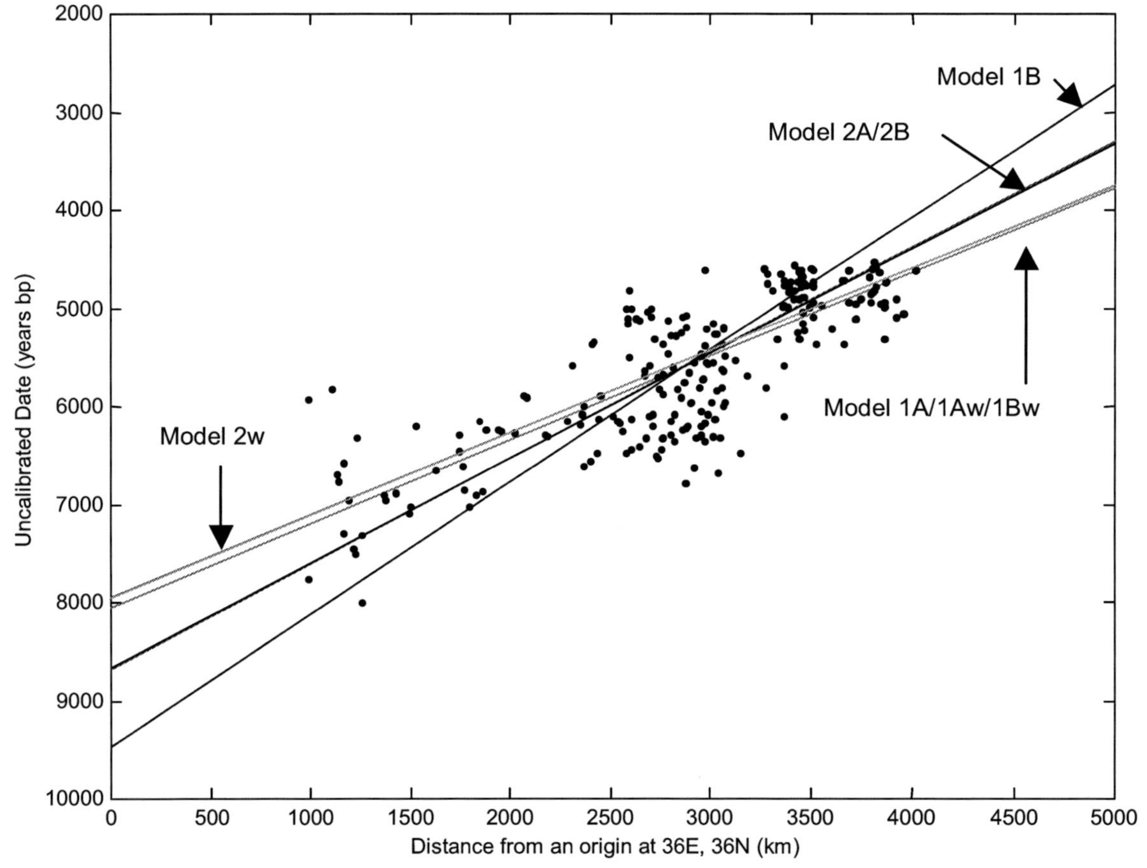

Figure 5.28 Comparison of the linear regression models using Dataset G.

Method	Least squares cubic (Model 2w)	Weighted model I regression (Model 1Aw(i)/1B w(i))	Least squares bisector (Model 2B)	Geometric mean (Model 2A)	Standard model I regression (Model 1A)	Inverted model I regression (Model 1B)	Weighted model 1 regression (Model 1Aw(ii) /1Bw(ii)
Correlation coefficient (r)	-0.8008	-0.7976	-0.7956	-0.7956	-0.7956	-0.7956	-0.7942
Rate of spread (km/year)	1.19	1.18	0.93	0.93	1.17	0.74	1.19
Determination at origin (years BP)	7953	8002	8675	8681	8052	9472	7953

Table 5.6 Rates of spread and time at the origin for Dataset G calculated using the different regression methods. They are ordered by the size of the correlation coefficient.

The different regressions give a tight range of results. The correlation coefficients range from a minimum of -0.7942 to a maximum of -0.8008. Rate measurements calculated from the slope of the regression line range from a minimum of 0.74 km/year to a maximum of 1.19 km/year. The time of departure from the origin at 36°E and 36°N ranges from a minimum of 7953 years before present to a maximum of 9472 years before present (Table 5.6).

5.4.2.2 The effect of the weight given to the variables on the linear regression results.

1. The effect of weighting the determination variable:
For the determinations, two weights were compared. These were 1/standard deviation and 1/variance (Model 1Aw (i), Model 1Aw (ii), Model 1Bw (i) and Model 1Bw (ii), Table 5.4). The Matlab script automatically calculates the weight as 1/variance but as the work in Section 5.2.2 used 1/standard deviation, the weight the comparison was necessary. The standard deviation used is that reported with the radiocarbon determination as calculated by the radiocarbon dating laboratory (namely ± 1 sigma range). Variance is simply the square of the standard deviation. As Table 5.4 shows the difference between the results for weighted determinations using the two alternative weighting factors have a small effect on the results. The correlation coefficients are 0.7976 for the weight 1/standard deviation and 0.7942 for the weight 1/variance.

2. The effect of weighting the distance variable:
For the distance measurement four weights were compared. The only way to weight the determinations was to apply a uniform weight to all the distance variables as the error on individual weights is unknown. Possible sources of error on the distance measurement are the calculation of geodesic distance that make the assumption that the earth is spherical with a radius of 6378. This is a reasonable approximation of the earth's radius for purposes that require accuracy of one kilometre or thereabouts (Longley *et al.* 2001). There are also errors in the measurement of the precise location of the archaeological site. Taking these factors into consideration four standard deviations for the distance measure were tested. These were 1 kilometre, 5 kilometres, 10 kilometres and 50 kilometres (Model 2w, Table 5.5). The weights used were 1/variance (as this is the automatic weight calculated by the Matlab scripts).
The highest correlation coefficients of the dataset G are found when distance has a standard deviation of 1 kilometre and 5 kilometres coupled with a weighting of 1/variance for the determinations (Table 5.5, Model 2w). The correlation coefficient for dataset G increases as the weight on the distance variable is increased (Figure 5.29).

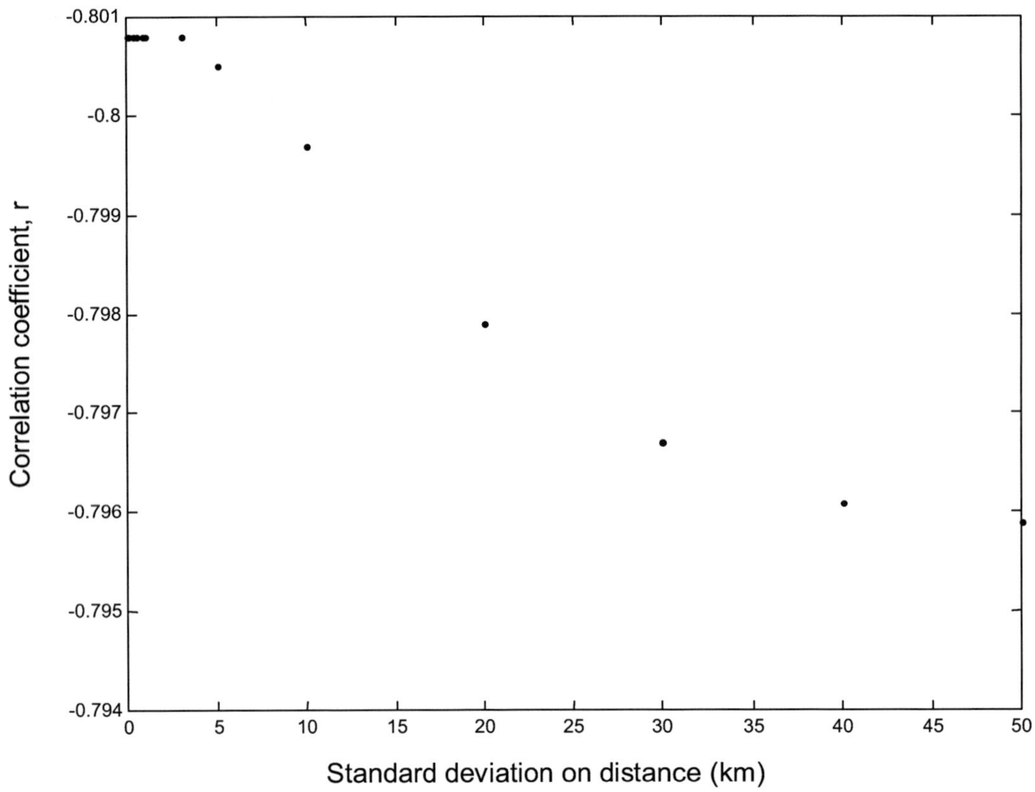

Figure 5.29 To show the effect of changing the weight ($\frac{1}{\sigma^2}$) on the distance measurement for the regression of Dataset G using the least squares cubic model. The weight on the determination remains constant ($\frac{1}{\sigma^2}$).

5.4.2.3 Choosing the most appropriate regression model for this dataset:

Ammerman and Cavalli-Sforza used the equivalent of Geometric mean regression to calculate the rate of spread of farming in Europe. Since they did not give a weight to either the radiocarbon determinations or the distance measurements this was the most suitable regression model to use. For the linear regression exercise that follows radiocarbon determinations were weighted (by $[\text{variance}]^{-1}$ where the variance is the square of the 1-sigma range error term as reported by the radiocarbon laboratory). Although the effect of distance weighting was explored using least squares cubic linear regression (Model 2w), the decision was made to run the analyses without a distance weighting. Currently there is no way of determining what weights each distance should have and to blanket them all with the same weight is not ideal (Steele *et al.* 2000).

The analyses in the following section used weighted model 1B regression (Table 5.2).

5.4.3 Rate measurements for the spread of the first farming calculated for each of the seven datasets A to G:

Rate measurements and the determination at the origin point were calculated for datasets A to G using weighted model 1B linear regression. In each case the origin used was that which gave the best correlation coefficient in the exercise to search for the origin. The distance variable is not weighted. Radiocarbon determinations are weighted.

5.4.3.1 Results for Dataset A:

Origin	36ºE, 30ºN
Number of sites	509
Weight on distance	none
Weight on determination	1/var (var = σ^2)
Correlation coeff., r	-0.7531
Slope (m)	-0.8308
Y intercept (c)	8379.6
Error on slope	0.0322
Error on intercept	111.8820
Equation of line	$Y_{determination} = (-0.83 \pm 0.03)\ x_{distance} + (8380 \pm 112)$
Rate (km/year) (1/m)	1.2 km/year
Determination at origin (yrs BP), i.e. $X_{dist.} = 0$	8380 ± 112

Table 5.7 Results of weighted model 1B linear regression on Dataset A.

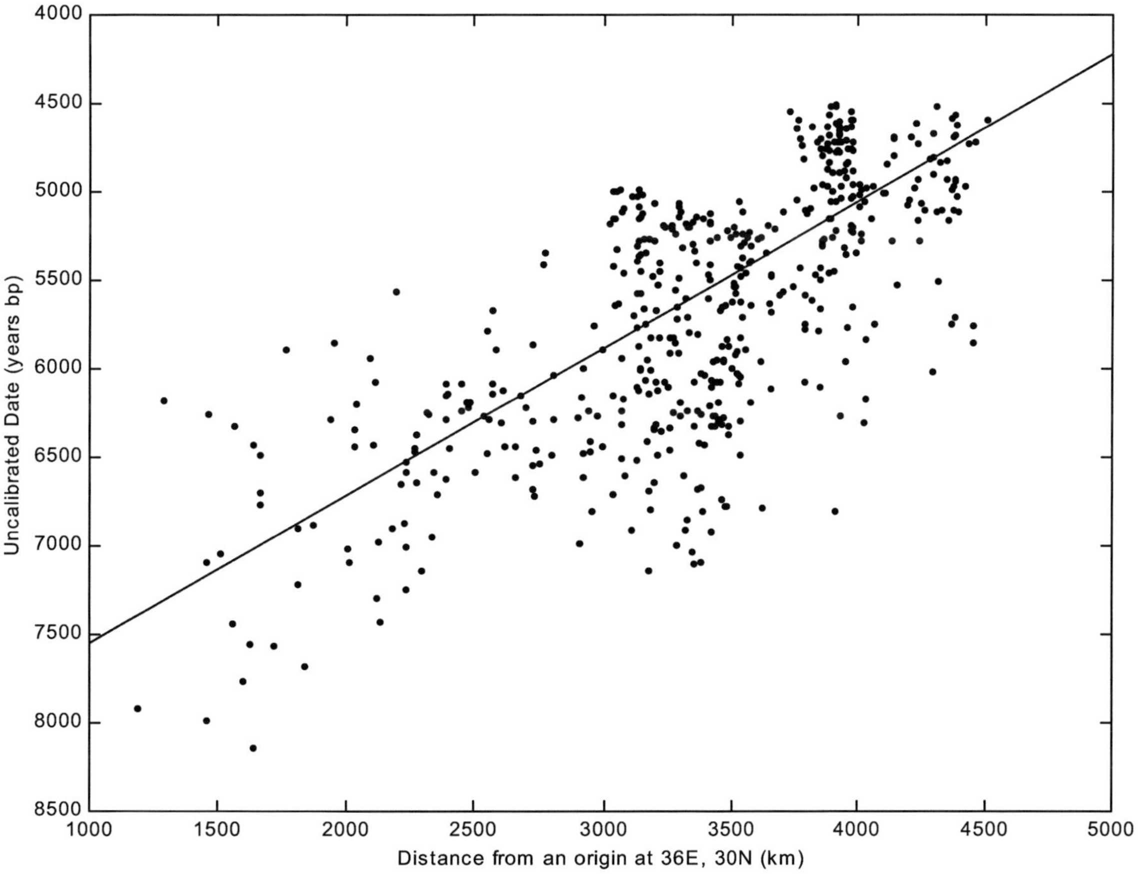

Figure 5.30 Linear regression of Dataset A.

5.4.3.2 Results for Dataset B:

Origin	24°E, 42°N
Number of sites	509
Weight on distance	none
Weight on determination	1/var (var = σ^2)
Correlation coeff., r	-0.7443
Slope (m)	-0.8290
Y intercept (c)	6879.7
Error on slope	0.0330
Error on intercept	61.5697
Equation of line	$Y_{determination} = (-0.83 \pm 0.03)\ x_{distance} + (6880 \pm 62)$
Rate (km/year) (1/m)	1.2
Determination at origin (yrs BP), i.e. $X_{dist.} = 0$	6880 ± 62

Table 5 Linear regression of Dataset B.

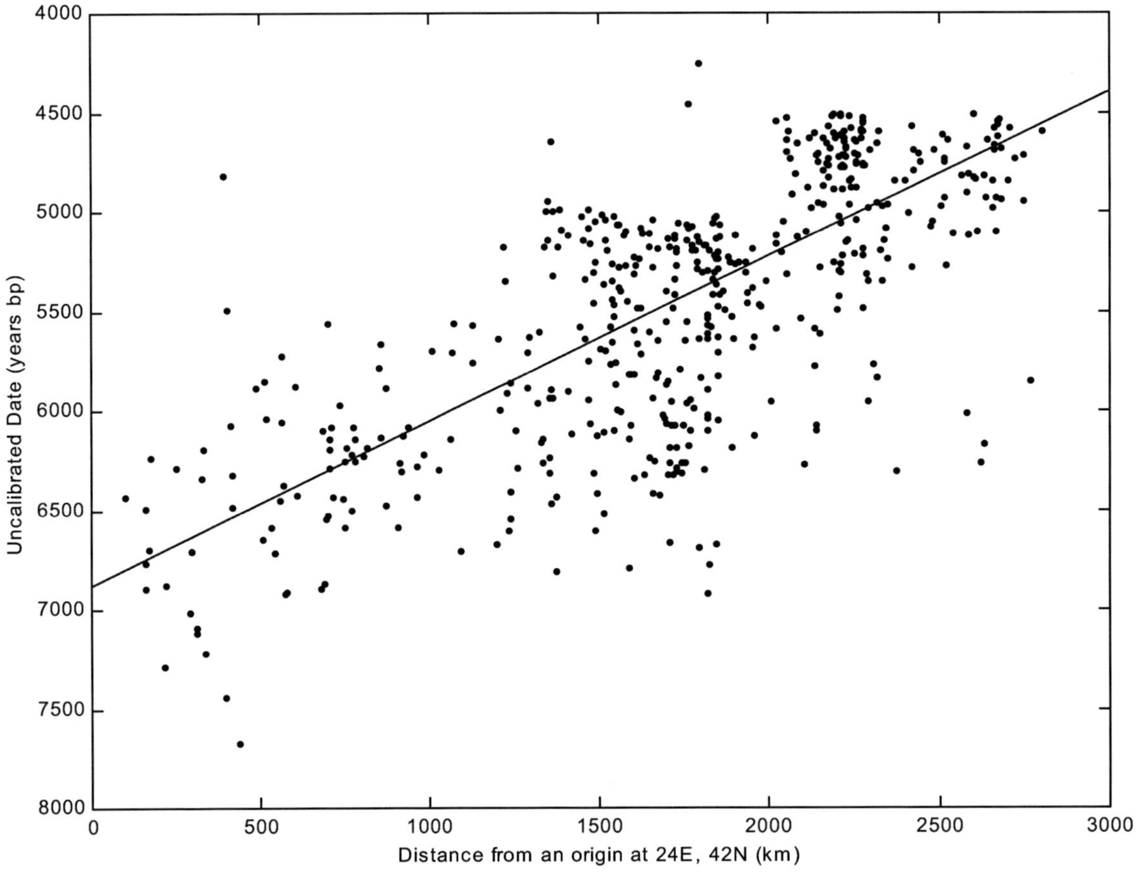

Figure 5.31 Linear regression of Dataset B.

5.4.3.3 Results for Dataset C:

Origin	36°E, 30°N
Number of sites	510
Weight on distance	none
Weight on determination	1/var (var = σ^2)
Correlation coeff., r	-0.7265
Slope (m)	-0.7662
Y intercept (c)	8243.7
Error on slope	0.0322
Error on intercept	109.75
Equation of line	Ydetermination = (-0.77 ± 0.03) xdistance + (8240 ± 110)
Rate (km/year) (1/m)	1.3 km/year
Determination at origin (yrs BP), i.e. $X_{dist.} = 0$	8240 ± 110

Table 5.8 Results of linear regression on Dataset C.

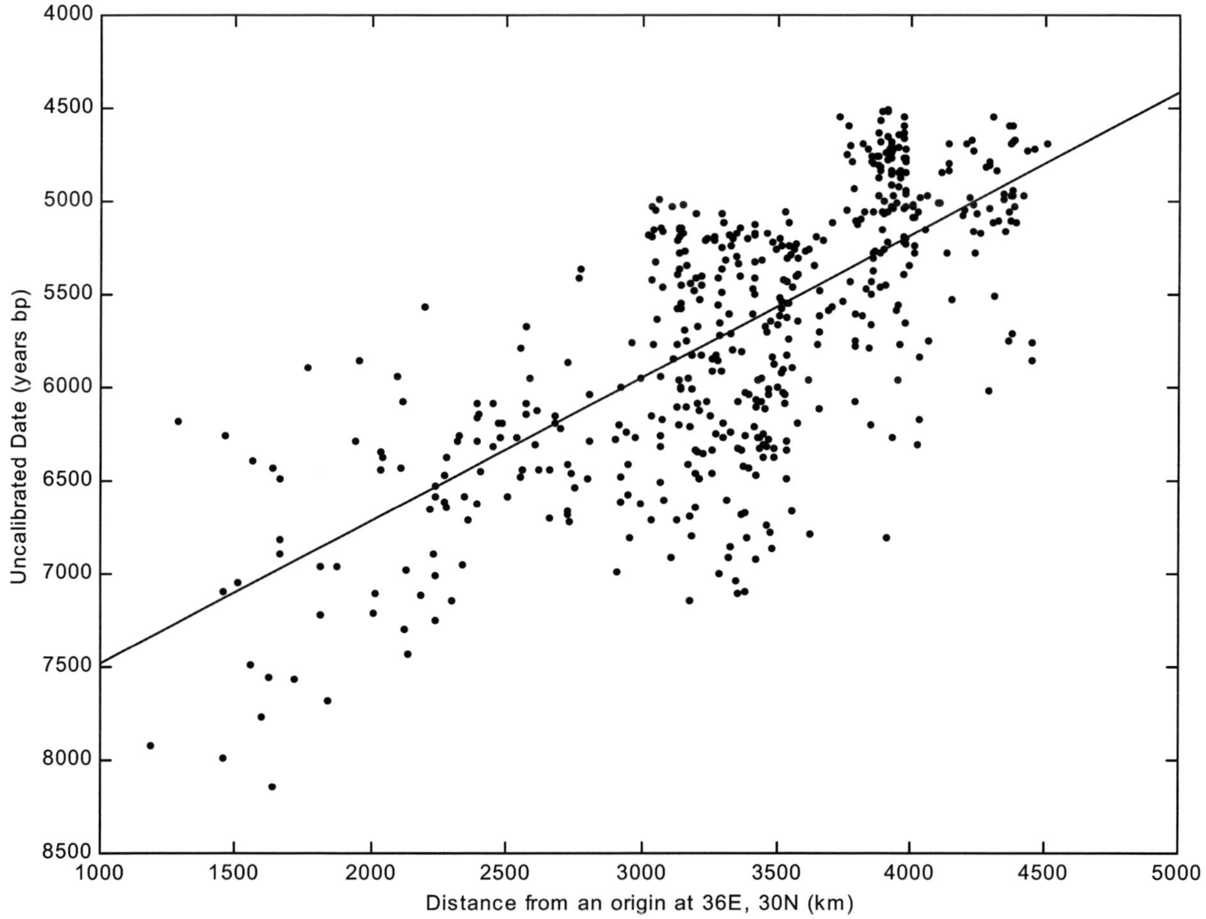

Figure 5.32 Linear regression of Dataset C.

5.4.3.4 Results for Dataset D:

Origin	24ºE, 42ºN
Number of sites	510
Weight on distance	none
Weight on determination	1/var (var = σ^2)
Correlation coeff., r	0.7102
Slope (m)	-0.7527
Y intercept (c)	6730.5
Error on slope	0.0331
Error on intercept	60.28
Equation of line	$Y_{determination} = (-0.75 \pm 0.03)\ x_{distance} + (6730 \pm 60)$
Rate (km/year) (1/m)	1.3 km/year
Determination at origin (yrs BP), i.e. $X_{dist.} = 0$	6730 ± 60

Table 5.9 Linear regression on Dataset D.

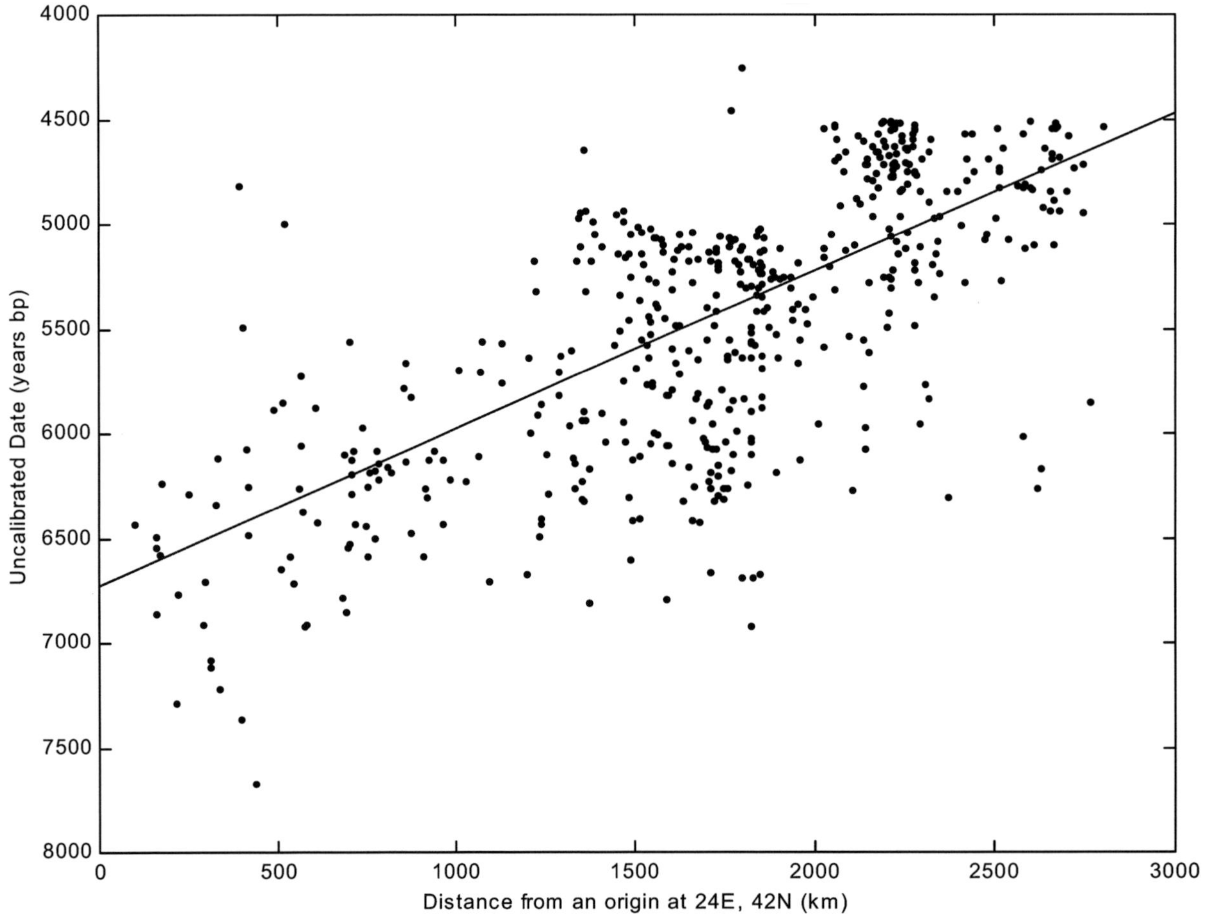

Figure 5.33 Linear regression of Dataset D.

5.4.3.5 Results for Dataset E:

Origin	24°E, 44°N
Number of sites	264
Weight on distance	none
Weight on determination	1/var (var = σ^2)
Correlation coeff., r	0.7638
Slope (m)	-0.8914
Y intercept (c)	6799.4
Error on slope	0.0465
Error on intercept	78.8162
Equation of line	$Y_{determination} = (-0.89 \pm 0.05)\ x_{distance} + (6800 \pm 80)$
Rate (km/year) (1/m)	1.1 km/year
Determination at origin (yrs BP), i.e. $X_{dist.} = 0$	6800 ± 80

Table 5.10 Results of linear regression on Dataset E.

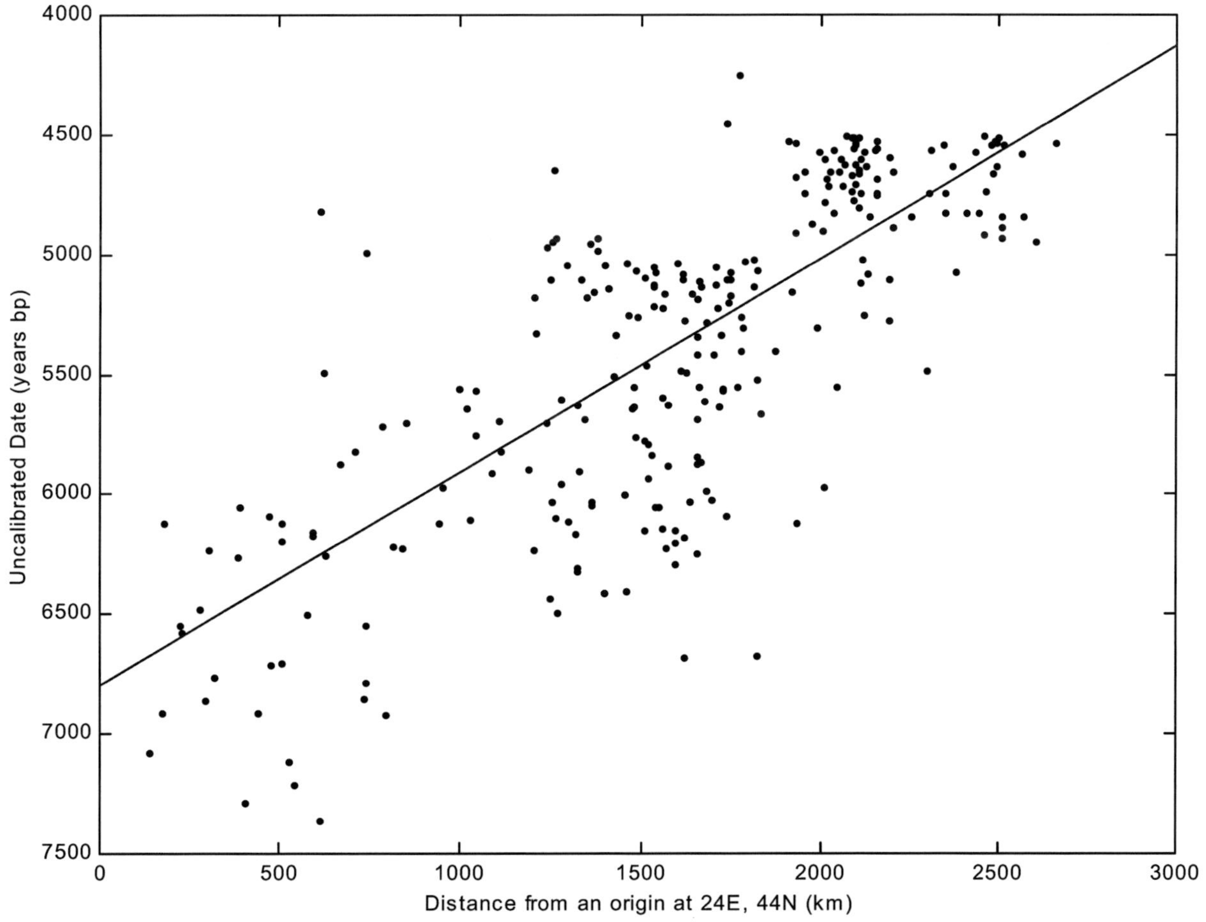

Figure 5.34 Linear regression of Dataset E.

5.4.3.6 Results for Dataset F:

Origin	36°E, 30°N
Number of sites	264
Weight on distance	none
Weight on determination	1/var (var = σ^2)
Correlation coeff., r	-0.7581
Slope (m)	-0.8053
Y intercept (c)	8483.8
Error on slope	0.0428
Error on intercept	143.9987
Equation of line	$Y_{determination} = (-0.81 \pm 0.04)\, x_{distance} + (8484 \pm 144)$
Rate (km/year) (1/m)	1.2 km/year
Determination at origin (yrs BP), i.e. $X_{dist.} = 0$	8484 ± 144

Table 5.11 Results of linear regression on Dataset F.

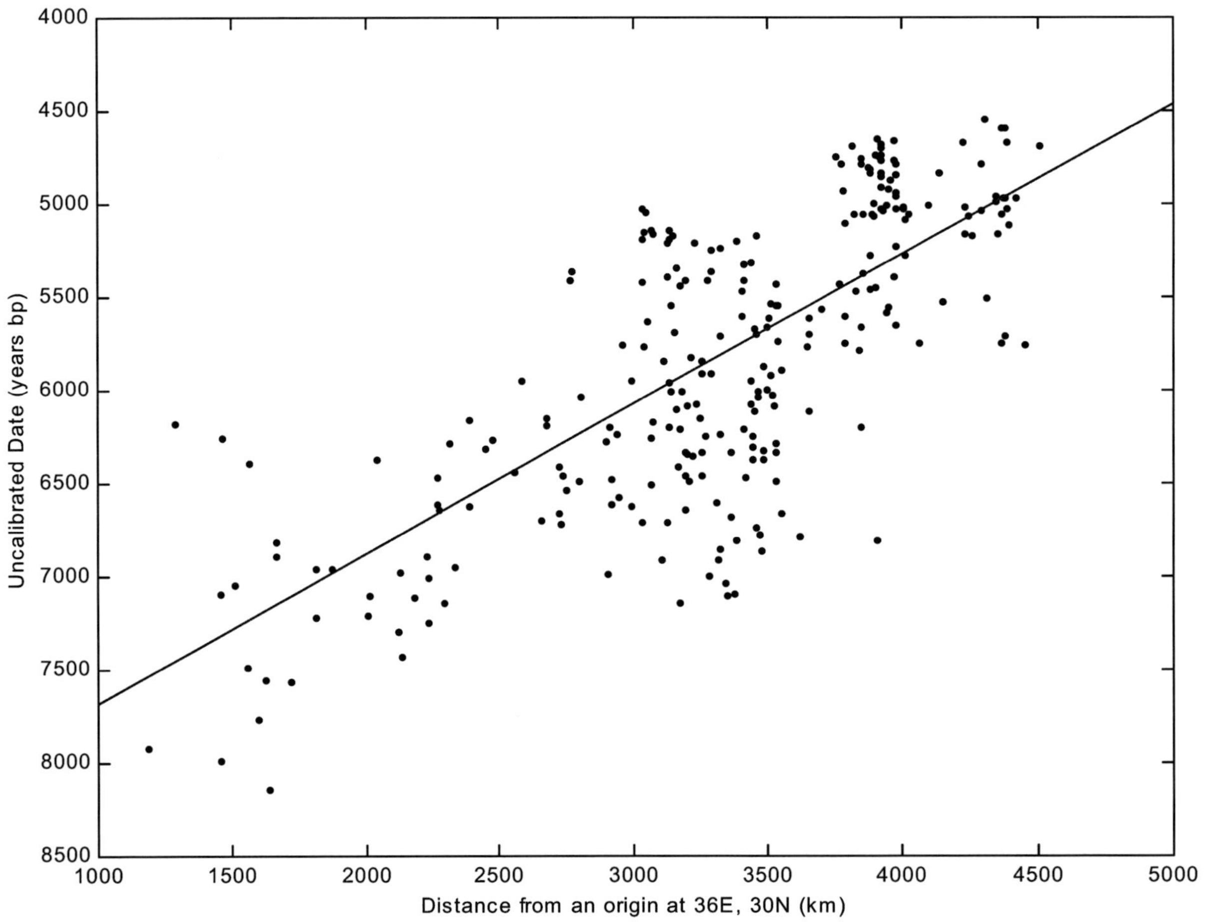

Figure 5.35 Linear regression of Dataset F.

5.4.3.7 Results for Dataset G:

Origin	36°E, 36°N
Number of sites	229
Weight on distance	none
Weight on determination	1/var (var = σ^2)
Correlation coeff., r	-0.7942
Slope (m)	-0.8426
Y intercept (c)	7952.5
Error on slope	0.0428
Error on intercept	130.5863
Equation of line	$Y_{determination} = (-0.84 \pm 0.04)\, x_{distance} + (7953 \pm 130)$
Rate (km/year) (1/m)	1.2 km/year
Determination at origin (yrs BP), i.e. $X_{dist.} = 0$	7953 ± 130

Table 5.12 Results of linear regression on Dataset G.

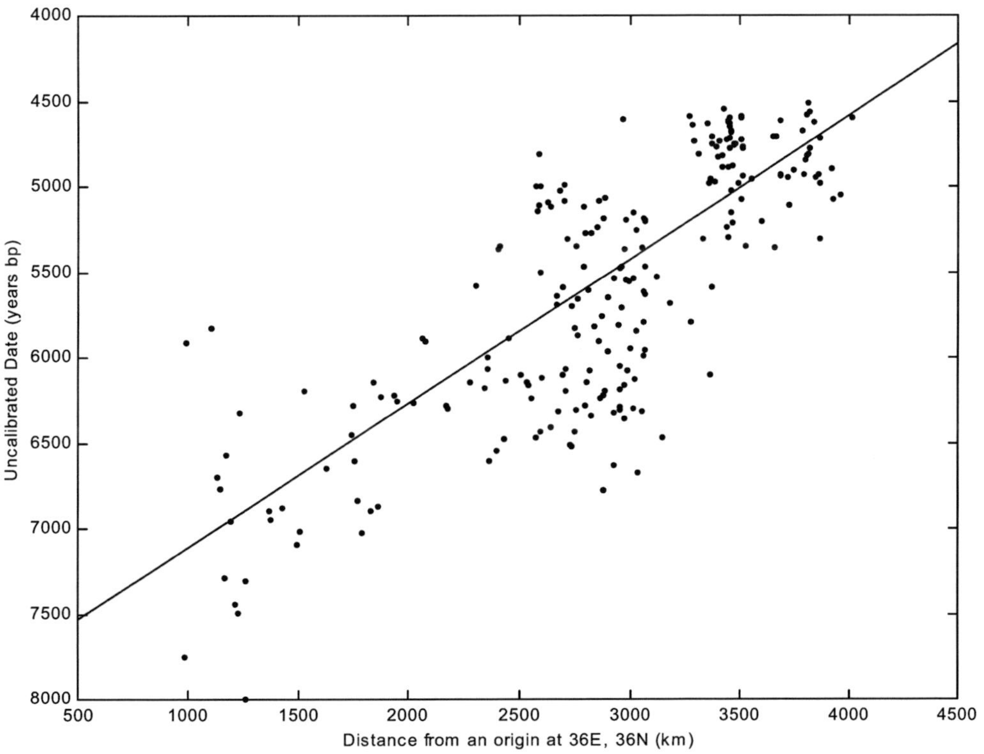

Figure 5.36 Linear regression of Dataset G.

5.4.4 *Discussion of rate and predicted departure times for Datasets A to G*

The predicted departure time from the origin is approximately 6800 years BP for origins in Bulgaria and Romania (Datasets B, D and E). When the departure point is in Jordan then the predicted departure time is approximately 8300 years BP (Datasets A, C and F). The origin in Turkey (Dataset G) has a predicted departure time of approximately 7900 BP. These predicted departure times can be used as a check on the model. To do this an independent dataset was used: that of Ammerman and Cavalli-Sforza (1971) (Table 5.13). These 1971 determinations have no overlap with those in the Gkiasta database as only post-1980 radiocarbon determinations were collected. The first ten sites as listed by Ammerman and Cavalli-Sforza are used to test the relationship between determination and distance as predicted by the linear regression results for Dataset C.

The distance had to be adjusted as the 1971 distances were from an origin at Jericho and the regression of Dataset C uses an origin that is approximately 260 km (geodesic) further south than Jericho. So 260 km was added to each distance above. The regression equation calculated for Dataset C was used to predict the age of the sites:

Predicted determination = (-0.77 ± 0.03) distance + (8240 ± 110)

These results were compared to the true age shown in the table above.

Apart for Konens Hoj and Azmak the predicted determinations are consistently ± 1000 years more recent than the measured radiocarbon determination. It is not possible to say how accurate the 1971 determinations are and this may be a factor in this discrepancy. However the result is not too unexpected. All the regression plots show that is a lot of scatter around the regression line (Figure 5.30 to Figure 5.36).

The rate of spread of farming across Europe as predicted from Datasets A to G varies between 1.1 km/year to 1.3 km/year. This is comparable to the rate of spread of 1.08 km/year that Ammerman and Cavalli-Sforza calculated in 1971.

Site name	Determination BP	Distance (km)
Neo Nikomedeia	8180	1815
Konens Hoj	5260	3366
Curnic	5980	4064
Ballynagilly	5625	4333
Knossos	7877	1452
Elateia	7480	1722
Argissa	7500	1770
Franchthi	7726	1670
Sidari	7670	1999
Azmak	6948	1631

Table 5.13 The first ten sites as listed in Ammerman and Cavalli's 1971 site list for early Neolithic sites in Europe.

Site name	Predicted Determination (BP) (Gkiasta Dataset C)	True Determination (BP) (Ammerman & Cavalli-Sforza 1971)
Neo Nikomedeia	6634	8180
Konens Hoj	5524	5260
Curnic	4910	5980
Ballynagilly	4703	5625
Knossos	6998	7877
Elateia	6790	7480
Argissa	6676	7500
Franchthi	6310	7726
Sidari	5980	7670
Azmak	6784	6948

Table 5.14 Comparison of 1971 determinations with those predicted by the regression equation calculated using Dataset C.

5.5 Summary:

The rate of spread of farming across the whole Europe is ±1 km/year as first calculated by Ammerman and Cavalli-Sforza using 53 early Neolithic sites in 1971. This figure is confirmed with a database ten times larger. The value of such an overall rate is discussed in the final chapter of this research.

A common criticism of rate calculations at this scale of analysis is that they mask underlying regional variation. Regional rates have been calculated for the spread of the Neolithic (Ammerman & Cavalli-Sforza 1971, Van Andel and Runnels 1995) and they have been used by proponents on both sides of the debate concerning the diffusion mode for farming, (most frequently to criticize the idea of demic diffusion in Ammerman and Cavalli-Sforza's 'wave of advance model'). To explore both the regional variation in rates using the Gkiasta's database and the notion that rates of spread will vary according to the mode of spread (i.e. by the movement of people or the movement of ideas), rate measurements are made using Zvelebil and Lillie (2000)'s work on the diffusion mode for farming in different regions of Europe. This analysis is presented in the following chapter.

Chapter 6 Cultural and demic diffusion mechanisms and regional variations in the rate of spread of farming.

6.1 Introduction

By 4000 BC the Neolithic economy based on the cultivation of wheat and barley and the breeding of cows, goats, sheep and pigs reached the British Isles and Scandinavia, completing the replacement of hunter-foraging by farming throughout Europe (Price 2000). With the exception of einkorn (a wheat species, *Triticum sp.*), which might have grown wild in the Balkans, the wild grasses from which domesticated crops have descended are not indigenous to temperate Europe. The Near East is their likely source (Ammerman and Cavalli-Sforza 1984, Bogucki 1995, 2000). In Central Europe the key domestic plant species and the two major livestock species have no wild progenitor. They were introduced from the Near East via south east Europe (Bogucki 1995).

In other parts of Europe the picture is varied; for example in the eastern Mediterranean there is evidence for the early exploitation of wild barley and lentils by foragers (Dennell 1985). However there is no such thing as a clear-cut case in archaeology. Recent suggestions for the presence of domesticated flax seed and wheat pollen from Mesolithic contexts in Switzerland, that predate the arrival of the Neolithic, must be considered (Price 2000). Wild sheep and goats are not thought to have existed north of the Mediterranean zone and are definitely exotic to the temperate forests of Europe first cultivated during the Neolithic (Bogucki 2000) (the Near East is their suggested source (Helmer 1984). However, wild cattle and pigs did exist in temperate Europe.

6.2 Search for the diffusion mode:

The debate has centred on how farming (i.e. the reliance on it by people as their chief food source) spread to Europe. The argument is whether its appearance was a consequence of the movement of people, or its indigenous adoption by hunter-gathering Mesolithic people, or a mixture of the two. Proponents of the two points of view are listed in Table 6.1. This list is by no means exhaustive, it is simply an illustration of the diversity of opinion that exist. Diffusions as a consequence of the movement of people are grouped as 'demic' diffusions. Diffusions that are a consequence of the exchange of ideas and their local indigenous adoption by groups already present in a landscape are grouped as 'cultural'diffusions.

	Demic diffusion (colonisation)	Cultural diffusion (indigenous adoption)
Greece & the Aegean:		
Mainland Greece	Papathanassopoulos 1996 Runnels 1995 Van Andel & Runnels 1995 Demoule & Perles 1993 Lewthwaite 1986	Douglas Price 2000 (for Peleponnese region) Tringham 2000 Halstead 1996 Whittle 1996 Chapman 1994 Dennell 1983 Jacobsen 1981 Theocaris 1973
Islands of the Aegean	Colonisation by sea: Douglas Price 2000 Jarman 1996 Broodbank & Strasser 1991 Cherry 1990 Sondaar 1971	

Table 6.1 **CONTINUED OVERPAGE**

Table continued	Demic diffusion (colonisation)	Cultural diffusion (indigenous adoption)
The Balkan Peninsula	Van Andel & Runnels 1995 Demoule & Perles 1993 Gimbutas 1991 Ammerman 1989 Renfrew 1989, 1987 Bogucki 1988 Barker 1985 Ammerman & Cavalli-Sforza 1984 Piggott 1965 Childe 1957	Tringham 2000 Zvelebil & Lillie 2000 Whittle 1996 Chapman 1993, 1994 Seferides 1993 Zvelebil 1986, 1986b Dennell 1983 Theocaris 1973
Central Europe:	Bogucki 2000, 1996 Keeley 1992 Zvelebil & Dolukhanov 1991 Dolukhanov *et al.* 1998	Jochim 2000 Whittle 1996 Hayden 1990
North central Europe	Bogucki 2000, 1995, 1988 Keeley 1992	Whittle 1996
Southern France and Italy	Binder 2000 Pluciennik 1998 (S. Italy)	
Iberian Peninsula	Zilhao 2000, 1993	Jackes *et al.* 1997
Ireland and Britain:	Cooney & Grogan 1994 Woodman 1985	Mercer 1986 Whittle 1985 Bradley 1984
Northern Europe:	Solberg 1989	Douglas Price 2000 Dolukhanov *et al.* 1998 Ahlfont *et al.* 1995 Price *et al.* 1995 Price 1991 Jennbert 1985, 1984
Eastern Europe		Dolukhanov *et al.* 1998

Table 6.1 Summary of the proponents of acculturation versus demic diffusion.

This debate has been a long one, but as Zvelebil and Lillie (2001) point out, it would be wrong to dismiss it as sterile, irrelevant or outdated. It has strong political connotations as it concerns the origin and ethnicity of European people today (see Harke 1998), and the issue is yet to be resolved.

Childe wrote in 1969, "The spread of a culture that can be traced by archaeological means, therefore, does not necessarily imply a spread of people who first created it. Nevertheless no other evidence is usually available to define folk movements" (1969:1). "Folk movements" are still extremely hard to determine archaeologically –as is evident in the many debates on this issue. In this section an attempt is made to distinguish between "folk movements" (demic diffusion) and the movement of ideas (acculturation/indigenous adoption) by looking at the radiocarbon determinations.

6.2.1 *Different types of diffusion*

6.2.1.1 Demic diffusion (the movement of people)

The concept of change brought by the migration of people rather then indigenous innovation suffers from its misuse and misinterpretation both historically and in the present day (Anthony 1997). Early proponents of migration hypotheses, for example Gordon Childe in his early work, interpreted change as the inevitable consequence of the replacement of the inferior by the superior. One of the difficulties with early migration hypotheses is that the indigenous people are seen as the passive recipients of whatever change came their way (Zvelebil and Dolukhanov 1991, Price 2000). There is now a better understanding of the interactions that occur when foragers and farmers meet. The consequences are varied. Both groups may coexist in a mutually beneficial system with neither side changing their subsistence base. Foragers may adopt farming; the contact may be hostile leading to the retreat of one or other group (Thorp 2000, Layton 2001).

The problem is how to identify the groups and their movements archaeologically. Bogucki (1997, 2000), who sees migration as the mechanism behind the arrival of farming in central Europe, uses Rouse's 1958 criteria for identifying migration. In very broad terms the evidence for the movement of people is the appearance of cultural traits that are different to those of the indigenous population and that have affinities with the origin area of the migration. In addition, and somewhat harder to see archeologically, Rouse's criteria require that conditions that favour migration are identified, and finally that the archaeological facts are not better explained by an alternative hypothesis (Bogucki 1997, 2000).

Other lines of evidence for migration are the absence of Mesolithic foragers in the areas where farming first appears (for example, in the in the eastern Mediterranean, (Van Andel and Runnels 1995) and in southern France and northern Italy (Binder 2000). The one area of Europe where there is general consensus amongst archaeologists that farming spread by migration is the region colonised by the Linear Bandkeramik culture of Central Europe (Bogucki 1997, Zvelebil 2000).

Anthony (1990) differentiates between two type of migration: short distance migration and long distance migration. These are key to understanding the way in which farming moved in areas of demic diffusion.

Short-distance migrations:

These are short distant local movements, for example the relocation to a new residence following marriage. Ammerman and Cavalli-Sforza's (1984) wave of advance model for the demic diffusion of farming across Europe is a stylised representation of this type of migration. The model was based on the relationship between the rate of population growth, rate of migratory activity and the velocity of spread at the edge of an advancing wave front. Migration direction is random. The relationship between these variables is described by the Fisher equation that was formulated to describe the spread of an advantageous gene in a linear habitat. In this model population growth at the frontier of expansion determines its expansion rate:

$$r = 2.04\sqrt{a.m}$$

Where,

r = the rate of advance of the wave front (km/year).

m = the migratory activity (km^2/generation).

$a =$ the population growth rate (%).

$a =$

Estimates of the population growth rate of a Neolithic population were made using archaeological evidence at Bandkeramik sites in Aldhoven, Germany (Ammerman & Cavalli-Sforza 1984). The number of houses (a total of 160 houses for a 500 year occupation period) excavated within a 1.3km stretch of the Merzbach stream was used to estimate population growth under the assumption that the growth was logistic. This produced an estimate of 1.9% growth per annum. An estimated range of possible variation of population growth, as compared to ethnographically known rates among modern people under various subsistence conditions, was 0.6% to 3% per annum.

The migratory activity (m) (mean square distance of movement per unit time) was estimated using ethnographic evidence based on the distance between birthplaces of husbands and wives in modern populations of farmers in Europe and Africa. The estimated range of possible values was between 300 to 2000 km^2/generation. Using all the possible combinations of the range of each of the two variables, the rate of advance was estimated (Figure 6.1).

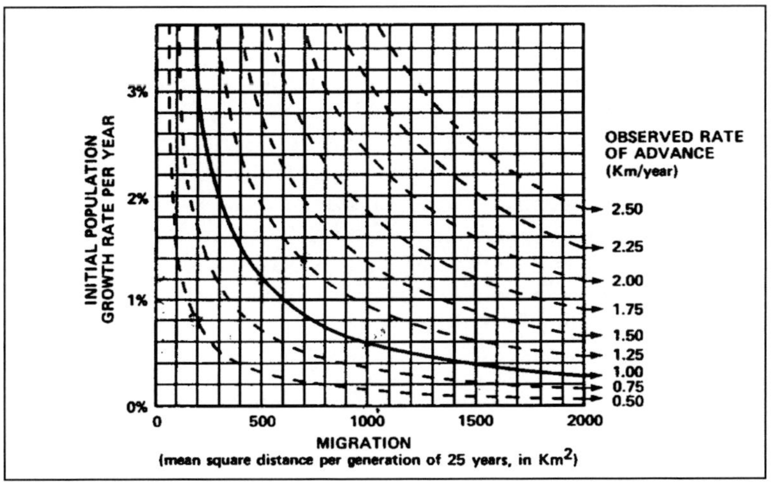

Figure 6.1 Evaluation of the wave of advance model (Ammerman and Cavalli-Sforza 1984:81)

The findings showed that a rate of advance of 1 km/year (as calculated from the linear regression of determination and distance from an origin at Jericho) was possible within the bounds of this model. Ammerman and Cavalli-Sforza (1984: 82) conclude, "The conclusion to be drawn is that the rates of advance predicted by the model when reasonable values of rates of growth and migratory activity are employed are essentially compatible with the observed rate of advance".

Long distance migrations:

These may be selective 'leap-frog' colonisations where groups of people move from one discrete location to the next, leaving intervening territory uninhabited. An example of this type of migration is the settlement of the Greek islands and the Balkans (Van Andel and Runnels 1995). They have proposed a revision of Ammerman and Cavalli-Sforza's wave of advance model (Figure 6.2). Rather than a continuous movement pushed forward by population growth at the front of the expansion (Figure 6.2, A), migration occurs in a discontinuous stepwise fashion with people 'leap-frogging' from one desirable location to the next (Figure 6.2, B). In Figure 6.2 B, the blue areas might represent water, mountains or some other barrier to settlement.

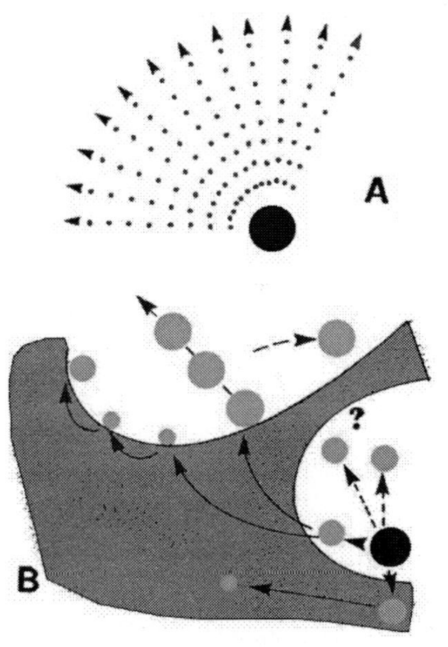

Figure 6.2 Modified wave of advance model for 'leap frog' colonisation of Greece and the Balkans (B) compared to Ammerman & Cavalli-Sforza's (1984) wave of advance model (Van Andel & Runnels 1995: 496, Fig.12)

In Greece and the Balkans, Van Andel and Runnels have shown that the preference of early farmers was for active floodplains and wet bottomland environments. This preference is also seen in the Great Hungarian basin, southern Yugoslavia, southern Bulgaria and Eastern Italy (*ibid.*). This is in agreement with Bogucki's (1997) finding in Central Europe, that early farmers showed a preference for settling in the upland loess basin and the lowlands of the Northern European Plain. Bogucki's interpretation of this follows from Beaton's (1991) idea of colonizing logic, that colonists unfamiliar with a new territory will only use the landscape at a gross level of resolution, locating their settlements in similar habitat types at the expense of equally fertile adjacent areas. Over time as familiarity with the landscape increases and the population grows, they spread to fill those areas that were initially avoided. Tringham (2000) describes a similar scenario for the catchment area of the middle and lower Danube basin in the northern Balkans. Here the settlement pattern is one of "tactical, short-term opportunism; a location is occupied according to an immediate need (water availability, a place to cross a river)" (*ibid.* 25). Leap-frog colonization as defined here can occur both across land and sea. It is the diffusion mode attributed to the spread of farming across the Mediterranean by sea-faring people (Price 2000). The leap-frogging between desirable habitats is not seen throughout Europe. In south western Macedonia early Neolithic settlements are found in a broad range of environments more consistent with gradual demic diffusion than leap-frogging selective migration (Wilkie & Savina 1997).

Anthony (1990) defines five different sub-categories of long distance migration. 'Leap frog migration' is summarised here, not only because it has been cited as a possible migration pattern for the European Neolithic, but also because it captures the essence of the others: this is that long-distance migration suggests either high risk taking or prior knowledge of destinations on the part of the migrating people.

6.2.1.2 Cultural diffusion (the movement of ideas)

This term places the emphasis on the indigenous adoption of farming by hunter-gatherers and its subsequent spread (Dennell 1983, Barker 1985, Whittle 1996). Europe was occupied by Mesolithic foragers prior to the first appearance of farming, and it is the extent of their role in the spread of farming that is debated. Few would argue that either indigenous adoption or demic diffusion is solely responsible for its spread. The disagreement is over the relative influence of each mechanism in different parts of Europe. The role of Mesolithic foragers in the change to a food producing economy is now widely recognized. Examples are found in northern Europe (Zvelebil & Rowley-Conwy 1986) and eastern Europe (Zvelebil & Dolukhanov 1991). Evidence from the last 60 000 years of hunter-gatherer history is that change is not necessarily caused by contact with other people (Layton 2001), nor brought by the migration of exotic people. The case for acculturation is built on theory and ethnography. Ethnographic studies of modern day foraging people provide examples of the interactions and the consequences of contact between the foragers and farming peoples. A summary of the findings is provided by Layton (2001: 294),

> "Comparative studies show that hunter-gatherers do normally interact with farmers or pastoralists when they come into contact. The interaction ranges from symbiosis to exploitation of hunter-gatherers by their neighbours. Moreover, hunter-gatherers can change their mode of subsistence and farmers or pastoralists can learn to become hunter-gatherers".

Between the two extremes from ethnography - the resistance to the adoption of farming on one hand and its appropriation on the other - there are many variations about the degree to which each group influences the other. It is this blurred area in the middle that is difficult for archaeologists. As the distinction between the two groups dissolves, it become harder to distinguish between them in the archaeological record (a problem that is addressed in detail in considering the appearance of the first sheep in South Africa, Chapter 8). Parallels with known ethnographic examples are one source for interpretation of archaeological evidence. Examples of other archaeological evidence that is used to identify the indigenous adoption of farming in Europe are the degree of overlap between Mesolithic and Neolithic sites on the landscape. Dennell (1992), a proponent of a hypothesis that emphasises indigenous adoption, argues that Neolithisation always occurs in area where Mesolithic hunter-gatherers are already present and that this suggests local innovation. This view is challenged in central Europe where the co-existence of the two groups is less evident (Bogucki 1987). Vencl (1986) uses palaeo-anthropological evidence to demonstrate that skeletal populations in northern Europe do not show the changes that would be predicted if a new immigrant population was bringing farming to northern Europe.

6.2.1.3 Demic and cultural diffusion (movement by a combination of the two mechanisms):

There is now general consensus amongst archaeologists that the arrival of farming throughout Europe was a result of a combination of demic and cultural diffusion.

This is what Zvelebil (2000) terms the 'integrationist' position. Zvelebil and Rowley Conwy (1984) suggest a mechanism of spread involving both the migration of farming people and the adoption of farming by indigenous people. The concept of the agricultural frontier as a boundary across which farmer-forager interactions take place is well documented (Alexander 1977, Green and Perlmen 1985, Moore 1985, Megaw 1997). Evidence of this is not restricted to the Neolithic of Europe. Frontier interactions between farmers and hunter-gatherers are described for southern Africa (Alexander 1984, Sampson 1984, Mazel 1989, Hall *et al.* 1994, Thorp 1996, 2000). Zvelebil and Rowley-Conwy make a strong case for the role of such a frontier in the spread of farming to Europe. The model has three stages: availability, substitution and consolidation. In the availability phase farming is known by foragers but not adopted. There is mutual cooperation between the two groups and exchange of items and information occurs. The substitution phase describes the stage when farming replaces hunter-gathering, although elements of foraging are maintained. In the final phase, consolidation, farming is adopted.

These stages are acted out across a frontier. The frontier is mobile during the spread of farmers into an area. Farming groups at the continually evolving frontier zone are in a situation of rapid population growth, extensive land-use patterns, frequent relocation, low subsistence and high labour costs, and a population structure with more males and young people than females and the elderly (Zvelebil & Lillie 2001).

> "…the frontier zone is expanded, contracted and relocated in space and through time and its characteristics change (i.e. as in the availability, substitution and consolidation phases [Zvelebil and Rowley-Conwy 1984, 1986]). These activities leave archaeological signatures which then have to be decoded by archaeologists and interpreted in terms of past human behaviour" (Zvelebil and Lillie 2001: 64).

The frontiers associated with the European Neolithic are shown in Figure 6.13.

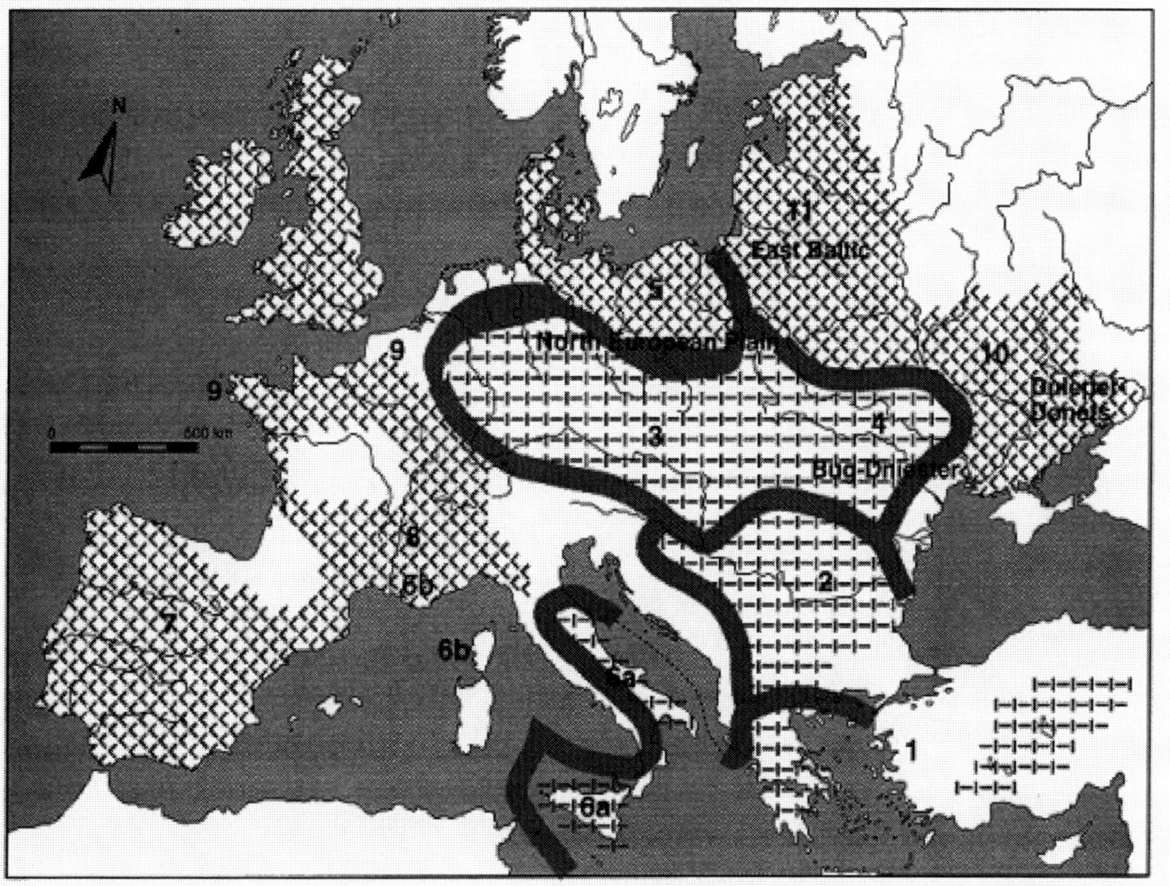

Figure 6.3 Major frontier zones (black) between early Neolithic cultures (Zvelebil & Lillie 2000:63, Figure 3.1). Cross hatching indicates those areas where farming spread by demic diffusion. Triangular hatching indicates the areas where farming spread by indigenous adoption.

At the frontier between the hunter-gatherers and the farmers interaction may be cooperative or disruptive (Hall 1987, Thorp 2000). The interaction may lead to the exchange of food and technology, information, marriage partners and diseases, to list a few. In Zvelebil and Rowley-Conwy's model the initial contact is one of cooperation. According to Zvelebil and Rowley-Conwy during this cooperative phase there is the exchange of:

1. Food. This reduces the risk of failure for both hunter-gatherers and farmers. This is particularly beneficial to farmers who have recently adopted farming and are in a new landscape (Zvelebil and Lillie 2000).

2. People. Kin relations may develop between the two groups (*ibid.*).

3. Goods and information: prestige items, raw materials and information may be exchanged for the mutual benefit of both groups. Archaeological evidence for contact and exchange across the frontier are found in, for instance in the Northern European Plain. Pottery, stone axes and shoe-last adzes and domestic fauna at forager sites are evidence of trade and exchange. Similar evidence of exchange of axes, pots, amber and seal fat is seen in the central and East Baltic (Zvelebil 2000).

4. Livestock. This may be passed to hunter-gatherers for social rather than economic reasons (*ibid.*).

This relationship of mutual cooperation deteriorates as the frontier endures. A new phase emerges characterized by the disruption to hunter-gatherer communities:

1. The prestige items and increased social competition cause breakdown within the social fabric of the hunter-gatherer society.

2. Farmers start to make direct use of the hunter-gatherer land. This disrupts the hunter-gatherer's by affecting their foraging strategies, information exchange networks and causing ecological change.

3. Farmers acquire raw materials and wild food directly by expanding into hunter-gatherer land.

4. Forager women are lost to farmers (hypergyny).

The rate of spread of farming can be used as an indication of how long a frontier zone remained in each of the three stages. For example the spread of farming from the Northern European Plain to Scandinavia was very slow. The availability phase is thought to have endured for over one thousand years whilst the spread of the Linear pottery culture to central Europe was rapid suggesting that the three phases did not last long (Zvelebil and Dolukhanov 1991). Frontier zones that stayed in the availability phase for hundreds of years should be picked up in the isochron maps (Figures 4.9 to 4.15, Chapter 4). Tightly packed isochrons show the slowness of spread (for instance the time lag in the Alpine region and the delay in the spread of farming to the British Isles), whereas in regions of rapid spread the isochrons are far apart (this is clear in the Balkans and central Europe).

6.2.2 A search for an origin for the spread of farming in 'demic diffusion' regions and 'cultural diffusion' regions and a comparison of the rate of spread in each of these regions.

The aim of this section is twofold. The first is to see if Zvelebil and Lillie's frontiers (Figure 6.3) can be identified by spatial analysis of the radiocarbon data. The second is to see whether there are differences in the rate of spread of farming in those areas that are attributed to demic diffusion and those attributed to cultural diffusion.

There was the expectation that the frontier hypothesis could be tested. Each frontier is equivalent to an origin. It is the point from which farming is thought to have spread from farmers to hunter-gatherers. If this was the case then dates for the conversion to farming should be older at the frontiers and there should be a correlation between the timing of the adoption of farming by indigenous groups and their distance from the frontier zone. There is another possibility, however. Whilst the expectation is that hunter-gatherers at the frontier zone would show the earliest conversion to farming in an area, the pattern may not be a simple trend of decreasing antiquity with distance away from the frontier. Hunter-gatherers adopting farming would not necessarily relocate and the adoption of farming might proceed along communication networks already established between different groups rather than as a function of their distance from the original point of influence.

The second expectation was that there would be a difference in the rate of spread in the two areas if they are indeed the result of different processes, as one is essentially a movement of people whilst the other is essentially a movement of ideas. The expectation was that demic diffusion might be faster than cultural diffusion. This is influenced by the theory that the switch from foraging is not one that is taken causally (Ingold 1980). Support for this can be found in the time lags that are seen in the spread in certain areas, for example, Northern Europe, where farming is known but not adopted by hunter-foragers for hundreds of years (Zvelebil and Rowley-Conwy 1986). A fast spread rate suggests that either an area is unoccupied, or that space can be shared (suggesting that there is no pressure on resources or that groups are occupying different niches) or that any resistance is easily overcome (that foragers are displaced, assimilated or killed) (Dennell 1985, Zvelebil 1986a). In many parts of Europe, authors who argue for the indigenous adoption of farming also emphasise the slowness of transition to farming (Zvelebil and Dolukhanov 1991). This lends support to the notion that cultural diffusion is slow. However, cultural diffusion may also be fast, after the approximate 1000 year halt in the spread of farming in North Europe it then spread extremely rapidly through Scandinavia, a spread most popularly attributed to indigenous adoption. However this was after many generation of hunter-gatherers knowing farming, and so the pattern seen here might be one that might not apply to situations of the first appearance of farming.

6.2.3 Spatial analysis to identify the origin and rate of spread in demic and demic diffusion areas

Radiocarbon determinations were selected according to the areas shown in Figure 6.3.. Two datasets were used in this analysis, Dataset G (grouped determination or median determination) as this gave the highest correlation coefficient in the previous analysis and Dataset A (grouped determination or oldest determination) as this represents the earliest dates for the spread of farming. Following Zvelebil and Lillie (Figure 6.1), the radiocarbon determinations in each dataset were sorted into areas representing demic diffusion and those representing acculturation. Further divisions were made to see the effect of excluding sites that lay on the periphery of these divisions or which were associated with possibly different modes of diffusion by other authors (namely, the spread westwards across the Mediterranean by sea).

The seven subdivisions for both Dataset A and Dataset G are as follows:

1. Zvelebil and Lillie's demic diffusion area (Areas A and B, Figure 6.6).

2. Zvelebil and Lillie's demic diffusion area without Italian sites (Area A, Figure 6.6).

3. Demic diffusion area using Italian sites only (Area B, Figure 6.6).

4. Demic diffusion Italy and enclave sites only (Areas B and C, Figure 6.6).

5. Zvelebil and Lillie's western cultural diffusion area (Areas D and C, Figure 6.6).

6. Zvelebil and Lillie's western cultural diffusion area excluding enclave[1] sites (Area D, (Figure 6.4, Figure 6.5 and Figure 6.6).

7. Zvelebil and Lillie's northern cultural diffusion area (Area E, Figure 6.6).

The Matlab script to search for an origin (described in section 6.3.1) was used in this exercise. For each of these subdivisions the script routine searched for the origin point with the best fit to the data by calculating the linear correlation coefficient for the regression of distance (from each of the 504 potential origins to each of the Neolithic sites) and the age of the Neolithic site. This was to see whether the 'best fit' origin coincided with the frontier zones and whether there was a linear relationship between distance and determination from this origin point. The 'best fit' origin to each of the subdivisions of data was used to calculate the rate of spread and the time at the origin (where the trend in the data is linear this is an accurate calculation).

[1] These are Mediterranean coastal sites (referred to as 'enclaves' (after Barnett 2000, Zilhao 1993, 2000)).

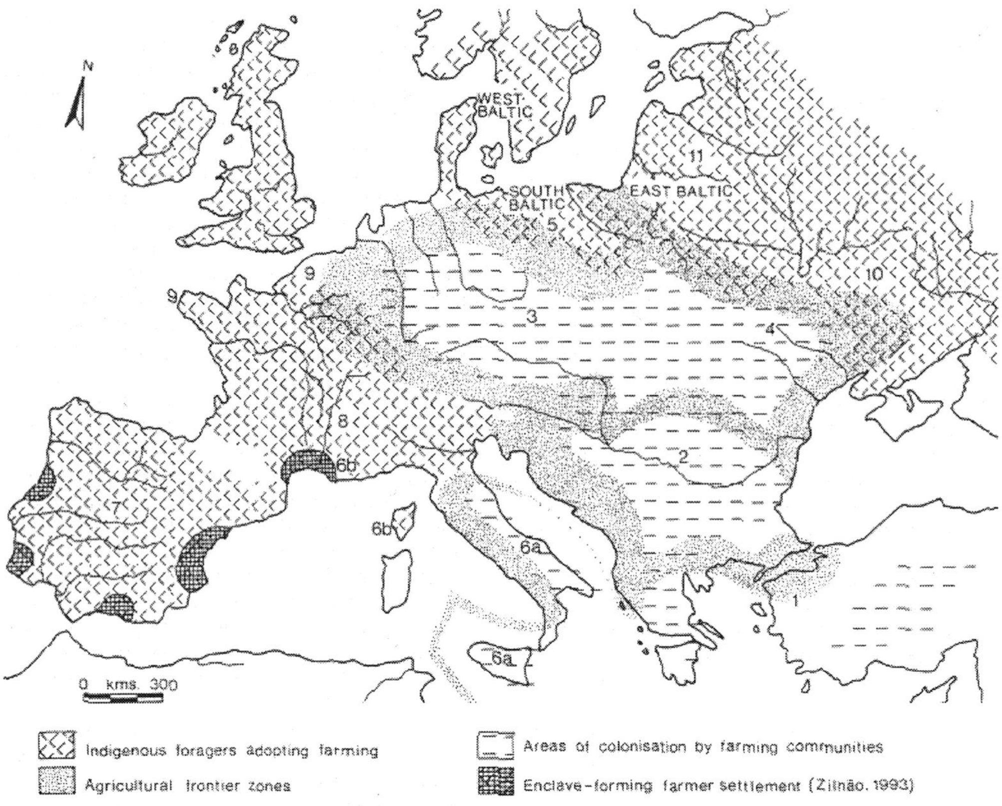

Figure 6.4 The map shows the enclave areas that were separated from the indigenous adoption area and analysed with Italian sites in this exercise in accordance with the hypothesis that they may be the result of migration by sea (Zvelebil 2000:61, Figure 7.1).

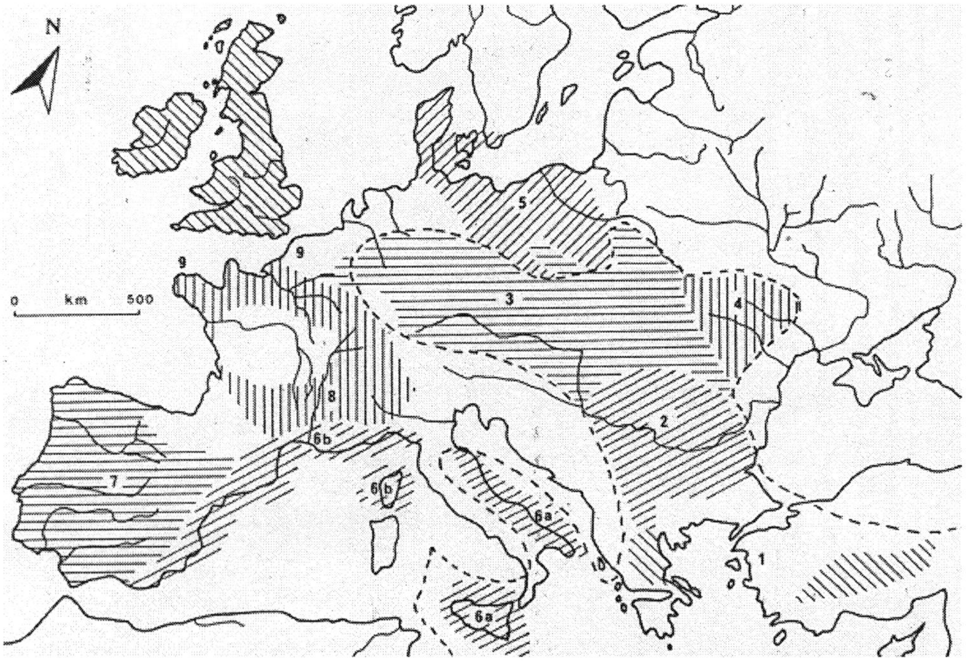

Figure 6.5 In an earlier map by Zvelebil and Dolukhanov (1991: 242, Figure 4) the area labelled as 6b (the Impressed ware culture of the eastern Mediterranean) is the area that in these analyses is called the enclaves following Barnett 2000 and Zilhao 1993, 2000. In these analyses it is coupled with area 6a (the Impressed Ware culture of the western Mediterranean) following the hypothesis that these two areas might represent the spread of farming by the demic migration of people by sea.

6.2.4 *Results for Dataset A:*

The radiocarbon determinations for Dataset A were divided into areas representing demic diffusion and those representing acculturation (Figure 6.6).

The results of the analysis of Dataset A are in Appendix 6.

6.2.4.1 Rate of spread and origin of farming for Zvelebil and Lillie's demic spread region:

Figure 6.7 shows sites that fall within the area that Zvelebil and Lillie attribute to the demic spread of farming. Using the Matlab script to search for the region that gave the best correlation between distance and determination, the result is an origin area in Greece and the Aegean (Figure 6.8).

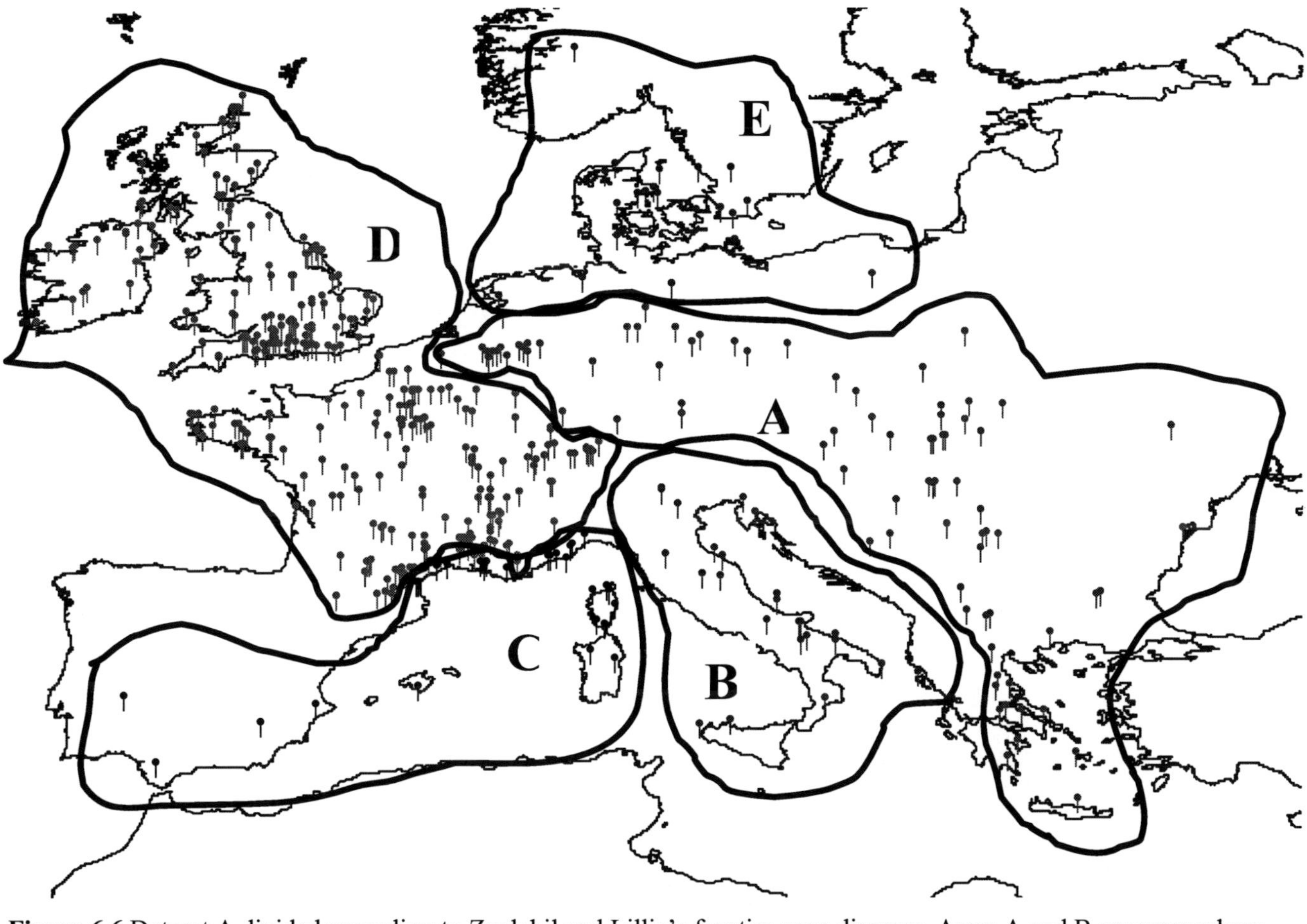

Figure 6.6 Dataset A divided according to Zvelebil and Lillie's frontier zone diagram. Areas A and B are areas where farming was introduced by demic diffusion. Areas D, C and E icons are areas where farming was introduced by its adoption by indigenous Mesolithic foragers. The subdivision of the demic and cultural diffusion areas is to test the effect of separating areas where the diffusion mechanism may have differed (i.e. travel by sea to Italy, Corsica and Sardinia).

Figure 6.7 Neolithic sites that fall in Zvelebil and Lillie's region of demic spread.

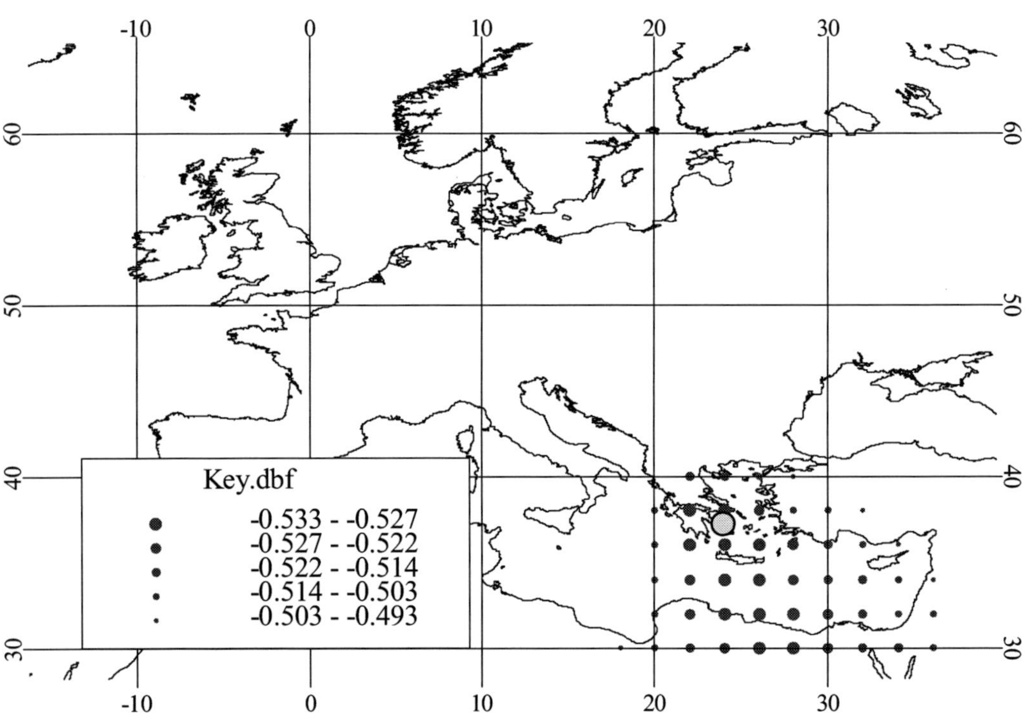

Figure 6.8 Map to show the region that produced the strongest correlation coefficient for the regression of determination and distance from an origin (using sites in Zvelebil and Lillie's demic spread region). The point with the strongest correlation coefficient is shown as pale circle.

The correlation coefficient gives an indication of the linearity of the relationship between date of archaeological site and distance from the origin. Low correlations show that the relationship is not linear. The correlation results shown in Figure 6.8 and Figure 6.9 were calculated using weighted model 1b regression (Section 6.4.1). This means that only the determinations are weighted and the regression line is fitted by minimising the Y (weighted determination) residuals only. Having identified the best-fit origin to the demic data, linear regression using weighted regression (model 1b) was used to find the rate of spread. The scatter plot also shows the degree to which the data deviates from the linear regression line.
The results of the linear regression of 121 Neolithic sites in the demic region with their distance from an origin in Greece (latitude 24°E, longitude 36°N) are shown in Figure 6.9.

The results are a correlation coefficient of –0.53. The equation of the regression line is:

$$y_{determination} = (-0.37 \pm 0.05)\, x_{distance} + (6897 \pm 74)$$

This gives an overall rate of spread for demic diffusion of between 2.4 km/year to 3 km/year and an estimated date of departure from the origin point between 6970 and 6820 years BP.

The correspondence between the calculated origin area and Zvelebil and Lillie's frontier zone between Greece and the Balkan Peninsula is good (Figure 6.9).

6.2.4.2 Rate of spread and origin of farming for Zvelebil and Lillie's demic spread region when the Italian sites are excluded from the analysis:

The decision was made to rerun the analysis without the Italian sites (Figure 6.10). Although these may be attributable to the demic mode of diffusion there is the

widespread view that the spread of farming in the Mediterranean was facilitated by sea travel. This would have implications for rate measurements, as far greater distances can be travelled in short lengths of time over water. As a consequence, when looking at rates of spread as a means to distinguish between demic and cultural spread, areas of diffusion across land must be separated from those across water.

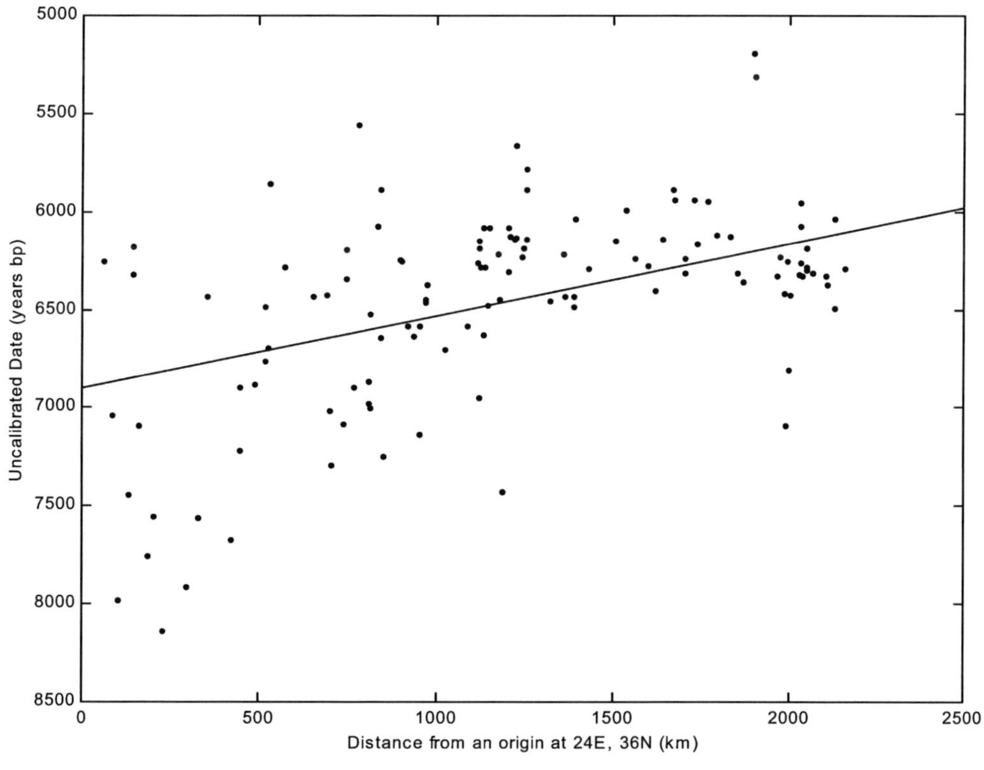

Figure 6.9 Linear regression of weighted determination and distance for 'demic' data. Origin at longitude 24°E and latitude 36°N.

Figure 6.10 Areas attributed to demic spread across land (Italian sites are excluded).

Once again a search was made for the best origin for the sites in this region. The strongest correlation coefficients show that the most likely origin is in a region that covers Greece, North Africa and western Turkey, are shown in (Figure 6.11).

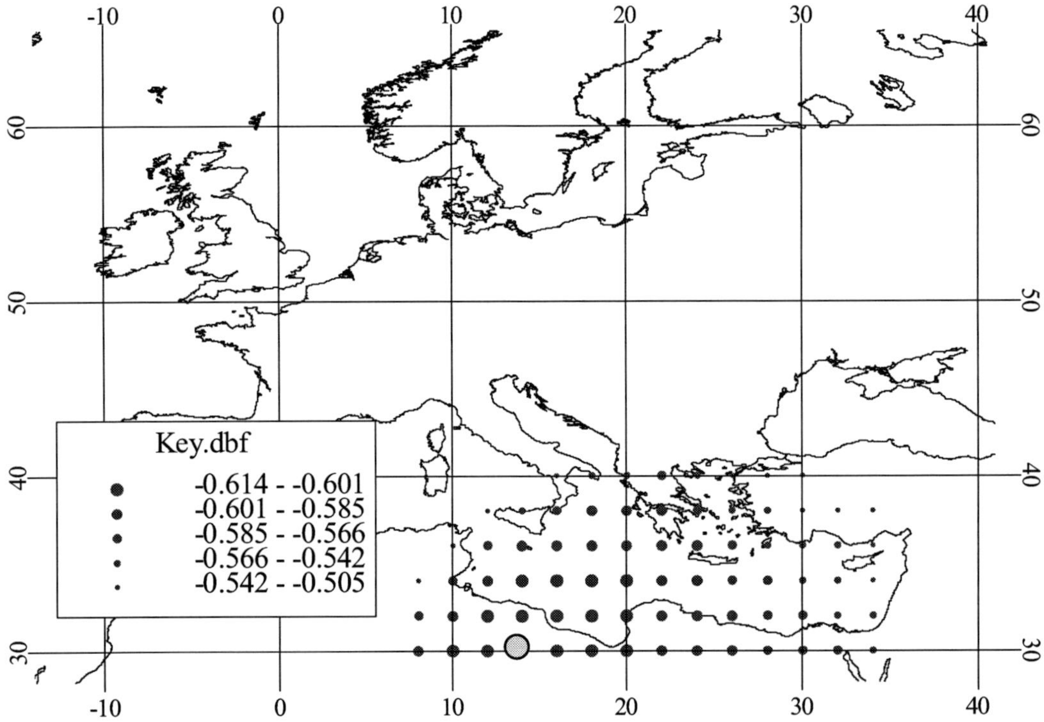

Figure 6.11 Origin points giving the strongest correlation coefficients for the spread of demic determinations excluding Italy. The point of strongest correlation is shown as pale circle.

Weighted regression model 1b was applied to the 96 Neolithic sites using the origin with the highest correlation coefficient (latitude 14°E, longitude 30°N, Figure 6.11). The results are shown in Figure 6.12. The correlation coefficient, r, when Italian sites are excluded, increases to –0.61, lending some support to the reasons for excluding Italian sites, namely that they may not produce a linear spread nor a similar rate to areas colonised by land.
The equation of the regression line (Figure 6.12) is:

$$y_{determination} = (-0.41 \pm 0.06) \text{ x }_{distance} + (8100 \pm 223)$$

Using this equation the rate of spread of farming in the area attributed to demic spread across land is 2.1 km/year to 2.9 km/year. The exclusion of Italian sites causes a drop in the rate of spread. The time of the departure from the origin is 7880 years BP to 8320 years BP.

Because archaeological evidence would not support a north Africa origin of spread the analysis was redone using one of the other origin points that had a strong correlation, namely 22°E, 38°N, the result is a correlation coefficient of –0.59 and the equation of the regression line is:

$$y_{determination} = (-0.43 \pm 0.06)\, x_{distance} + (7000 \pm 80)$$

This gives a rate of spread in the range 2 to 2.7 km/year and a predicted determination of departure from the origin in the range 6920 to 7080 BP.

The origin region for these sites (Figure 6.11) is in agreement with Zvelebil and Lillie's frontier between Greece and the Balkan Peninsula.

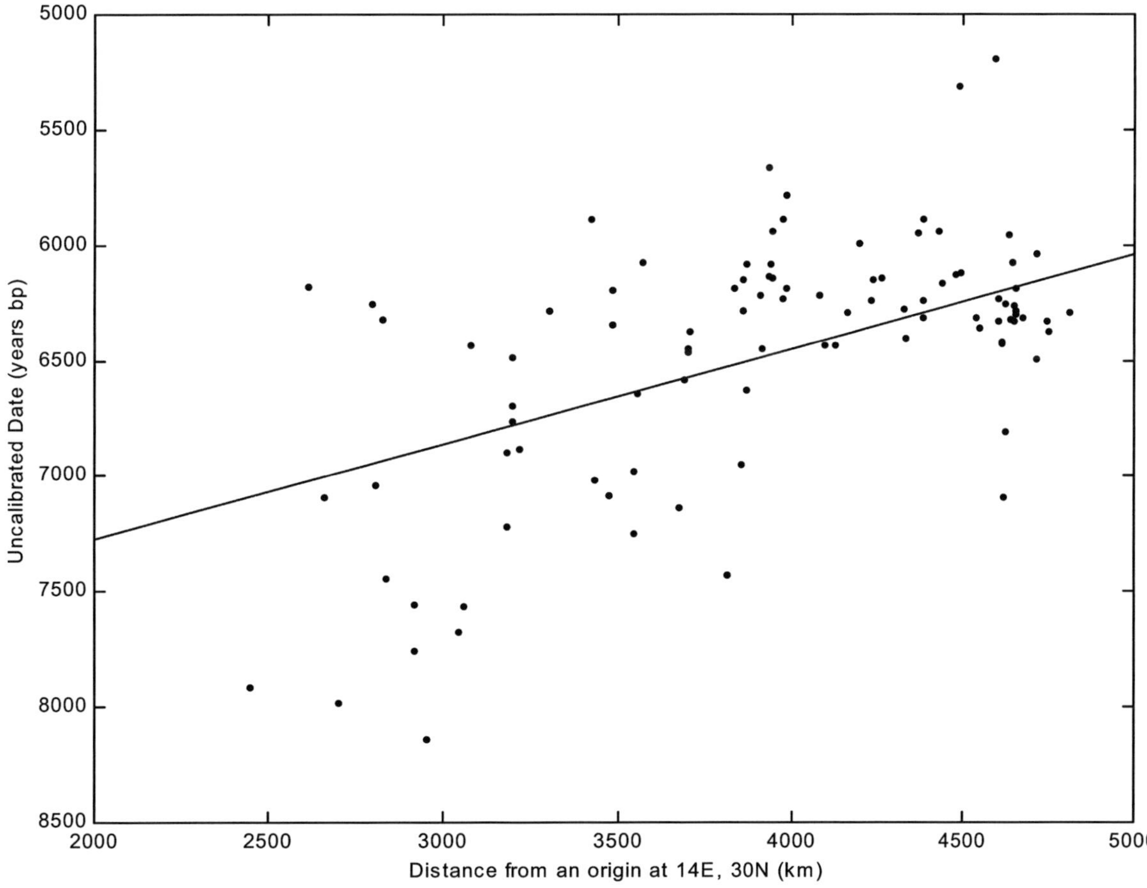

Figure 6.12 Linear regression of weighted determination and distance of 'demic' data, excluding Italian sites, from an origin at 14°E, 30°N.

6.2.4.3 Rate of spread and origin calculated for the Italian sites only

The rate of spread and the origin point was calculated for the 25 Italian sites. This was to see whether the rate of spread was noticeably higher and less linear than that for central Europe as expected if movement includes travel by sea. The Italian sites used in the analysis are shown in Figure 6.13. The origin region that showed the best fit to the Italian Neolithic sites is shown in Figure 6.14.

The correlation coefficients are extremely low (the strongest is -0.29), showing that there is not a linear trend between determination and distance for these sites. This is clear on the scatter plot (Figure 6.15).

Figure 6.13 The Italian sites that fall in Zvelebil and Lillie's demic spread region. The spread of farming to these sites may involve travel by sea.

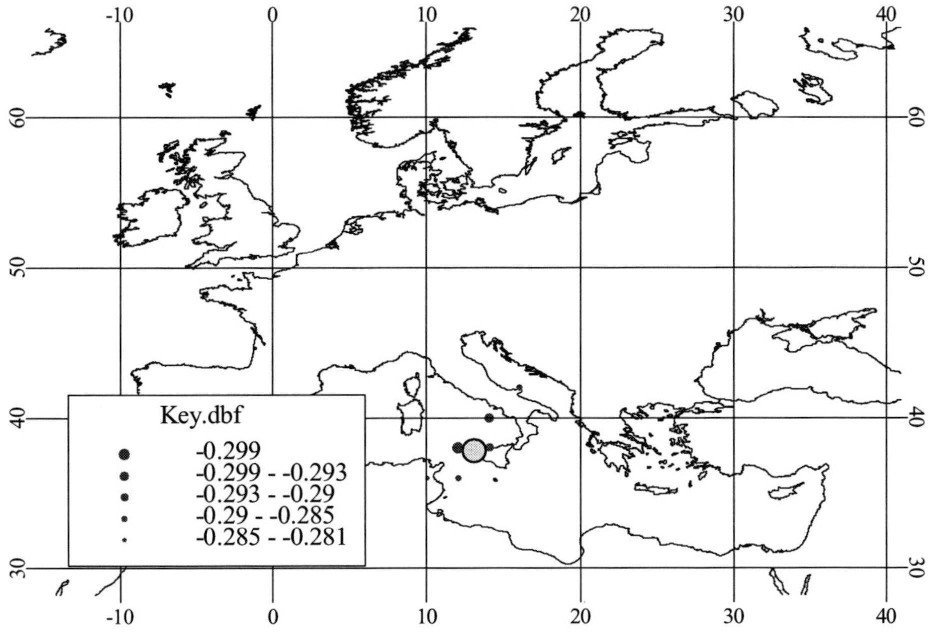

Figure 6.14 The origin points with the best correlation coefficients for the Italian Neolithic sites. The point with the strongest correlation coefficient is shown as pale circle.

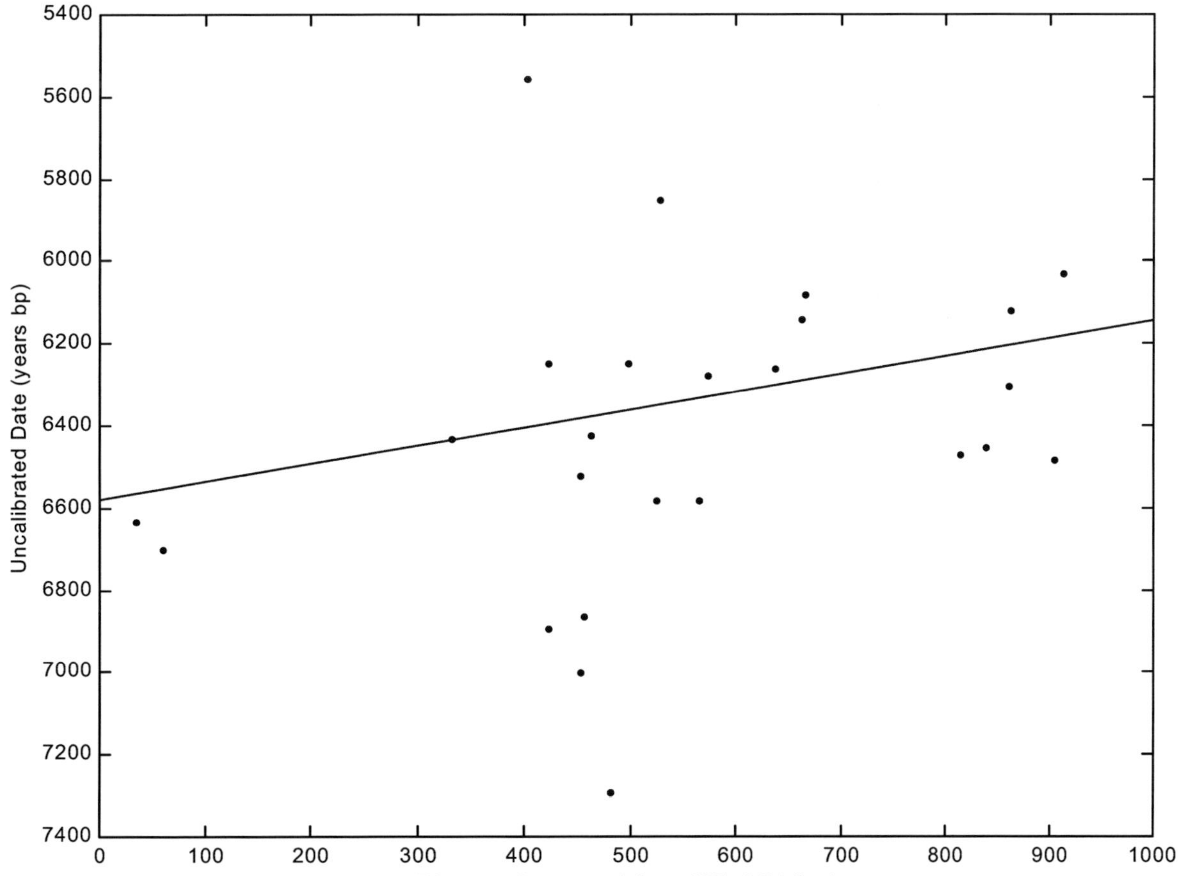

Figure 6.15 Linear regression of weighted determination against distance for Italian sites from an origin at 13°E, 38°N.

Using distance from an origin at 13ºE, 38ºN, the equation for the regression line:

$$y_{determination} = (-0.43 \pm 0.31)\, x_{distance} + (6578 \pm 202)$$

This gives a rate of spread of the range 1.4 km/year to 8.3 km/year and an origin departure date in the range 6380 to 6780 years BP. As mentioned this sort of rate calculation is only accurate when there is a linear trend between variables, but it is useful as an approximation of the rate of spread. The error terms associated with the slope and intercept are a consequence of departure of the determinations from a linear trend. The calculation was included here, nevertheless, because previous authors ((Ammerman and Cavalli-Sforza 1971, Housley *et al.* 1997) have calculated the rate of spread to sites in the Mediterranean; this is included as a comparison.

6.2.4.4 Rate of spread and origin calculated for Italy and the enclaves:

The next step was to add other Mediterranean coastal sites (referred to as 'enclaves' (after Barnett 2000, Zilhao 1993, 2000)) to the group containing the Italian Neolithic sites. To do this Neolithic sites that fall within Zvelebil and Lillie's acculturation group must be 'borrowed' (Figure 6.16). The hypothesis was that the sites shown in Figure 6.16 might be a result of the demic spread of farming by a sea-faring people.

The results for the most likely origin region for these sites are shown in Figure 6.17. The strongest correlation coefficient is found at latitude 30ºN and Longitude 20ºE, on the north coast of Libya.

Figure 6.16 Italy and enclaves: sites that might be representative of the spread of a sea-faring people.

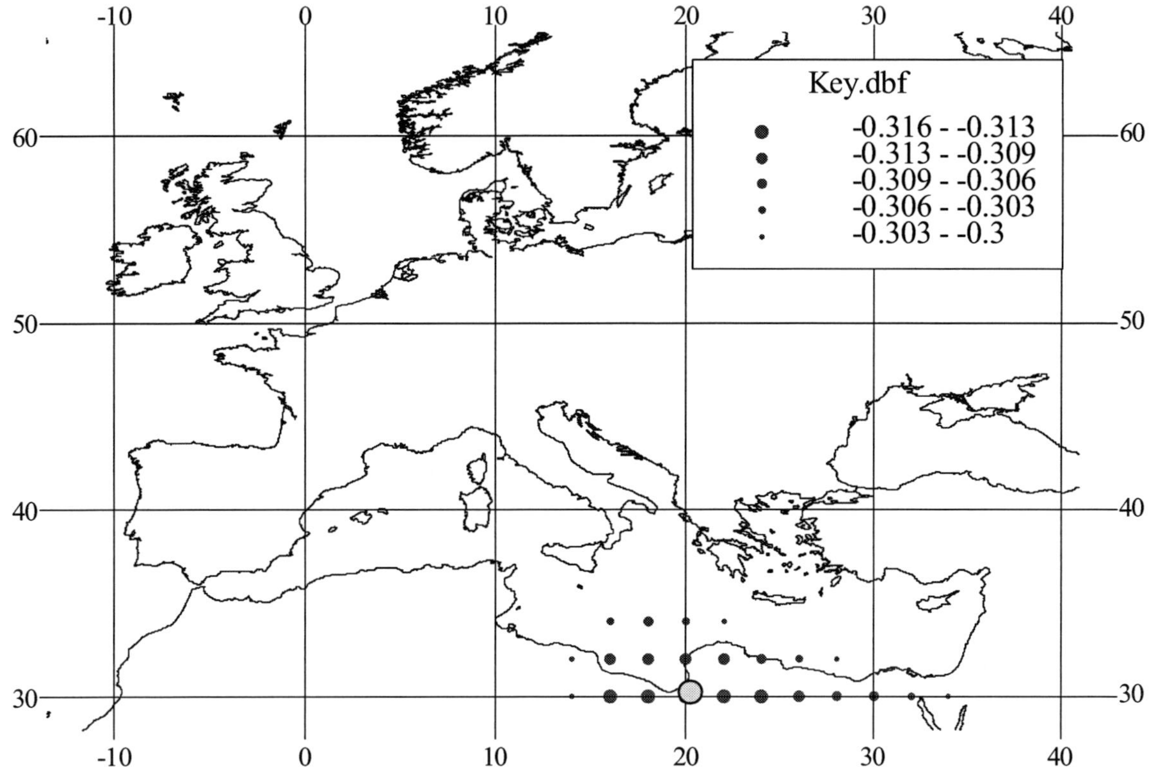

Figure 6.17 Origin points that gave the strongest correlation coefficients to the Neolithic sites in Italy and enclaves. The point with the strongest correlation coefficient is shown as a pale circle.

The origin point on the Libyan coast is used to find the rate of spread of farming and the most probable date of departure from this origin point. Once again the correlation coefficients are low (the strongest correlation coefficient is –0.32), an indication that as with the Italian sites, the trend is not linear. The scatter plot of the 55 Neolithic determinations used in the analysis and the best-fit line are shown in Figure 6.18. The equation of the line is:

$$y_{determination} = (-0.46 \pm 0.20)\, x_{distance} + (6898 \pm 304)$$

Using this equation, the rate of spread is in the range 1.5 to 3.8 km/year and the time of departure from the origin is in the range 6590 to 7202 years BP.

Although Neolithic sites are known on the Libyan coast (Chapter 4, Figure 4.1 Clark's 1965 map), this is not an accepted point of departure for the spread. The decision was made to rerun the analysis using an origin as close as possible to the Libyan coast but within Europe. An origin at 16ºE, 40ºN on the Italian coast was used. The correlation coefficient is –0.26 and the equation of the regression line is:

$$y_{determination} = (-0.40 \pm 0.18)\, x_{distance} + (6457 \pm 129)$$

This gives a spread rate in the range 1.7 to 4.5 km/year and a predicted date of departure from the origin in the time range 6330 to 6590 years BP.

6.2.4.5 Rate of spread and origin for the region that Zvelebil and Lillie attribute to cultural diffusion/indigenous adoption mechanisms.

Figure 6.19 shows the 355 Neolithic sites that fall within the western acculturation region (see inset).
The region that gave the strongest correlation coefficients for the linear regression of age of site and distance from 504 potential origin points across Europe is shown in Figure 6.20.

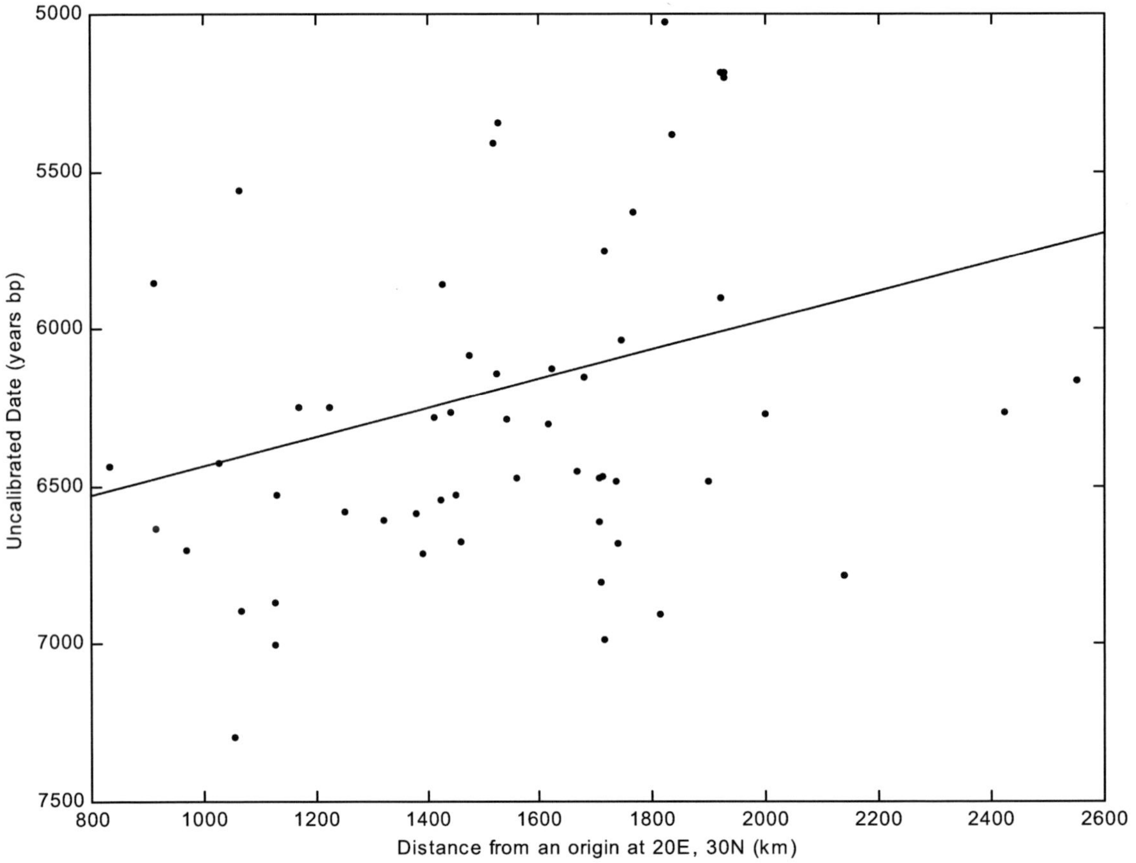

Figure 6.18 Linear regression of distance and weighted determination for Italy and enclaves Neolithic sites, from an origin at 20ºE, 30º N.

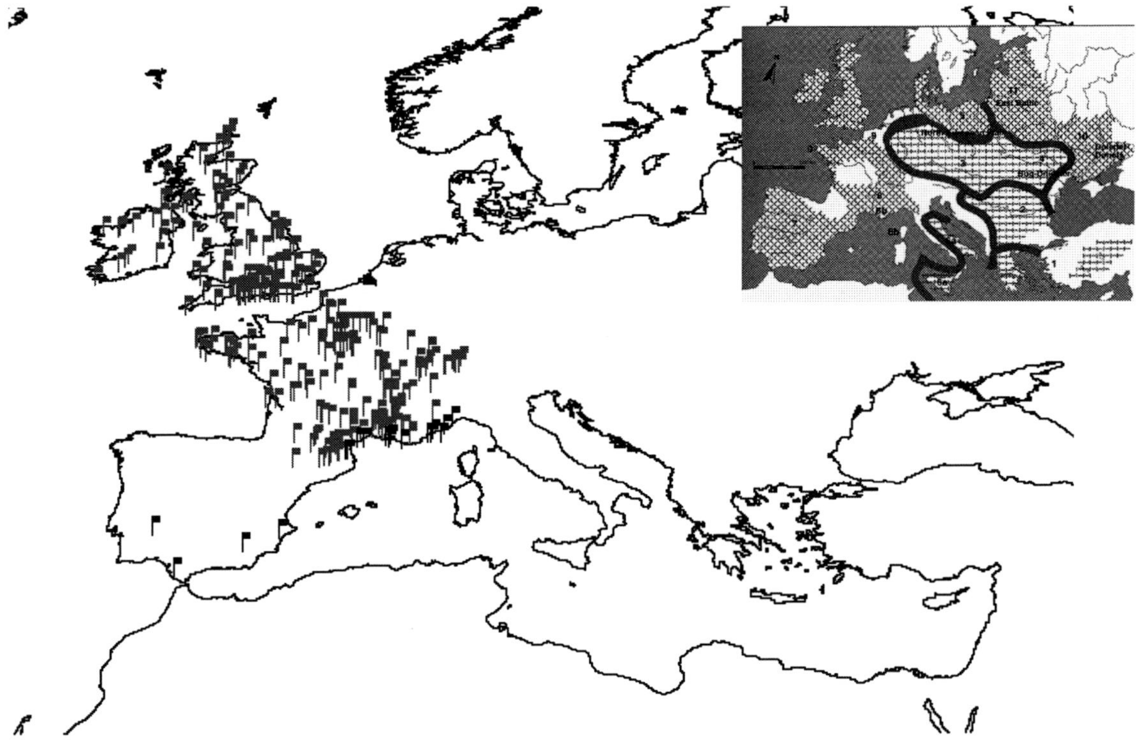

Figure 6.19 Zvelebil and Lillie's western acculturation area.

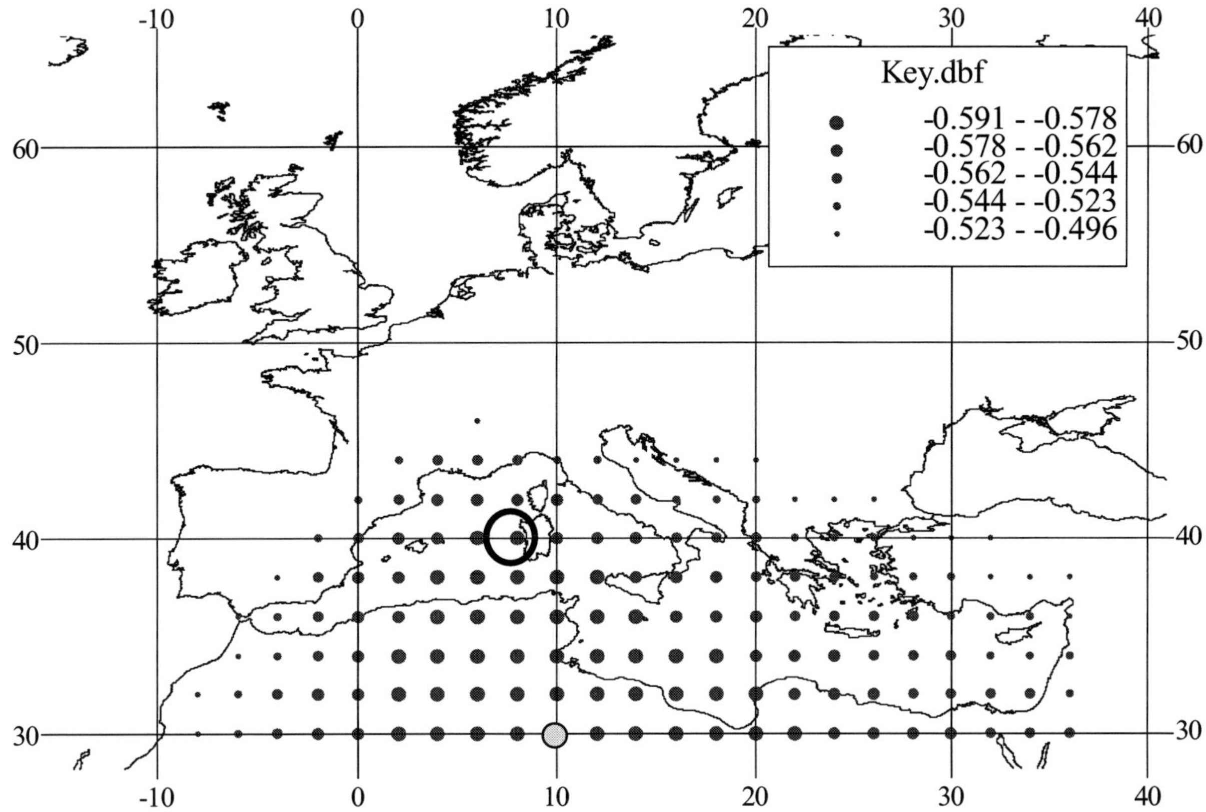

Figure 6.20 The map shows the origin points that gave the strongest correlation coefficients for the Neolithic sites that Zvelebil and Lillie attribute to an acculturation process. The origin with the strongest correlation coefficient is shown as a pale circle. That used in calculations is ringed.

Although the strongest correlation coefficient was for an origin in North Africa the analysis was run using one of the European origins with a strong correlation (9°E, 40°N) (Figure 6.19). This origin point was used to create the regression plot shown in Figure 6.21.

The correlation coefficient is –0.59. The equation of the line is:

$$y_{determination} = (-0.66 \pm 0.05)\, x_{distance} + (6058 \pm 68)$$

Using this equation, the rate of spread is in the range 1.4 to 1.6 km/year and the time of departure from the origin is 5990 to 6130 years BP. Interestingly the origin area for the western acculturation data does not correspond to the Zvelebil and Lillie's frontier region to the west of Germany, instead the pattern in the radiocarbon determinations suggest more influence from the frontier region in Italy and the Balkan Peninsula. The direction of spread of radiocarbon determinations, indicated by their decreasing antiquity with distance from the calculated origin, is from the Mediterranean north west to France and the British Isles (Figure 6.20). This corresponds with the direction of spread shown in Renfrew's (1987) map of the language dispersals in Europe (Figure 6.22).

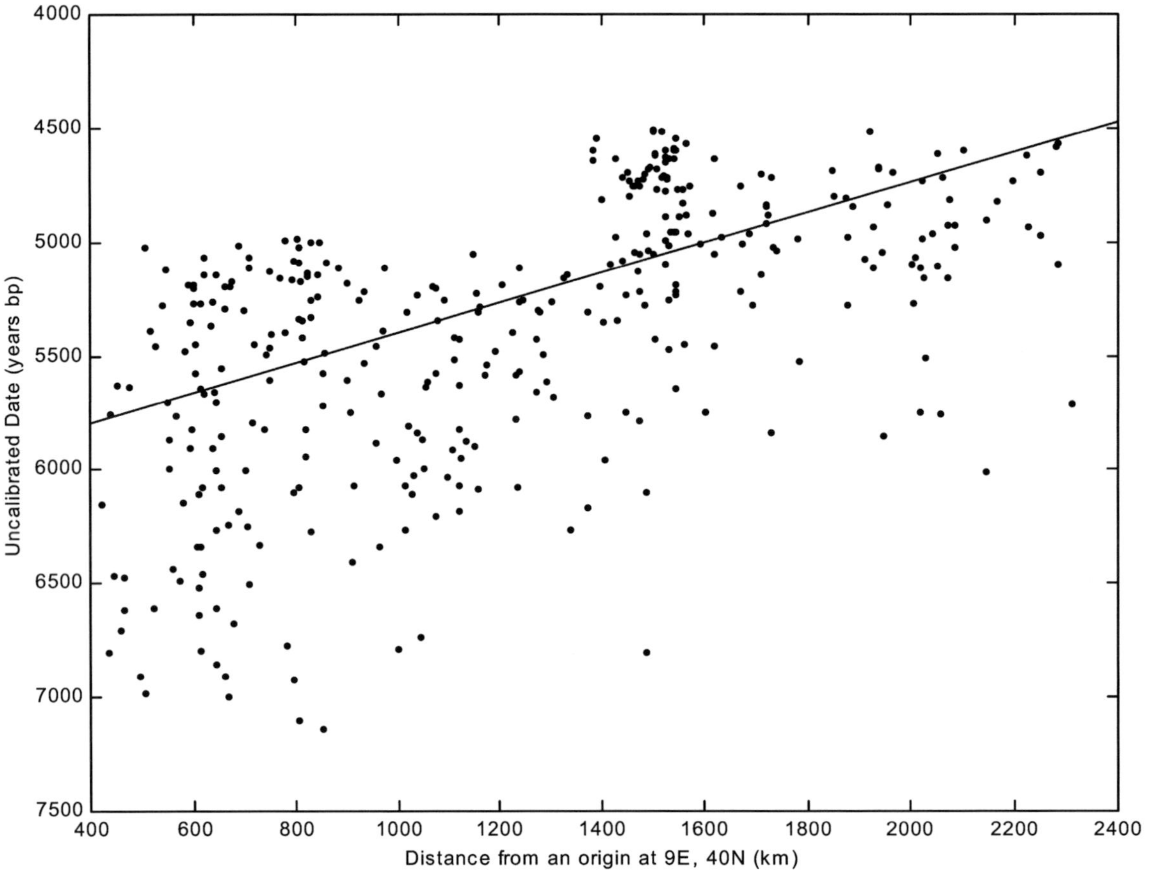

Figure 6.21 Linear regression of distance and weighted determination of 'acculturation' data from an origin at 9ºE, 40ºN.

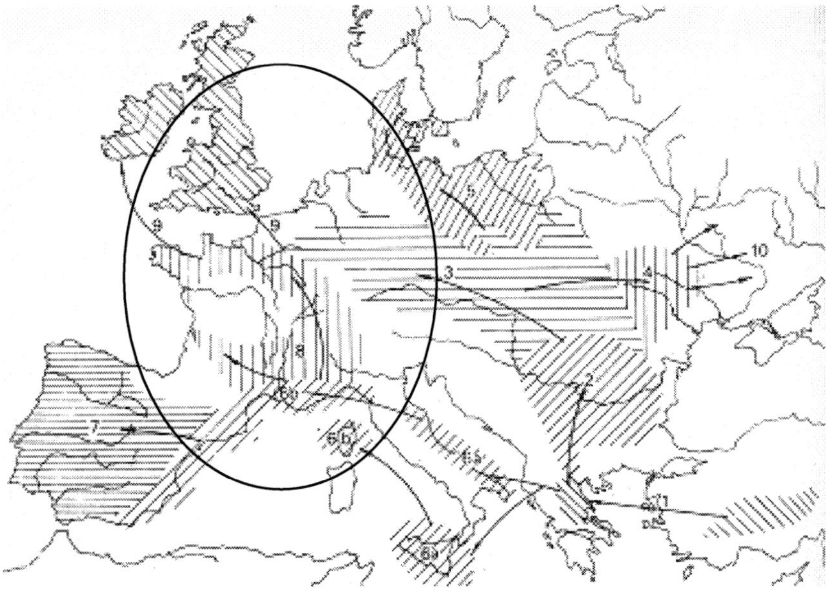

Figure 6.22 Renfrew's (1987:160, Figure 7.7) map to show the hypothetical sequence of cultural and linguistic transformations during the spread of the first farming to Europe. The circled area is the area where Renfrew's suggested direction of spread shows very clearly in the spatial pattern of radiocarbon determination that fall within Zvelebil and Lillie's western acculturation area (Figure 6.19).

6.2.4.6 Rate of spread and origin for western acculturation sites excluding enclaves:

For consistency the analysis for the acculturation sites was rerun without the enclave sites (which might be the result of demic spread by sea). The 335 Neolithic sites used in the rerun are shown in Figure 6.23. The origin region with the strongest correlation coefficient for the regression of determination and distance for these sites was calculated. The results show a wide zone of potential origins stretching across the Mediterranean (Figure 6.24).

The origin point with the highest correlation coefficient (Figure 6.24) was a sea site so the strongest correlation coefficient on land (6°E, 44°N) was used to calculate the rate of spread and time at origin for this dataset. The results of this analysis are shown in Figure 6.25.

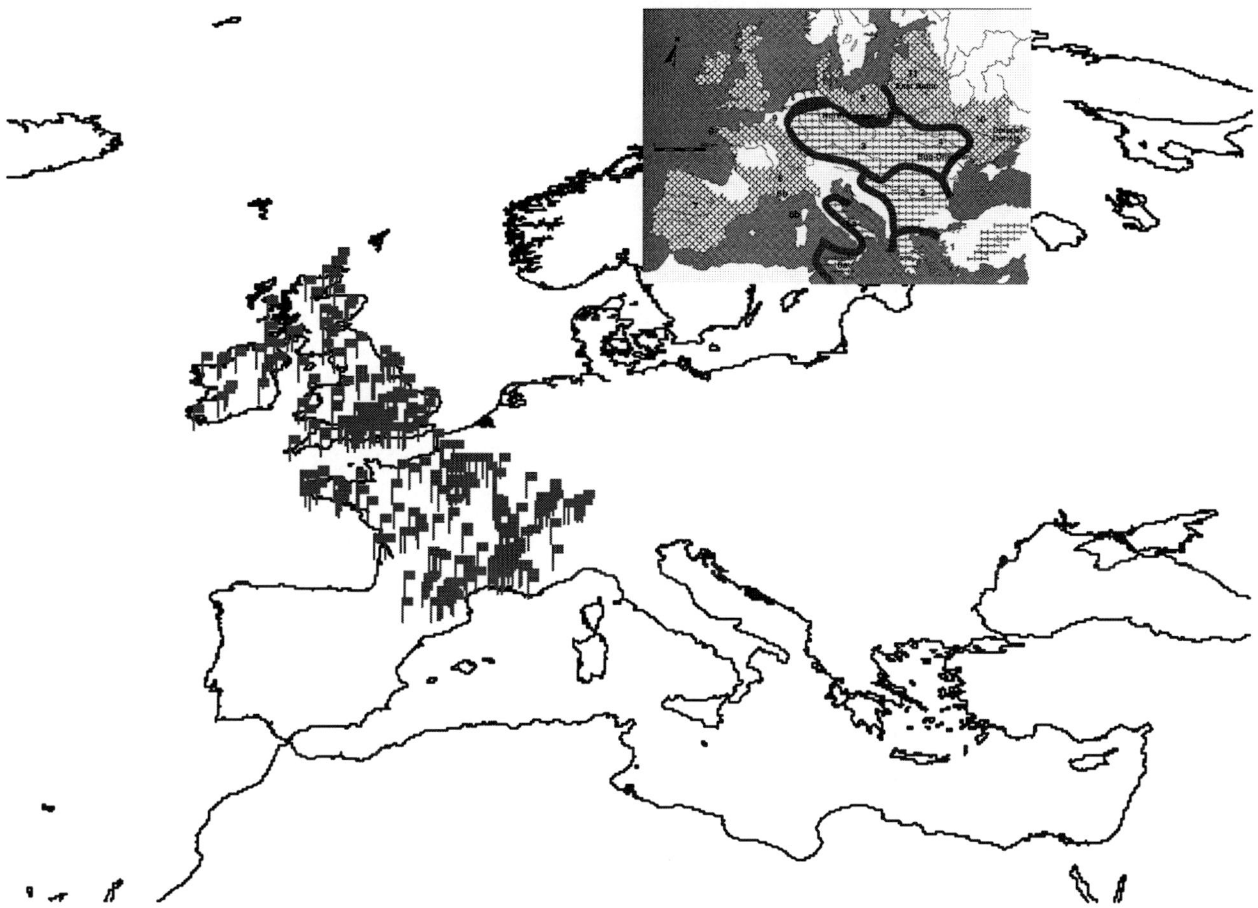

Figure 6.23: Zvelebil and Lillie's western acculturation sites with the exclusion of enclave sites.

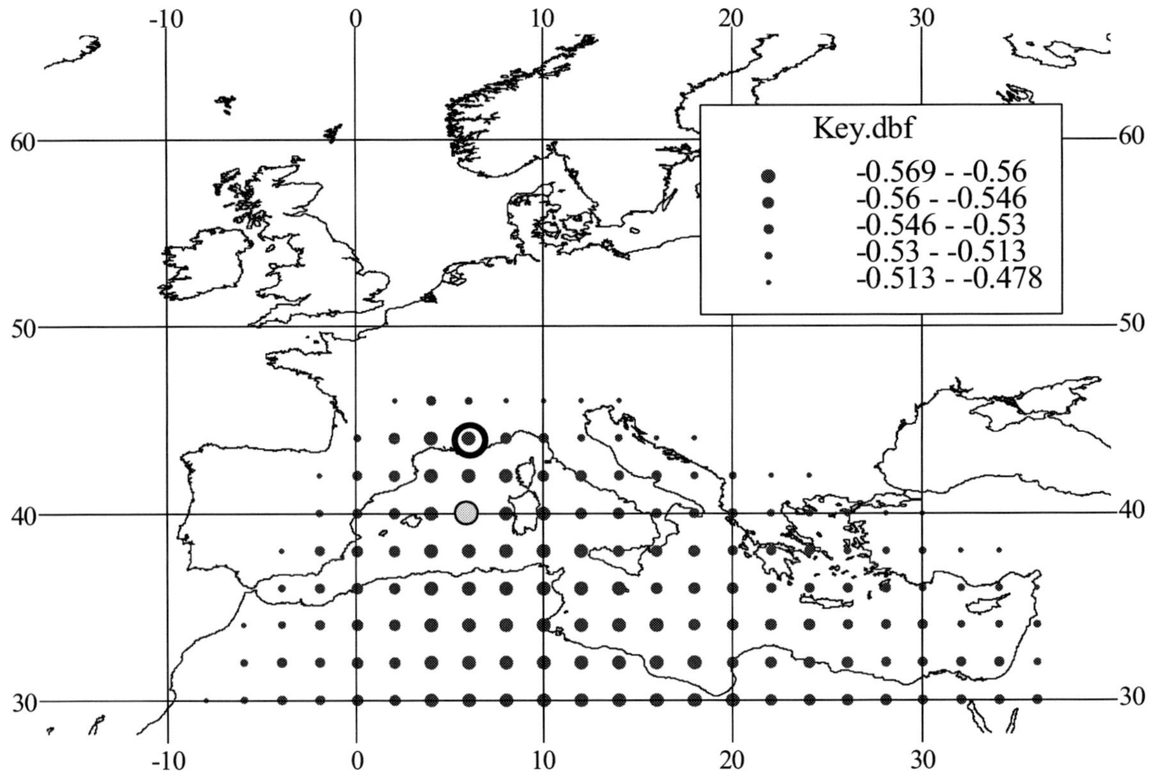

Figure 6.24 Points of origin that gave the strongest correlation of determination and distance to the Neolithic western acculturation sites when the enclave sites are excluded. The origin with the strongest correlation coefficient is shown as a pale circle. That used in calculations is ringed.

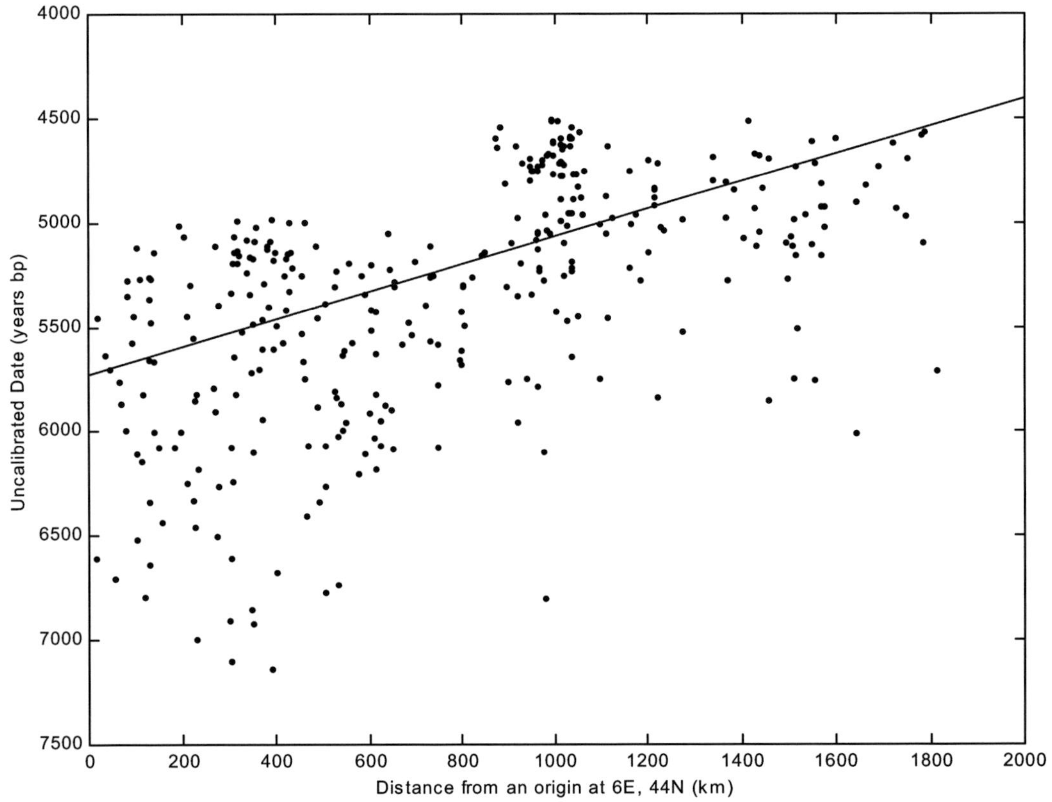

Figure 6.25 Linear regression of distance and weighted determination of 'acculturation' data, excluding enclaves, from an origin at 6°E, 44°N.

The correlation coefficient is –0.57 and the equation of the regression line is:

$$y_{determination} = (-0.66 \pm 0.05)\, x_{distance} + (5724 \pm 51)$$

This equation was used to calculate a rate of spread, this falls in the range: 1.4 to 1.6 km/year and a time for the departure from the origin falling in the range: 5670 to 5770 years BP.

As with the western acculturation zone (including enclave sites), the origin area and the direction of spread of the appearance of the first farming sites correspond to a frontier in the Mediterranean rather than in Central Europe (Figure 6.24).

6.2.4.7 The rate of spread and the origin for northern sites that Zvelebil and Lillie attribute to spread by cultural diffusion.

The 23 Neolithic sites used in this analysis are shown in Figure 6.26. As before the most likely origin for these sites was calculated by comparing the regression of determination and distance from 504 potential origin points across Europe. The results are shown in Figure 6.27.

Figure 6.26 Northern European Neolithic sites at which farming was adopted by indigenous foragers according to Zvelebil and Lillie.

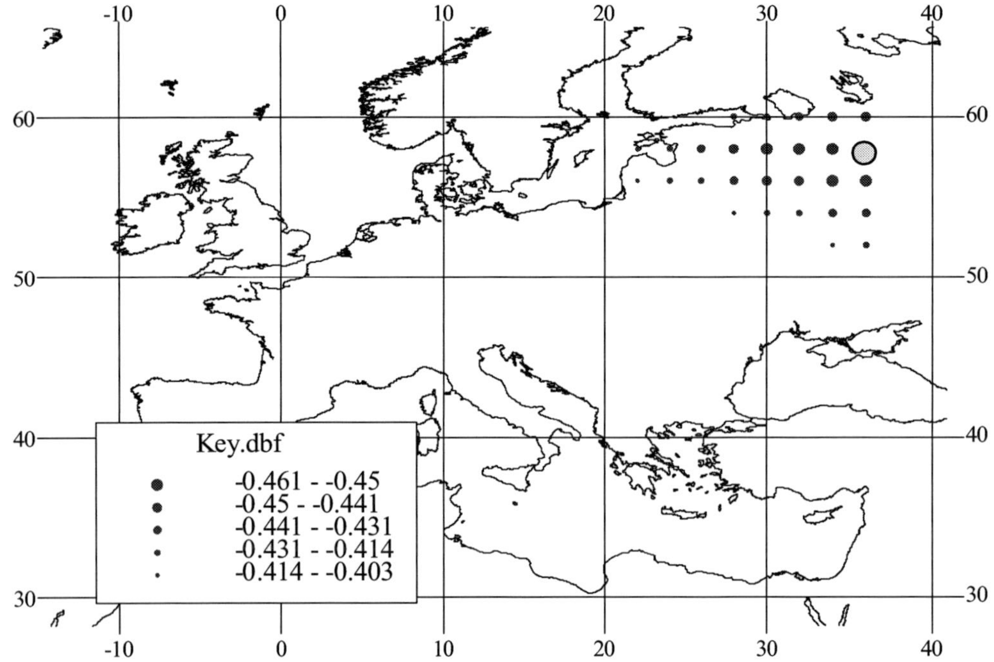

Figure 6.27 The origin points for the Neolithic acculturation north sites that had the strongest correlation coefficients. The strongest correlation coefficient is shown in as a pale circle.

The origin point with the strongest correlation point at latitude 58°N and longitude 36°E was used in linear regression analysis to calculate the rate of spread and time at origin. The results are shown in Figure 6.28. The correlation coefficient is -0.46 and the equation of the regression line is:

$$y_{determination} = (-2.12 \pm 0.74)\, x_{distance} + (9094 \pm 1140)$$

The equation gives a rate of spread in the range: 0.35 to 0.72 km/year and a time of departure from the origin in the range 7950 to 10 230 years BP.

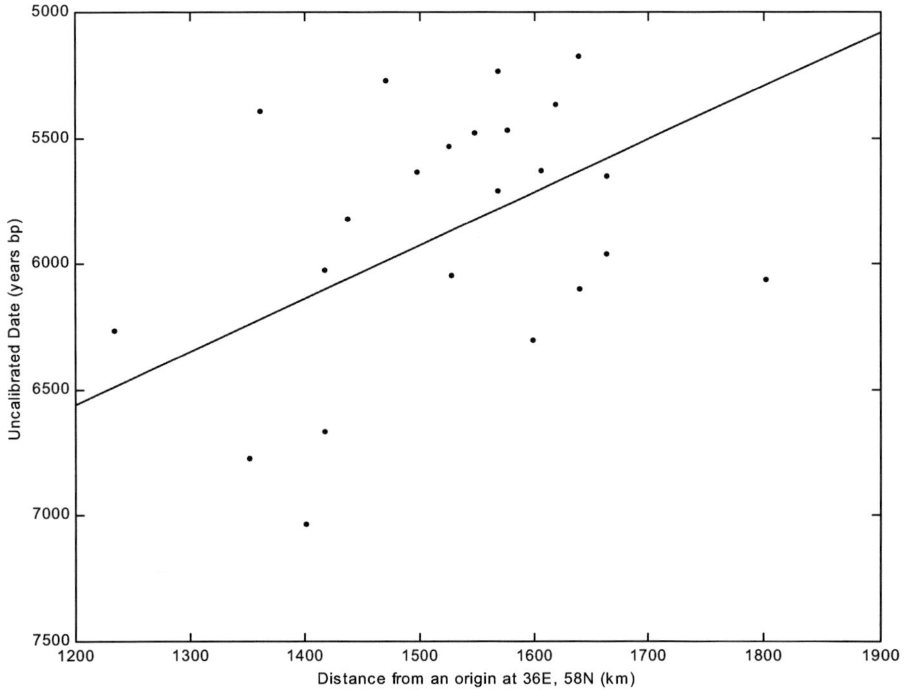

Figure 6.28 Linear regression of distance and weighted determination of 'acculturation' north Neolithic sites from an origin at 36°E, 58°N.

The correspondence between frontier region and origin is discussed in the following section.

6.2.5 Discussion of results for Dataset A:

The purpose of this exercise was threefold:

1. To see if rates of spread differed between demic and cultural diffusion areas.
2. To see if the most likely origin point for each subset of data coincided with the frontier positions suggested by Zvelebil and Lillie (2000).
3. To see the extent of regional variation in spread rate and the extent of its departure from the overall spread rate of 1.1 – 1.3 km/year calculated for continent wide spread (Chapter 5).

The results are discussed below. In addition the predicted origin areas and the predicted departure times from the origins were examined to see that they are archaeologically plausible.

6.2.5.1 Predicted origin areas: their agreement with the archaeological evidence and with the position of frontier zones as posited by Zvelebil and Lillie.

The origin point and date of departure predicted by the model are an independent check on the model and a means by which the plausibility of the model can be checked against archaeological data (Ammerman and Cavalli-Sforza 1971). To run this check the results were compared to the actual location and age of sites using Ammerman and Cavalli-Sforza's 1984 site list (as this list is independent of the Gkiasta database) (Table 6.2). Experimental results were also checked against the actual location and age of sites in the Gkiasta European database (Table 6.2).

Region	Predicted origin	Predicted departure time (years BP)	Archaeological check using Ammerman & Cavalli-Sforza's 1984 site list	Check using new European database
Demic	24°E, 36°N, Greece	6820 - 6970	Knossos 7875 BP (25.33°E, 35.52°N)	Theopetra cave 6911 BP (23.23°E, 38.5°N)
Demic excluding Italy	14°E, 30°N, North Africa	7880 - 8320	None	None
Demic excluding Italy revised origin	22°E, 38°N, Greece	6920 – 7080	Elateia 7480 BP (22.73°E, 38.62°N)	Platia Magoula Zakrou 7083 (22.8°E, 37°N)
Italian sites	13°E, 38°N, Mediterranean	6380 – 6780	Grotta dell'Uzzo 6750 BP (12.5°E, 38.12°N)	Lagnano da Piede 6700 BP (12.3°E, 38°N)
Italy & enclaves	20°E, 30°N, Libya	6590 – 7202	None	None
Italy & enclaves revised origin	16°E, 40°N, Italy	6330 – 6590	Piana di Curinga 6930 BP (16.25°E, 38.82°N)	Grotta Scaloria 6720 (16.9°E, 41.65°N)
Western Acculturation (origin revised)	9°E, 40°N, Sardinia	5990 – 6130	Arene Candide 6680 BP (8.5°E, 44.5°N)	Grotta Corbeddu 6690 BP (9.48°E, 40.25°N)
Western acculturation area (excluding enclaves)(origin revised)	6°E, 44°N, southern France	5670 – 5770	Le Capitaine 6050 BP (6.17°E, 43.75°N)	Baume de Fontbregoua 5690 BP (6.23°E, 43.55°N)
Northern acculturation[2]	36°E, 58°N, Russia	7950 – 10 230	None	None

Table 6.2 Comparison of predicted origins and predicted origin departure times against archaeological data from Ammerman & Cavalli-Sforza's 1984 Neolithic site list and the current European database.

[2] Ertebolle (see text for discussion)

When the origins in north Africa are replaced with more probable origin points (with slightly lower correlation coefficients to the data) then as a group all of the predicted origin points are found close to archaeological sites from the Neolithic period. The predicted determinations at the origins are also comparable to the radiocarbon ages of sites that are in close proximity to them. In fact the archaeological sites always either predate or are approximately equal to the predicted departure date: the predicted determination is never found to be of greater antiquity than the real archaeological evidence. This ties in well with the concept of a time lag at the frontier between the presence of farming on a landscape and its adoption by hunter-gatherers. It fits equally well with the demic hypothesis, the expectation is that people would settle and remain in an area for some time before migrating to a new location.
The 'best fit' origin for the demic diffusion area was in Greece (when Italian sites are included) and in North Africa when Italian sites are excluded. The origin shows the point from which there is the best linear relationship of decreasing age with increasing distance from the origin. It is useful when used to show the most likely origin region and, even when it points to an unlikely origin (i.e. north Africa) it is still useful as a tool to find the general direction and trend in the spread data. In this case the result shows that the origin for the central European data is south of these sites.

For the western acculturation area the origin for the data lies to its south-east in the direction of the Mediterranean and North Africa, suggesting that the change to farming originated here and spread in a north westerly direction. This would correlate with the Zvelebil and Lillie's frontier zone in Italy. The results suggest that this frontier zone could be extended to encompass parts of the southern French coast where the settlement of sea-faring farmers might have brought the frontier closer to Mesolithic forager groups. Interestingly the influence of the frontier zone to the east (that encompasses the central European sites in Zvelebil and Lillie's demic diffusion area) is not picked up as a possible region of origin for this dataset.

The determinations that fall within the northern acculturation zone in the European database are nearly all sites associated with the Ertebolle culture which, it is agreed, represents a local Mesolithic population which adopted various attributes such as pottery from fully agricultural populations to the south (this frontier is shown on Zvelebil and Lillie's map). What is unexpected here is that the origin is to the east. It is impossible to say whether this should be interpreted as influence from farming communities that are known from western Russia and Ukraine or whether the trend that is being picked up is the spread of the Ertebolle culture westwards and thus not correlated with the spread of farming.

6.2.5.2 Predicted rates of spread: their departure from the overall rate of spread for Europe and their usefulness in distinguishing between different mechanisms of spread.

For Zvelebil and Lillie's demic diffusion area in central Europe the decision was made to exclude Italian sites under the hypothesis that they represent at least some travel by sea and so were not directly comparable to areas of demic diffusion over land in terms of the rate of spread or even possibly linearity of the spread. The result of excluding Italian sites was an improvement in the correlation coefficient for central Europe (-0.53 to –0.61) and a decrease in the rate of spread to 2.1 to 2.9 km/year (Table 6.3).

Region	Demic rate of spread (km/year)	Cultural diffusion of spread (km/year)
Central Europe	2.4 to 3 (r = -0.53)	-
Central Europe (excluding Italy)	2.1 to 2.9 (r = -0.61)	-
Italian sites only	1.4 to 8.3 (r = -0.30)	-
Italian and enclave sites	1.5 to 3.8 (r = 0.32)	-
South western Europe	-	1.4 to 1.6 (-0.59)
South western Europe no enclaves	-	1.4 to 1.6 (-0.57)
Northern Europe	-	0.32 to 0.72 (-0.46)

Table 6.3 Rate of spread and linear correlation coefficients in demic and cultural diffusion areas of Europe.

When Italian sites are analysed as a separate group there is no linear pattern in the data (the best correlation coefficient is –0.30 for an origin near Sicily). Due to the non-linearity in the pattern of spread the possible rates vary widely (1.4 to 8.3 km/year – the highest spread rate overall for any region analysed). This result: the absence of a linear trend and a varying but possibly high rate of spread all meet the expectations of a pattern reflecting movement across water for the Italian sites. This coupled with the increase in correlation coefficient and fall in rate of spread for the central European sites once Italy was excluded all lends support to the hypothesis that Italian sites are a result of sea travel and should be treated differently in this sort of analysis.

To test the hypothesis that the whole of the Mediterranean coast was colonised by a demic diffusion by people across water, the coastal sites in France were also included in a group with the Italian sites (Italy and enclaves, Table 6.3). As with Italy the correlation is very low and the rate of spread, although lower than that when Italy is analysed alone, is higher than that calculated for areas where demic spread occurred across land.

For the sake of completeness the western acculturation region was analysed with and without the coastal enclave sites. This had a minimal affect on the results. For this region the rate of spread is lower than for any of the areas where farming is attributed to demic spread. This is interesting as it supports the idea that spread by indigenous adoption diffusion would occur at a slower rate than spread by migrating people. At the start of the analysis the degree to which the rate of cultural diffusion would follow a linear trend was uncertain. Multiple factors would affect this rate besides proximity to farming groups. The results show a linear correlation that is only marginally smaller than that for spread in demic areas.

In summary, rates of spread are variable and reach a maximum spread rate in the Mediterranean region, 1.4 to 8.3 km/year. Here the earliest Neolithic determinations might represent the demic movement of a sea-faring people. In the other demic spread region in central Europe the rate of spread is between 2.1 to 2.9 km/year. Both of these rates are higher than those of the western acculturation zone where the spread is 1.4 to 1.6 km/year. The northern acculturation zone is excluded from discussion here for reasons mentioned above.

There is regional variation in the rates of spread, as was expected from earlier work by Ammerman and Cavalli-Sforza (1971) and Housley *et al.* (1997), thus confirming that the spread rate calculated for the whole of Europe does indeed mask a great deal of underlying variability. Their results are presented in Table 6.4 and Table 6.5 for comparative purposes. Ammerman and Cavalli-Sforza caution that these are very tentative results as the data is so sparse (their total data set contains 51 dates, so for regional rates of spread the measurements are based on very small samples). In Ammerman and Cavalli-Sforza's words, " The rate of the Bandkeramik culture is probably too high due to less than complete documentation in central than in western Europe" (*ibid.* 1971: 684).

Detailed comparison is not possible between rates of spread calculated using the Gkiasta database and those of Ammerman & Cavalli-Sforza as the rates are calculated from different origin points and regional breakdowns. However there can be broad comparison. Ammerman and Cavalli-Sforza's prediction that the Bandkeramik rate was too high seems to be confirmed by the rate calculated for Zvelebil and Lillie's demic spread area (excluding Italian sites). The rate for this area is 2.1 to 2.9 km/year using the Gkiasta database, half the rate calculated by Ammerman and Cavalli-Sforza. For the Mediterranean, Ammerman and Cavalli-Sforza calculate a rate of spread of 1.52 to 2.08 km/year, with a correlation coefficient of +0.9. They draw particular attention to the sparseness of the data in this group. The equivalent area using the Gkiasta database gave a rate that falls in the range 1.4 to 8.3 km/year and a correlation coefficient of ± 0.3.

Housley *et al.* (1997) calculate rates of spread using the map that Van Andel and Runnel's show in their 1995 paper (Van Andel & Runnels 1995: 495, Figure 11).

Sites	Correlation coefficient	Diffusion Rate
Mediterranean	0.975	1.52
Western Mediterranean	0.915	2.08
Balkans	0.458	0.70
Bandkeramik	0.494	5.59
All of Europe	0.892	1.08

Table 6.4 Ammerman & Cavalli-Sforza's regional diffusion rates measured from an origin at Jericho (1971:684)

Sites	Diffusion rate (km/year)
Jericho to Tavoliere	0.87
Jericho to Sava Valley	0.64
Cayonu to Tavoliere	0.83
Cayonu to Sava Valley	0.57

Table 6.5 Housley *et al.*'s rates of spread of agriculture calculated from Van Andel & Runnels (1995) (Housley *et al.* 1997:49).

These results are much lower than the rates of any of the areas analysed using the Gkiasta database, possibly a further example of the variation of rates at a smaller scale of analysis. However the equivalent area was not analysed here so a direct comparison is impossible. Interestingly these results for the rate of spread from Jericho to Tavoliere would include travel by sea but here the rate of spread is far slower than that for the western Mediterranean (Italy and the enclaves).

The divisions of Europe presented here follow those laid out by Zvelebil and Lillie, although Table 6.1 shows the extent of the debate surrounding which mechanism was dominant in the spread of farming in different areas. The areas in Zvelebil and Lillie's map are still very large and more refined breakdowns would most probably reveal more variability. The question raised was whether rates of spread might differ if the underlying mechanism were cultural or demic diffusion. The rates of spread differ but it would be impossible to say with any conviction that these reflect a difference in the mechanism of spread. Table 6.1 shows how much disagreement there is about the mechanisms of spread across Europe. What would need to be confirmed is that a). Zvelebil and Lillie have correctly identified areas of cultural and demic diffusion; and b). The regions themselves are not still too large and therefore still masking a great deal more variation.

It is hard to draw conclusion as rates of spread may be affected by many variables besides the diffusion mechanism, for example a fast rate may also be suggestive of the effect of the environmental on movement: travel might be facilitated by water, an inhospitable terrain might increase the rate of spread in the rush to get to the other side, or alternatively it might act as a barrier to movement causing delay and detour (van Andel and Runnels 1995). A leap-frog pattern of migration is now favoured by many authors to describe the pattern of movement of farming. Farmers 'leap' from one desirable niche to the next (Zvelebil 1986a). This mechanism of spread should be rapid as large areas are skipped in the need to reach the next desirable location.

In 1971 Ammerman and Cavalli-Sforza used an origin at Jericho to calculate the rate of spread to Bandkeramik sites of central Europe of 5.59 km/year (r = 0.494). For comparison their result was compared to that from the demic diffusion region of central Europe. Taking an origin at 14°E, 30°N, the rate of spread is 2.1 to 2.9 km/year, much slower than Ammerman and Cavalli-Sforza's calculation, and the linear correlation coefficient is stronger (r = -0.61). For comparative purposes, the central European data excluding Italian sites was tested from an origin at Jericho. The result was a rate of spread of 2.5 to 3.2 km/year and a correlation coefficient of –0.57, confirming that the rate of spread calculated here is slower and more linear than that of Ammerman and Cavalli-Sforza.

6.2.6 *Repeat of the regional analysis using Dataset G:*

One of the aims of this thesis was to understand the effect of data selection on results. In this vein the whole exercise was repeated using a different subset of the European database. In this section the results of the analysis on Dataset G are presented (Figure 6.29). This is the dataset that was compiled by taking either the average determination at a site or the median determination at the site (see section 2.1.1, Chapter 2). The method followed is identical to that for Dataset A so the results will be presented over the following pages without describing the methodology. The comparability of results between the two datasets is of interest here. A full list of results is given in Appendix 7.

6.2.6.1 Rate of spread and origin for Neolithic sites that lie within Zvelebil and Lillie's demic spread region

Figure 6.30 shows the 43 Neolithic sites used. Of the 504 potential origins, those shown in Figure 6.31 have the strongest correlation for the linear regression of determination and distance to the sites. This origin point was used in linear regression to calculate the rate of spread and time of origin (Figure 6.32).

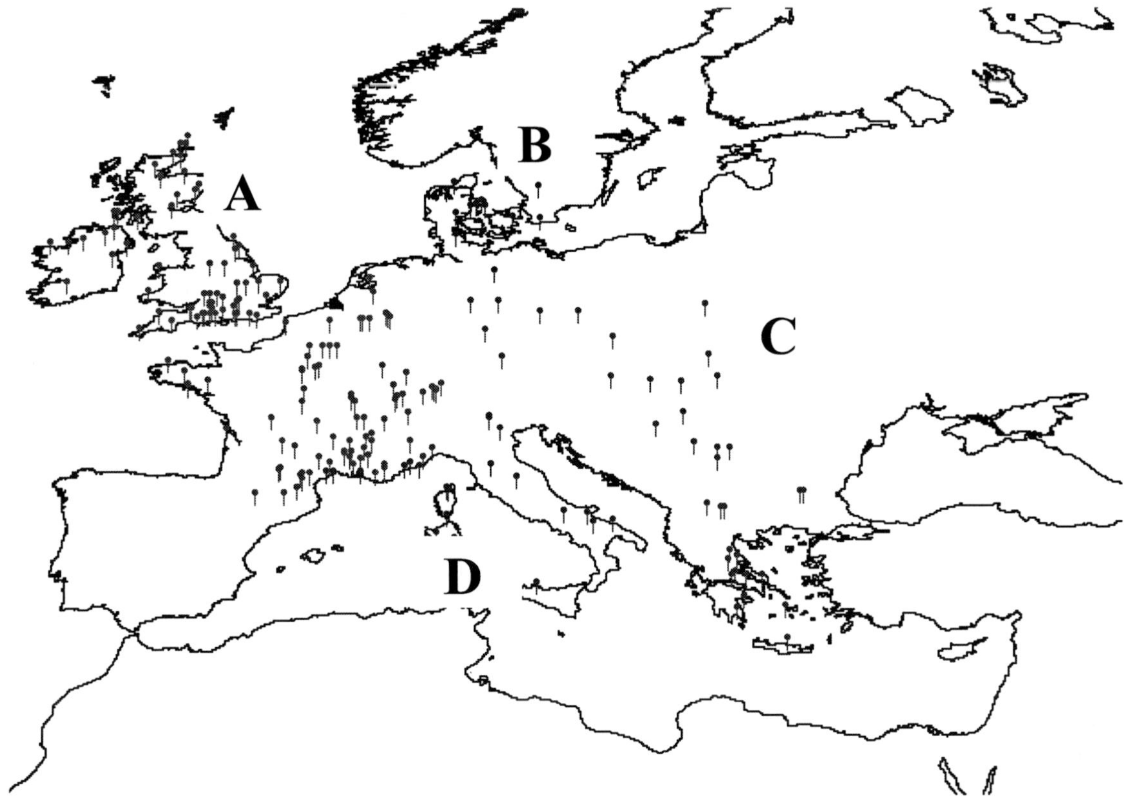

Figure 6.29 Dataset G divided to reflect areas of acculturation and demic spread after Zvelebil and Lillie. Icons in areas A and B are areas attributed to spread by acculturation/indigenous adoption whilst icons in areas C and D represent areas of demic diffusion. The further subdivision of the regions is to see the effect of analysing separately the areas that may be attributed to a different mechanism of spread (namely sea travel to Italy and land travel to the Northern European Plain).

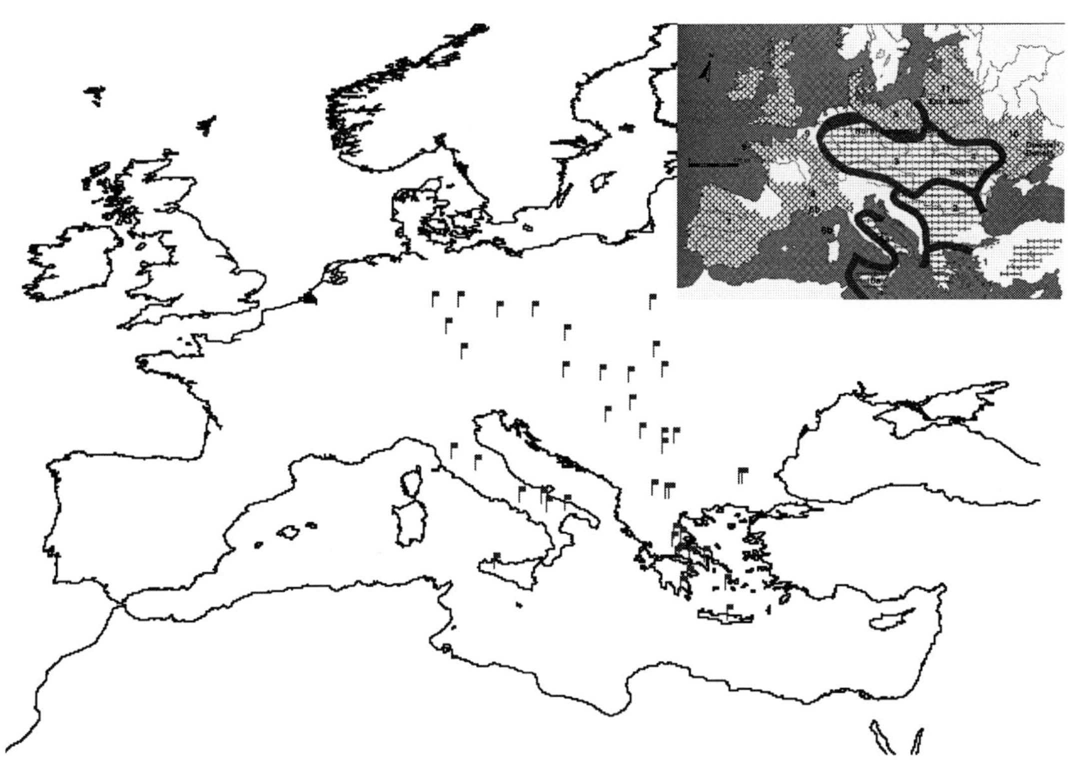

Figure 6.30 Neolithic sites that lie within Zvelebil and Lillie's demic spread region.

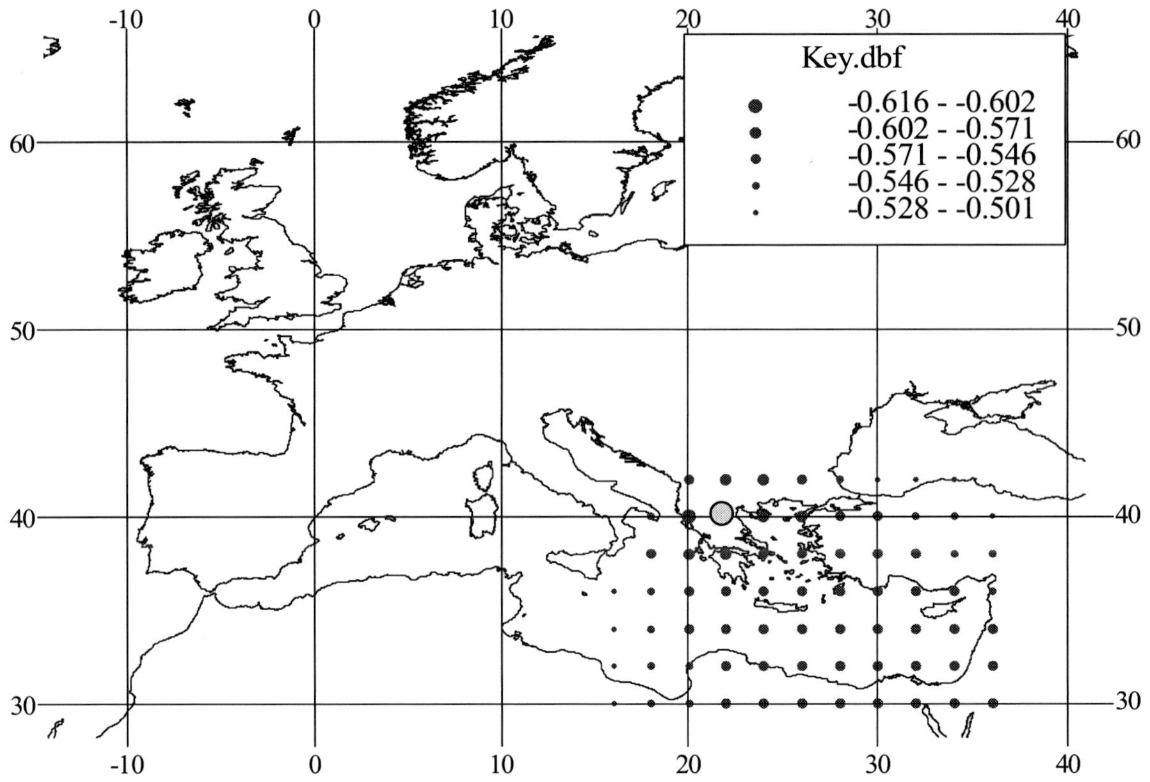

Figure 6.31: Origins that showed the strongest correlation coefficients with Neolithic sites falling within Zvelebil and Lillie's demic diffusion region. The origin with the strongest correlation is shown as a pale circle.

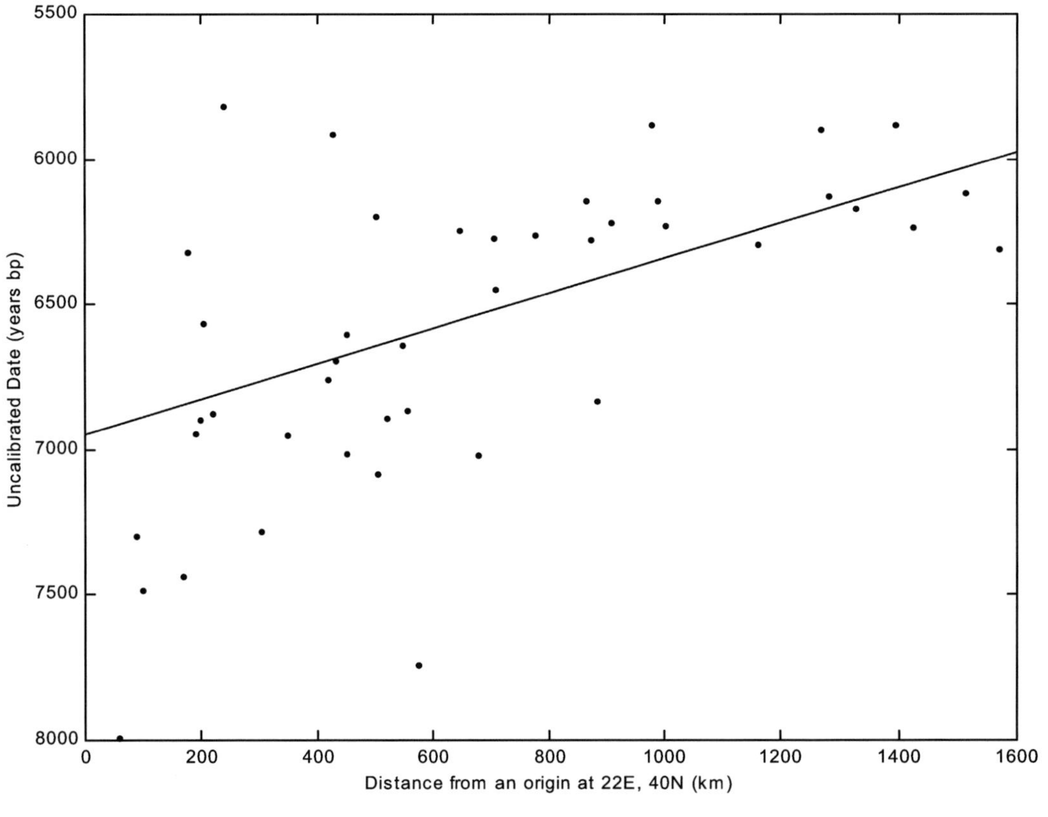

Figure 6.32 Linear regression of weighted determination and distance of 'demic' data from an origin at 22°E, 40°N.

6.2.6.2 Rate of spread and origin to Neolithic sites within Zvelebil and Lillie's demic diffusion area with the exclusion of the Italian Neolithic sites.

The Italian sites were excluded from the demic analysis for reasons already discussed (Section 6.1.3.2). This leaves 36 Neolithic sites within the region attributed to demic spread (Figure 6.33). The origin points that gave the best correlation coefficients to this dataset are shown in Figure 6.34.

The origin point (40ºN, 20ºE) was used in linear regression of determination of Neolithic site with distance from the origin (Figure 6.35).

Figure 6.33 Neolithic sites that fall within Zvelebil and Lillie's demic diffusion area when Italian sites are excluded.

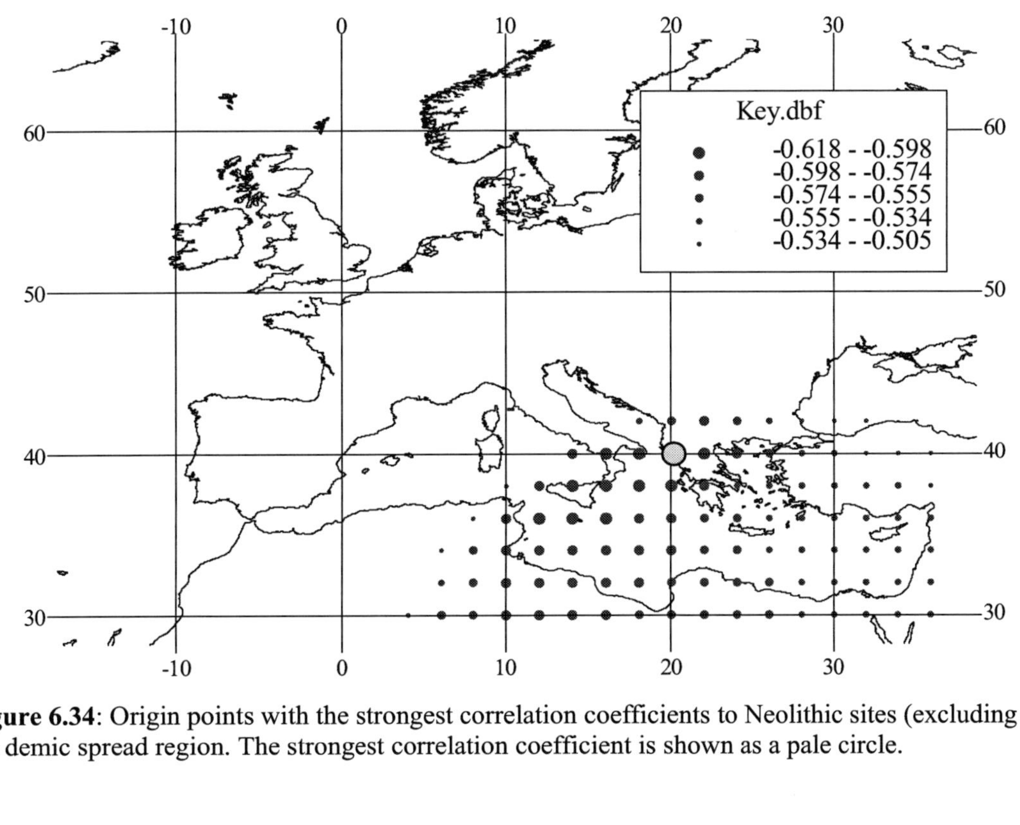

Figure 6.34: Origin points with the strongest correlation coefficients to Neolithic sites (excluding Italy) that fall within the demic spread region. The strongest correlation coefficient is shown as a pale circle.

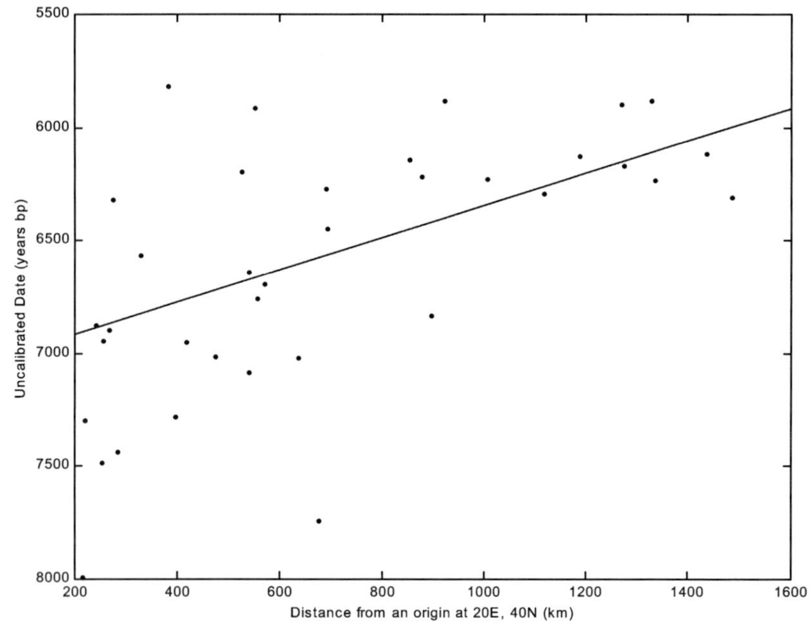

Figure 6.35 Linear regression of distance and weighted determination of 'demic' data, excluding Italy.

The correlation coefficient is –0.61. The equation of the regression line is:

$$y_{determination} = (-0.71 \pm 0.16)\, x_{distance} + (7057 \pm 140)$$

This gives a rate of spread in the range: 1.1 km to 1.8 km/year and a time for departure from the origin in the range: 6920 to 7200 years BP.

6.2.6.3 Rate of spread and origin for the Italian Neolithic sites (demic diffusion).

The 10 Neolithic sites are shown in Figure 6.36. The results of the search for the most likely origin area for this set of sites is shown in Figure 6.37.

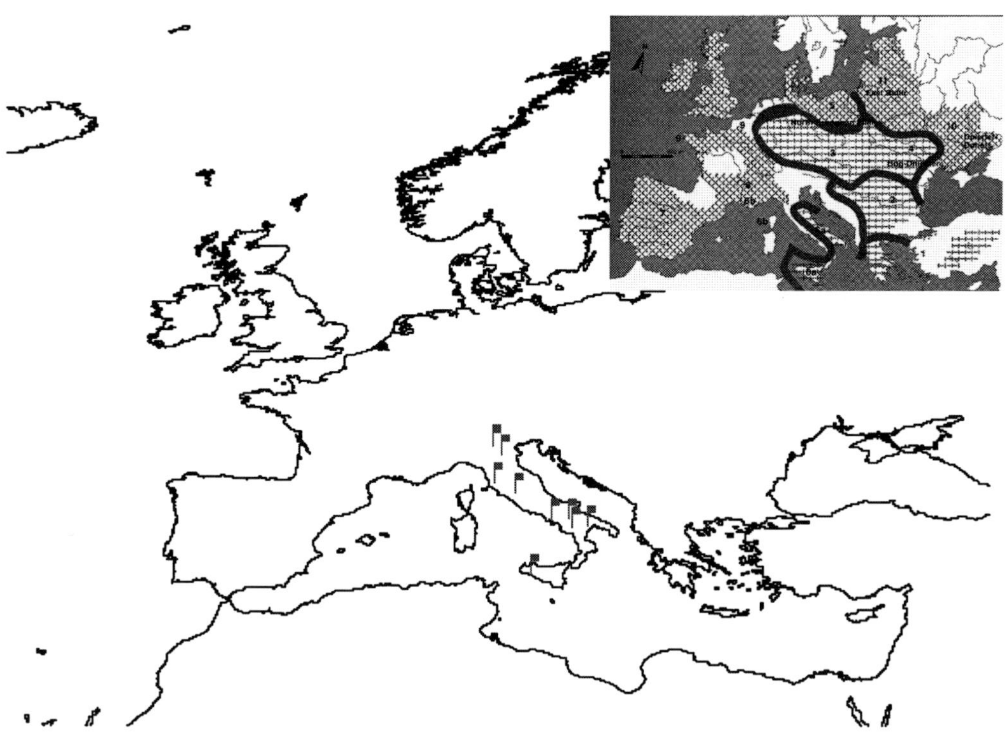

Figure 6.36 The Italian Neolithic sites that fall within Zvelebil and Lillie's demic diffusion region.

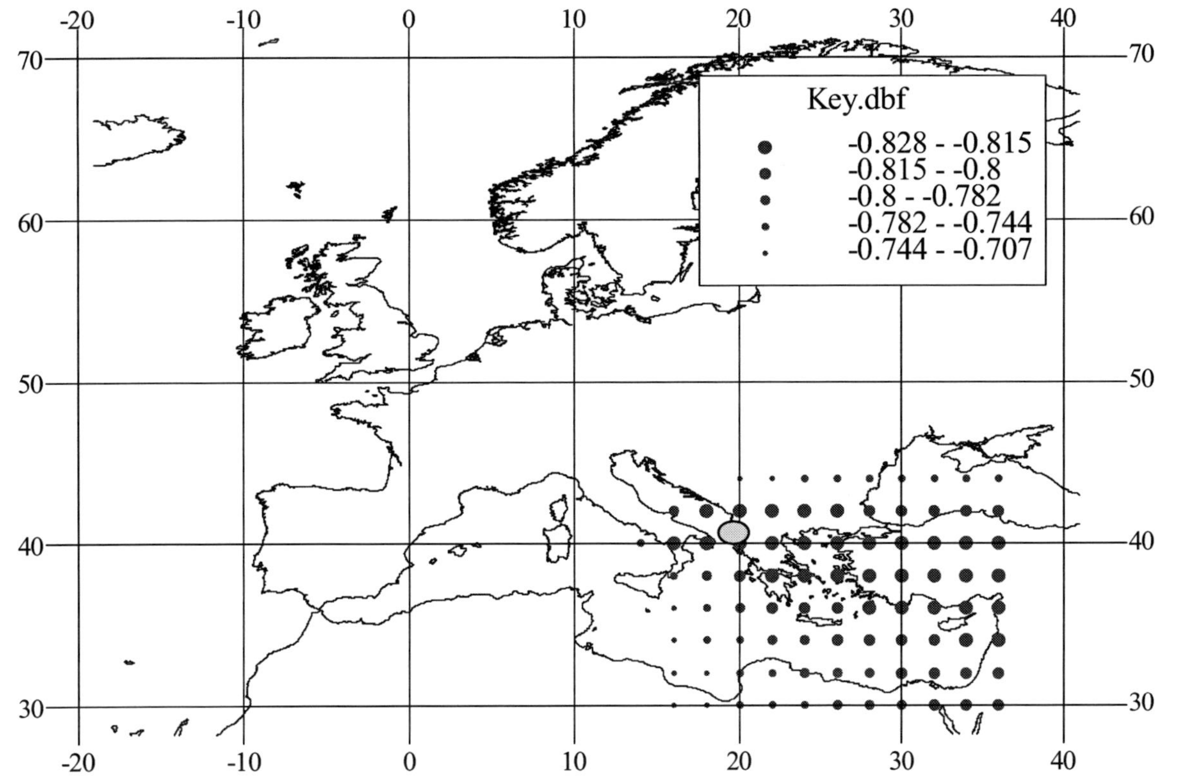

Figure 6.37 Origin points that showed the strongest correlation to the Italian Neolithic sites. The origin with the strongest correlation coefficient is shown as a pale circle.

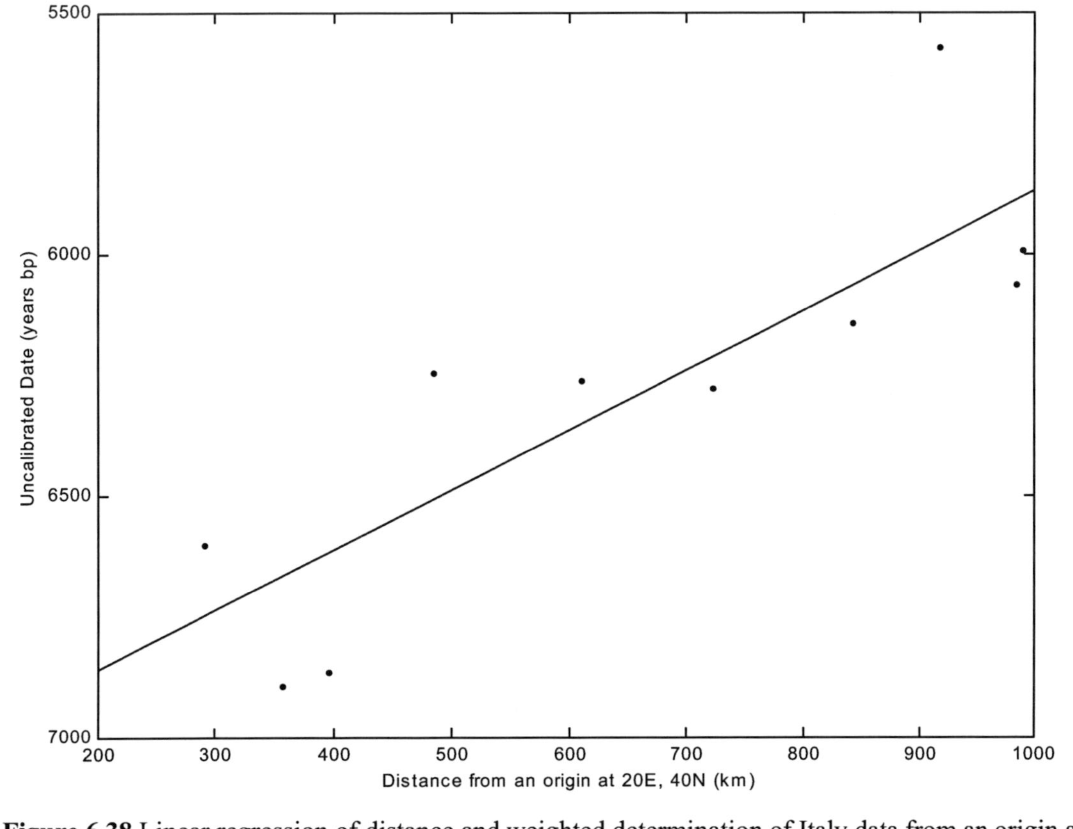

Figure 6.38 Linear regression of distance and weighted determination of Italy data from an origin at 20°E, 40°N.

The origin in Greece was used to calculate the rate of spread of farming to the Italian sites and the time of departure from this origin point (Figure 6.38).

The correlation coefficient is –0.83. The equation of the regression line is:

$$y_{determination} = (-1.24 \pm 0.32)\, x_{distance} + (7106 \pm 252)$$

From this equation, the rate of spread falls within the range: 0.64 km/year to 1.09 km/year and the time of departure from the origin is in the range: 6850 to 7360 years BP.

6.2.6.4 Rate of spread and origin for Italian Neolithic and enclave sites:

The hypothesis is these 21 Neolithic sites may fall within an area where the spread of farming was a consequence of the movement of a colonising sea-faring people (Figure 6.39).

Of the 504 potential origin points that were tested for their fit to the Italian and enclave sites, those giving the strongest correlation coefficients are shown Figure 6.40.

The origin point with the strongest correlation coefficient was used to calculate the rate of spread of farming to these sites and the time of departure from the origin (Figure 6.41).

Figure 6.39 Neolithic sites in Italy and the enclaves.

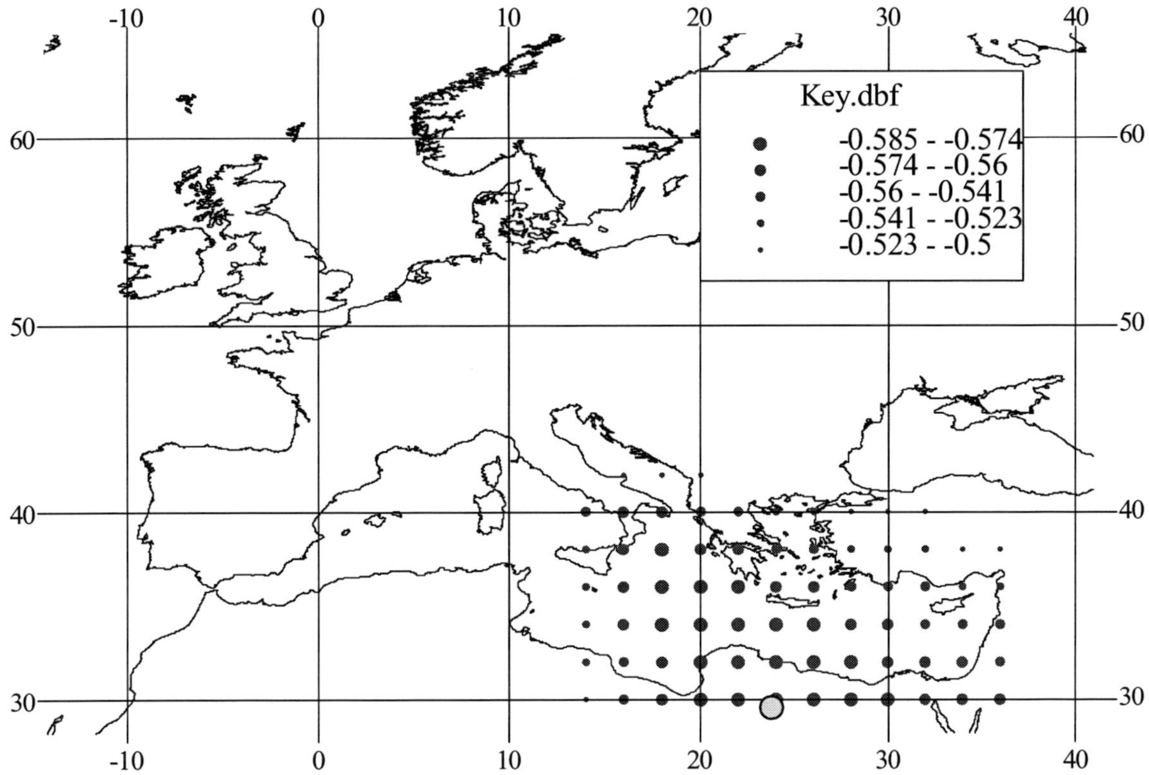

Figure 6.40 The origin points that give the strongest correlation coefficients for the regression of determination and distance to the Neolithic sites in Italy and the enclaves (Dataset G). The strongest correlation coefficient is shown as a pale circle.

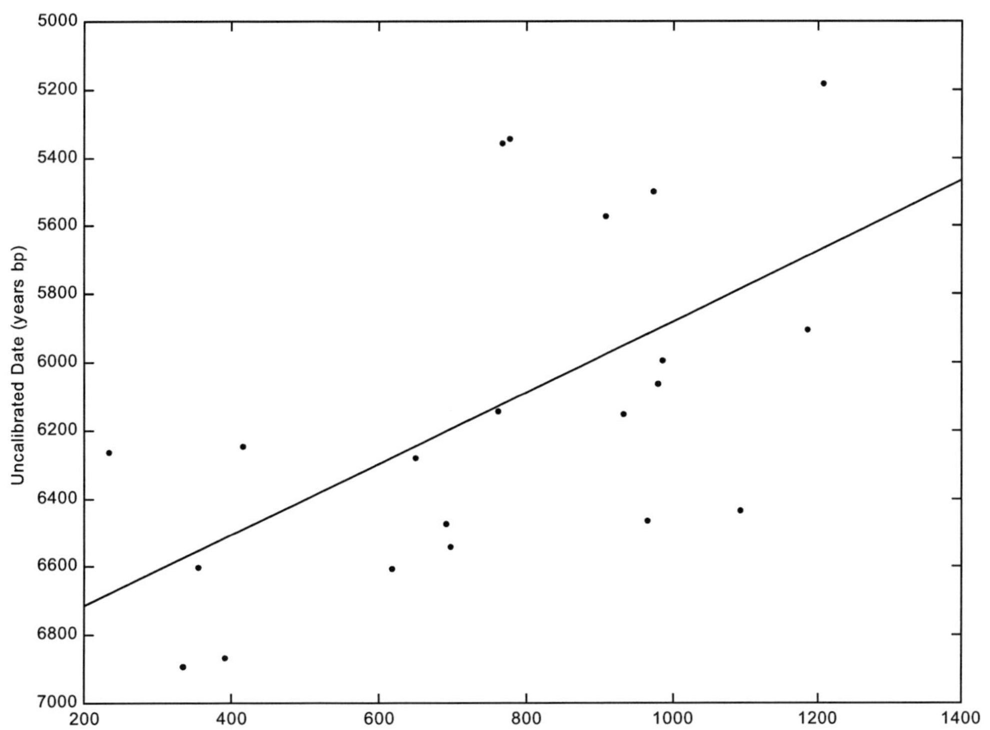

Figure 6.41 Linear regression of distance and weighted determination of 'demic' data, Italy and enclaves, from an origin at 24°E, 30°N.

The correlation coefficient is -0.59. The equation of the regression line is:

$y_{determination} = (-1.04 \pm 0.37)\, x_{distance} + (6920 \pm 299)$

Using this equation, the rate of spread falls within the range: 0.7 to 1.5 km/year. The time of departure from the origin is within the time range: 6620 to 7220 years BP.

6.2.6.5 Rate of spread and origin for Neolithic sites within Zvelebil and Lillie's western acculturation zone.

The 165 Neolithic sites used here are shown in Figure 6.42. Of the 504 potential origin points those with the best correlation coefficients for the linear regression of determination and distance to the sites are shown in Figure 6.43.

The origin in Turkey was used to calculate the rate of spread of farming to the Neolithic sites and the time of departure from the origin (Figure 6.44).

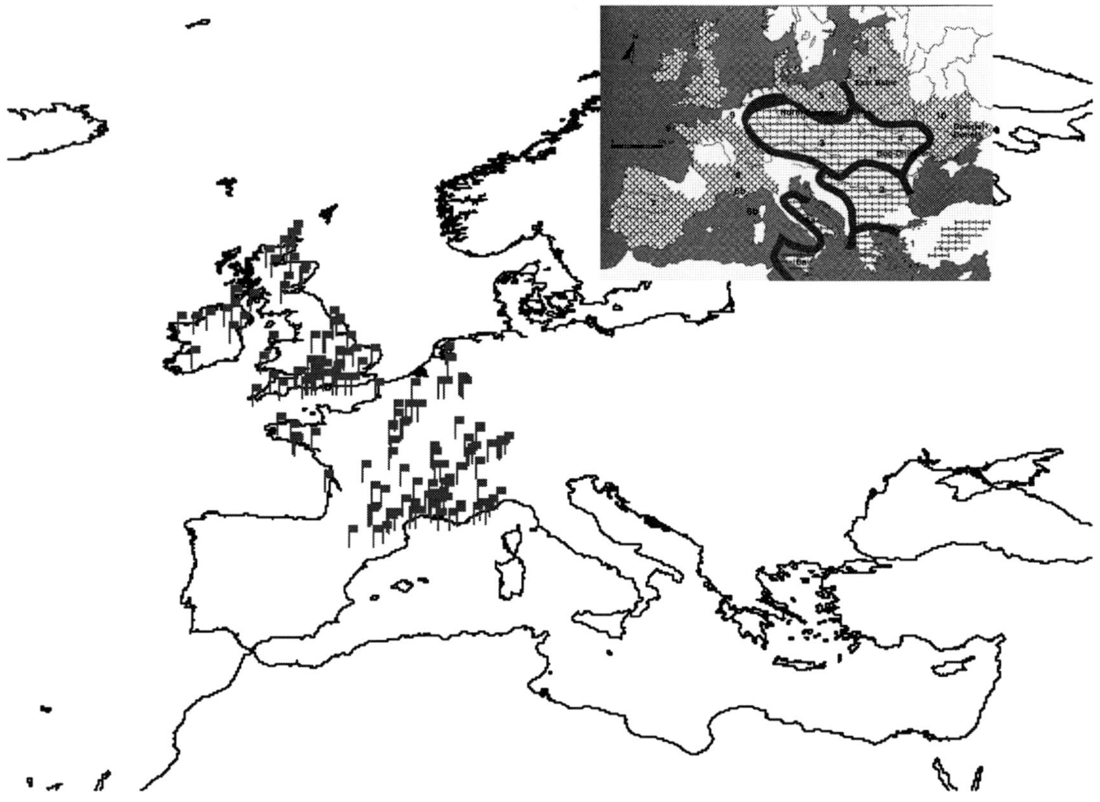

Figure 6.42: Neolithic sites that fall within Zvelebil and Lillie's western acculturation region (Dataset G).

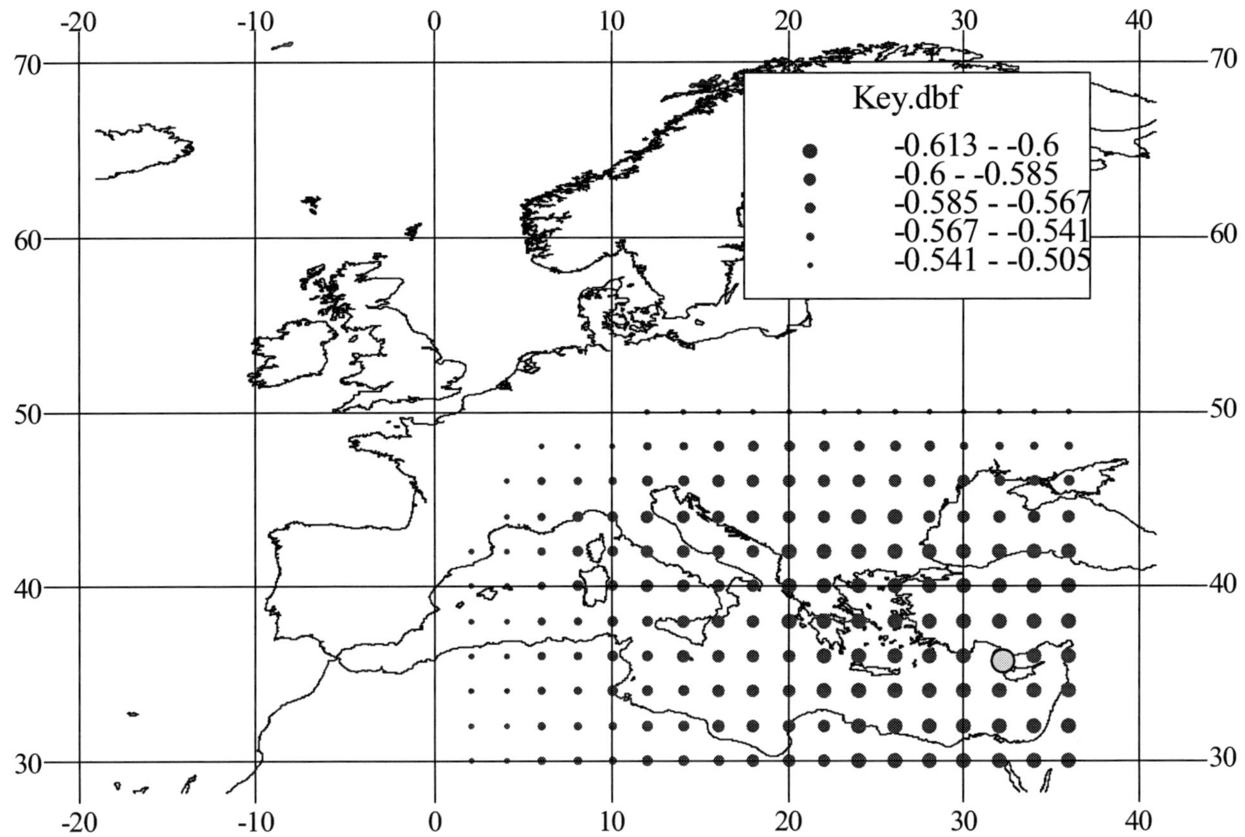

Figure 6.43 Origin points that showed the strongest correlation with Neolithic sites in Zvelebil and Lillie's southern acculturation region. The strongest correlation coefficient is shown as a pale circle.

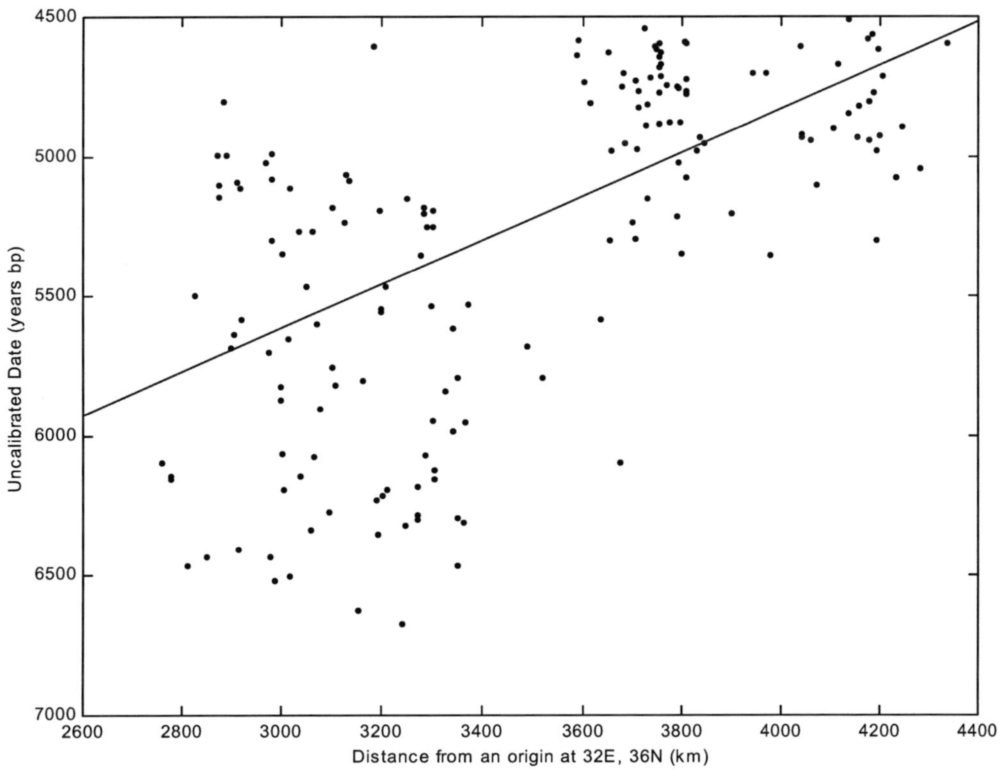

Figure 6.44 Linear regression of distance and weighted determination of western 'acculturation' data.

The correlation coefficient is –0.61. The equation of the regression line is:

$$y_{determination} = (-0.79 \pm 0.09)\, x_{distance} + (7973 \pm 309)$$

Using this equation, the rate of spread falls within the range: 1.1 to 1.4 km/year. The time range for the departure from the origin is within the range: 7660 to 8280 years BP.

6.2.6.6 Rate of spread and the origin for the Neolithic site falling within Zvelebil and Lillie's western acculturation region when the enclave sites are excluded:

The 159 Neolithic sites in this group are shown in Figure 6.45. Of the 504 potential origin points those that produced the strongest correlation coefficients for the regression of determination and distance to these sites are shown in Figure 6.46.

The origin in southern Turkey is used to calculate the rate of spread of farming and the time at the departure from the origin (Figure 6.47).

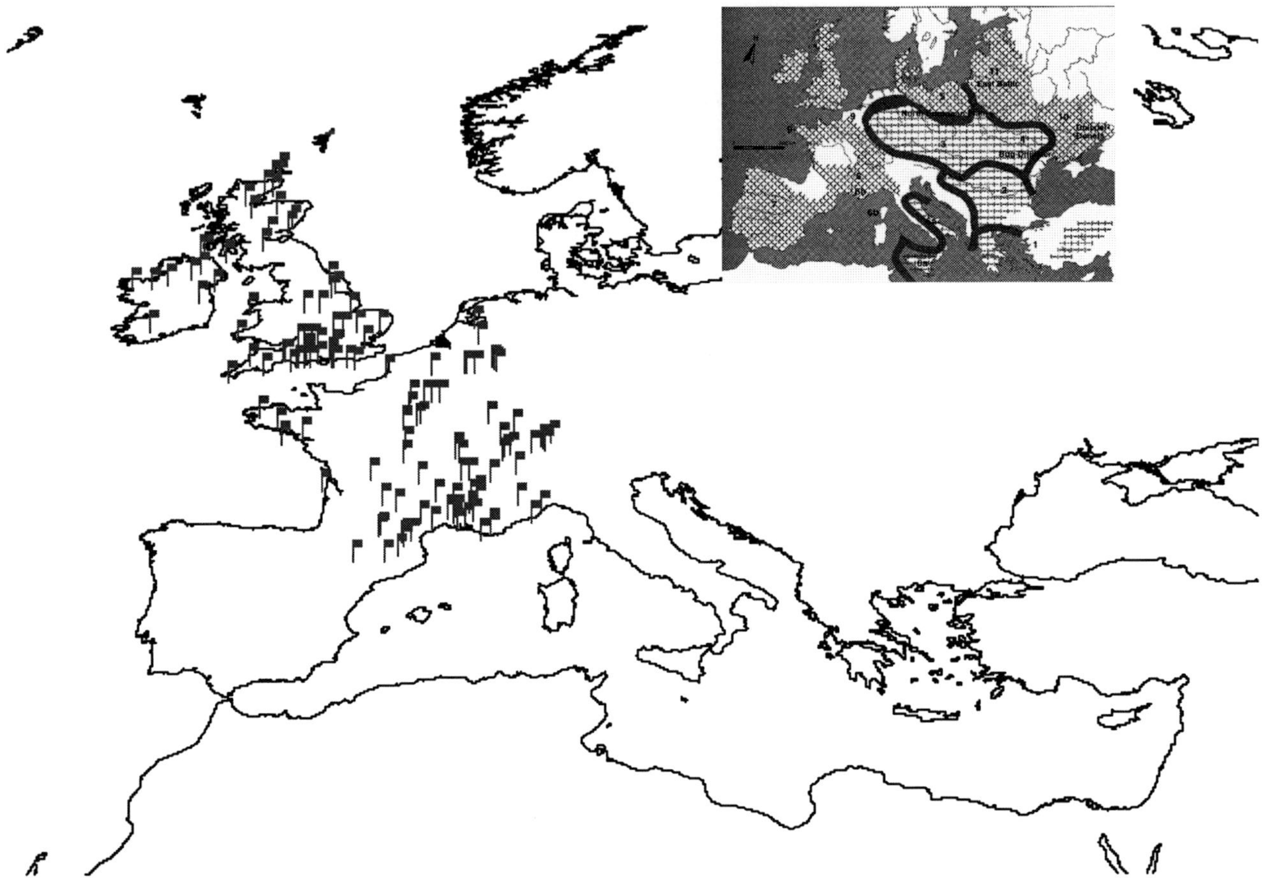

Figure 6.45 Neolithic sites (with the exclusion of enclave sites) that fall within Zvelebil and Lillie's acculturation region (Dataset G).

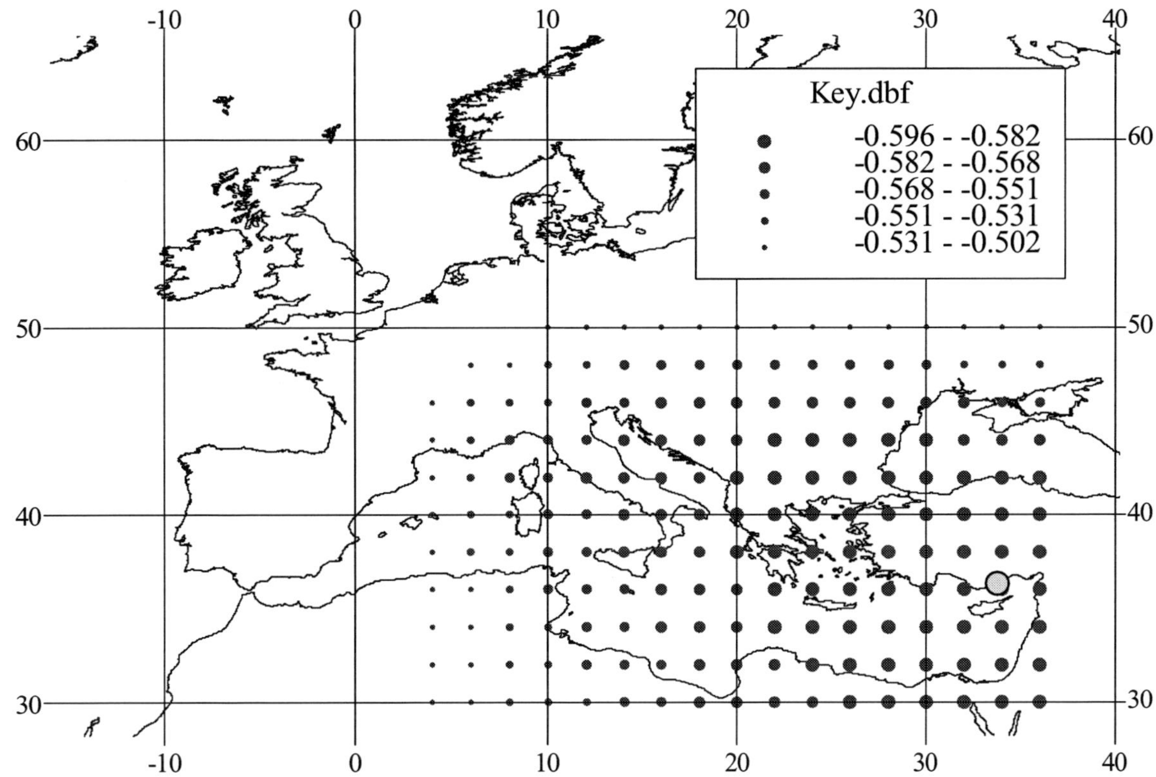

Figure 6.46 Origin points that showed the strongest correlation to the Neolithic sites that fall within Zvelebil and Lillie's western acculturation region (excluding enclave sites). The origin with the strongest correlation coefficient is shown as a pale circle.

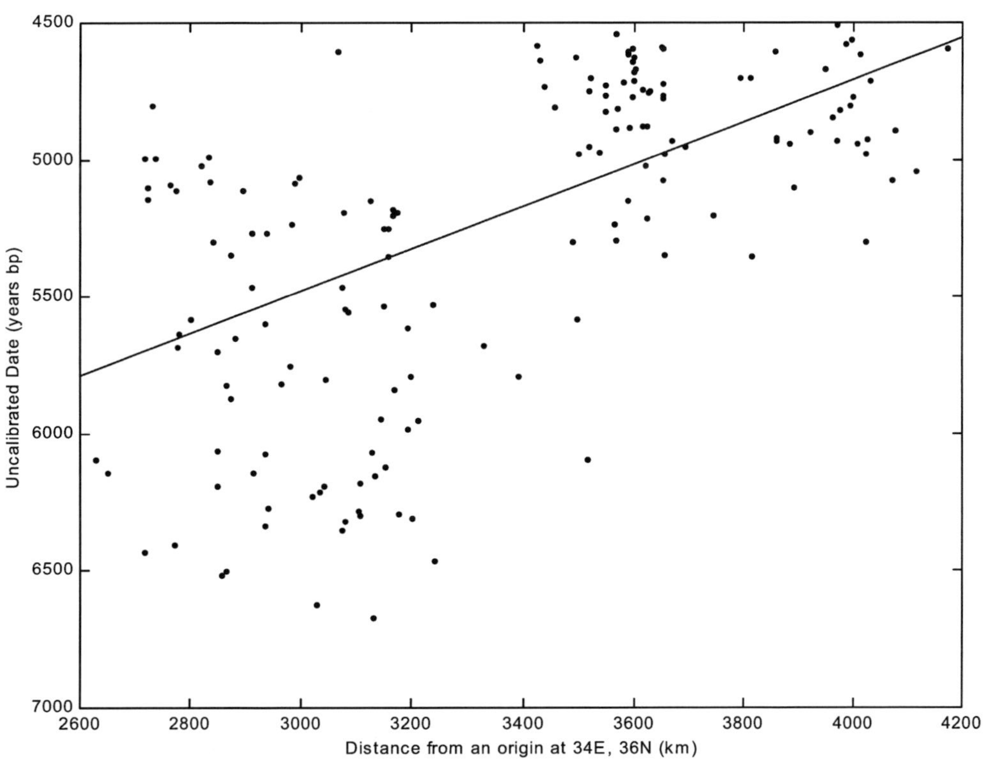

Figure 6.47 Linear regression of weighted determination and distance of 'acculturation' data, excluding enclaves from an origin at 34°E, 36°N.

The correlation coefficient is -0.60. The equation of the linear regression line is:

$$y_{determination} = (-0.77 \pm 0.09)\, x_{distance} + (7804 \pm 314)$$

Using the equation, the rate of spread is in the range: 1.2 to 1.5 km/year. The time of departure from the origin is within the range: 7490 to 8120 years BP.

6.2.6.7 Rate of spread and origin for Neolithic sites that fall within Zvelebil and Lillie's northern region of acculturation.

The 13 Neolithic sites are shown in Figure 6.48. Of the 504 potential origin points those that gave the strongest correlation coefficients for the linear regression of determination and distance to the sites are shown in Figure 6.49.

The origin in Russia (52°N, 36°E), with the strongest correlation coefficient, was used to calculate the rate of spread of farming to the Neolithic sites and the time of departure from the origin (Figure 6.50). The correlation coefficient is –0.65. The equation of the regression line is:

$$y_{determination} = (-3.54 \pm 1.12)\, x_{distance} + (11\,170 \pm 1941)$$

Using the equation, the rate of spread of farming falls within the range: 0.2 to 0.4 km/year. The time of departure from the origin is in the range: 9230 to 13 100 years BP.

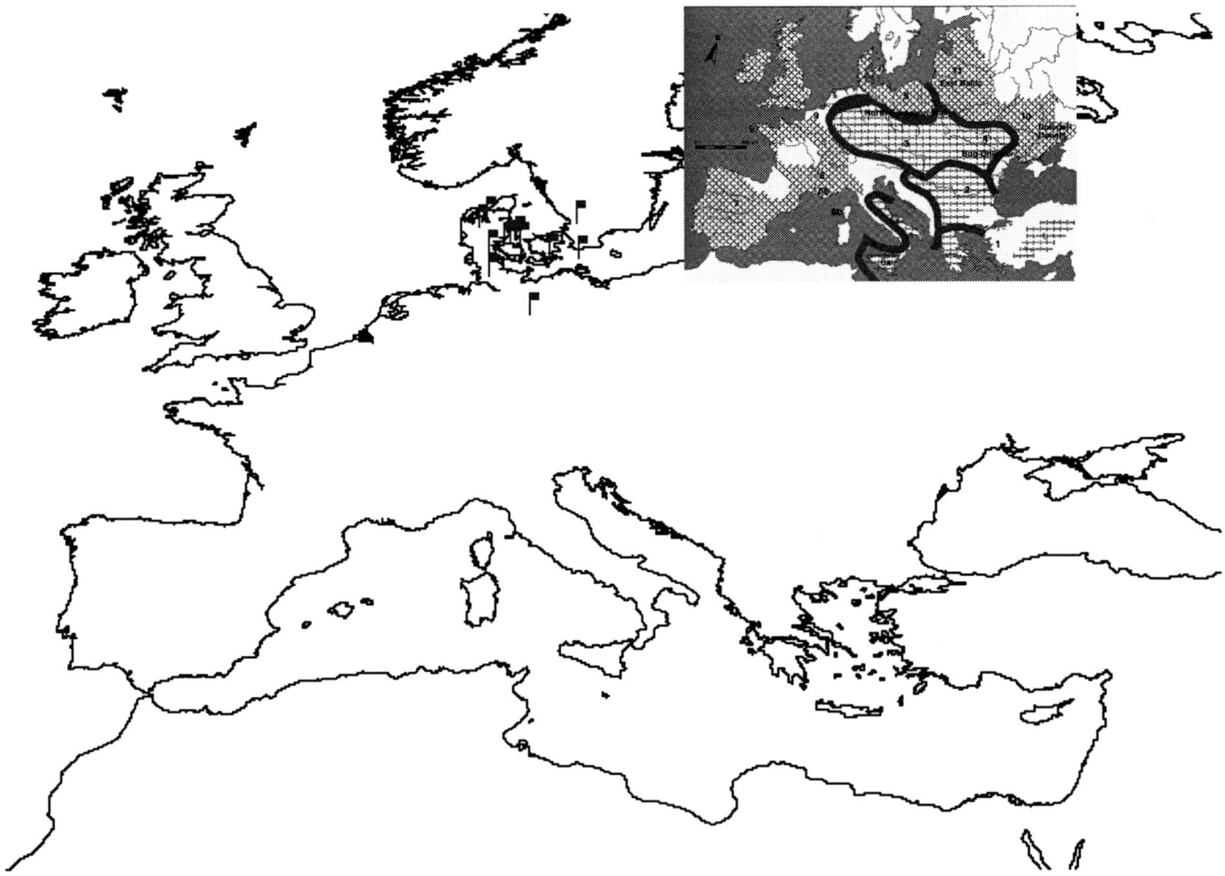

Figure 6.48 Neolithic sites that lie with the region of Europe that Zvelebil and Lillie attribute to the spread of farming by acculturation (Dataset G).

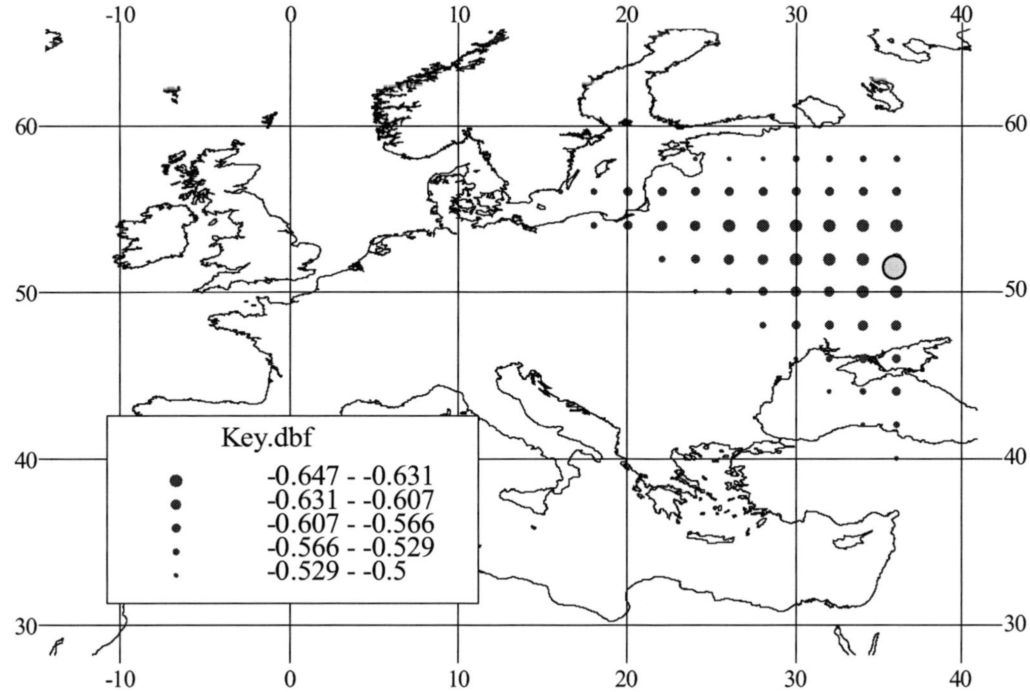

Figure 6.49 Origin points that gave the strongest correlation coefficients to the Neolithic sites within the northern acculturation region. The strongest correlation coefficient is shown as a pale circle.

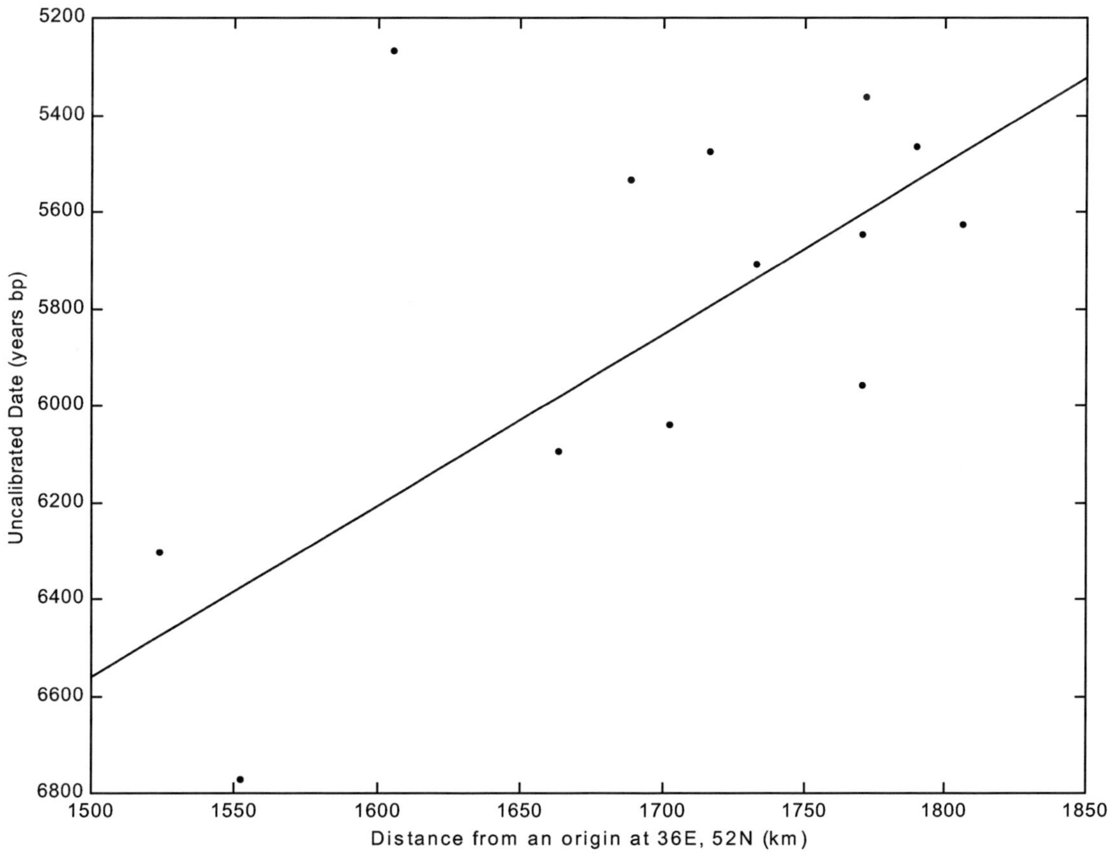

Figure 6.50 Linear regression of distance and weighted determination of acculturation north data from an origin at 36°E and 52°N.

6.2.7 Comparison of Dataset A results to Dataset G results.

The results of predicted origin, rate of spread and departure time at the origin for each dataset are summarised in Table 6.6.

For Dataset G only the origin for Italy and the exclaves was outside of Europe and the Near East. All origins (except Italy and the enclaves and the northern data – which is unlikely to be linked to the spread of the Neolithic as discussed previously) are in the same general area of Greece and Turkey. The predicted departure times fit with the archaeological evidence for Neolithic settlement in the origin areas. The rates of spread are lower than those calculated for Dataset A. All the correlation coefficients are high. The main area of difference between the two datasets is in the results for the areas of Italy, and Italy and the enclaves. These two groups show a high rate of spread and a poor correlation coefficient for Dataset A and the opposite, namely low rate of spread and a strong correlation coefficient for Dataset G. The difference can be explained by examining the underlying data. Dataset A contains the average determination at each site (when the Chi-square test allowed) or the oldest determination (when the Chi-square test proved that the range of determinations at a site were too great to average). Dataset G contains the average determination at each site, as with Dataset A, but instead of taking the oldest determination the median is selected. The consequence of this is that Dataset G has neither one-determination sites nor two-determination sites (as the decision was made not to average to dates to find the median but simply take the date that is closest to the central point of the chronological sequence). Additionally the difference between the oldest determination and the median determination at a site can be large, especially at sites with a long sequence of Neolithic determinations. For example at the Neolithic site of Molino Casarotto the determination selected for Dataset A is the oldest determination at the site (6450 ± 110 BP) whilst that selected for Dataset G, the median determination at the site is 5570 ± 50 BP. Similarly at the site of Scamuso, Dataset A selects the oldest determination (7290 ± 110 BP) and Dataset G selects the median determination 6600 ± 120 BP.

In summary, the trends reflected in Dataset G are associated with more recent Neolithic determinations. Whilst the trends with Dataset A are associated with the oldest Neolithic determinations and are thus arguably more representative of the spread of the first farming.

6.3 Summary

The chapter set out to test Zvelebil and Lillie's frontier hypothesis by searching for the presence of frontiers in the pattern of the spread of radiocarbon dates for the spread of the first farmers. The suggestion was that each frontier could be treated as an origin at which points dates for farming would be old, and that from the frontier one might expect to see a trend of decreasing antiquity with increasing distance from the frontier. The results did not contradict the frontier map, more interesting is the possibility that by examining the radiocarbon dates in this way one might be able to see which frontiers were most influential on areas of indigenous adoption.

The second aim of the chapter rested on the validity of Zvelebil and Lillie's divisions. Taking their map as correct, rates of measurement were made to see whether there was a difference between rates of spread in cultural diffusion areas and those in demic diffusion areas. Rates were calculated and they are very similar, although marginally higher for areas labelled as demic expansion areas (and possibly considerably higher for spread to the western Mediterranean although this calculation is subject to large error margins). The rates provide a comparative dataset for further research.

In this chapter analysis used radiocarbon determinations weighted by their one-sigma error range. In Chapter 7 techniques are presented that use calibrated dates in analysis to see the effect that this might have on results.

		Dataset A				Dataset G			
Diffusion mode	Area	Predicted origin	Predicted time of departure from origin (BP)	Rate of spread (km/year)	Correlation coefficient	Predicted origin	Predicted time of departure from origin (BP)	Rate of spread (km/year)	Correlation coefficient
Demic	Central Europe	Greece (22ºE, 38ºN)	6820 – 6970	2.4 – 3.0	-0.53	Greece (22ºE, 40ºN)	6840 – 7060	1.4 – 2.1	-0.61
Demic	Central Europe – no Italy	*Greece (22ºE, 38ºN)	6920 –7080	2.0 – 2.7	-0.59	Greece (20ºE, 40ºN)	6920 – 7200	1.1 – 1.8	-0.62
Demic	Italy only	Sicily (13ºE, 38ºN)	6380 – 6780	1.4 – 8.3	-0.29	Greece (20ºE, 40ºN)	6850 – 7360	0.6 – 1.1	-0.83
Demic	Italy and enclaves	*Italy (16ºE, 40ºN)	6330 – 6540	1.7 – 4.5	-0.26	Libya (24ºE, 30ºN)	6620 –7720	0.7 – 1.5	-0.59
Acculturation	Western region	*Sardinia (9ºE, 40ºN)	5990 – 6130	1.4 – 1.6	-0.59	Turkey (32ºE, 36ºN)	7660 – 8280	1.1 – 1.4	-0.61
Acculturation	Western region – no enclaves	*France (6ºE, 44ºN)	5670 – 5770	1.4 – 1.6	-0.57	Turkey (34ºE, 36ºN)	7490 – 8120	1.2 – 1.5	-0.60
Acculturation	Northern region	Russia (36ºE, 58ºN)	7950 – 10 230	0.32 – 0.72	-0.46	Russia (36ºE, 52ºN)	9230 – 13 100	0.2 – 0.4	-0.65

Table 6.6 Comparison of the results of Dataset A to those of Dataset G. Origin points that were revised when they were found to lie in north Africa are marked with an asterisk (*).

Chapter 7: Analyses using calibrated radiocarbon dates

7.1 Introduction

In Chapter 4 two visualisation methods were presented: isochron maps and chronological maps. One of the problems with these methods is that they use uncalibrated radiocarbon determinations. In this chapter a technique of visualisation that uses calibrated dates is presented.

7.2 Visualisation technique using calibrated radiocarbon date.

A subset of 508 Neolithic dates and 207 Mesolithic dates were used in these analyses. At sites where there are multiple radiocarbon dates only the oldest Neolithic date and/or the most recent Mesolithic date were used. The methodologies discussed in Chapter 4 used uncalibrated dates and treated the date's modal value as a point value. The irregular shape of many of the calibrated probability distributions of radiocarbon dates makes statistical treatment that requires point values problematic. Most commonly this is overcome by treating the modal value of the radiocarbon probability distribution as a point value and using weighting to minimize the influence of dates with a large associated error (Blockley *et al.* 2000, Steele *et al.* 2000, Glass *et al.* 1999).

To assess how problematic the use of the modal value might be, particularly in instances when the calibrated radiocarbon distribution deviates from a normal distribution (Figure 7.1) a visualisation method was devised to show the whole of the probability range for a calibrated date.

To do this dates were calibrated in OxCal Version 3.5 (Bronk Ramsey 2000). The area under the calibration curve was used to calculate the probability that a site was occupied throughout the whole period of the calibrated date range (Figure 7.2). The probability that a site was occupied at successive 100-year intervals is reflected by the rise and fall in the area of the circle that marks a site's location.

A sample of this analysis is shown in below (Figure 7.3).

The full animation is available as a PowerPoint presentation on CD (Russell 2002). When the whole of the calibrated date distribution is used, rather than the modal value of the uncalibrated date, the pattern of spread is far less obvious. The value in this approach is that it captures more of the fluidity of the spread than does the use of point values in isochron and chronological maps and it is also useful to assess the extent to which Mesolithic and Neolithic sites may have overlapped both in space and time. Of course, the successive maps make it much harder to see any overall pattern than the scatter diagram. A method to quantify this was required, one that still used the information held in the probability distribution of the calibrated radiocarbon date.

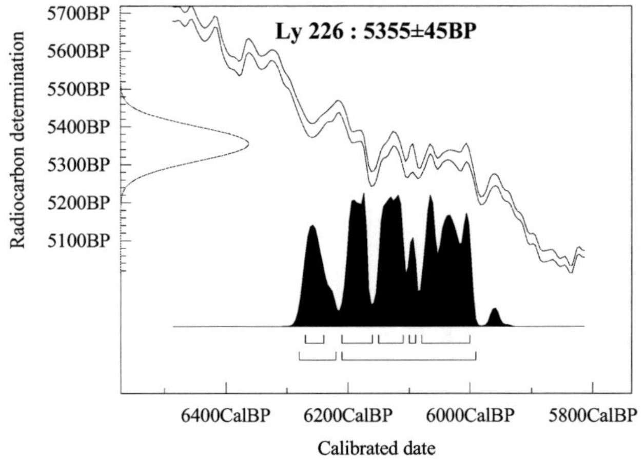

Figure 7.1 The calibrated probability distribution of a radiocarbon determination from the Neolithic site of Camp del Ginebre, France (from the Gkiasta database). The calibration used OxCal version 3.5 (Bronk Ramsey 2000)

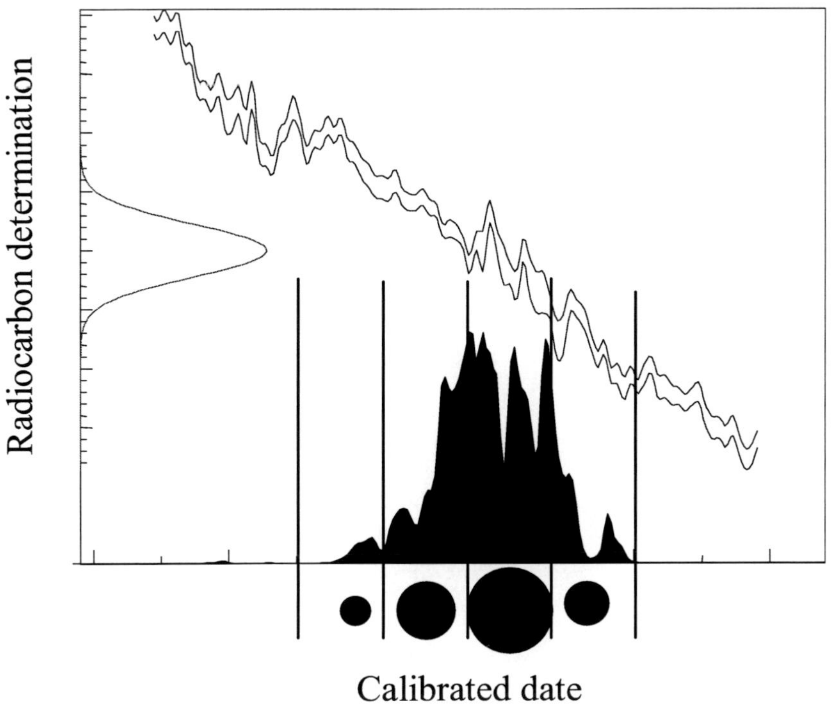

Figure 7.2 The calibrated date's probability distribution is divided into 100-year intervals; the area within these time bins (shown as circles of different area) reflects the varying probability that a site is occupied throughout the calibrated date range.

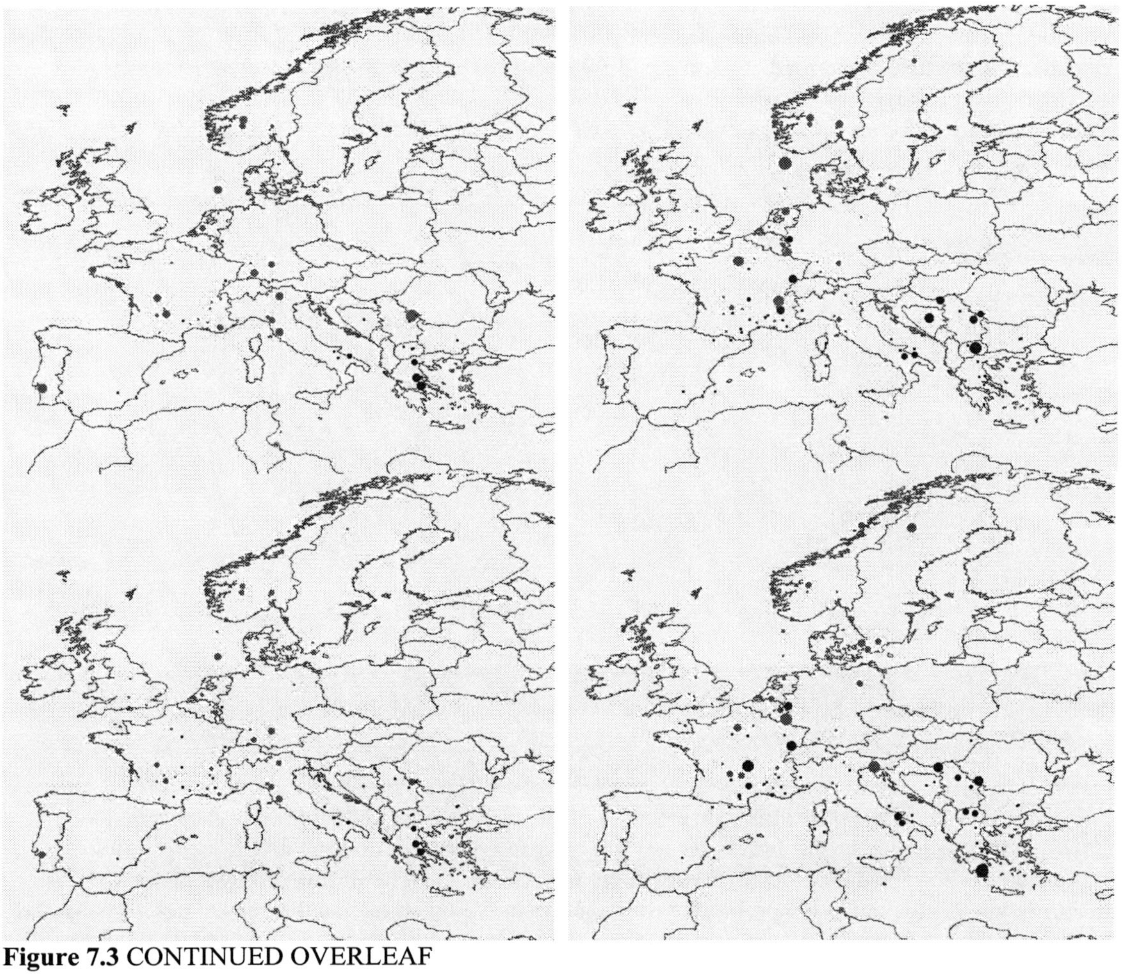

Figure 7.3 CONTINUED OVERLEAF

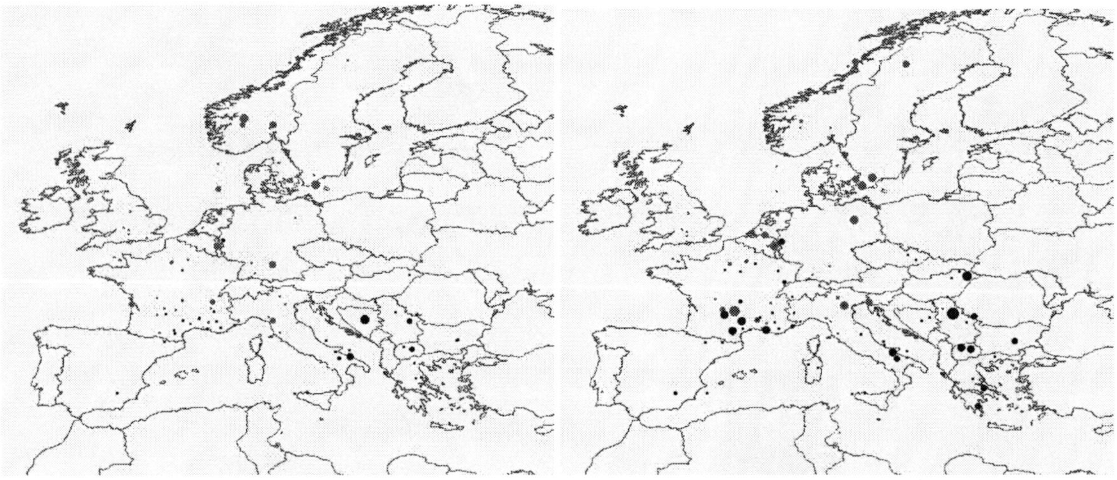

Figure 7.3: A sequence of probability maps for Mesolithic (grey) and Neolithic (black) site occupation for the period 6400 BC to 5800 BC (top to bottom). Each map represents a 100-year period. The full animation is enclosed on CD found with the thesis (Russell 2002).

7.3 Probability weighted time bins: a method by which calibrated radiocarbon dates can be used in statistical analysis

In this section a method is described that is similar in principle to the visualisation method described: it too makes use of the total area under the probability distribution of the calibrated date. The aim was to devise a method by which the whole of the calibrated date range could be used in statistical analyses rather than a single modal point value. The result is best illustrated with a stepwise example.

Step 1. The radiocarbon determinations were calibrated in OxCal version 3.5. OxCal generates plot date files (extension .14d) that contain the information required to build the probability distribution of the calibrated date (Figure 7.4). This is the data that OxCal uses to generate the probability distribution of the calibrated date (Figure 7.5).

```
!   95.4% probability
!     6270BP (2.0%) 6240BP
_2 -4320 30
!     6210BP (93.4%) 5780BP
_2 -4260 430
^0.00498582
-4240  0.251029
-4236  0.239696
-4232  0.234754
-4228  0.231662
-4224  0.304042
-4220  0.570751
-4216  0.893598
-4212  1.000000
-4208  0.975068
-4204  0.884679
-4200  0.700329
-4196  0.519283
-4192  0.440954
-4188  0.412500
-4184  0.370979
```

Figure 7.4 An example of a portion of the probability distribution of the radiocarbon date distribution as generated in OxCal Version 3.5. Of the two columns the left lists dates BC in 4 year intervals, the column on the right shows the relative probability associated with each date. Each of these probability values must be multiplied by a factor (underlined) and by four to get an area under the probability distribution curve equal to 1 as the values are normalised to the maximum of the date's calibration curve (shown here at –4212 BC). The highlighted area shows the probability distribution of the date.

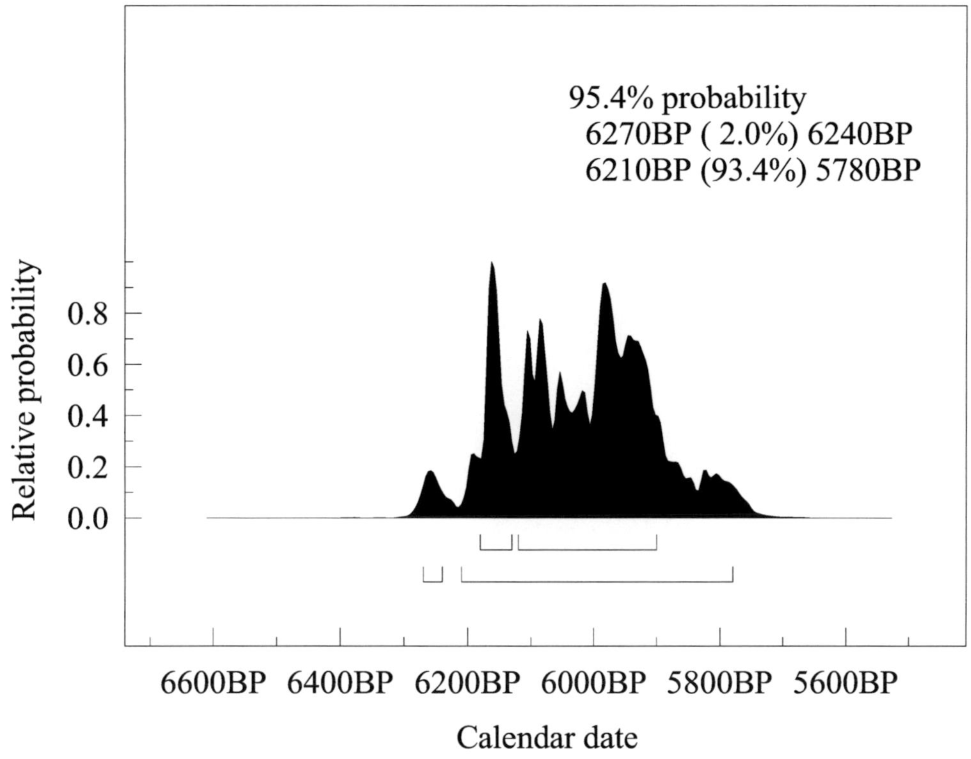

Figure 7.5 Calibrated probability distribution. The highest peak (relative probability 1) is at 6162 BP (this corresponds to the –4212BC in Figure 7.4).

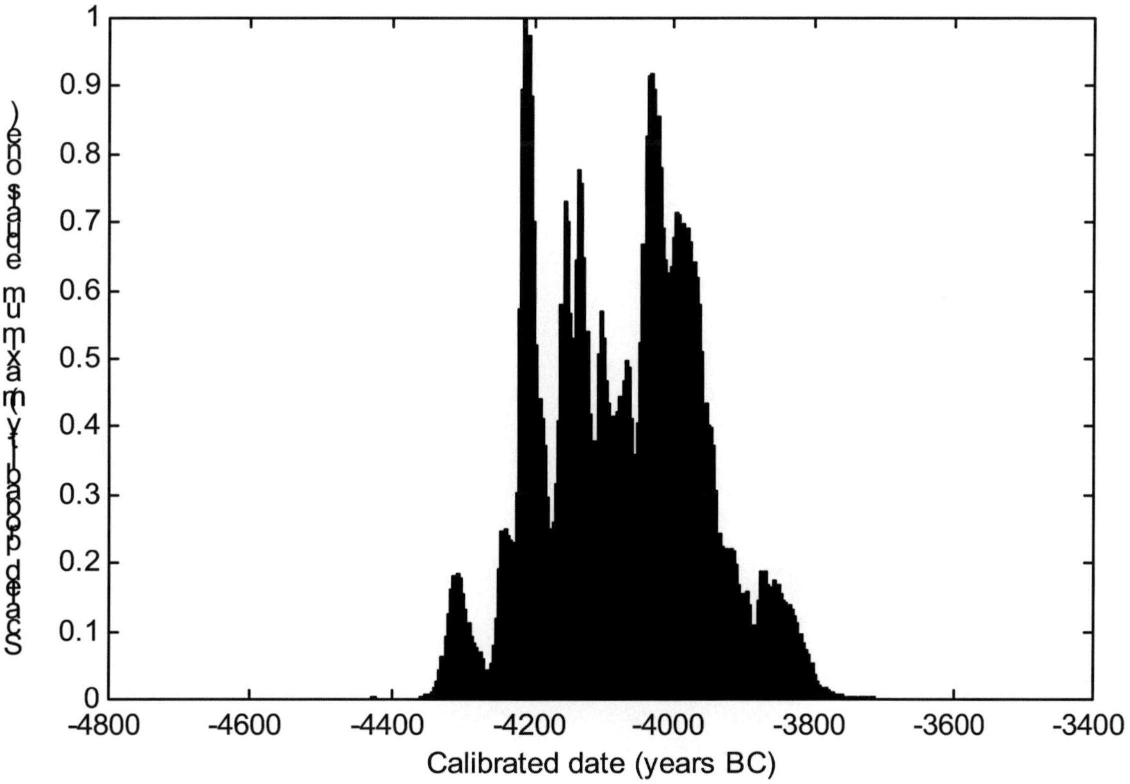

Figure 7.6 Probability distribution of the date rebuilt using the plot data from OxCal

Step 2. As a cross check the calibration curve was rebuilt using the OxCal plot data file. Matlab was used to reproduce the curve (Figure 7.6).
Step 3. The area under the calibrated probability distribution curve was regrouped from 4 -year intervals to 20- year intervals (Figure 7.7).
Step 4. Rather than seeing the calibrated range as a single measurement, the date at the mid-point of each 20-year time bin is treated as a possible date for the event. In Figure 7.7 above this results in 24 dates (24 columns), only one of which is the real date of this event. It is impossible to say which one is correct, but the probability associated with each date can be used to suggest which is the most likely to be correct. To do this each of the dates was given a weight determined by the probability that it is correct: this is equivalent to the height of the column. The result is that dates associated with high probabilities (high columns) are given more influence in the analysis. To amplify the effective weight, the probabilities were squared (this means that very small probabilities become negligible (i.e. 1^{-3} becomes $1^{-6)}$, whilst the larger probabilities remain large (i.e. 9^{-2} becomes 8.1^{-3}) (Figure 7.8). This is analogous to using the inverse square of the standard deviation when weighting radiocarbon determinations by their standard error.

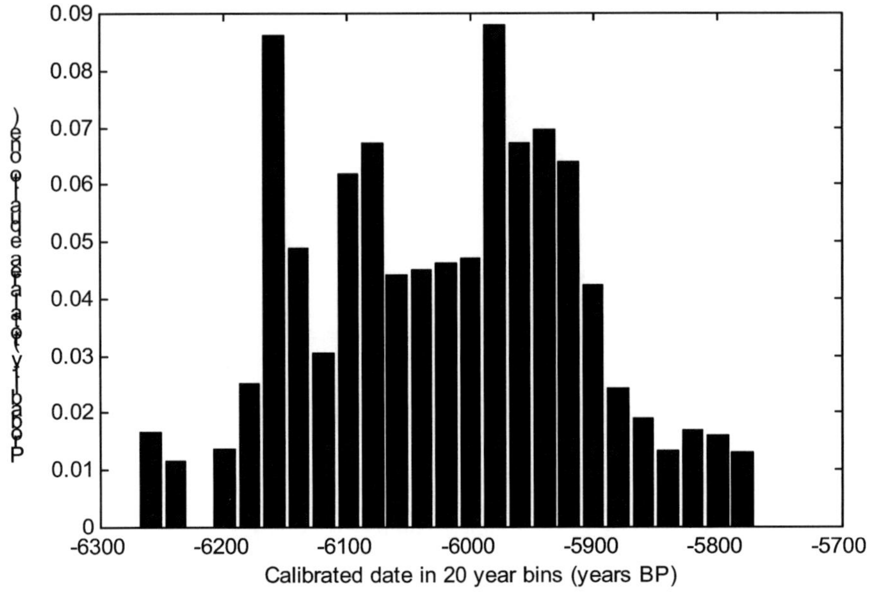

Figure 7.7 Calibrated date probability distribution with the probability range regrouped in 20 years time intervals.

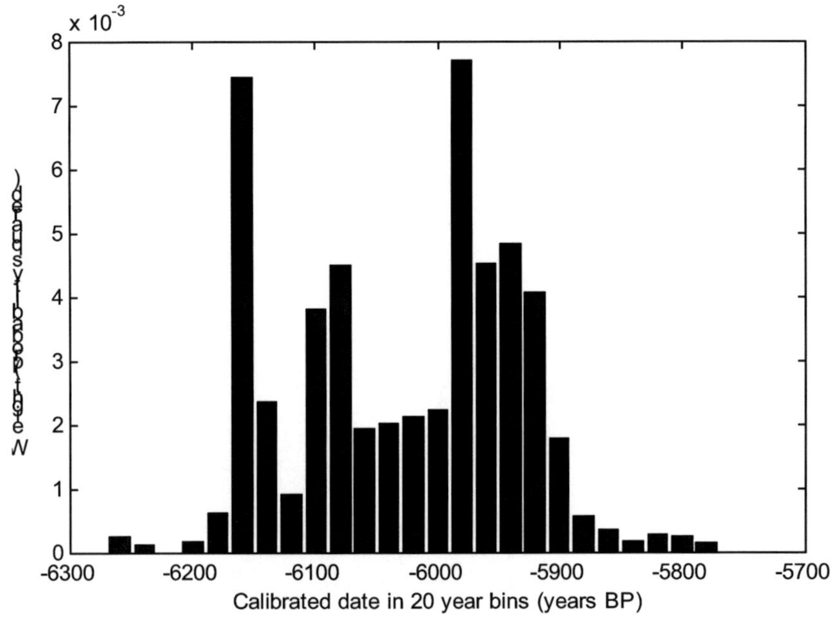

Figure 7.8 The probability associated with each dated 20-year time interval is squared to amplify large probabilities whilst minimising negligible probabilities.

The result is that in spatial-temporal analyses, at each archaeological site instead of using one radiocarbon determination and its standard deviation, the analysis uses multiple dates (generated from a determination's calibrated probability distribution) each weighted with their associated probability.

7.4 Re-analysis of the European dataset using whole of the calibrated date range and the 'first' event rather than the oldest date.

The method outlined was applied to explore the effect of using calibrated dates rather than radiocarbon determinations in analysis. In addition, rather than choosing the 'oldest' date or one of the other combinations of date selection (Chapter 2) use was made of the OxCal tool that uses Bayesian statistics to calculate the most probable date for an event (namely, the first arrival of farming) that is undated but which is thought to have occurred prior to the oldest date in a sequence (Bronk Ramsey 1995, 1998). The reasoning is that there is very little chance that the oldest radiocarbon date in a sequence actually dates the first arrival of farming. This method uses all the dates that are linked to this first event to approximate when it occurred. The date of the 'first' event is calculated on the basis of the radiocarbon probability distributions of the dates that are archaeologically linked to it (Figure 7.9). In OxCal the command 'first' returns the probability distribution for the estimated 'first' event based on the group of radiocarbon determinations linked to it (Appendix 8). The application of this analytical method to the European database was inspired by a presentation by Caitlin Buck, University of Sheffield at a meeting held at the University of Southampton, 2002 (Buck *et al* 1994, 1996).

Using this method, the 'first' event was calculated for each of the Neolithic sites in the European database that had more than one radiocarbon determination: a total of 264 sites. The probability weighted time bins method (section 7.3) was applied to each of the 'first' event distributions. The probability weighted times bins were then used in analysis to search for the best possible origin to the data. The result is a correlation coefficient of –0.76 and an origin at 30°N, 36°E (south west Jordan).
The best-fit origin was used to calculate the geodesic distance to each site. Linear regression using weighted regression model 1B (Chapter 5) is shown in Figure 7.10.

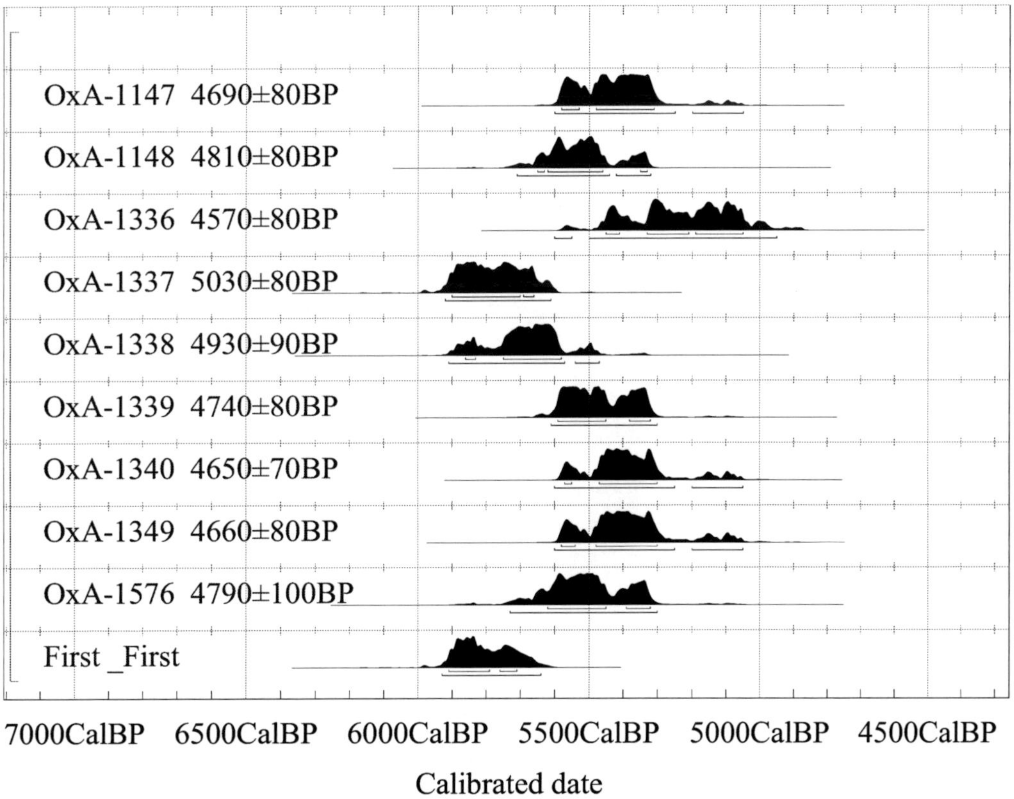

Figure 7.9 Diagram to show the probability distribution for the 'first' event calculated using the 9 Neolithic radiocarbon dates from Maiden Castle, Great Britain.

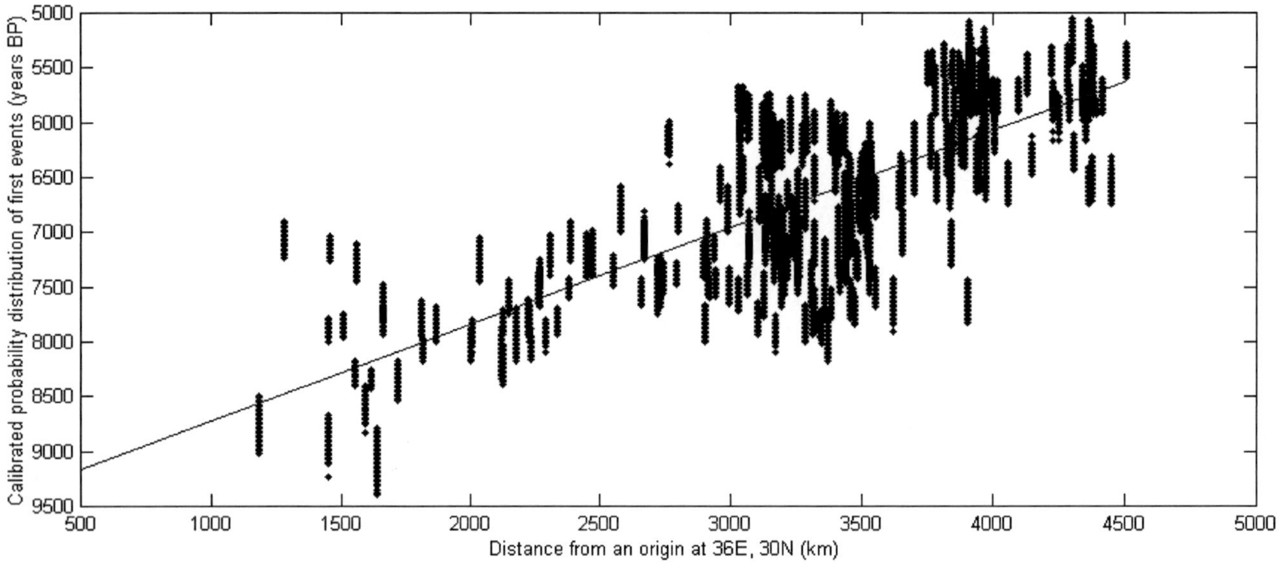

Figure 7.10 Regression analysis of date and distance using the probability weighted time bins as the dates at each of 264 Neolithic sites.

The equation of the regression line is:

$y_{date} = (-0.88 \pm 0.01)\ x_{distance} + (9613 \pm 37)$

This gives a rate of spread of 1.1 km/year and a time of departure from the origin in the time range: 9570 to 9650 years calBP.

7.4.1 Comparison of analysis using calibrated dates to those using radiocarbon determinations:

The dataset that is most directly comparable to the one used here is Dataset F, which contains the oldest date at multiple dated sites (264 sites)(Chapter 5). This dataset uses an identical site list to the dataset for the 'first' event. The analysis of Dataset F used uncalibrated radiocarbon determinations. The best-fit origin to the Dataset F is also at 36ºE, 30ºN. The equation of the regression line for Dataset F is:

$y_{determination} = (-0.81 \pm 0.04)\ x_{distance} + (8484 \pm 144)$

This gives a rate of spread of 1.2 km/year and a time of departure from the origin between 9000 to 9900 calBP (2-sigma range, calibrated using OxCal version 3.5).
The results are similar between datasets and with or without calibrated dates (Table 7.1). This result is explained, in part, by the method of date selection (Figure 7.11).

	Predicted origin CalBP (2-sigma range)	Rate of spread (km/year)	Origin
'First' Dataset	9650 - 9570	1.1	South west Jordan
Dataset A	9550 – 9000	1.2	South west Jordan
Dataset B	7840 – 7580	1.2	Southern Bulgaria
Dataset C	9500 – 8990	1.3	South west Jordan
Dataset D	7690 – 7970	1.3	Southern Bulgaria
Dataset E	7800 – 7480	1.1	Southern Romania
Dataset F	9900 – 9000	1.2	South west Jordan
Dataset G	9250 – 8400	1.2	Turkey

Table 7.1 Comparison of the results of the calibrated dataset to those that were analysed using radiocarbon determinations (Chapter 5).

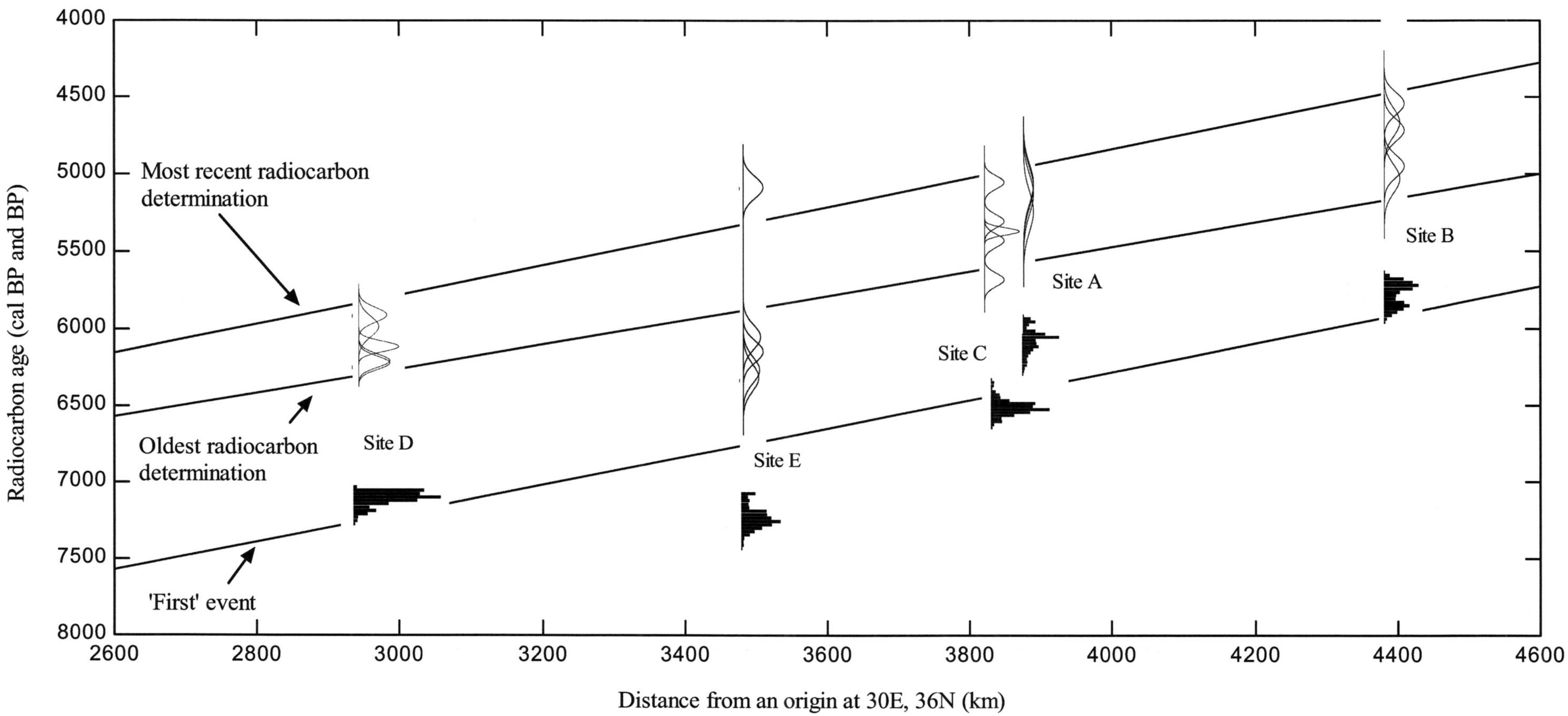

Figure 7.11 The effect of data selection. The line distributions are the normal distributions of radiocarbon determinations. The bar diagrams are the calibrated distribution of 'first' events using the probability weighted time bin method. The regression line that is fit to the date will shift (vertically up and down) according to the whether the most recent radiocarbon determination is used, the oldest radiocarbon determination or the probability distribution of the 'first' event.

Over a long time range and with a big sample the effect of the calibration seems minimal. To look at the effect of using this probability weighted time bin approach on a smaller dataset and a smaller time range, the origin and rate measurement for the 13 sites from Dataset G that were associated with the northern acculturation zone of Europe (Chapter 6) were re-analysed. The diagram from the previous section is reproduced here for ease of comparison (Figure 7.12).

Using radiocarbon determinations (a modal value weighted by its error term) the origin point for the thirteen sites is at 36°E and 52°N, the correlation coefficient is
-0.65 and the equation line for the regression is:

$$y_{date} = (-3.54 \pm 1.12)\ x_{distance} + (11\ 170 \pm 1941) \quad \text{(equation 1)}$$

The analysis was redone using the probability weighted time bin method, so that the whole of the calibrated dates range was used. The origin is still at 36°E and 52°N but the correlation coefficient is stronger, r = – 0.74. The regression analysis uses the whole dates as illustrated in Figure 7.13.

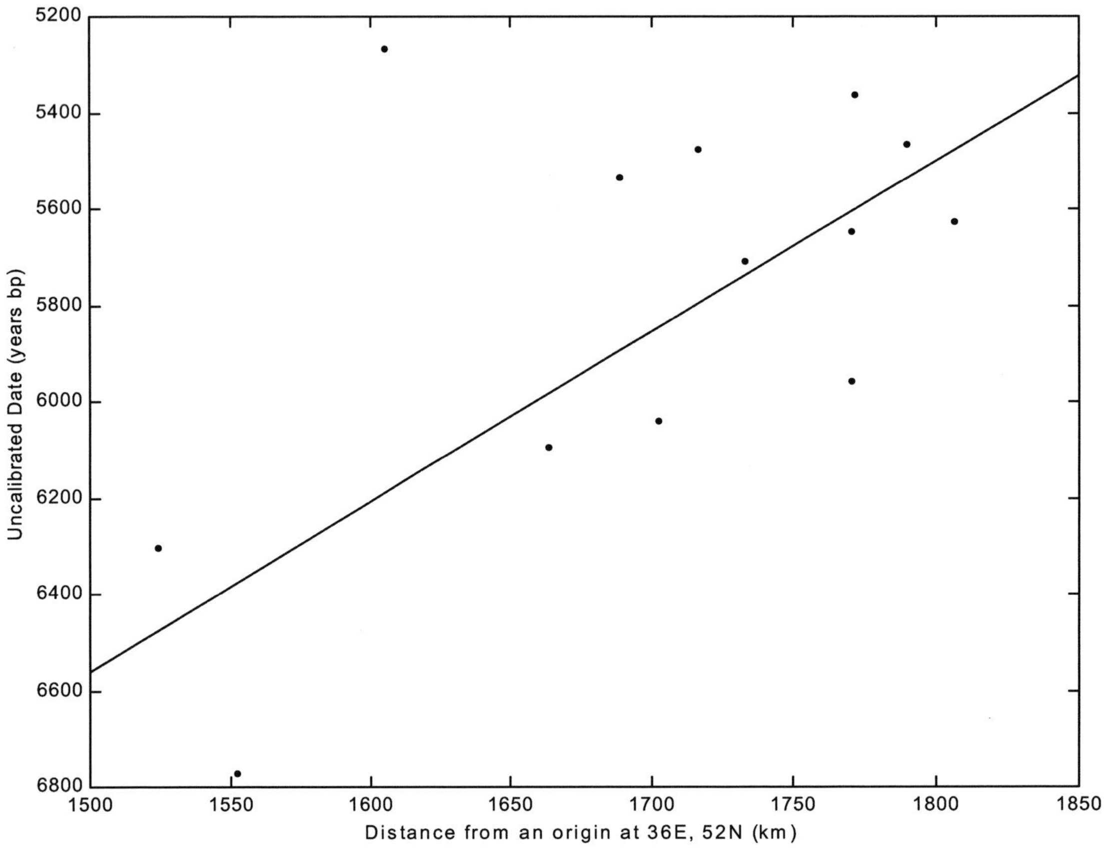

Figure 7.12 Linear regression of distance and weighted determination for acculturation north data from an origin at 36°E and 52°N.

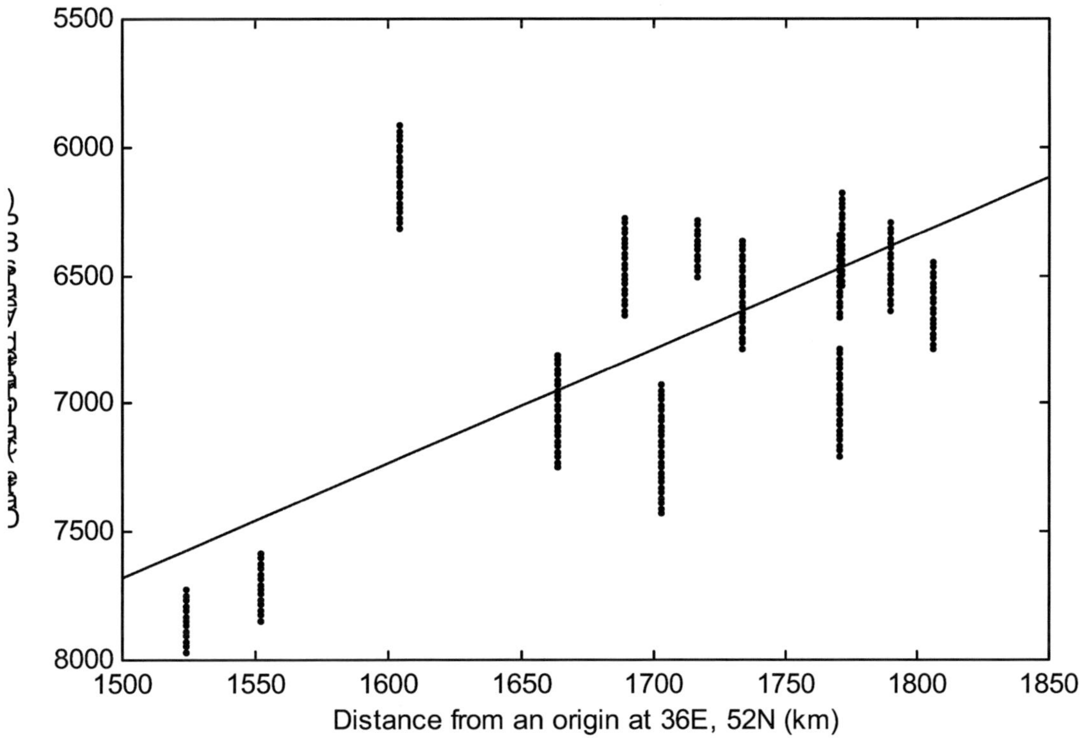

Figure 7.13 Regression analysis with the probability weighted time bin method so that the whole of the calibrated date's range is used. The sites used are identical to those shown in Figure 7.13.

The equation line for the regression is:

$$y_{date} = (-4.5 \pm 0.26)\, x_{distance} + (14\ 351 \pm 434) \quad \text{(equation 2)}$$

The most noticeable difference between equations 1 and 2 is that the probability weighted time bin method gives reduced error terms associated with the y-intercept and the slope of the line. This means that the predicted date of departure from the origin point and the rate of spread will differ to those calculated using equation 1. However the large error terms in equation 1 mean that there is overlap between the results (note that the predicted date of departure in equation 1, 11 170 ± 1941 is an uncalibrated determination and calibration will bring it within the same time range as the calibrated date of departure, 14 351 ± 434, in equation 2). The different methods produce similar results even over a smaller time range and reduced sample.

7.5 Summary

This chapter described a technique by which calibrated dates can be used in statistical analysis. More comparative analysis is required to understand the difference between analysis that uses radiocarbon determinations and those using calibrated dates, for instance, the shape of the calibration curve will cause the effect to be more or less significant depending on where along the calibration curve a set of determinations are calibrated. So that for small samples, in particular, the similarity between results using calibrated and results using uncalibrated dates may well be determined by where they lie on the calibration curve.
It would also be interesting to compare results from the probability weighted time bin method to results from analyses that treat radiocarbon dates as point values, using the modal value of the probability distribution as a point datum and weighting this datum inversely to the square of the standard deviation (Steele *et al.* 2000).

Chapter 8: The spread of the first sheep to southern Africa

8.1 Introduction

The following two chapters report on the exploratory application of the techniques developed with the Gkiasta Database to a much smaller database for the spread of sheep in South Africa[1]. There are parallels between the questions that are asked about the spread of the first farming into Europe and those that are asked for a similar process in southern Africa. Much debate surrounds the origin, timing, route and method of introduction of the first domestic stock in to the western half of what is now South Africa. A lot has been written on the theory of the change from hunter-gatherer to food-producer/stock-keeper and the means by which the two groups can be separated on the landscape, but this has not yet brought resolution to the debate. There has not been a thorough spatial analysis of the radiocarbon data. The aim of this work was to see whether this approach would lend anything new to the discussion. In this chapter the debate is introduced. Only the spread of the first sheep into the western half of southern Africa is analysed here but mention of the other broadly contemporaneous processes (i.e. the spread of the Iron Age farmers) is made below, because although distinct in the geographical area under study it would be incorrect to think of them as isolated processes: they have connections further back in time and further north in Africa. In Chapter 9 the geographically referenced radiocarbon dataset for the spread of domestic sheep is analysed using the statistical methods developed in Chapters 3 to 7.

8.1.1 The first stock-keepers and the first agro-pastoralists in southern Africa:

Until the first century AD southern Africa was inhabited solely by Stone Age hunter-gathering people whose descendants are known as the Bushman, San or Soaqua (parallels can be drawn with the Mesolithic inhabitants of Europe). This changed at *circa* 2000 years ago with the arrival of the first Bantu agro-pastoralists who migrated into the eastern part of southern Africa bringing with them cattle, sheep and goats, iron working technology, a new language and the ability to grow sorghum and millet crops (Holl 1998, Huffman 1996, Huffman & Herbert 1996). There is some disagreement as to the order of appearance of this group, the agro-pastoralists, as distinct from another, the non-agricultural stock-keepers (Maggs 1996). At approximately the same time (or perhaps a few centuries earlier) the first sheep appeared in the western half of southern Africa (Robertshaw 1978). Sheep predate cattle in this area (Klein & Cruz Uribe 1989, Sealy & Yates 1994, Clutton-Brock 1996, Sadr 1998); it is solely the spread of the first sheep that is explored here. Since sheep do not have wild progenitors in southern Africa (Robertshaw 1978, Sealy and Yates 1994, 1996, Smith 1997, Sadr 1998,) they must have been introduced into this area from further north.

There is disagreement about the mechanisms by which sheep spread. As with the Neolithic in Europe, two models currently compete: a cultural diffusion model with the indigenous adoption of sheep-keeping by San hunter-gatherers (Kinahan (1991, 1996) and a demic diffusion model with the migration of Khoi[2] pastoralists (Smith 1997, 1998a). As with the European Neolithic, there is also a growing body of thought that suggests both processes took place (Sadr 1998). In the words of Klein (1986:9), "...broad continuity in stone artefact traditions and in hunting-gathering patterns before and after the introduction of stock and pottery to the Cape suggest that acculturation (diffusion) was at least as important as population movement in promoting the spread of pastoralism". It is to avoid this controversy that the analysis is primarily focussed on tracing the first appearance of sheep in the archaeological record rather than of a particular group of people.

[1] My previous work has been in South Africa (see Russell 2000) where my familiarity with the landscape, the debates and some of the sites (Steenbokfontein Cave, Smitswinkelbaai, Elands Bay Cave, Tortoise Cave, Kasteelberg A/B and Witklip) has facilitated the finer grained approach adopted in this section.

[2] The relationship and distinction between the Khoi and the San (also known as Bushman or Soaqua) is a difficult one (Smith 1998b: 208). This is reflected in the fact that they are commonly referred to as the Khoisan in discussions of South African prehistory. " The term 'Khoisan' was coined by Schultze (1928) by combining Khoi (part of the term KhoiKhoi which was used by the KhoiKhoi or 'Hottentots' to refer to themselves) with San, which was the term the KhoiKhoi used to refer to the hunter-gatherers or 'Bushmen'." (Soodyall & Jenkins 1997:1). Although the San and KhoiKhoi are not always distinguishable in physical appearance, certain cultural and linguistic differences are apparent. Whereas the San are classically the hunter-gatherers and keep no animals for their own use, the KhoiKhoi were until recently pastoralists who kept large herds of cattle, sheep and goats (Ehret 1982).

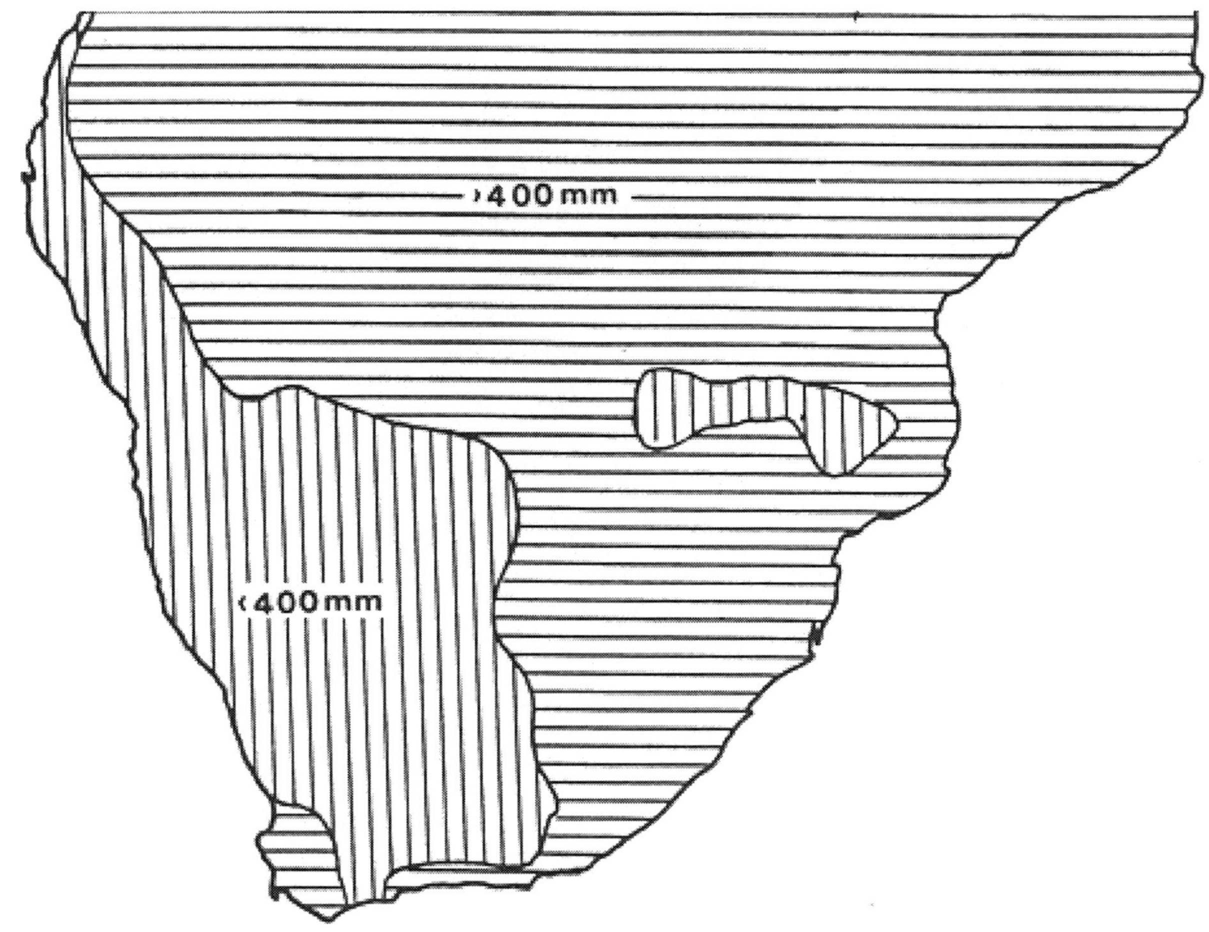

Figure 8.1 Annual rainfall in Southern Africa. The spread of agropastoralists is limited to areas with summer rainfall patterns with >400mm per annum. This excludes south western corner of southern Africa as it is a winter rainfall area and this is unsuitable to growing sorghum and millet (Hall and Smith 1986:3).

The geographic separation of stock-keeping and agro-pastoralism, west and east, is in part explained by the rainfall patterns in southern Africa. Rainfall is high in the east and, apart from a pocket of high winter rainfall in the south western corner at the Cape, low in the west (Figure 8.1). This imposed an environmental restriction on the westward and southward expansion of the agro-pastoralists; the Bantu never made any attempt to colonise the western area of southern Africa. They were limited to settling in areas with reliable summer rainfall (Kinahan 1996, Vogel and Fuls 1999). There is, however, archaeological evidence for the interaction of the Bantu farmers and the Khoi pastoralists at the Orange River in the central region of South Africa (Smith 1997)(Figure 8.2).

8.2 The sheep: identification and description.

Sheep bones are found at Later Stone Age sites from central Namibia south as far as the Cape, and east as far as the eastern end of the Cape Fold Belt mountains (Parkington 1984) (Figure 8.2).

The breed of sheep most commonly associated with their first occurrence is the fat tailed sheep[3] (*Ovis Aries*) (Figure 8.3). This is based on several independent lines of evidence. Historic accounts by the first Europeans to sail along the shores of South Africa in the 16th to 18th centuries describe the indigenous sheep as hairy, non-woolly, fat tailed and floppy eared (Raven-Hart 1967, Robertshaw 1978). The paintings in the rock shelters of southern Africa similarly depict fat-tailed floppy eared sheep (Figure 8.4)(Robertshaw 1978, Manhire *et al.* 1986, Jerardino 1999). Thackeray & Humphreys (1983:94) working at Wonderwerk cave report "hair samples from layers 2b and 3a identified as sheep may indicate the presence of the Afrikaner (fat-tailed) breed in prehistoric context of the site".

[3] On presentation of this work at the Southern African Archaeologists biennial meeting 200, I learnt that genetic analysis of sheep bone is bringing into question the commonly held belief that all sheep bones at archaeological sites for this period are those of the fat-tailed sheep.

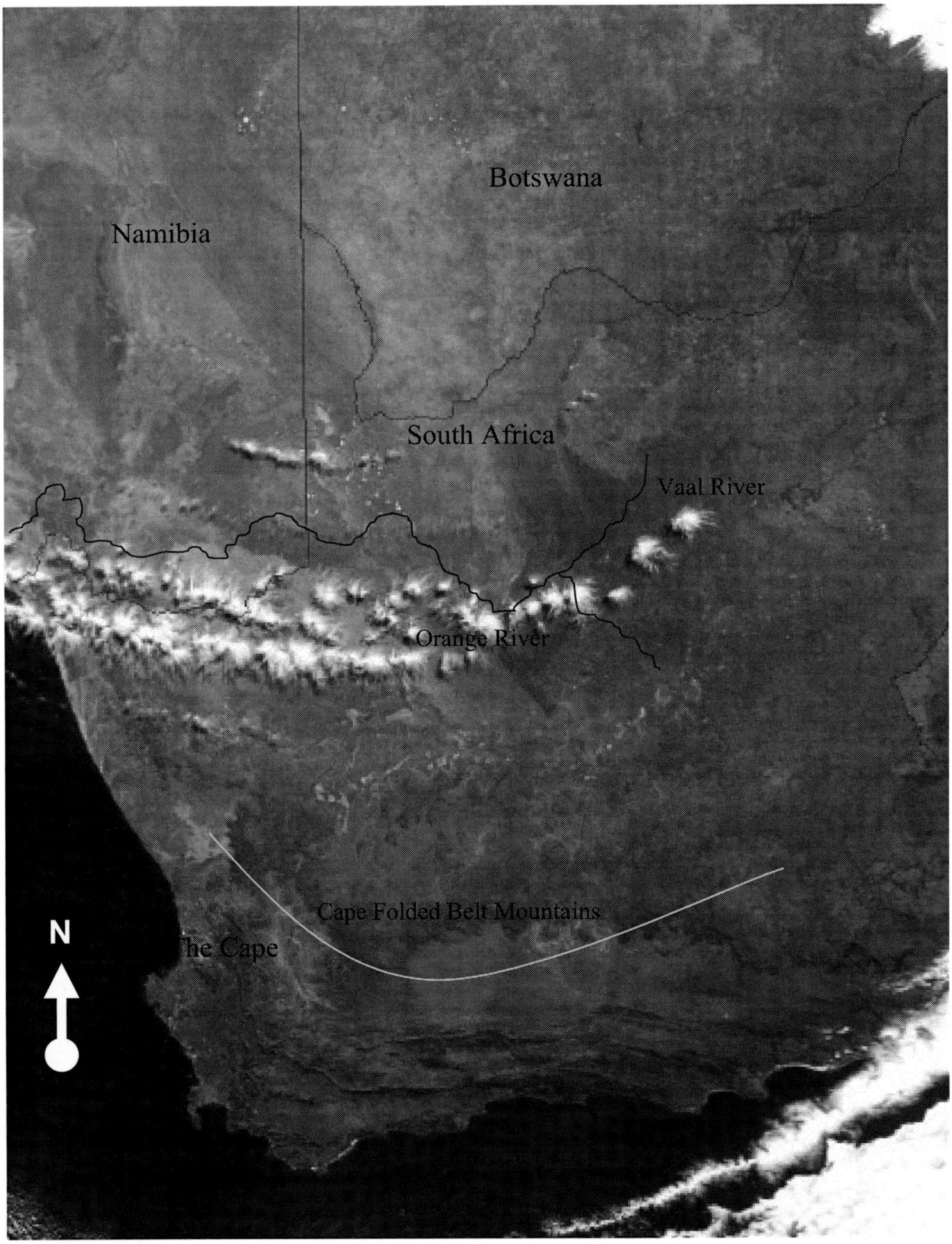

Figure 8.2 Map of the western half of southern Africa to show the locations of the major rivers and mountains mentioned in the text.

Figure 8.3 Fat tailed sheep.

Figure 8.4 An example of rock painting of fat-tailed sheep from Mazoe, Zimbabwe (Anati 1993: 131).

8.3 Pottery: Part of a sheep-herder package or an earlier introduction?

For completeness pottery must be mentioned. Accounts of the first spread of sheep frequently include pottery as part of a migrating system. Initially archaeologists believed that livestock and pottery arrived as a package in the western half of South Africa, as both seemed to occur together (Smith & Jacobson 1995). Indeed, until recently, pottery and domestic stock were seen as the archaeological signature of pastoralism, with the presence of one being inferred from the other (Kinahan 1996). Thus Robertshaw (1978:120) …"it seems reasonable to suggest that pottery and domestic stock make their appearance almost simultaneously and that the spread of these two elements in the Cape is linked". However, the direct dating of sheep bones (Sealy & Yates 1994, 1996), and new archaeological excavations at sites such as Spoegrivier (Webley 1992a), Blombos (Henshilwood

1996) and sites in Namibia (Kinahan 1991), all show indisputably that pottery is of greater antiquity. Pottery regularly pre-dates the arrival of sheep by as much as 500 years (Smith & Jacobson 1995).

8.4 Different hypotheses about the diffusion mechanism that spread the first sheep:

The evidence for and against both cultural diffusion and demic diffusion are discussed in the section that follows.

8.4.1 Evidence for the cultural diffusion of sheep between hunter-gatherer groups.

Kinahan (1991, 1996) is the main proponent of this view. His observations at archaeological excavations in Namibia have led him to conclude that, ..."the pastoral economy of Namibia was in all likelihood a local development resulting from an internal transformation of the indigenous hunting economy" (*ibid.* 1996:218).
Recent work by Sadr (1998) argues convincingly for at least an initial diffusion of livestock into southern Africa. He argues that the unpopularity of the diffusion hypothesis is largely due to an assumption that the change from hunter-gatherer to pastoralist is not an easy one, as argued by people like Smith (1997) who suggests that the egalitarian nature of hunter-gatherer societies and their ethic of sharing would have prevented them from easily adopting a pastoralist way of life. Sadr (1998) cites ethnographic case studies to show that hunter-gatherers are far less rigid to change than has been previously argued. "Kalahari ethnography shows that hunters with even the strictest ethos of sharing can keep livestock" (*ibid.*117). Smith (1990) has vehemently opposed this view. Like Ingold (1980) he suggests that the difference between hunter-gatherers and herders lie in their fundamentally different social, ideological and economic mindsets. Further, he argues that these differences are so divisive that the transition from one to the other would involve a revolutionary change in lifestyle and mindset; it could not happen casually.
A second argument against diffusion is the commonly held belief that the rapidity at which livestock appear throughout the southern African landscape excludes diffusion (Robertshaw 1978). Against this Sadr (1998:117) argues that domestic stock could have diffused from Zimbabwe to the western Cape in just 200 years with a diffusion rate of 10 kilometres a year. The regional diversity in the styles of the earliest pottery and the variability in the order of appearance of pottery and sheep are taken as evidence for diffusion approximately 2000 years ago (*ibid.*).

Sadr agrees that there was some demic diffusion but that this was a much later event towards the end of the 1st millennium AD. His evidence for the demic migration is a later stylistic chain of pottery running from northern Botswana to the Cape and other sudden changes in the archaeological record interpreted as indicative of an immigrant population either replacing or displacing the indigenous people.

8.4.2 Evidence for the spread of sheep by demic diffusion

The second hypothesis is that livestock were introduced to southern Africa by the migration of stock-keepers (Robertshaw 1978, Deacon *et al.* 1978, Walker 1983, Parkington 1984, Smith 1997). The majority of archaeologists currently favour the migration hypothesis over the diffusion hypothesis, although few would see them as exclusive processes. In support of the migration hypothesis is the apparent rapidity of the spread of sheep from sites in the northern Cape to the southern Cape. The linguistic similarities between the language spoken by the Khoi at the Cape and the people living in northern Botswana are taken as further evidence for the movement of people rather than livestock. "As both Khoi-speakers and the earliest livestock had to have come from the north, parsimony favours the interpretation that they came together" (Sadr 1998:103).

8.4.2.1 Stow and Cooke's demic diffusion routes:

Stow (1905) and Cooke (1965) are early proponents of the migration hypothesis. Both suggested that pre-Iron Age Khoisan groups brought sheep into southern Africa. The route they propose originates in east Africa and moves southwards to Zimbabwe and Zambia and then westwards to Namibia and finally southwards down the western coast of southern Africa to the Cape (the final part of this journey is shown in Figure 8.5). This route was proposed on the basis of oral traditions and the distribution of rock paintings of sheep that are known from Zimbabwe and the south-western Cape (Sealy & Yates 1994). Cooke (1965) argued that the harsh and arid semi-desert environment adjacent to the route would have kept the pastoralist confined to the proposed corridor.

At present the Cooke's western route is favoured as the most likely route for a migration of domestic stock (Webley 1992a, Sealy & Yates 1994, Avery 1995, Bousman 1998). Amongst its proponents are Deacon *et al.* (1978), Robertshaw (1978), and Webley (1997).

8.4.2.2 Elphick's demic diffusion route:

Another proponent of the demic diffusion hypothesis is Elphick who uses linguistic data as evidence for his migration route (Elphick 1977, 1985)(Figure 8.6). Ehret's (1982) research to trace the 'core vocabulary' of dialect chains supports Elphick's route. Other support for this route is found in historic accounts by early European travellers to the Cape, people with stock encountered by Europeans placed their affinities with people in the east not the north (Hall 1987). Hall (1987) favours Elphick and Ehret's route.

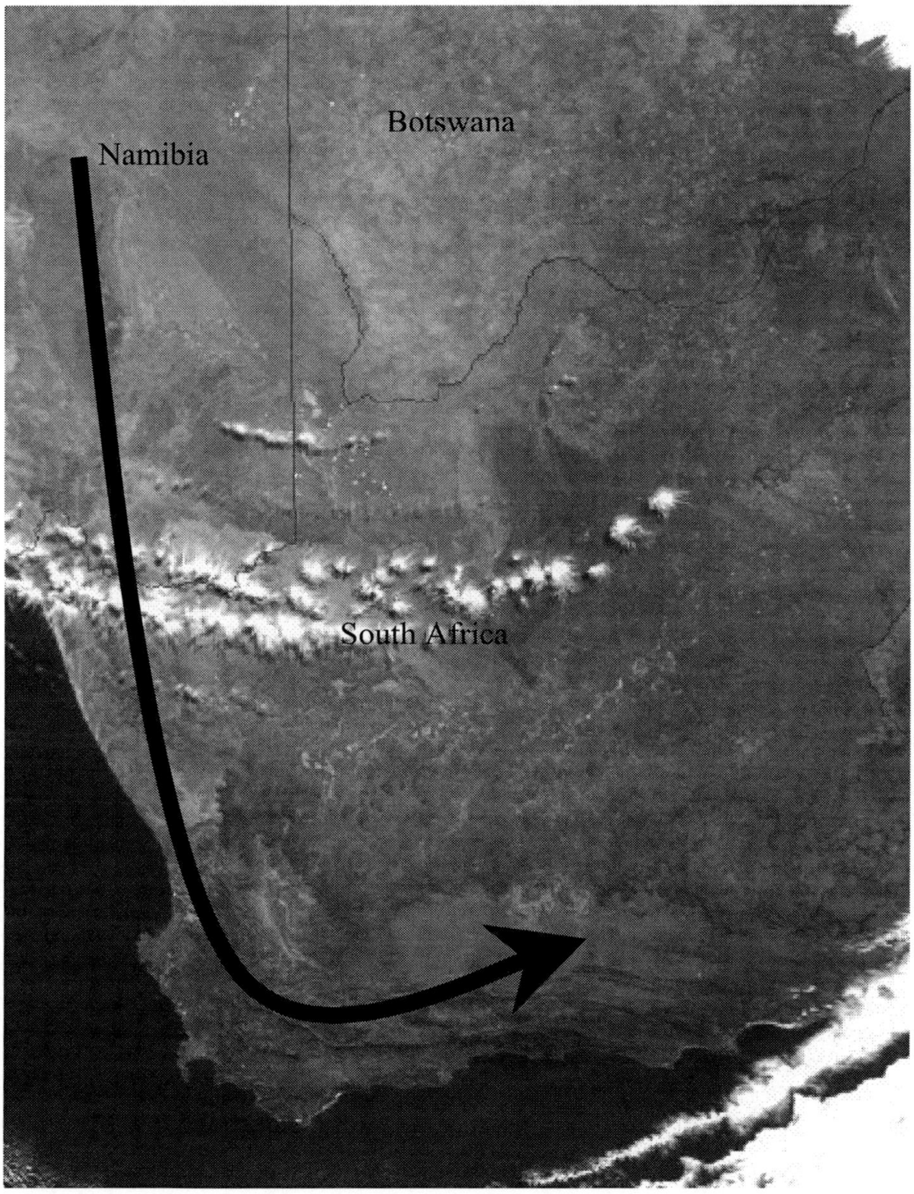

Figure 8.5 Stow and Cooke's routes for the introduction of domestic stock into southern Africa.

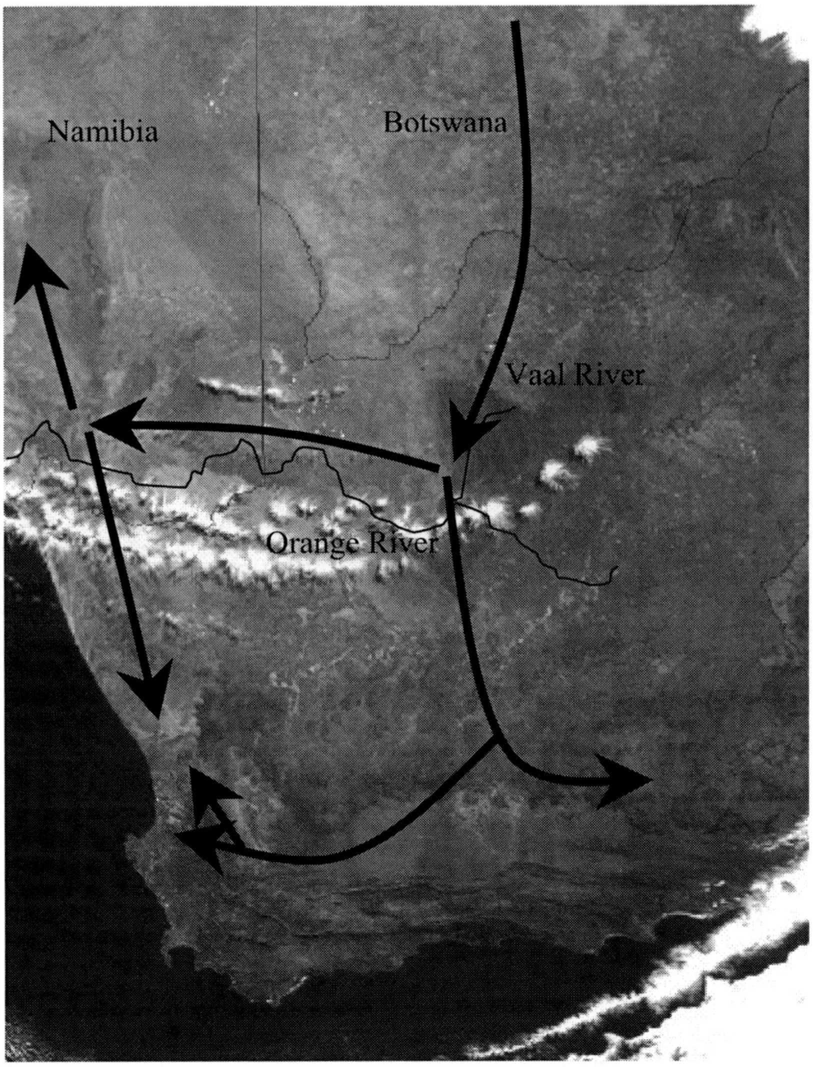

Figure 8.6 Elphick's route for the introduction of domestic stock.

Elphick's route proposes that a pre-Iron Age Khoisan migration took place from northern Botswana southwards to the confluence of the Orange and Vaal rivers where there was a split into two branches, one of which moved westwards along the Orange river into Namibia where it split further, north and south along the coast. The second branch moved southwards through the central Karoo following rivers and the western edge of the interior grasslands. When it reached the south-western Cape it split into two, with one group moving eastwards along the south coast and the other moving north up the west coast (Sealy & Yates 1994, Bousman 1998).

Neither of these routes have been either confirmed or discredited by the archaeological evidence. If one of the routes is indeed correct then the expectation is that the radiocarbon dates for the first occurrence of sheep might show a chronological order along the route with increasing antiquity from the origin southwards. This has not yet been satisfactorily tested. The most recent work is by Bousman (1998) who presents the radiocarbon data associated with the earliest evidence of domestic stock in southern Africa in an attempt to review the competing diffusion hypotheses. Bousman's study is problematic in several ways and it was this that provided the impetus for the analysis of the radiocarbon data (Chapter 9).

8.4.2.3 Bousman's demic diffusion route:

Bousman's study area is larger than the one presented in Chapter 9, it includes both the south eastern spread of the Iron Age farmers and the south western spread of livestock. It includes radiocarbon date covering a greater area from southern Zambian southwards (Figure 8.7)(Bousman 1998). For simplification Bousman's model is reproduced excluding the Iron Age sites and the sites outside of the study area of the research (Figure 8.8).

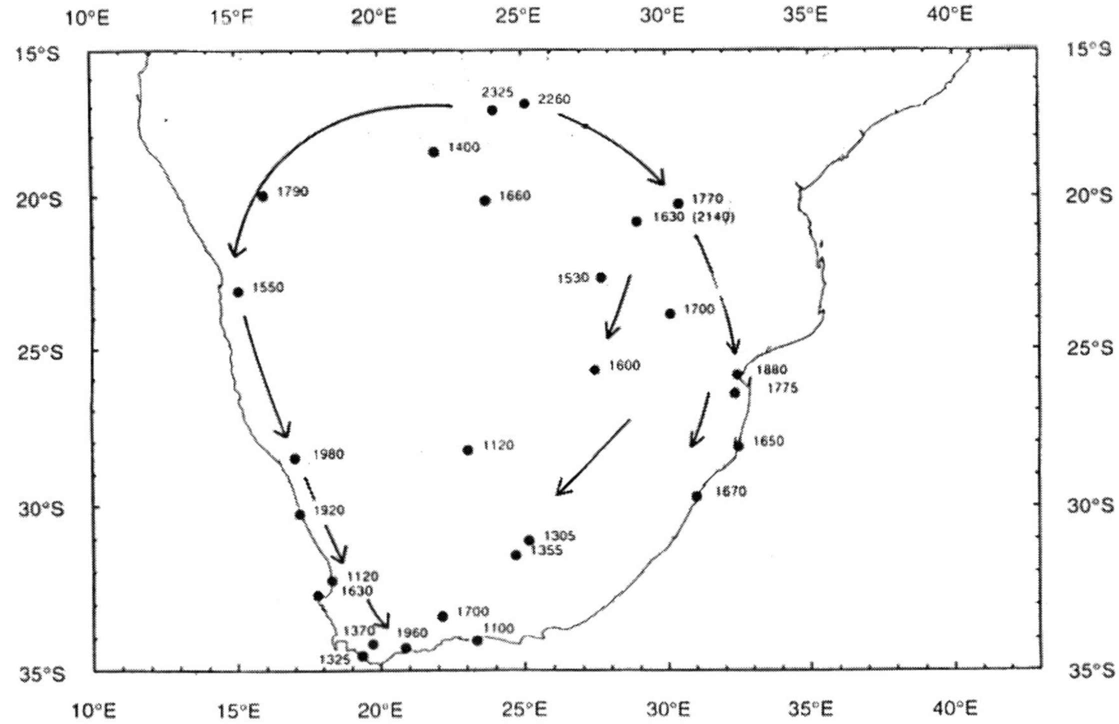

Figure 8.7 Bousman's model for the introduction of domestic stock into southern Africa (Bousman 1998:145, Figure 4). The numbers are the modal values of radiocarbon determinations. The arrows follow the presumed direction of spread. Those to the east associated with Iron Age agropastoralist sites and those to the west with the non-agricultural stock-keeper sites.

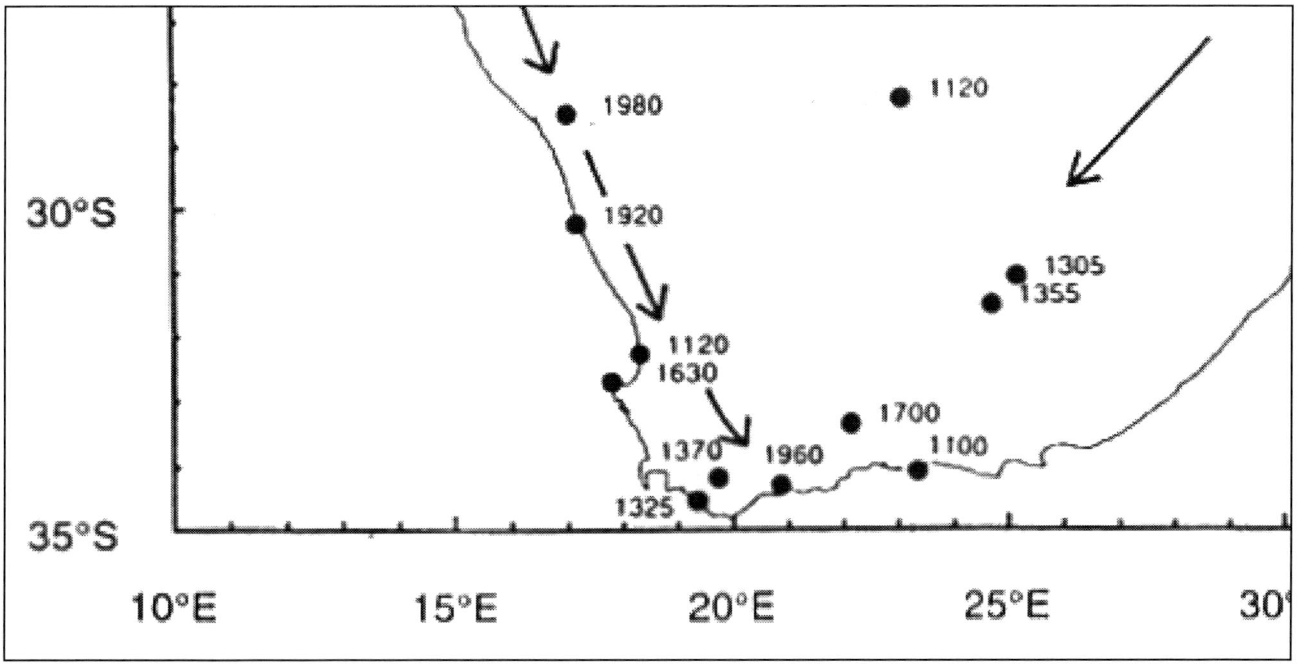

Figure 8.8 The section of Bousman's model that mirrors the study area of this research. The numbers represent the modal value of uncalibrated radiocarbon determinations in years before present for Later Stone Age sites with evidence for domestic stock are shown. The arrows show the suggested direction of spread.

There are 4 criticisms of Bousman' work:

1. He dismisses the use of calibrated dates without providing the evidence in support of this decision. To quote Bousman, "...Tree-ring calibrations do not increase the resolution nor, in fact, change the general patterns observed with the uncalibrated radiocarbon ages" (Bousman 1998:135).
2. He uses the modal value of the radiocarbon determination without its associated error term.
3. He does not explain the arrows shown on the map (Figure 8.8): arrows on the map join dates that have no apparent chronological order.
4. He does not explain his date selection criterion. Sites shown in Figure 8.7 have more than one date associated the first sheep. Bousman does not appear to be selecting the earliest date for the sheep nor is the selection consistently favouring AMS dated sheep bone over charcoal dates or any other obvious parameter (Figure 8.9).

The analysis presented in the following chapter presents a more thorough analysis of the patterns in the radiocarbon dates. The aim was to try to improve on Bousman's model.

The remainder of this chapter contains a discussion of two problems that were encountered during the database compilation:

1. The separation of herder from hunter-gatherer sites.
2. The discrepancies between direct dates on sheep bone and those on carbon-containing samples associated with sheep bones.

8.4.3 Separating herder sites from hunter-gatherer sites:

While compiling the radiocarbon database it became clear that many of the sites that contained sheep bones were not recognised as herder sites by their excavators but rather as hunter-gatherer sites with access to sheep. As Smith (1997: 210) states, " It is important to recognise that at any time when there are domestic animals in the landscape, all people of different economic persuasions have access to them, either by exchange or theft". It was unclear whether ignoring this distinction would have any effect on the spatial analysis of the first occurrence of sheep. It seemed reasonable to suggest that hunter-gatherer sites with sheep would be found in close proximity to those of herder and that the timing of the appearance of sheep in hunter-gatherer sites would be closely coupled to their first appearance on the landscape in herder sites, making the distinction unimportant for this type of analysis. To keep all options open sites were classified as either herder or hunter-gatherer with access to stock. This distinction can be hard to make and the following is a discussion of the different approaches that may be taken.

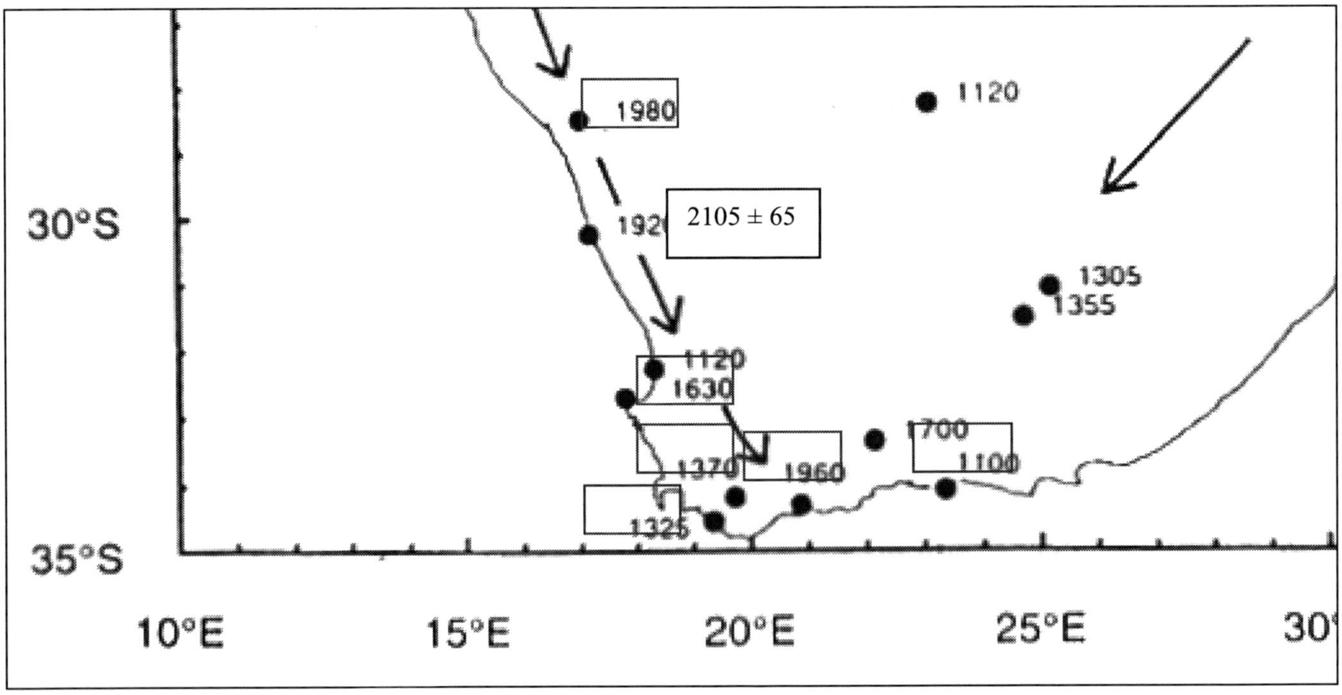

Figure 8.9 Unexplained inconsistencies in Bousman's map. Red boxes ring AMS dates on sheep bone. The blue box shows an instance where the radiocarbon determination on charcoal is mapped (1920) when there is a direct date on sheep bone (2105 ± 65 years bp). Determinations without boxes are conventional radiocarbon dates on charcoal.

Three criteria have been used to separate herder sites and hunter-gatherer-sheep sites:

1. Evidence from trends in the artefact assemblages and the faunal composition:
Smith (1997, 1998a) has identified two contemporaneous yet culturally and economically distinct groups of people living simultaneously within close proximity of each other in the Cape. Two sites typify the distinction. One site, Kasteelberg, a large open-air site in the western Cape is clearly pastoralist whilst the other, Witklip, a small rock shelter 12 kilometres south of Kasteelberg, has a hunter-gatherer signature (Table 8.1).

Witklip exhibits cultural continuity before and after the introduction of sheep and pottery to the site (Smith 1998a) as measured in the continuity in the proportion of formal stone tools in the deposit and the size of the ostrich eggshell bead[4]. This view supports earlier work by Parkington (1984). In a discussion of San hunter-gatherer sites post-2000 BP, Parkington (1984:123) states that, "...the technological traditions of previous millennia are retained and developed in the post herder period and some form of continuity must be assumed. Our current working hypothesis is that the increase in occupied sites and their numbers in areas with low grazing potential indicate a shift in settlement and site choice in the face of competing herders". Smith suggests that the patterns in the assemblages at Witklip and Kasteelberg can be used to classify other sites containing domestic stock.

Unfortunately Smith *et al*'s (1991) criteria are rarely seen as distinctly at other sites, as they are at their type-sites. There are instances of hunter-gatherer sites with no domestic stock and large ostrich eggshell beads: Smith *et al* (1991) interpret this as an indication of the acceptance of 'norms of manufacture' of the herder communities. Similarly hunter-gatherer sites have been found to have a formal tool component that is closer to the herder type-site. This is not a criticism of Smith *et al*'s work but an illustration of the complexity of the problem. By clearly listing a set of criteria Smith *et al* (1991) have provided archaeologists with a checklist for future excavations and a common framework from which the debate can grow.

Webley (1992,1997) excavated a number of sites in the northern Cape in order to test whether there is in fact archaeological evidence for two separate economic groups at the time when domestic stock is first introduced to South Africa. She concludes, "Sites in Namaqualand occupied during the period after 1000 BP exhibit a great deal of variability and cannot easily be characterised as either hunter-gatherer or herder using Smith *et al*'s (1991) criteria" (Webley 1997:6).

In an attempt to separate herder and hunter-gatherer sites in this database, Smith's criterion for faunal assemblage composition was tested at 20 sites that contain sheep bone. Smith (1997) suggested that sites with a high percentage of sheep bone (measured at the herder type-site of Kasteelberg as 23.4%) and a low percentage of hunted *Raphicerus sp.*[5] bone (measured as 5.8 % at Kasteelberg) were herder sites. Similarly sites with a high percentage of *Raphicerus sp.* bone (37% of the total assemblage at Witklip) and low numbers of domestic sheep (percentage unspecified) were hunter-gatherer sites.

The number of sheep and *Raphicerus sp.* at each site was calculated as the percentage MNI of the total faunal assemblage. The total faunal assemblage used is the total MNI of all the species present excluding those species that were classified as too small to be of a comparable body size to the sheep and *Raphicerus sp.* Those excluded are mole rats, hyraxes, badgers, mongoose, tortoise, porcupine and hares. These calculations were made (a) at the time when sheep first occur at the site and (b) when they were at their peak. The results were as follows:

Faunal composition at the first appearance of sheep:

1. At the first occurrence of sheep, ten of the twenty sites had a percentage MNI of sheep > 23.4%, thus fitting them to Smith's herder category (Figure 8.10).
   However when the percentage MNI of *Raphicerus sp.* at each of these sites is analysed for the same time period, 13 sites fit the herder criteria (<37% *Raphicerus sp.*) (Figure 8.11). Eight sites fit both categories depending on whether the sheep or the wild fauna are being analysed.

2. Faunal composition when sheep numbers reach their maximum in the faunal assemblage at each site:
   When sheep are at their peak 11 of the 20 sites fit Smith's herder criterion based on the count of sheep (Figure 8.12). Four have no peak (they are single occupation sites, or sheep only occur once in the stratigraphy i.e. sites 2, 16, 17 and 18). When the percentage of wild fauna is analysed for this period then eight sites are found to fit both categories (Figure 8.13).

---

[4] Analysis of the size of ostrich eggshell beads has shown that they can be used as a stylistic marker (Jacobson 1987, Smith & Jacobson 1995). Large beads are associated with sites that have high proportions of domestic stock (i.e. pastoralist sites) whilst smaller beads are markers of hunter-gatherer occupation (Smith *et al.* 1991).

[5] *Raphicerus sp.* fit into the category, by body size, of small bovid (examples are the steenbok and the grysbok). For comparison sheep are in the category small-medium bovid (Klein & Cruz-Uribe 1989)

Herder Site Type (e.g. Kasteelberg)	Hunter-gatherer Site Type (e.g. Witklip)
A high M.N.I.[6] sheep bone count	A low M.N.I sheep bone count
Large ostrich egg shell beads	Small ostrich egg shell beads
Large quantities of pottery	Low quantities of pottery
Negligible numbers of formal retouched stone tools	High numbers of formal retouched stone tools
Low count of antelope	High count of antelope

Table 8.1 Smith's criteria for the identifying herder and hunter-gatherer sites.

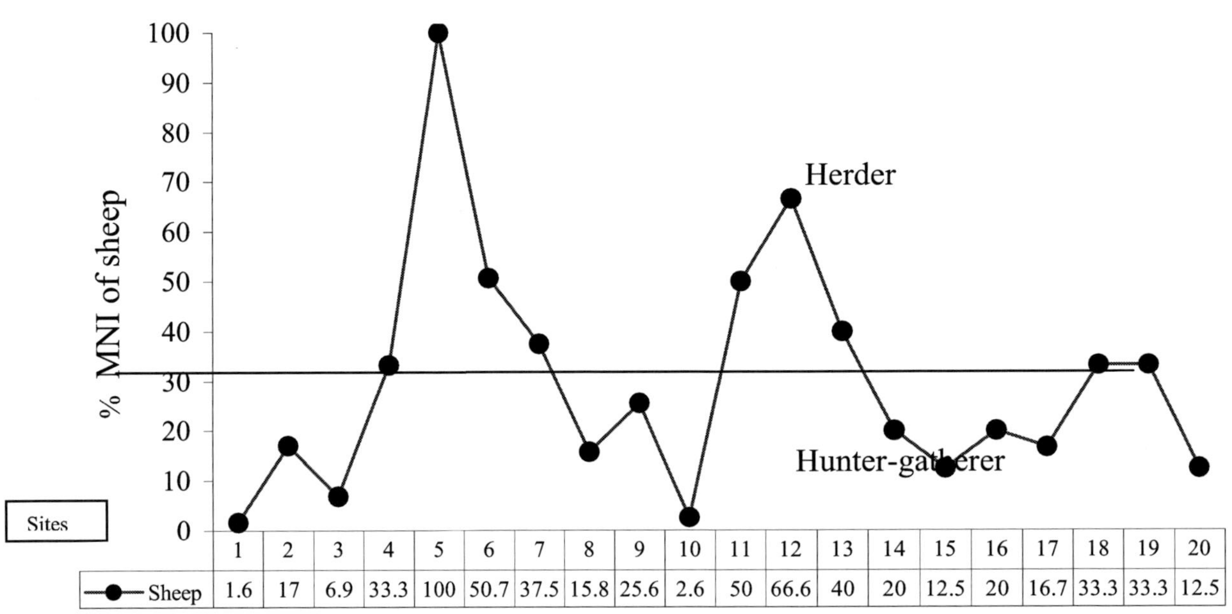

Figure 8.10 Graph to show the sites that fit Smith's herder category (>23.4% sheep) at the first occurrence of sheep for 20 archaeological sites in South Africa. The horizontal line is the 23.4% mark; those above are herder and those below hunter-gatherer. Sheep is given as the percentage MNI of the total faunal assemblage.

[6] Minimum number of individuals

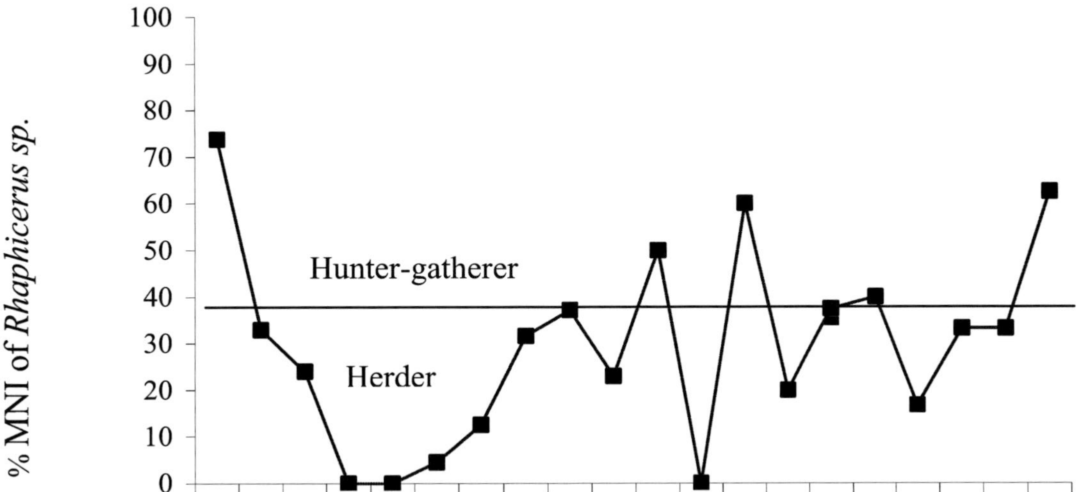

Sites	1	2	3	4	5	6	7	8	9	10	11	12	13	14	15	16	17	18	19	20
Raphicerus sp.	74	33	24	0	0	5	13	32	37	23	50	0	60	20	38	40	17	33	33	63

Figure 8.11 Graph to show sites that have a wild faunal component matching Smith's criteria for classification as hunter-gatherer or herder when sheep first appear at the sites. The horizontal line is 37% marker, above which sites are hunter-gatherer, below which sites are herder.

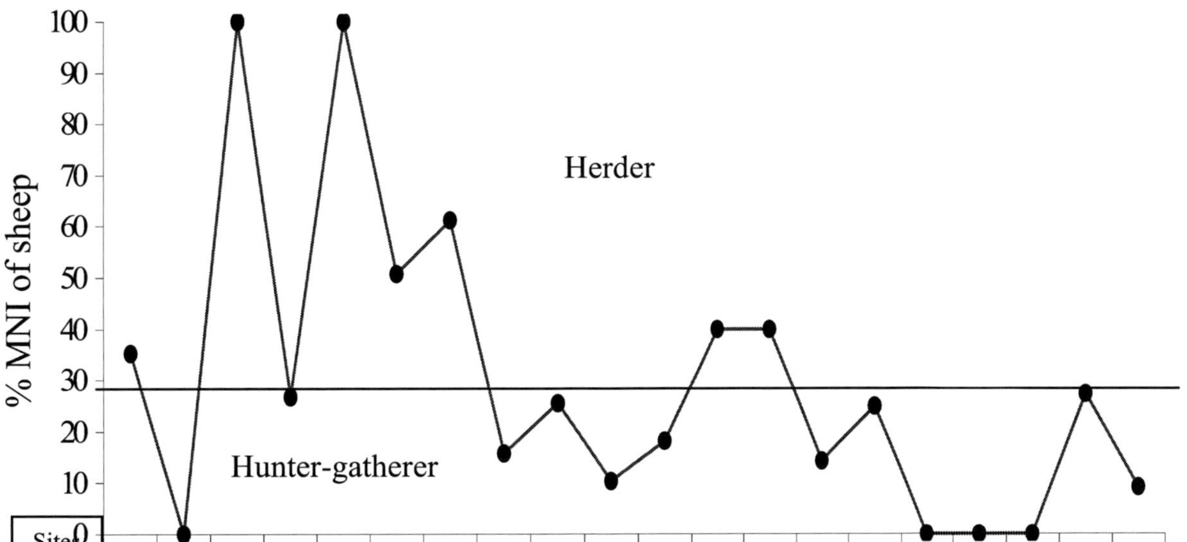

Sites	1	2	3	4	5	6	7	8	9	10	11	12	13	14	15	16	17	18	19	20
Sheep	35.3	0	100	26.7	100	50.7	61.3	15.8	25.6	10.3	18.2	40	40	14.3	25	0	0	0	27.3	9.1

Figure 8.12 Sites that fit Smith's criterion based on the count of sheep at the site when sheep are at a peak. The horizontal line is the 23.4% threshold between the two classifications.

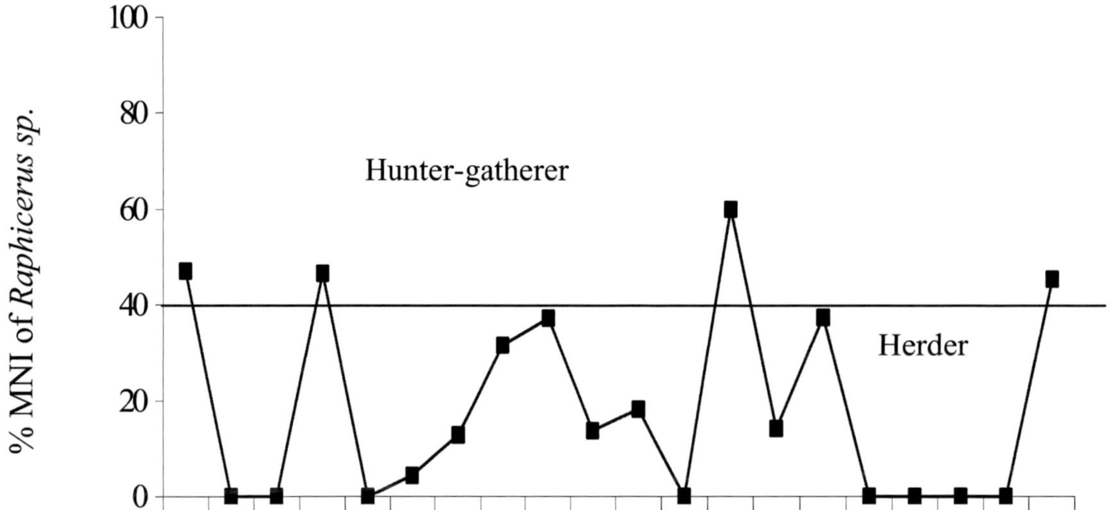

	1	2	3	4	5	6	7	8	9	10	11	12	13	14	15	16	17	18	19	20
Raphicerus sp.	47	0	0	47	0	4.5	13	32	37	14	18	0	60	14	38	0	0	0	0	45

Figure 8.13 Sites that fit Smith's criterion based on the count of Raphicerus when sheep are at a peak. Sites above the 37% threshold (the horizontal line) are hunter-gatherer, those below the threshold are herder.

This exercise shows the practical difficulty of using Smith's apparently straightforward faunal criteria for distinguishing between herders and hunters. The results are inconclusive. This does not mean that the percentage faunal contribution of species is not important but rather that their ratio at each site must be considered individually. The sample size and the location of the site can influence the results. Small samples falsely amplify the percentage calculations. At certain sites the hunted fauna is made up of species other than *Raphicerus sp.* Webley (1997) suggests that at the site of Spoegrivier although domestic stock does not make up a great percentage of the total faunal assemblage this cannot be taken as an indication of the lack of economic and social importance that domestic stock might have had to the site's inhabitants. As Klein (1986:10) states, "It is tempting to conclude that hunter-gatherers occupied a site when domestic bones are rare…even in Iron Age sites, where context argues against occupation by hunter-gatherers, domestic animals can be substantially less numerous then indigenous ones". The rarity of domestic stock does not rule out herder occupation but the abundance of stock suggests it, especially abundance through successive levels.

It was following this analysis that the decision was made to use the presence of sheep at a site, irrespective of their numbers, as the criterion for including sites in this analysis. The classification of sites as either herder or hunter-gatherer with stock access is made using a combination of the viewpoints of the archaeologists in published reports and the presence and increase of sheep bones in the faunal assemblage through a number of stratigraphic layers.

2. Evidence from dung accumulations:
Large accumulations of dung in the same area that show its repetitive use, are taken as evidence of the presence of stock-keepers. Webley (1997) and Deacon (1978) have both identified Boomplaas as a herder site on such evidence. "That the Boomplaas people were also herders is suggested by the fact that the matrix enclosing sheep bones is largely calcined sheep dung. This means that the human inhabitants were kraaling sheep in the cave" (Webley 1997:62).

3. Evidence from age profiles of fauna:
Maggs & Whitelaw (1991) argue that the presence of domestic animal bones is insufficient evidence for pastoralism; that more important is the age profile of the animals at the time of slaughter. Klein & Cruz-Uribe (1989) suggest that a herder assemblage will reflect the selective culling of male lambs and older female sheep that are past their productive prime; in order to classify a site as herder the assemblage must consist of the bones of these two groups. This method was used by Klein, "The age profiles of the sheep at Die Kelders could be determined from the jaws and frontlets and the conclusion was that they represented selective culling and were herders not thieves" (Klein1978: 68). In practice this method of identifying herder sites is hindered by poor bone preservation and inconsistencies in the

preservation of particular bone parts (Klein & Cruz-Uribe 1984,1989). Male sheep might be more readily identifiable at a site, not because they were culled in greater numbers but because the male frontlets used in sex determination are stronger and thus more likely to be preserved than the same cranial bone in females (*ibid*). At the overwhelming majority of southern African sites with domestic sheep, both the sample size and poor preservation make it difficult to deduce much about the age or sex profile of the original stock population.

It seems unwise to exclude the possibility that a site is herder because of either an indeterminable age/sex profile or a paucity of domestic stock remains. The faunal profiles suggested by Klein & Cruz-Uribe (1984,1989) assumes a very specific method and reason for stock keeping, namely, that animals were being kept for maximum meat and milk returns, a purpose which contemporary ethnography suggests is rare amongst herders in Africa (Herskovits 1926, Deng 1973).

8.4.4 Discussion of the discrepancies between the results of AMS dates on sheep bone and conventional dates on charcoal associated with the first sheep:

The radiocarbon dates in this database have overwhelmingly been made on charcoal. So in the majority of cases the date of the first occurrence of sheep in the archaeological stratigraphy is only as good as the association of the charcoal (and its age) with the sheep bones (Klein 1986). The development of AMS dating has facilitated the direct dating of small fragments of sheep bone. Sealy & Yates (1994, 1996) have compared these dates with conventional dates on charcoal associated with sheep bone. They found that in all but one of the seven cases the direct date on the sheep bone was significantly younger than that of the stratigraphy with which it was associated (Table 8.2).

Sealy & Yates (1994, 1996) attribute this to the post-depositional movement of artefacts within archaeological deposits. Humphreys (1974) had anticipated their results from his own excavations twenty years earlier. He cautioned on the reliability of his own findings: "final confirmation must however await further work at the site, because in a context like Dikbosch 1 where the deposits accumulated very slowly the possibility of slight but significant downward movement of a single tooth cannot be ruled out" (*ibid*: 117). A similar caution has been expressed by others at sites where there has been no AMS dating, but that show very early dates associated with sheep bones. Most archaeologists have interpreted such early dates as reflecting the post-depositional movement of the deposit rather than the early occurrence of sheep. For instance, at Tortoise Cave, "The presence of sheep, *Ovis aries,* in the pre-pottery layers is attributed to disturbance & vertical movement" (Robey 1987:319). A similar explanation is given for early sheep dates at Steenbokfontein (Jerardino & Yates 1996). This sort of reassessment has made it clear that the direct dating of early domestic stock remains is essential to the understanding of the chronology of the introduction of stock into South Africa.

Site	AMS Date* (bone) (bp)	Radiocarbon Date (charcoal)(bp)
Spoegrivier	2105 ± 65	1920 ± 40
Kasteelberg	1630 ± 60 1430 ± 55	1860 ± 60 1790 ± 60
Nelson bay cave	1100 ± 8	1930 ± 60
Byneskranskop	1370 ± 60	1880 ± 50
Die Kelders	1325 ± 60 1290 ± 60	1960 ± 85 1960 ± 95
Wonderwerk	460 ± 120	1890 ± 50
Jubilee Cave	310 ± 50	1840 ± 50

Table 8.2 Comparison of AMS dates on sheep bone to conventional radiocarbon dates on charcoal associated with sheep bones (Sealy & Yates 1994,1996).

8.5 Summary:

The different views as to how sheep may have spread in to southern Africa have been outlined. Possible routes of introduction have been proposed using linguistic evidence, the distribution of rock paintings of sheep, oral traditions and more recently the radiocarbon evidence but there is yet no confirmation that any of the routes are correct.

The aim of this exercise was to gather radiocarbon determinations and geographic coordinates for all sites containing the earliest evidence of sheep in what is now South Africa and to apply statistical and spatial analysis to see whether either of the proposed routes could be confirmed or refuted. Whilst compiling the dataset problems were encountered in site classification (this mirrors the problem encountered in the Europe concerning the criteria by which a site can be labelled Neolithic) and this led to the faunal analysis that was described in this chapter. The other problem encountered was that recent direct dates on the sheep bone give more recent dates to their first appearance than the traditional conventional radiocarbon dating by association.
In the next chapter an attempt to find the routes of the introduction of the sheep, in the light of the radiocarbon evidence, is presented.

Chapter 9: PART ONE - Spatial and statistical analysis of the southern African sheep database

9.1 Introduction

In this chapter, the statistical method of handling calibrated dates in linear regression analysis and the technique to find the most probable origin for a set of dates, as developed in Chapters 5 to 7, are applied to the dataset for the first appearance of sheep in the western half of what is now South Africa. The analysis is divided into two parts. The first explores the dataset by seeing the effect of excluding and including certain subsets of sites and dates in the analysis. The second attempts to test the models of Stow (1905), Cooke (1965), Elphick (1977) and Bousman (1998) for the route of the first introduction of sheep by spatial analysis of the radiocarbon dates.

9.2 The sites and the grading system:

Twenty-seven sites (72 determinations) are used in this analysis (Appendix 9). Sites were graded 1-3 to track the effect of different dating methods, as follows:

Grade 1: A direct AMS on sheep bone.
Grade 2: A series of consistent charcoal determinations in good stratigraphic association with sheep bone.
Grade 3: A single charcoal determination or multiple discrepant determinations in poor association with sheep bone.

Sites were also separated into three categories on the basis of the strength of evidence as to their proper classification, as:

1. Herder sites (Figure 9.1).

2. Hunter-gatherer sites with access to sheep (referred to as non-herder from this point on) (Figure 9.2).

3. Sites that could not be confidently classified as either herder or non-herder (Figure 9.3).

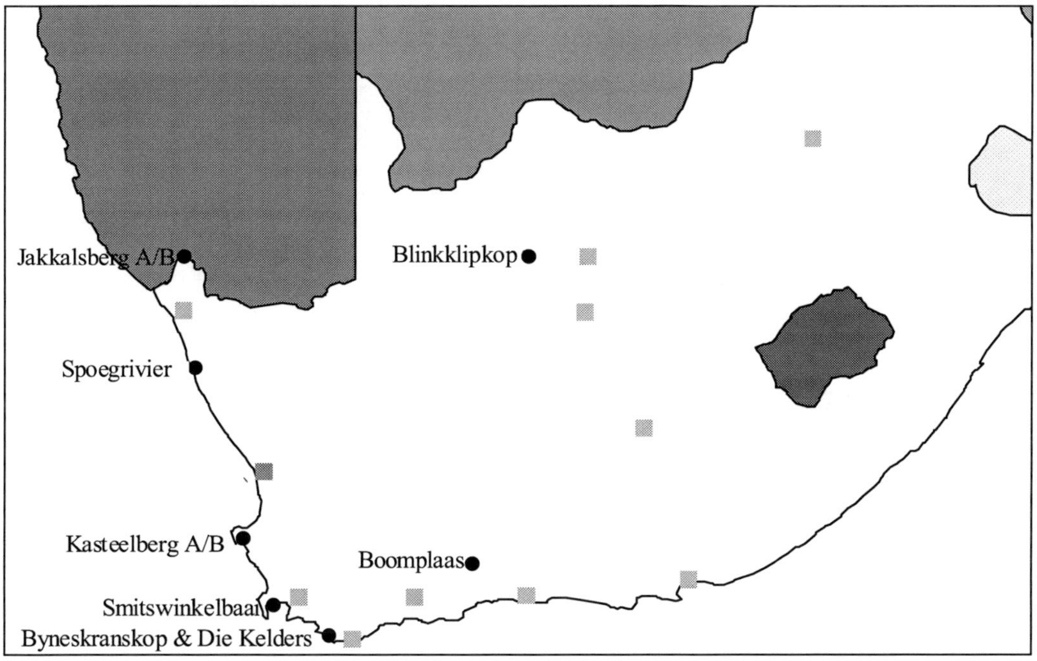

Figure 9.1 Sites that are classified as herder are shown with their names and a black circle. The grey squares show the position of non-herder and unclassified sites.

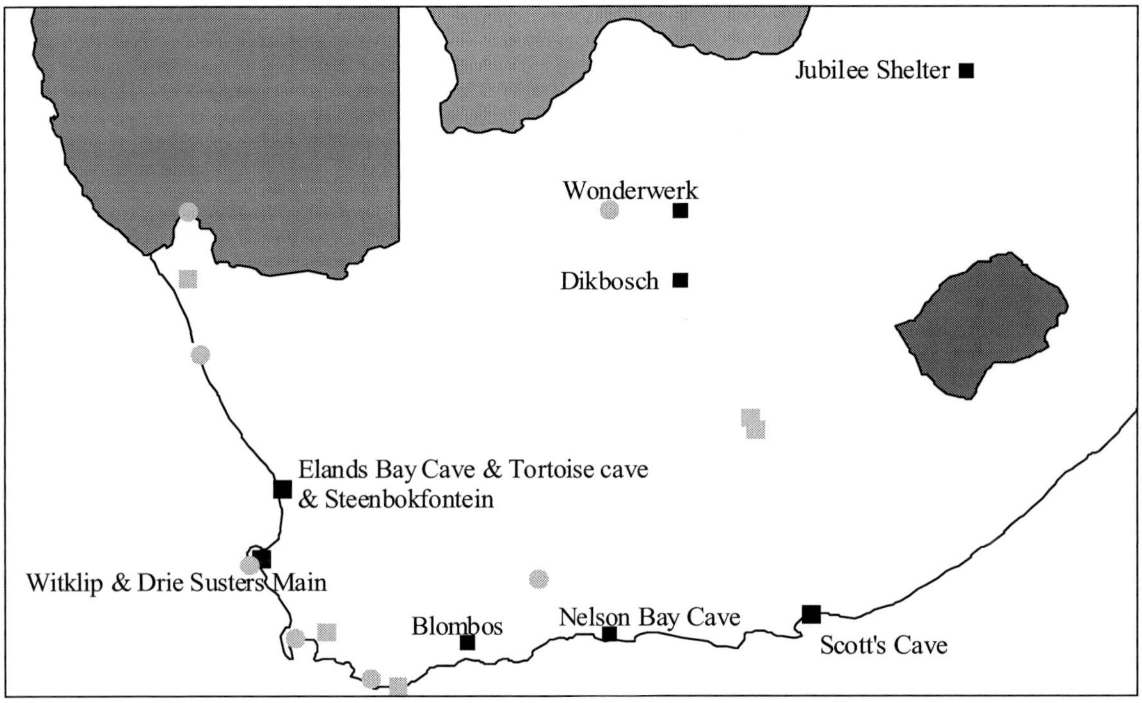

Figure 9.2 Sites that are classified as hunter-gather sites with access to stock (non-herder sites) are shown with their names and a black square. Grey squares show the location of unclassified sites and grey circles show the location of herder sites.

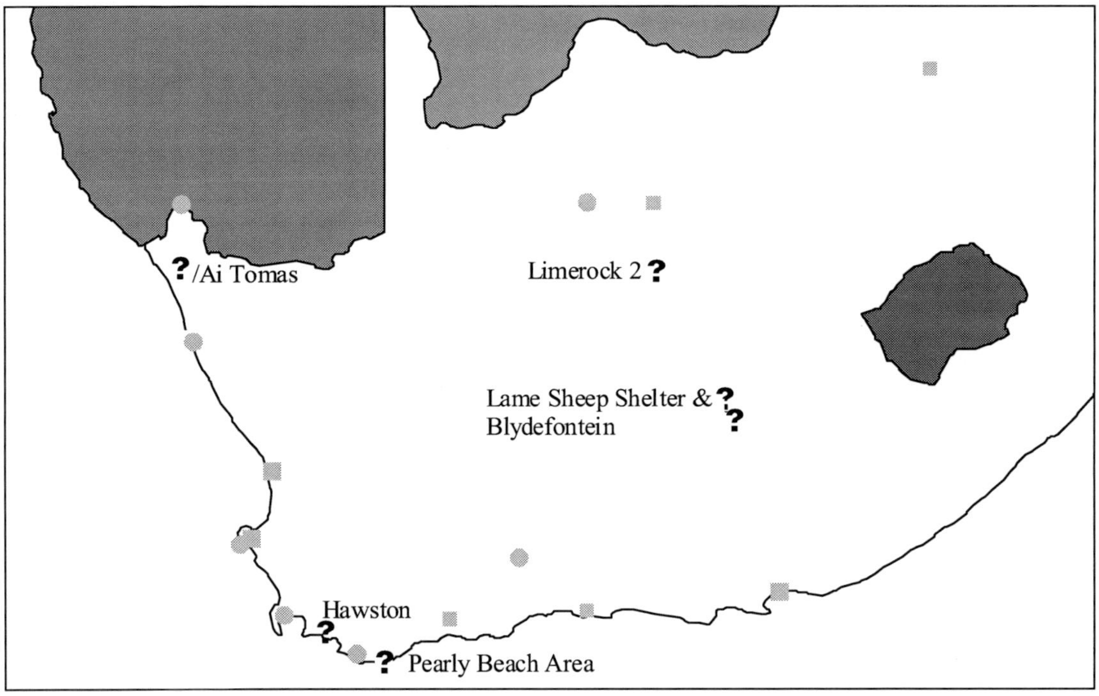

Figure 9.3 Sites that could not be confidently classified as either herder or non-herder sites are shown by their names and a question mark. The position of herder sites (grey circles) and non-herder sites (grey squares) are also shown.

9.3 Analysis 1: To see the effect of excluding and including certain subsets of sites and dates on the analysis.

The database was divided into 9 groups for analysis based on the site classification and the dating method.

Group 1: Herder sites, grade 1.
Group 2: Herder and non-herder sites, grade 1.
Group 3: Herder sites, grade 1 and 2.
Group 4: Herder sites, grade 2 and 3.
Group 5: Herder and non-herder sites, grades 1 and 2.
Group 6: All herder sites.
Group 7: All non-herder sites.
Group 8: Group 6 and Group 7.
Group 9: Group 6 and Group 7 and unclassifiable sites.

In all groups at sites that are dated by both methods, direct AMS determinations take precedence over conventional radiocarbon determinations.
For each group the best-fit origin was found using the method outlined in Chapter 4. The parameters were adjusted; the potential origin points were tested at one-degree intervals southwards from latitude 25°S to 35°S and from longitude 15°W to 33°W. The resultant grid contains 200 potential origin points (Figure 9.4).

Rather than selecting the oldest date at each site or an average of the dates, the 'first' event was calculated at each site. This gives a probabilistic estimate of when the sheep might have first appeared at a site based on all the radiocarbon determinations that are associated with the earliest sheep at a site (Chapter 7). Because of the discrepancy between direct dates on sheep bone and conventional dates on charcoal associated with sheep, as discussed (Chapter 8), AMS dates were separated from conventional radiocarbon dates at each site: the first event was calculated separately for each of these subdivisions. As with analysis on the European dataset, the origin point that gave the best fit to the data was used to calculate the rate of spread and the time at the origin. The results were as follows.

9.3.1 Results for Group 1

This consists of the four herder sites with direct AMS determinations on sheep bone. The origin point with the best fit is 25°E, 28°S (Figure 9.5).

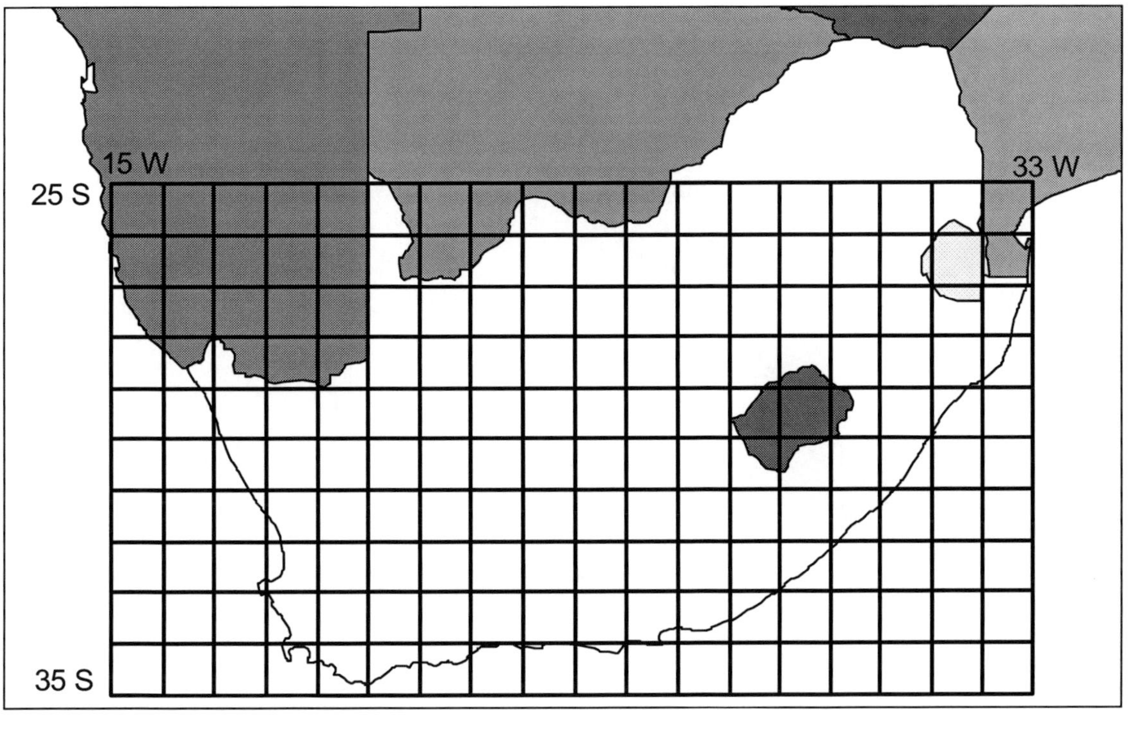

Figure 9.4 The position of 200 potential origin points at each vertex of latitude and longitude at one decimal degree intervals west and south.

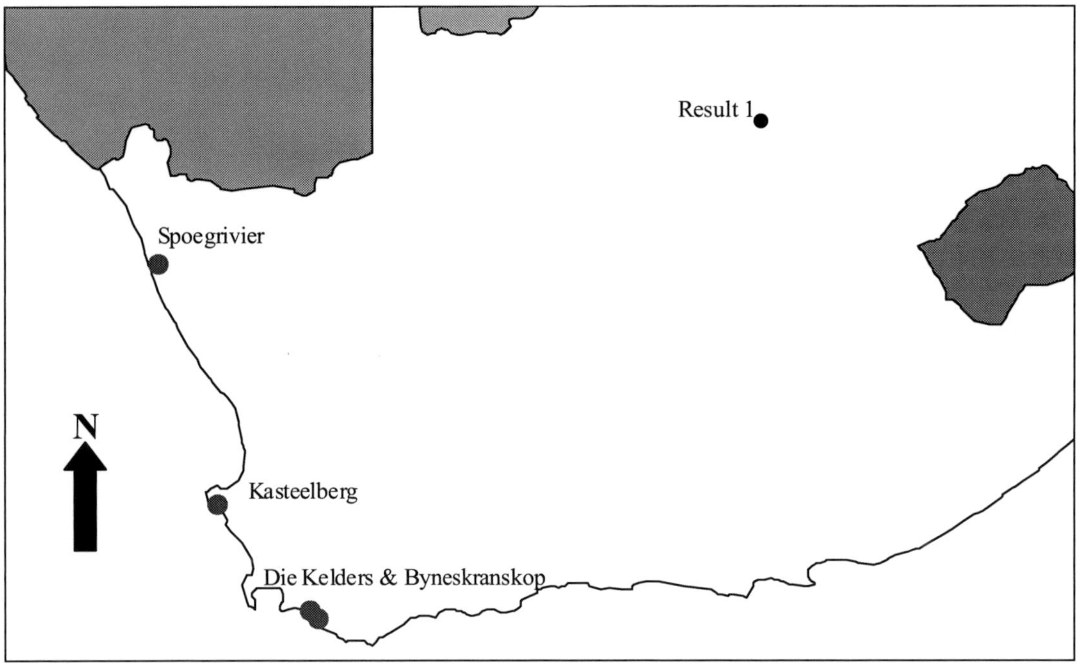

Figure 9.5 Group 1: Herder sites with a direct AMS date on sheep bone. Result 1 is the origin point with the strongest correlation coefficient.

With such a small sample size the most that can be said of the results is that they show the most probable area of origin to be in the north east of the searched area. This trend is clearly shown in Figure 9.6.

From Figure 9.6 it is clear that correlation coefficients are high and negative for half of the searched area and high and positive for the other half of the searched area. This is illustrated in Figure 9.7.

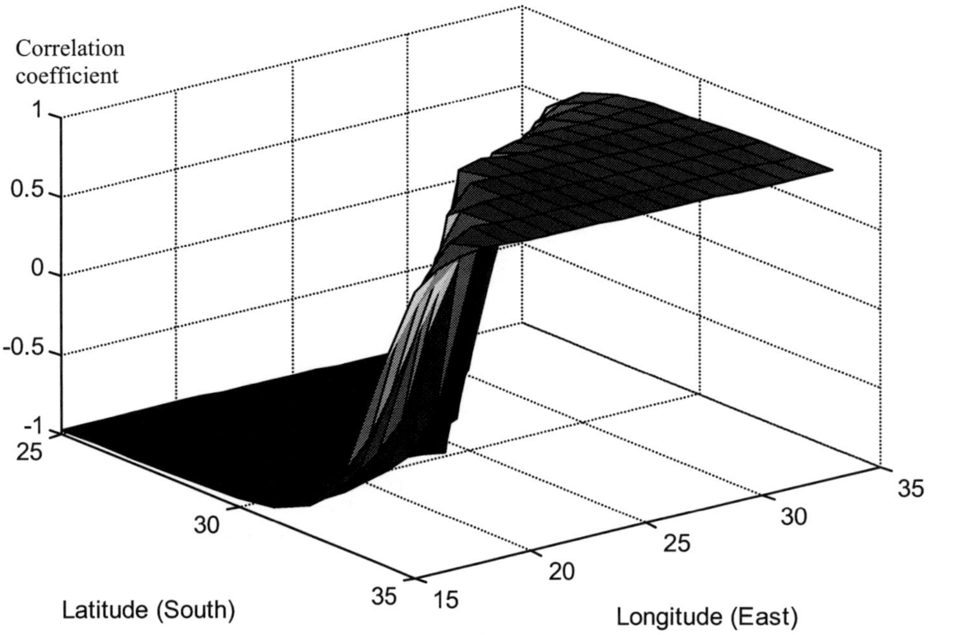

Figure 9.6 Correlation coefficients calculated from each of 200 potential origin points to the sites in Group 1. The dark flat area (bottom) shows the area of most likely origin for Group 1.

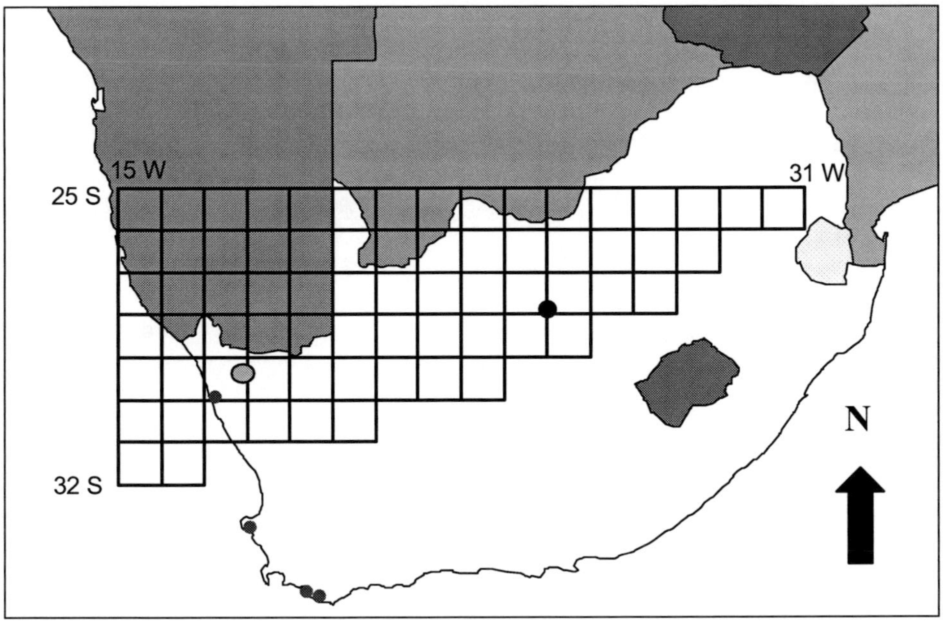

Figure 9.7 The grid shows all origin points for which there is a strong negative correlation (>-0.9), on the other side of a imaginary diagonal line running south-west to north-east all the correlation coefficients are strong and positive. The black circle (25°E, 28°S) shows the origin point that gave the best correlation coefficient. The grey circle shows the origin point at 17°E, 28°S that was also analysed.

The plot of date and distance from the best-fit origin is shown in Figure 9.8. The correlation coefficient is –0.99 and the equation of the regression line is:

$$y_{date} = (-5.85 \pm 0.12)\, x_{dist} + (9125 \pm 158)$$

The predicted departure time from the origin is approximately 9000 years calBP. This result is far too old and cannot be reconciled with the archaeological evidence that the spread of sheep occurred within the last 2000 years.

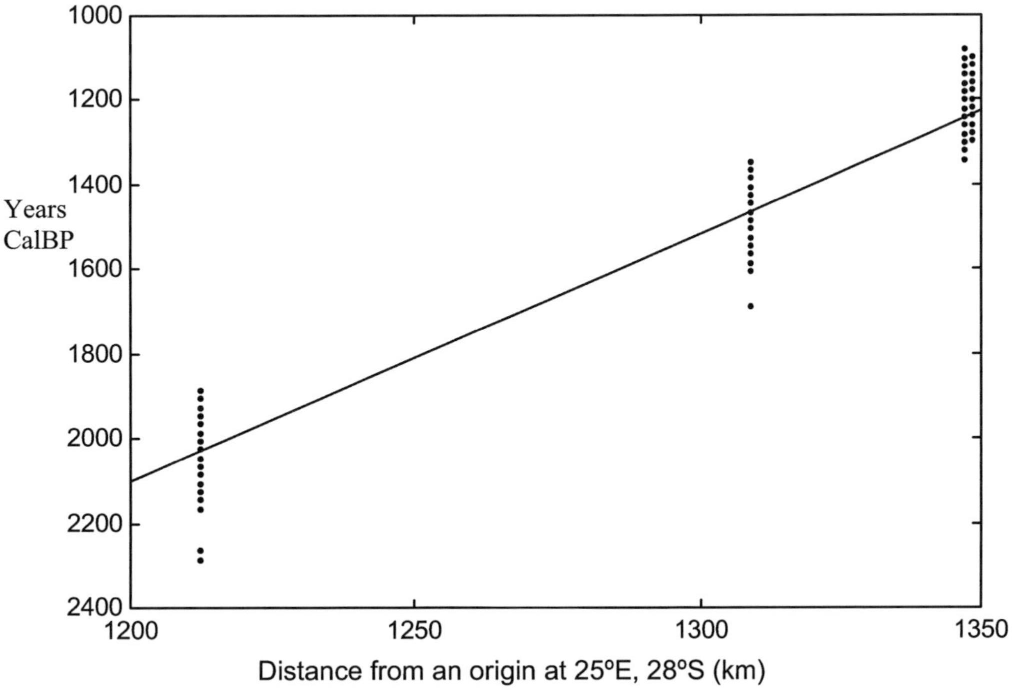

Figure 9.8 Linear regression of weighted calibrated probability distribution of date and distance for Group 1.

Because the correlation coefficients are strong over a wide area, another origin point was chosen to fit in with current archaeological evidence (Webley 1997) that is interpreted as pointing to an origin north towards Namibia. The regression analysis was rerun using an origin at 17°E, 28°S (Figure 9.7). The correlation coefficient is 0.97 and the regression line is:

$Y_{date} = (-1.56 \pm 0.05) \times dist + (2410 \pm 33)$

This gives a predicted departure time from the origin at approximately 2400 calBP and a predicted rate of spread of 0.62 km/year to 0.66 km/year. This result is far more plausible archaeologically. Whilst there is no archaeologically known site of this antiquity at this origin point, sites to the north in Namibia (e.g. Geduld) date to around 1880 calBP (Smith and Jacobson 1995) whilst sites to the south of this origin have varying antiquity. Spoegrivier is of greatest antiquity with a date with a 2-sigma range of 2310 – 1870 calBP. This illustrates that whilst the origin with the best-fit to the data might not give a result that is plausible archaeologically, the general region of origin to which the model points provides an area within which to search for an origin point more in line with archaeological evidence.

9.3.2 Results for Group 2.

This group includes both herder and non-herder sites with direct AMS determinations on bone. There are 6 sites in this group (Figure 9.9).

The origin point with the strongest correlation coefficient is at 21°E, 31°S. The spread of correlation coefficients throughout the searched area is shown in Figure 9.10.

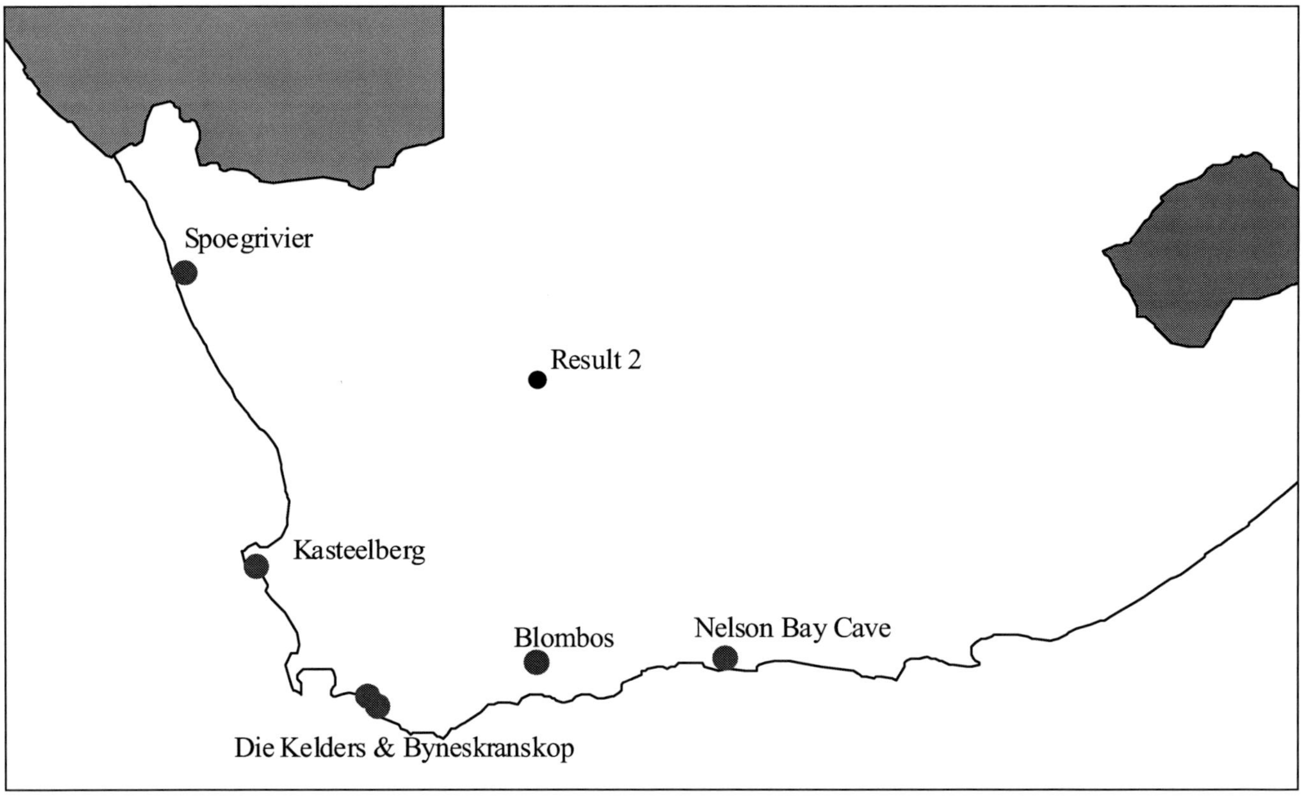

Figure 9.9 Group 2: herder and non-herder sites with direct AMS dates on sheep bone. Result 2 shows the most probable origin point for this group.

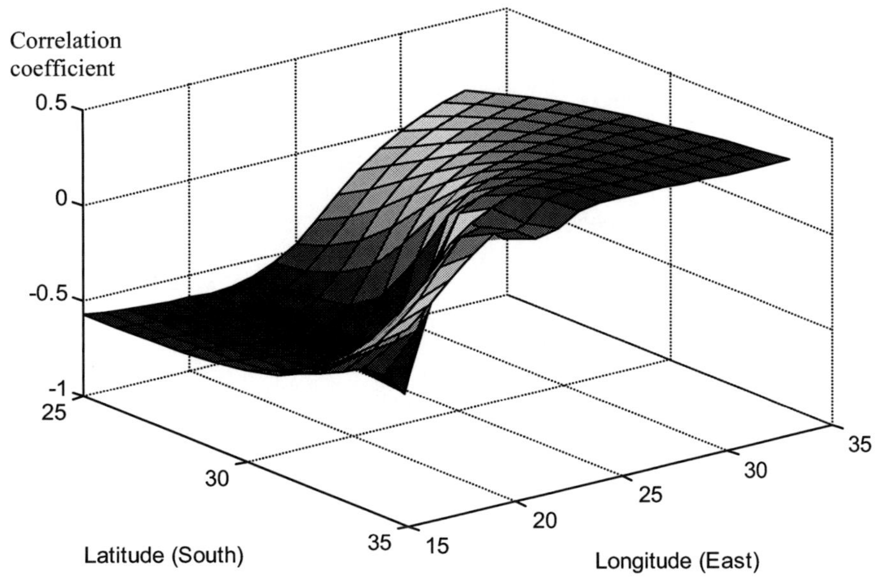

Figure 9.10 Correlation coefficients for the most probable origin for Group 2 are shown in the darkest area (bottom).

The trend in the data points once again to the north-west for the most likely origin. The plot of date (calBP) against distance from the origin with the strongest correlation coefficient (r = 0.74) is shown in Figure 9.11 below. The equation of the regression line is:

$$y_{date} = (-8.55 \pm 0.84)x_{dist} + (4828 \pm 332)$$

This gives a predicted departure date from the origin of more than 4000 years calBP. As with Group 1 the antiquity of the date does not fit in with current archaeological evidence. Once again the analysis was re-run using an origin point that is favoured by most archaeologists (Webley 1997). Using an origin at 17ºE, 28ºS gave a correlation coefficient of –0.59. The equation of the regression line is:

$$y_{date} = (-1.09 \pm 0.16)x_{dist} + (2222 \pm 116)$$

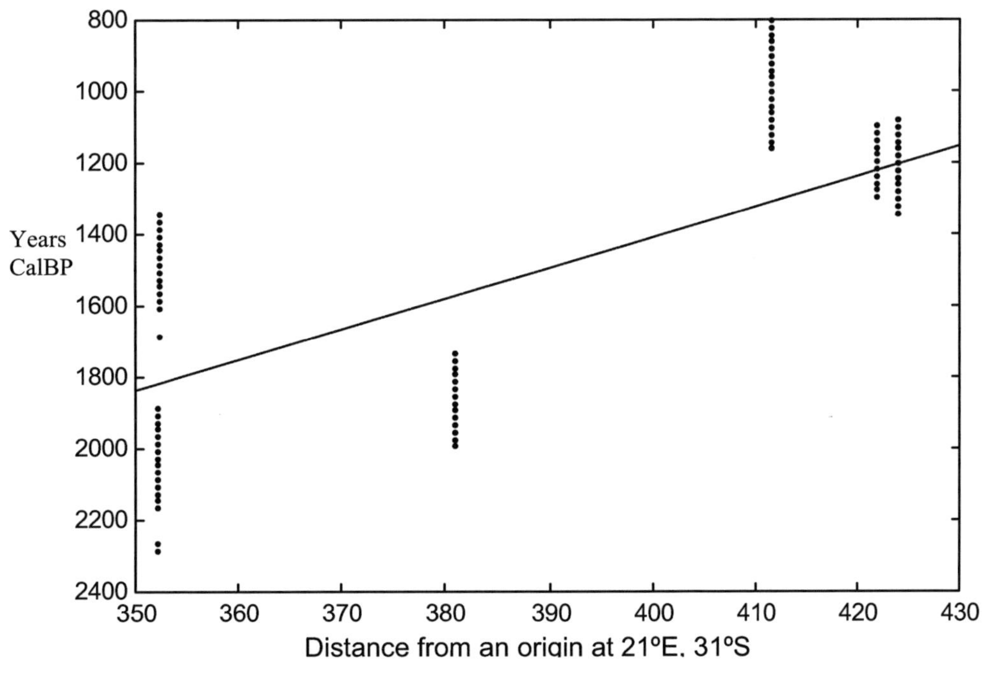

Figure 9.11 Linear regression of weighted calibrated probability distribution of date with distance from the origin for Group 2.

This gives a predicted rate of spread in the range 0.8 to 1.1 km/year and a predicted time of departure from this origin in the time range 2110 to 2340 calBP. This would correlate well with the archaeological evidence from a site just south of this origin at Spoegrivier which has a 2-sigma calibrated date range of 2310 – 1870 calBP.

9.3.3 Results for Group 3.

This group contains only herder sites with reliable radiocarbon dates. The sites and the best-fit origin are shown in Figure 9.12 below. The most probable origin point for Group 3 is found at 19°E, 31°S with a correlation coefficient of –0.63. The correlation coefficients for the whole origin search area are shown in Figure 9.13.

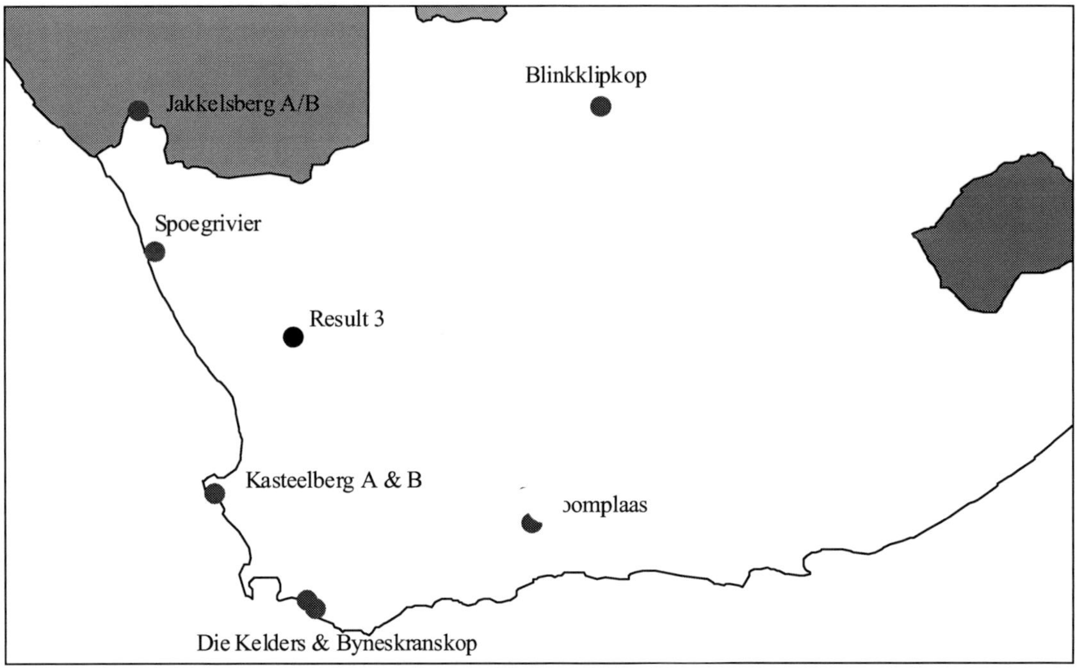

Figure 9.12 Group 3: herder sites with either AMS dates on bone or reliable conventional radiocarbon dates on charcoal. Result 3 shows the origin point that showed the highest correlation to the group.

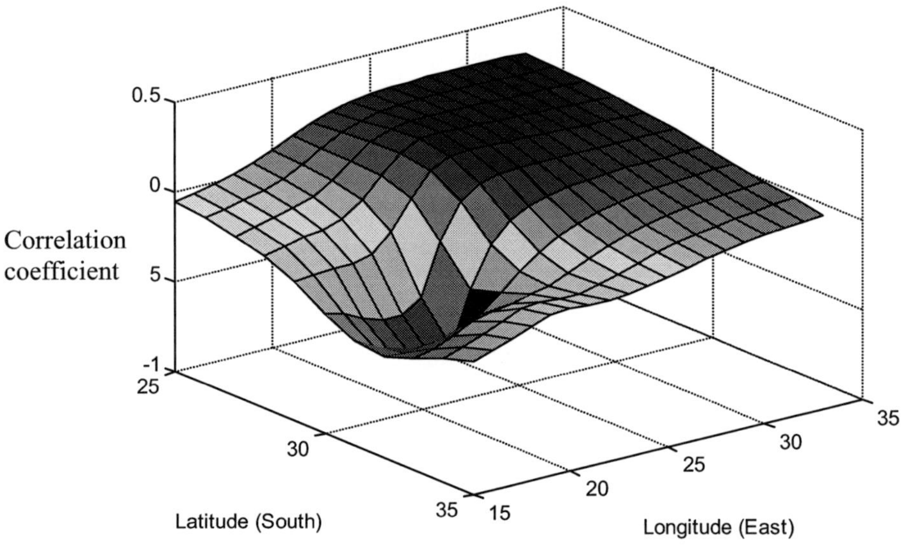

Figure 9.13 Correlation coefficients for Group 3. The most likely origin point is the dip (bottom).

The range of correlation coefficients that show good correlation for Group 3 is much smaller than those for Groups 1 and 2. This shows as the very definite dip in Figure 9.13. The origin was used in the regression of date and distance (Figure 9.14).
The equation of the regression line is:

$$y_{date} = (-1.57 \pm 0.2)\ \text{xdist} + (1887 \pm 73)$$

Using this equation the predicted departure date from the origin is in the range 1810 – 1960 calBP. The predicted rate of spread is 0.6 to 0.7 km/year. This fits well with the available archaeological evidence.

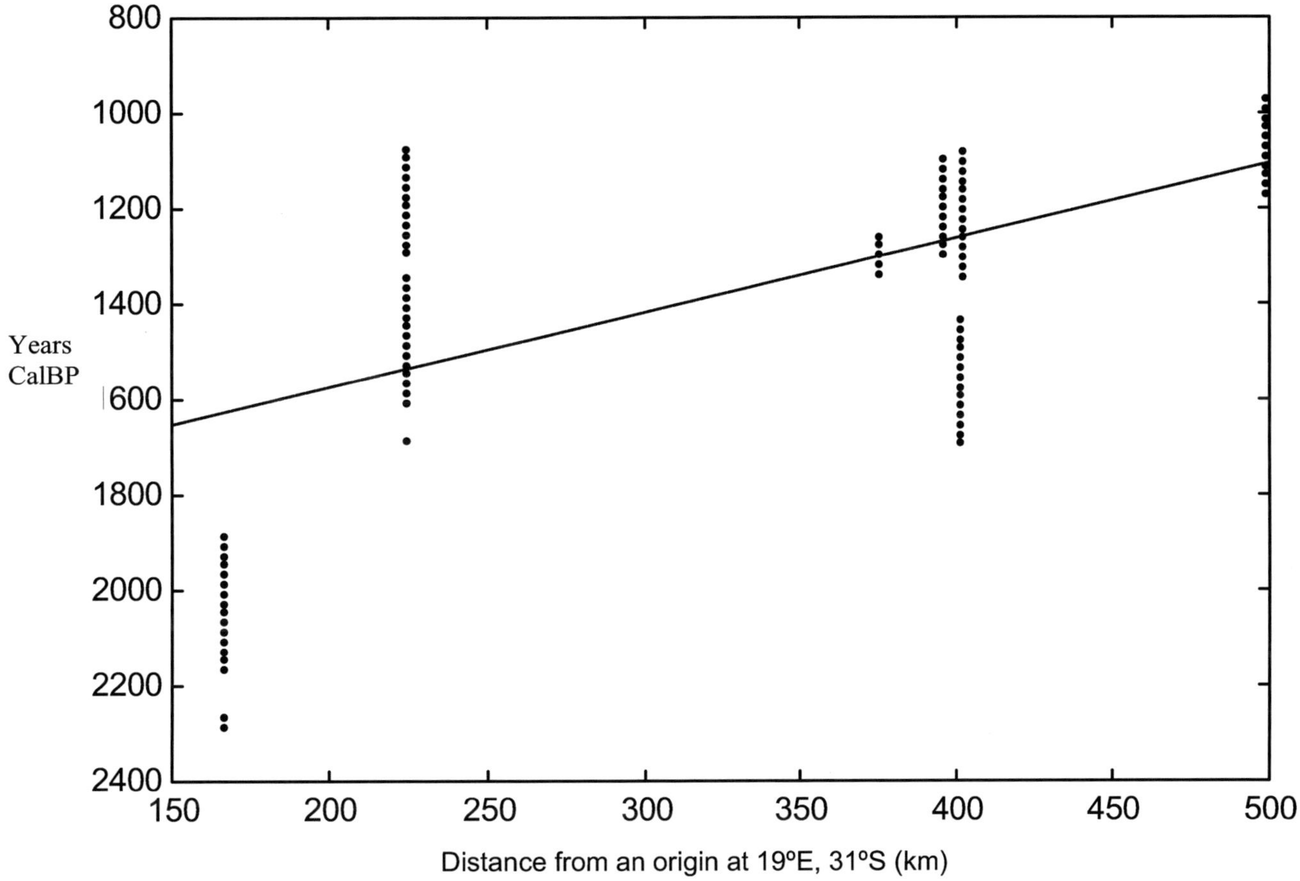

Figure 9.14 Linear regression of weighted calibrated probability distribution of dates with distance for Group 3.

9.3.4 Results for Group 4.

This group contains herder sites with less reliable radiocarbon dates. The 7 sites in are shown in Figure 9.15. The correlation coefficients for the whole of the searched surface are shown in Figure 9.16. Once again the general trend for the most probable origin in the north-west of South Africa is maintained. There is a dip at 20ºE, 30ºS, this is the origin point with the strongest correlation to Group 4. The correlation coefficient is – 0.82 at this point.

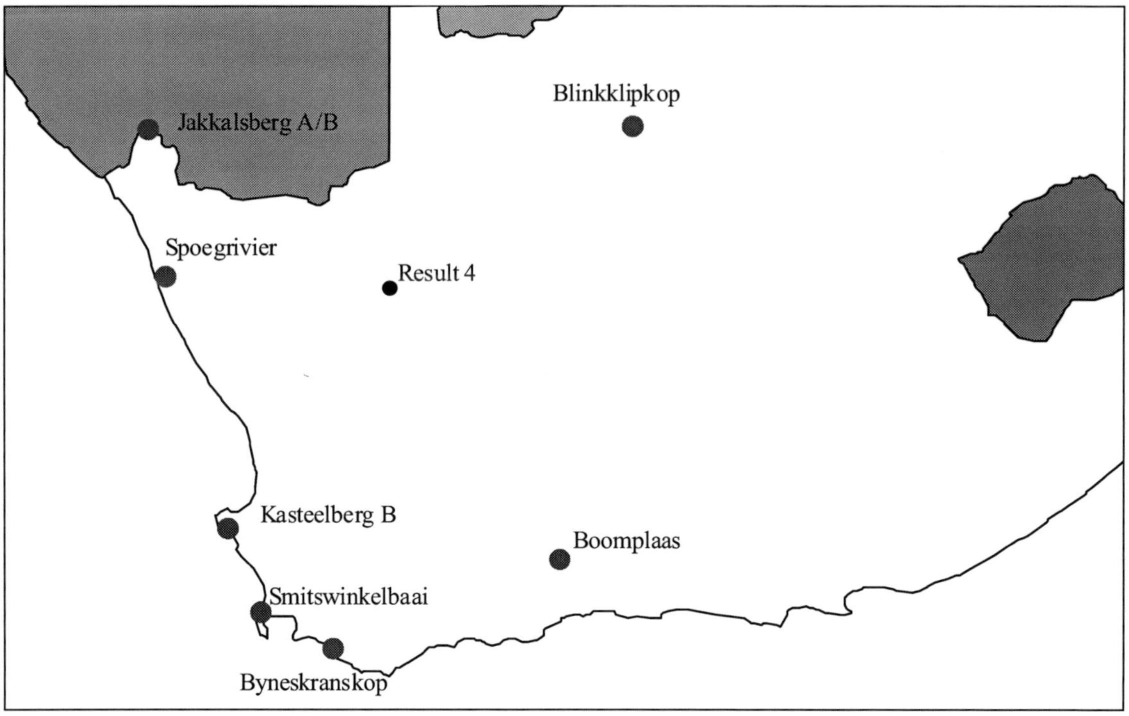

Figure 9.15 Group 4: herder sites with less reliable radiocarbon dates (grade 2 or 3). Result 4 shows the origin point with the optimal fit to the data.

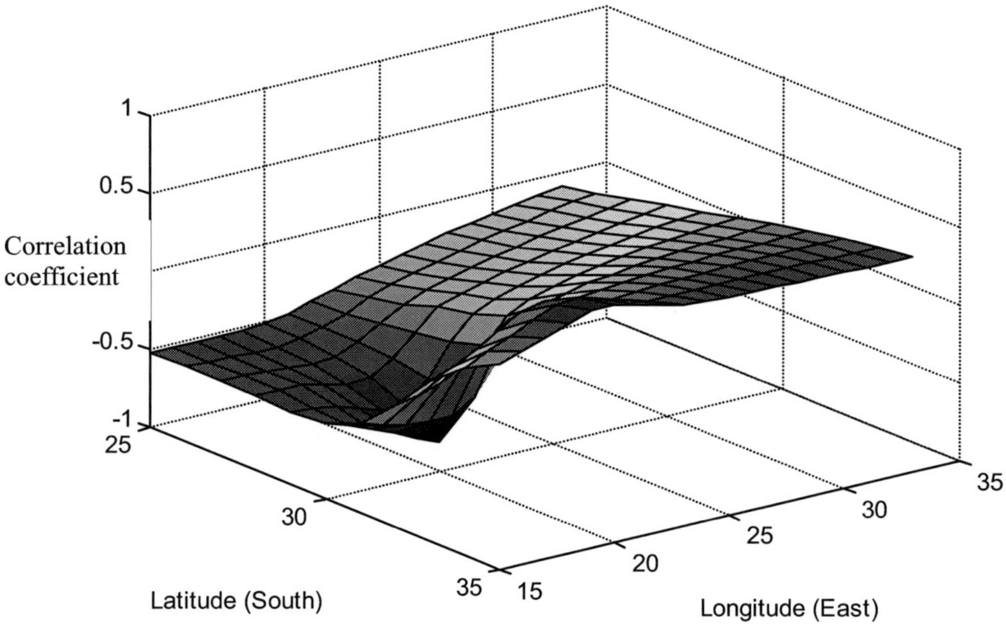

Figure 9.16 Correlation coefficients for Group 4. The best-fit origin points are those in the dip (bottom).

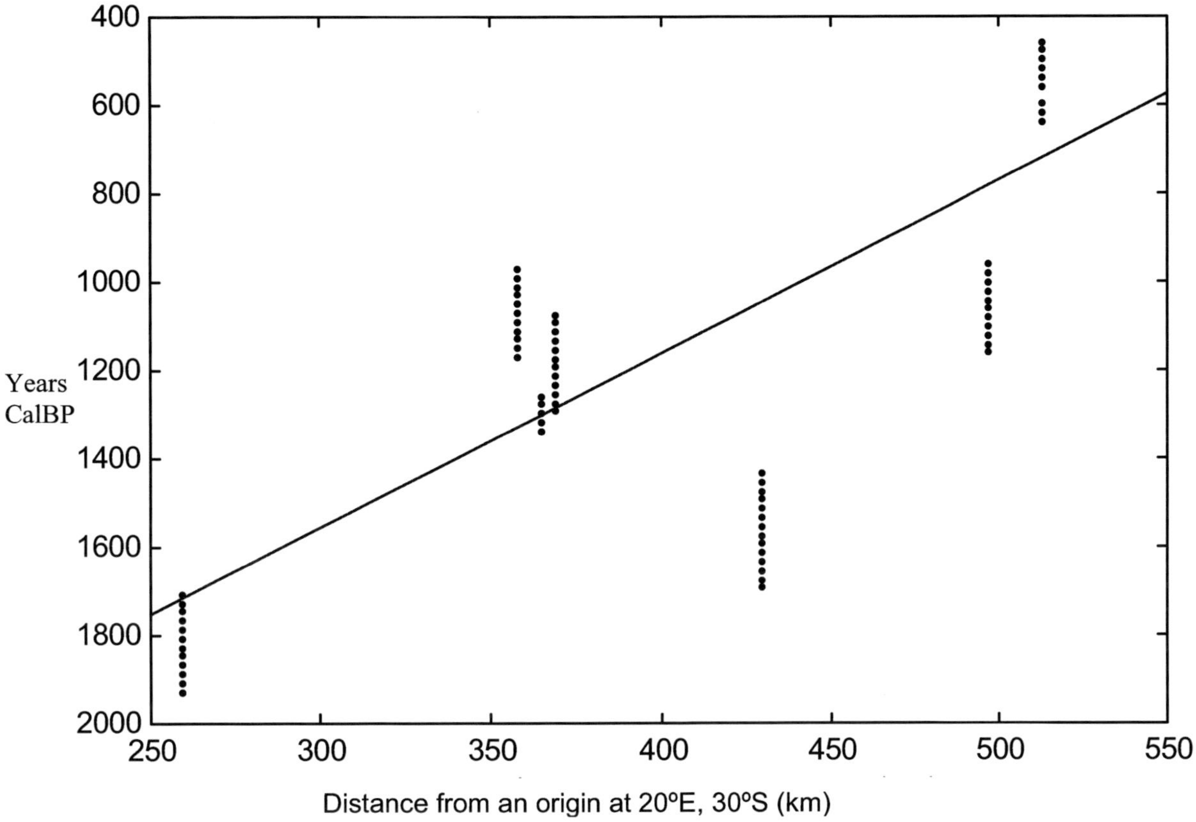

Figure 9.17 Linear regression of weighted calibrated probability distributions of dates with distance from an origin for Group 4.

The origin at 20°E, 30°S was used to draw the regression plot for Group 4 (Figure 9.17).
The equation for the regression line is:

$$y_{date} = (-3.9 \pm 0.33)\, x_{dist} + (2730 \pm 133)$$

This gives a predicted departure time from the origin of 2600 – 2830 calBP and a predicted rate of spread in the range 0.2 to 0.3 km/year. The date of departure is older than would be expected given other archaeological evidence for the spread of sheep in this region.

9.3.5 Results of Group 5.

This group, like Group 2, contains herder and non-herder sites, but whereas Group 2 was restricted to sites with AMS dates, Group 5 also includes sites with secure conventional dates. Sites are shown in Figure 9.18. There are 12 sites in this group.
The correlation coefficient is poor, r = 0.35 at the optimal origin, 20°E, 31°S. The correlation coefficients for the whole of the searched landscape are shown in Figure 9.19. This is the first time that the general trend is altered slightly, the area with the strongest correlation coefficients have shifted southwards. The best-fit origin point was used to create the linear regression plot of Group 5 (Figure 9.20).

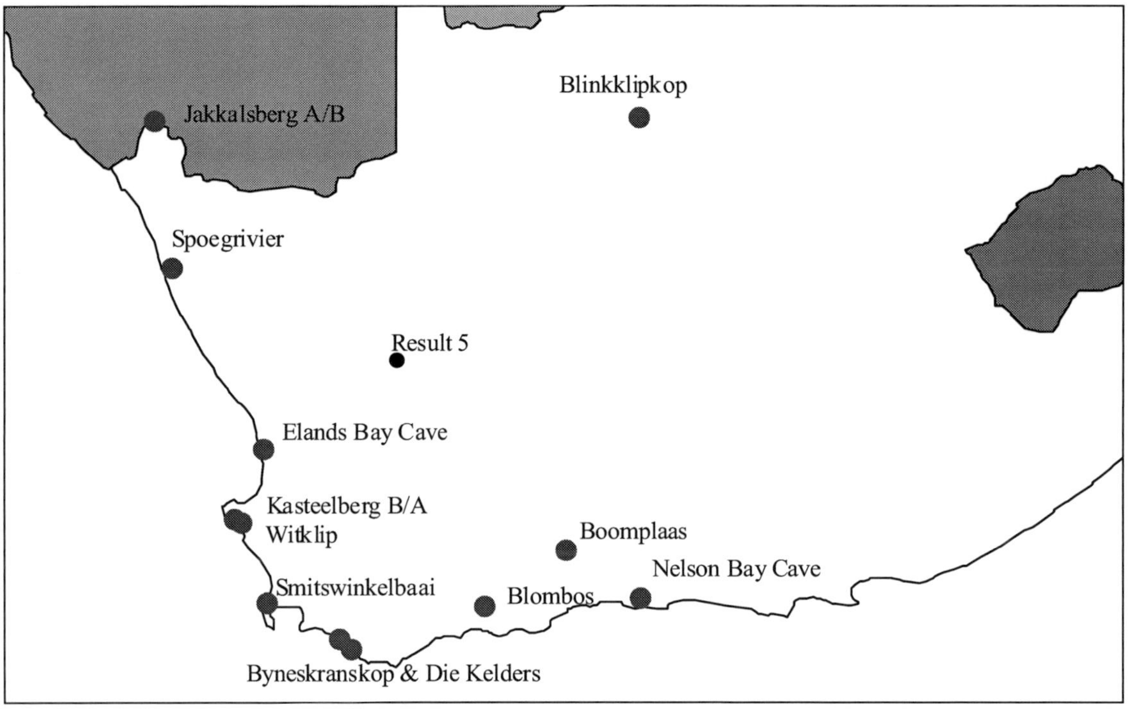

Figure 9.18 Group 5: herder and non-herder sites with AMS direct dates on sheep bone or conventional radiocarbon dates with a grading of 2 (secure). Result 5 is the most probable origin point for Group 5.

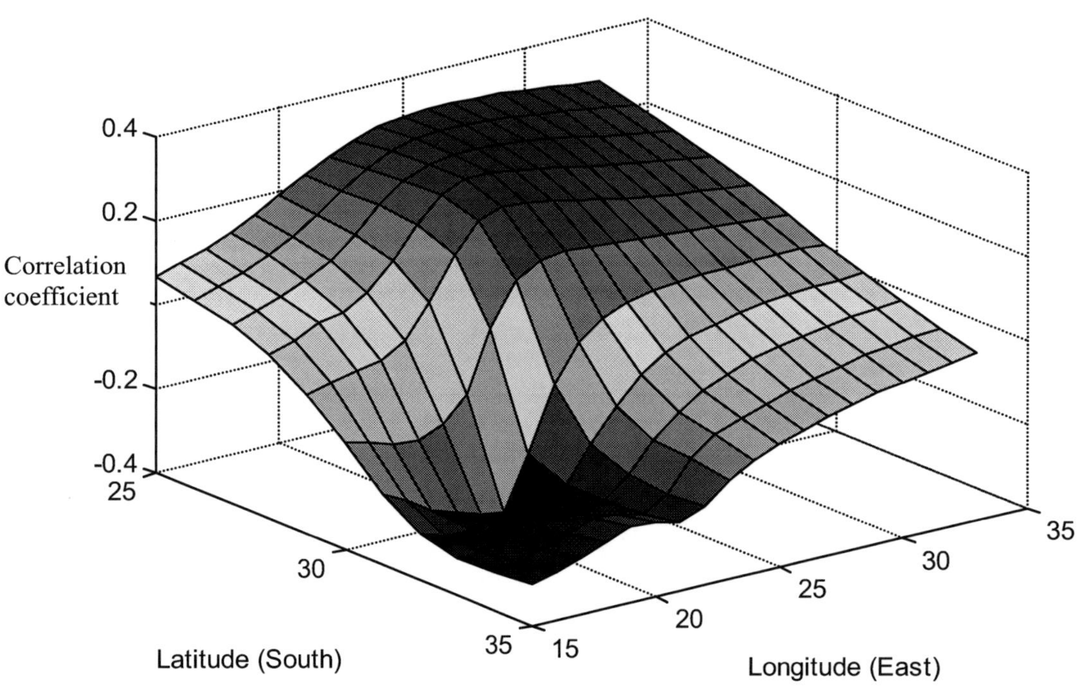

Figure 9.19 Correlation coefficients for possible origins to Group 5. The most probable origin is at the dip (bottom).

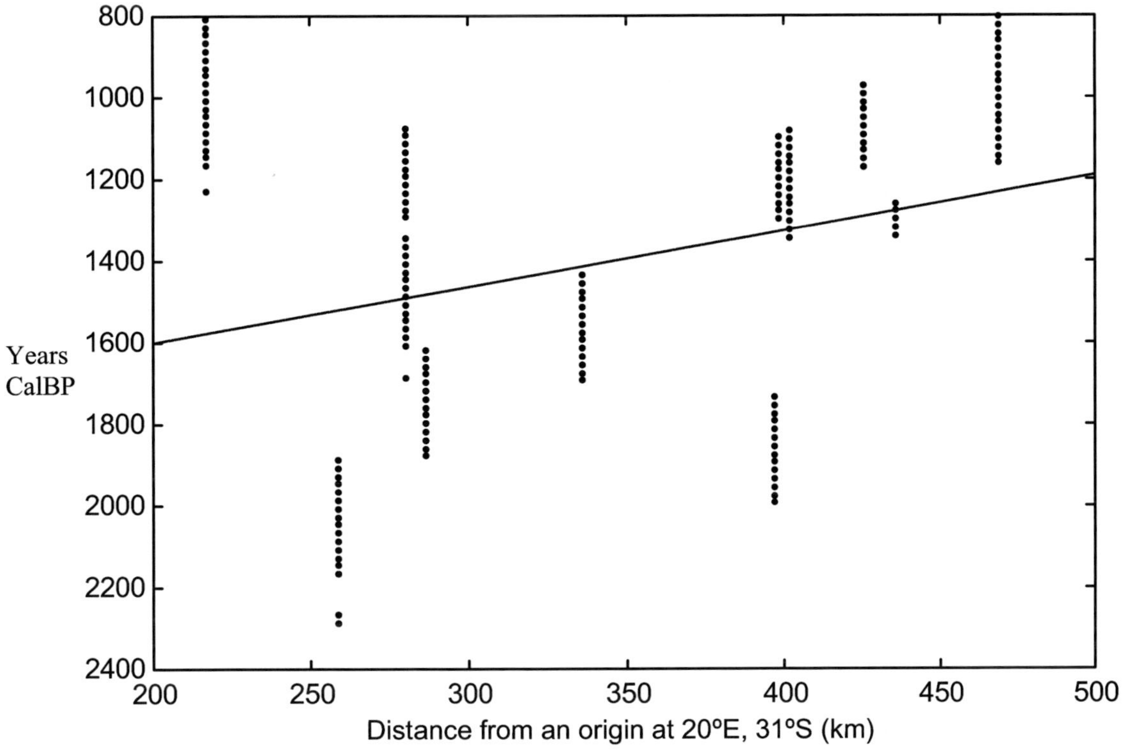

Figure 9.20 Linear regression of the weighted calibrated probability distribution of dates with distance from an origin for Group 5.

The equation of the regression line is:

$$y_{date} = (-1.36 \pm 0.29)\, y_{dist} + (1870 \pm 110)$$

This gives a predicted date of departure from the origin in the range 1760 to 1980 calBP. The predicted rate of spread is 0.6 to 0.9 km/year.

9.3.6 Results for Group 6.

This group contains only herder sites with reliable dates, whether conventional or AMS dates). There are 9 sites in this group (Figure 9.21).
The optimal origin point is found at 19°E, 31°S. The correlation coefficient at this point is –0.59. The correlation coefficients for the whole landscape are shown in Figure 9.22. The range of possible correlation coefficients is reduced dramatically (compare Figure 9.22 to Figure 9.6).
Using this origin point the linear regression plot was drawn for Group 6 (Figure 9.23).

The equation of the regression line is:

$$y_{date} = (-1.6 \pm 0.21)\, x_{dist} + (1870 \pm 80)$$

This gives a predicted departure date from the origin of 1790 to 1950 calBP. The predicted rate of spread is in the range 0.6 to 0.7 km/year.

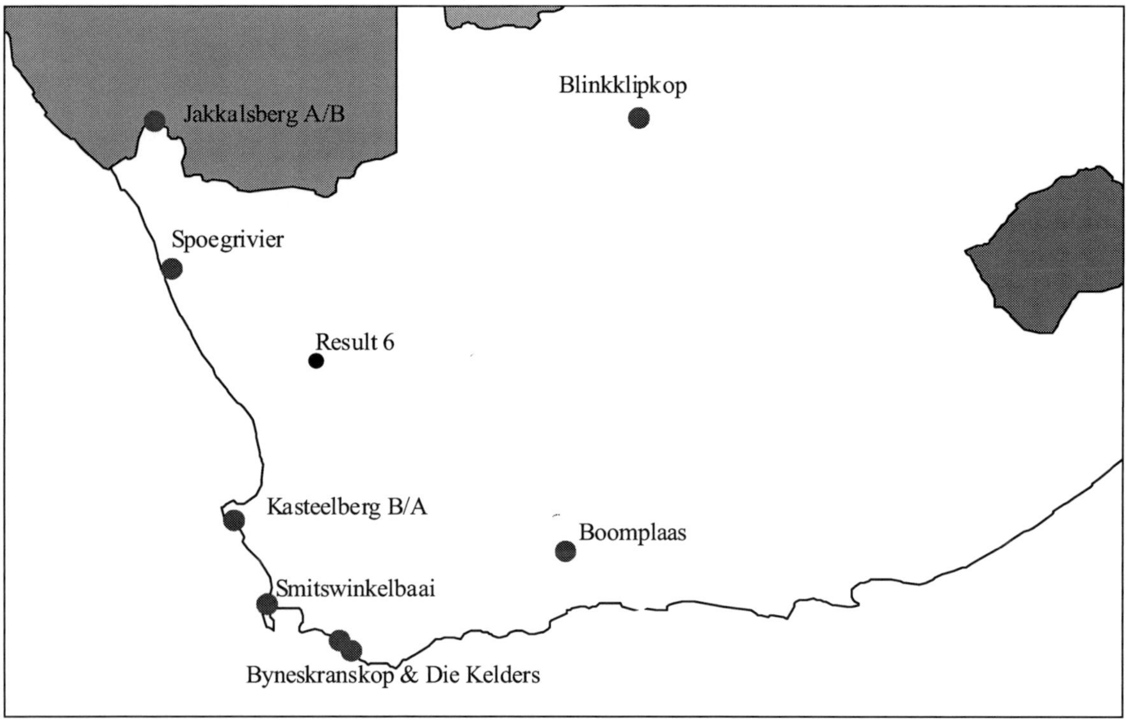

Figure 9.21 Group 6: All herder sites (both types of dating and all grades of date). Result 6 is the most probable origin point for these sites.

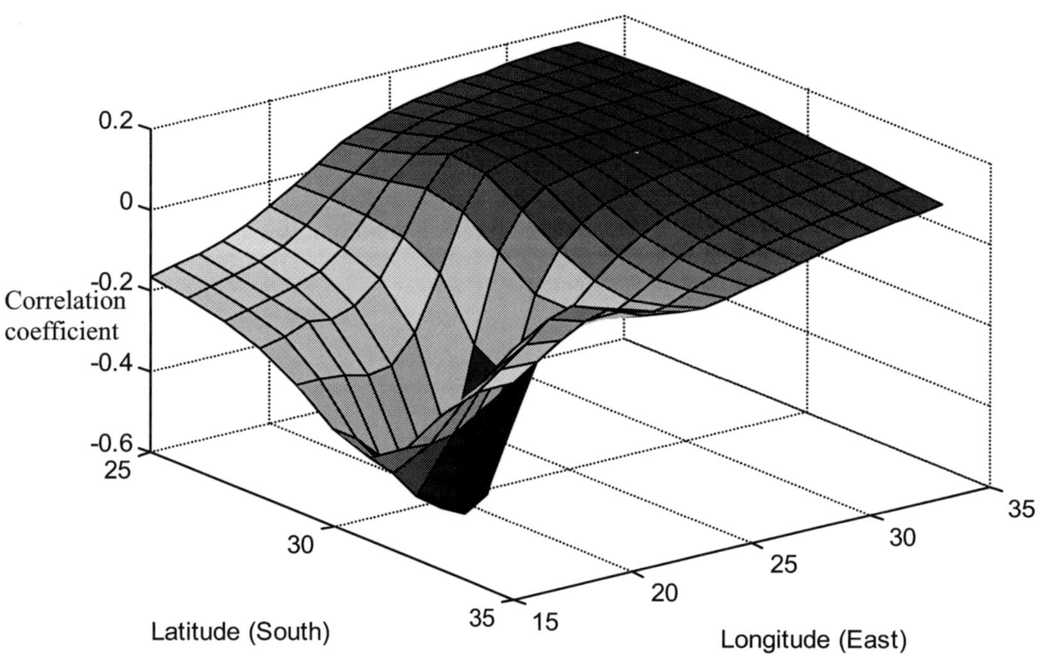

Figure 9.22 Correlation coefficients for Group 6. The best-fit correlation and most probable origin for Group 6 are shown by the dip (bottom).

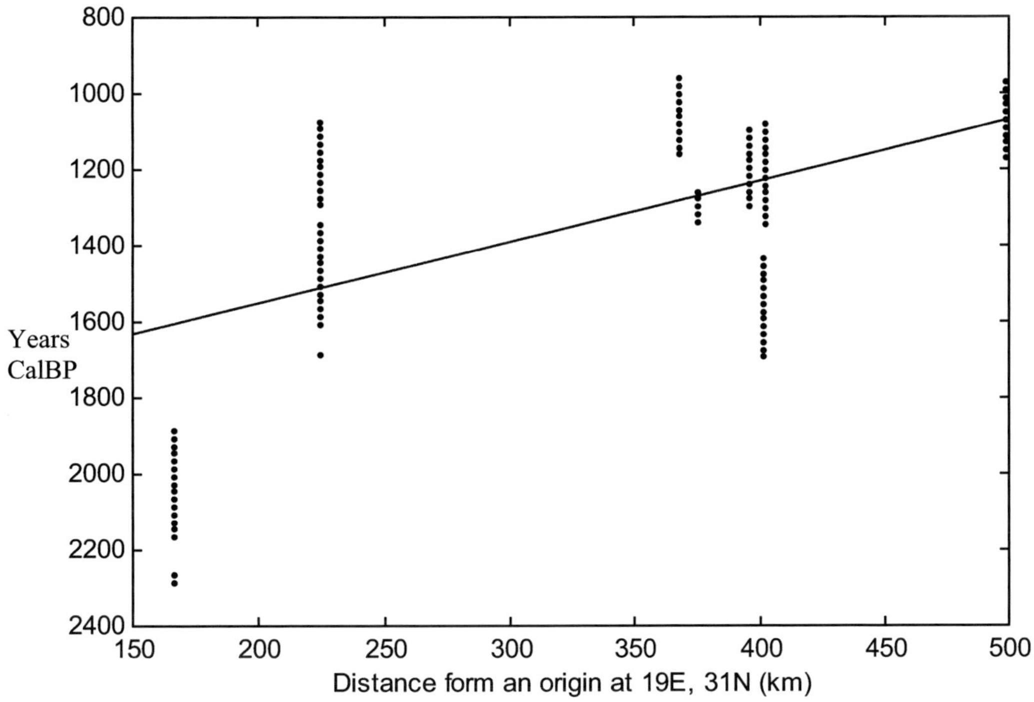

Figure 9.23 Linear regression of the weighted calibrated probability distributions of dates with distance from an origin for Group 6.

9.3.7 Results for Group 7

This group contain all non-herder sites regardless of reliability of radiocarbon dates. There are 8 sites (Figure 9.24).

The highest correlation coefficient, r = -0.46, is found at 21°E, 32°S (Figure 9.24).
The correlation coefficients for the whole searched surface are shown in Figure 9.25. The origin was used to create the linear regression plot (Figure 9.26).

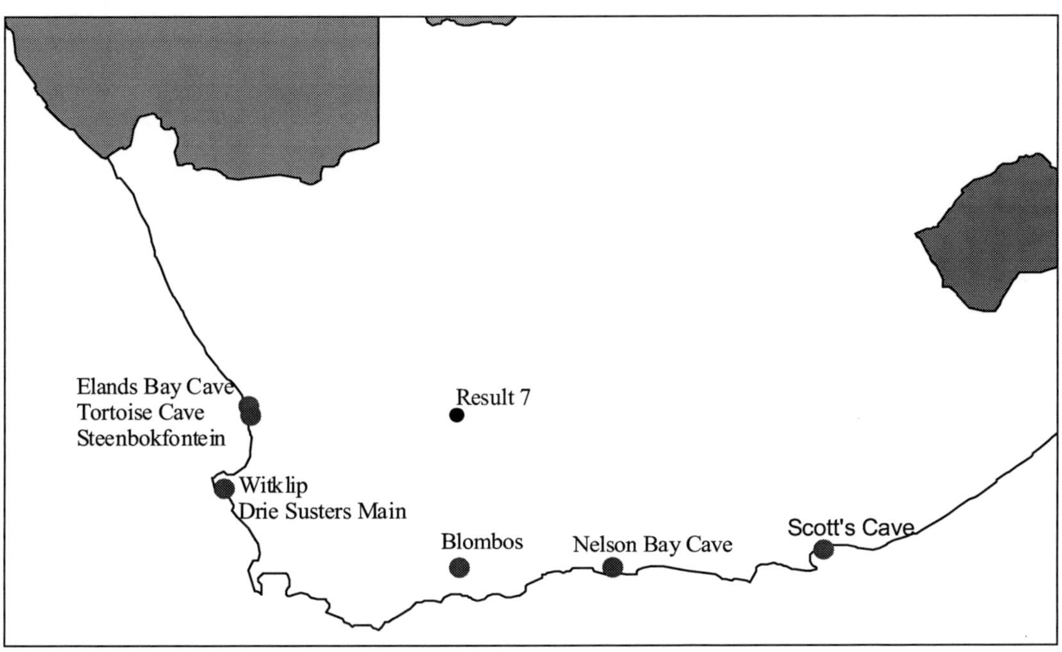

Figure 9.24 Group 7: non-herder sites (all dating methods and gradings). Result 7 is the origin point that had the strongest correlation coefficient to Group 7.

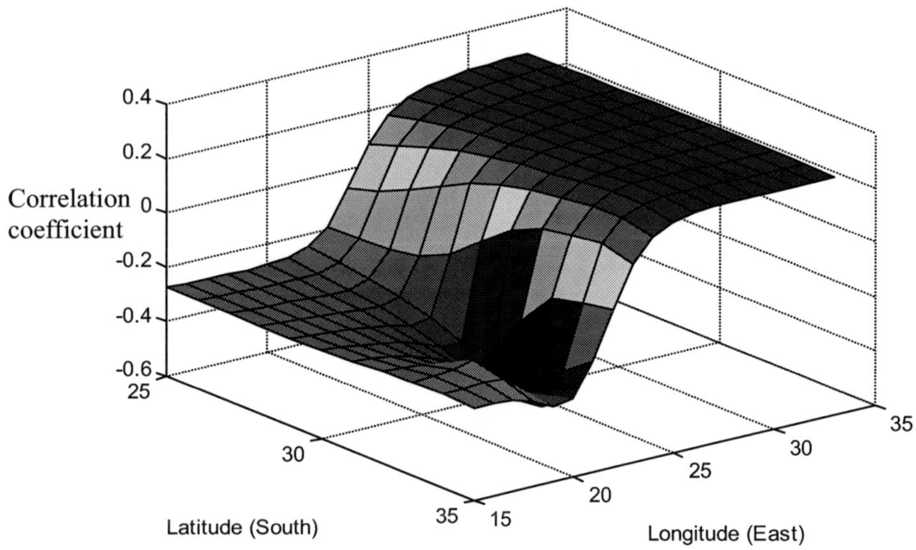

Figure 9.25 Correlation coefficients for Group 7. The dip area (bottom) shows the most probable origin area for the group.

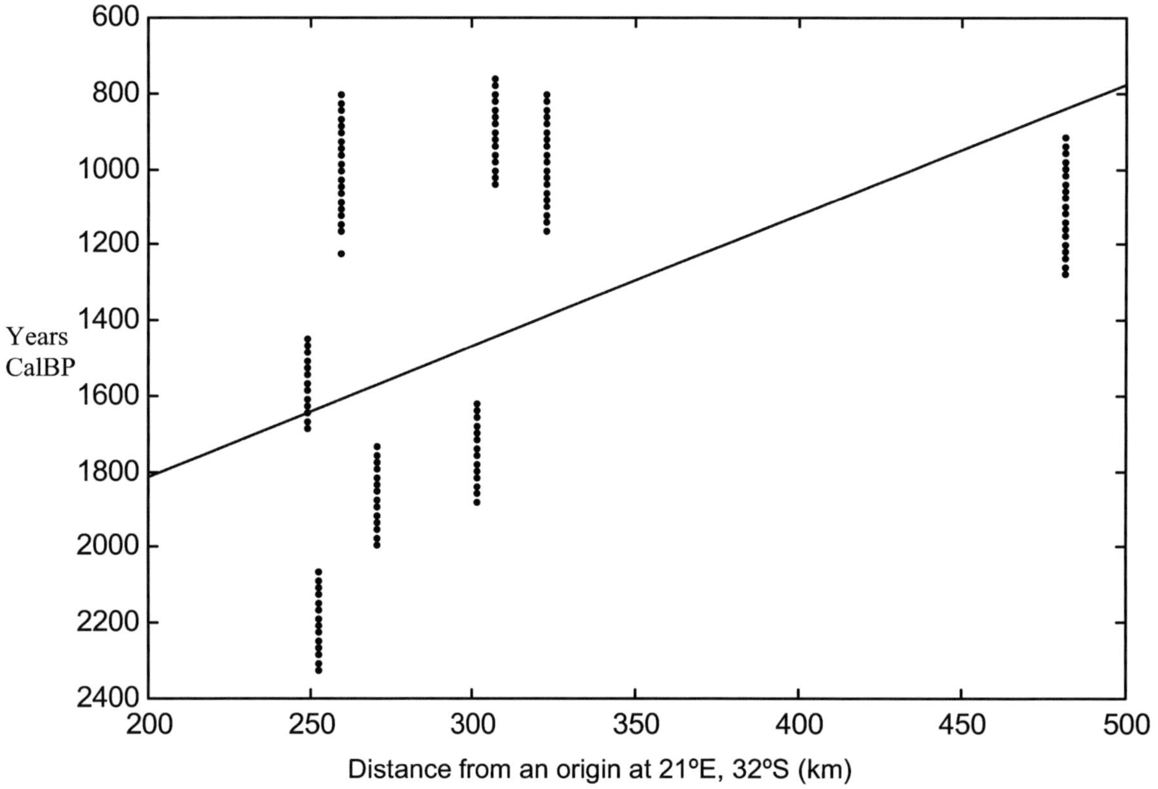

Figure 9.26 Linear regression of the weighted calibrated probability distributions of dates and distance from an origin for Group 7 sites.

The equation of the regression line is:

$y_{date} = (-3.47 \pm 0.60)\, x_{dist} + (2510 \pm 180)$

This gives a predicted departure date from the origin in the time range 2330 – 2690 calBP. The predicted rate of spread is in the range 0.2 to 0.3 km/year.

9.3.8 Results for Group 8.

This group combines the sites from Group 6 and Group 7. It contains all herder and non-herder sites (all dating methods and gradings). There are 17 sites in this group (Figure 9.27). The best correlation coefficient, -0.47, is found at 20°E, 31°S (Figure 9.27). The correlation coefficients to Group 8 as calculated for the whole of the searched landscape are shown in Figure 9.28. The best-fit origin point was used to draw the linear regression plot for Group 8 (Figure 9.29).

The equation of the regression line is:

$y_{date} = (-1.73 \pm 0.21)\, x_{dist} + (1990 \pm 80)$

This gives a predicted date of departure from the origin in the time range 1910 to 2070 calBP. The predicted rate of spread from the origin is 0.5 to 0.7 km/year.

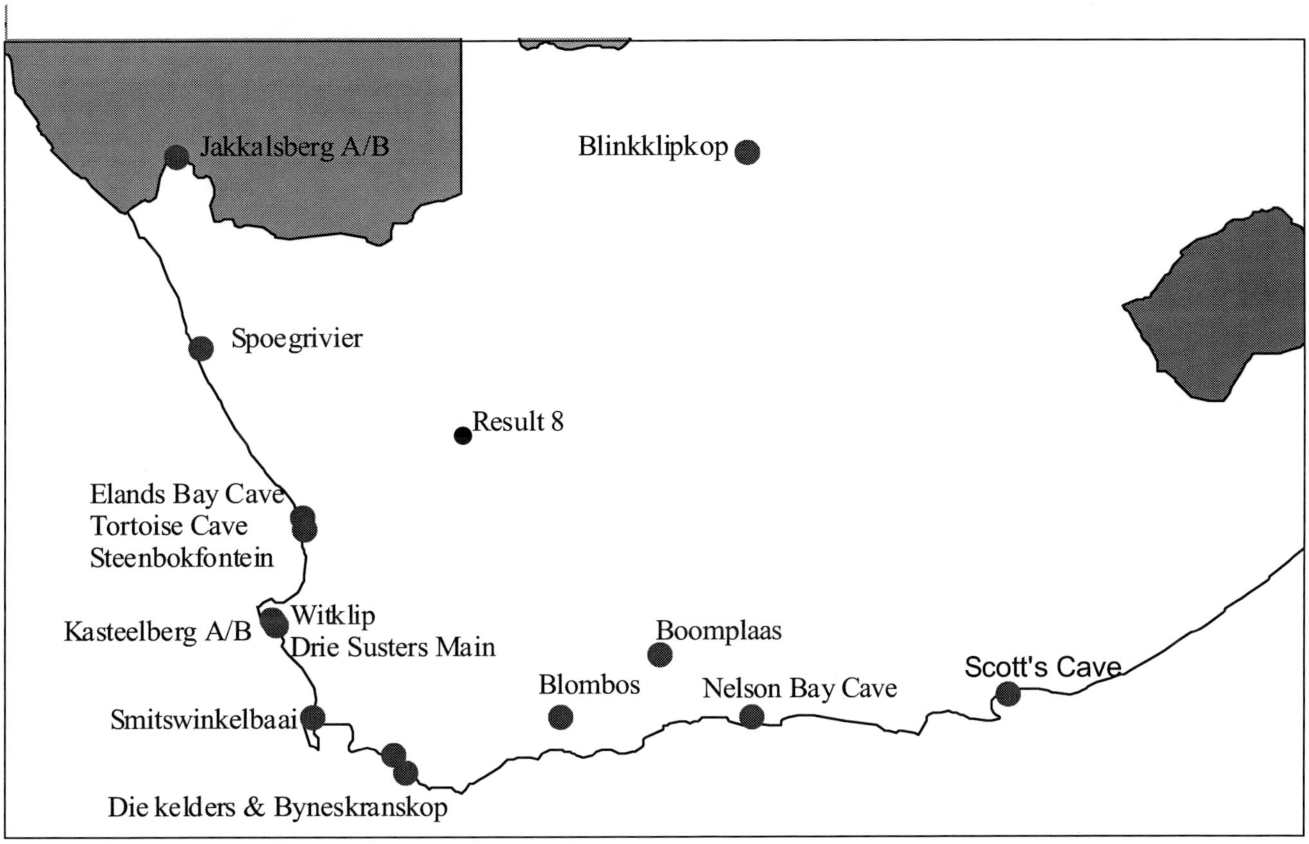

Figure 9.27 Group 8: All herder and non-herder sites (all dating methods and date gradings). Result 8 shows the origin point with the best correlation coefficient to group 8.

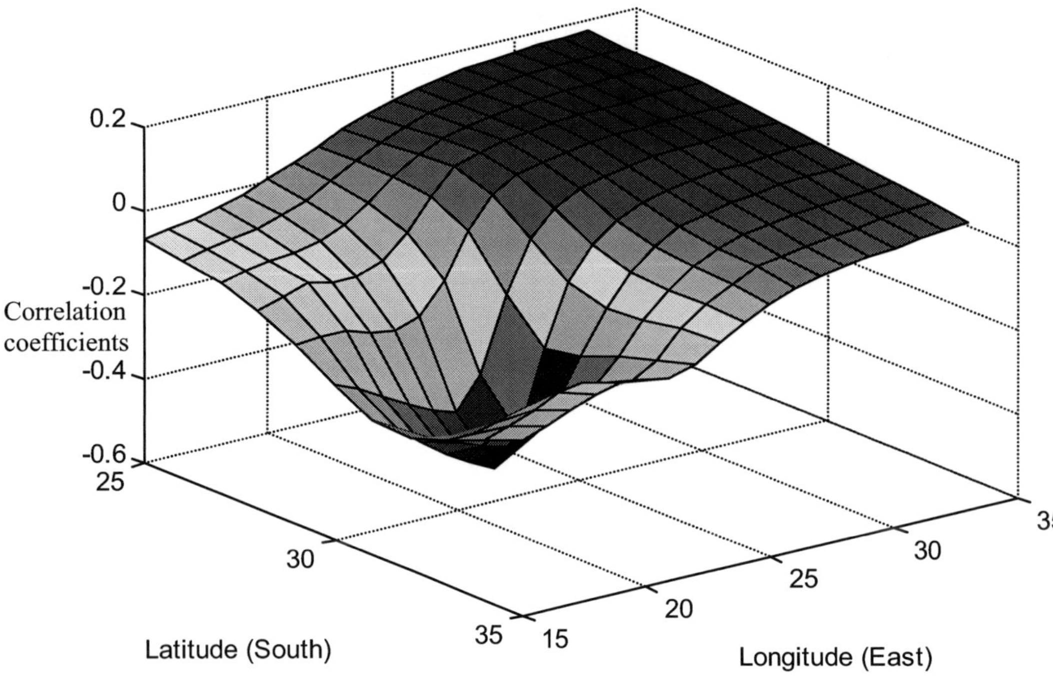

Figure 9.28 Correlation coefficients for Group 8. The dip (bottom) shows the most probable origin point (the highest correlation coefficient) for this data.

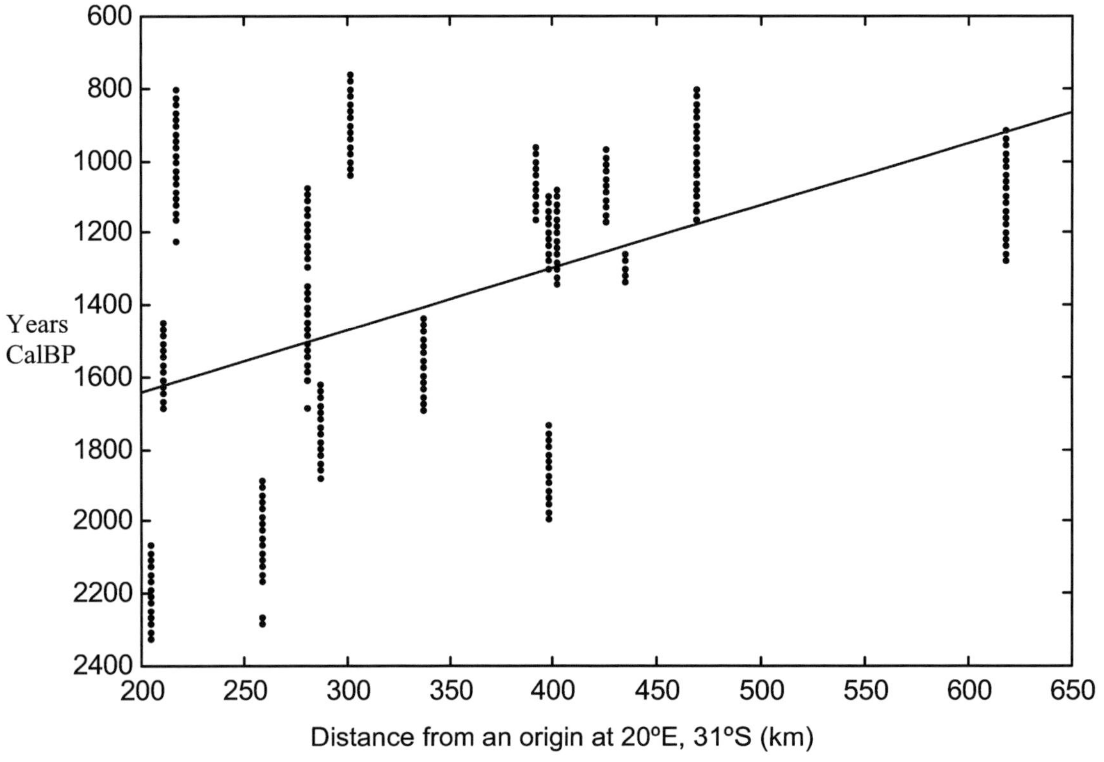

Figure 9.29 Linear regression of the weighted calibrated probability distributions of dates and distance for Group 8.

9.3.9 Results for Group 9.

This group contains all of the sites in Group 8 and additionally all sites with evidence of sheep that could not be classified as securely as either herder or non-herder sites. There are 22 sites in this group (Figure 9.30). The origin point with the strongest correlation coefficient, r = -0.29, is at 20°E, 31°S (Figure 9.30). The trend in correlation coefficients over the entire searched area is shown in Figure 9.31. The origin with the strongest correlation coefficient was used to draw the regression plot (Figure 9.32).

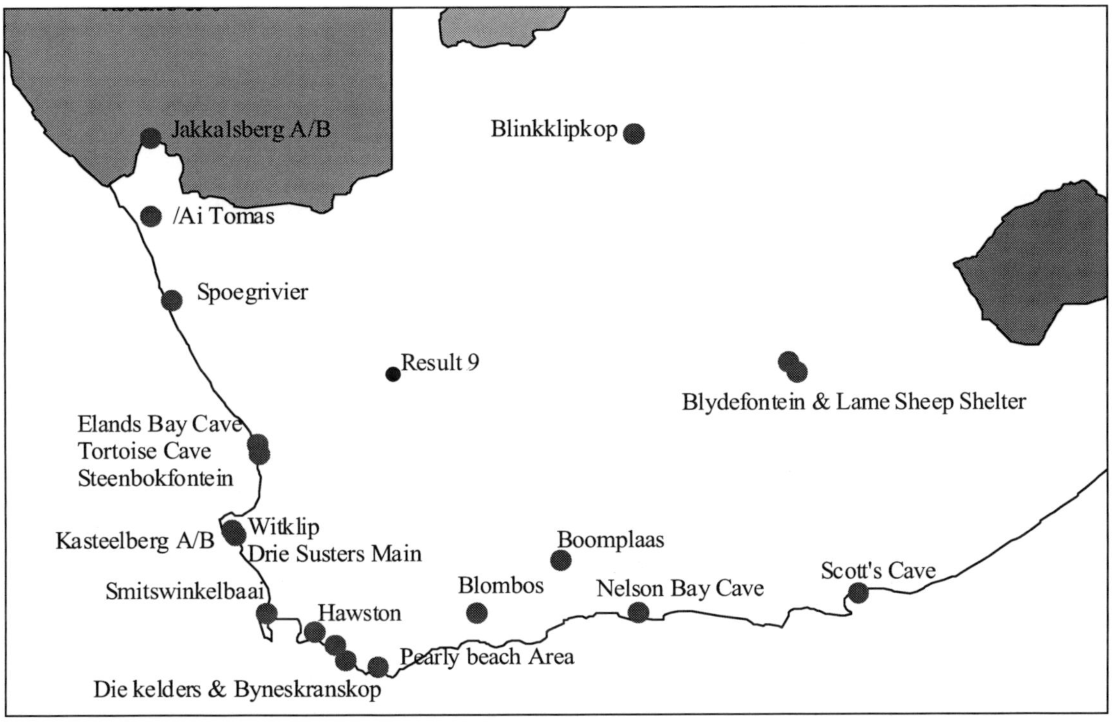

Figure 9.30 Group 9: All sites (herder, non-herder and unsure) and all dating methods. Result 9 is the most optimal origin for Group 9.

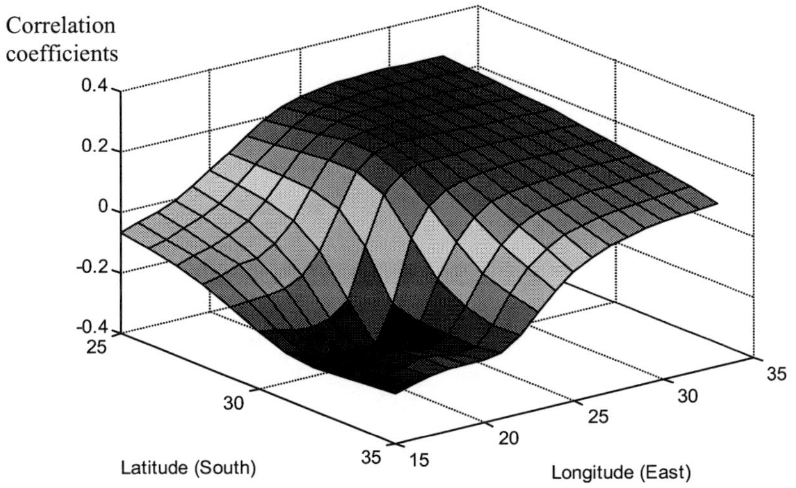

Figure 9.31 Correlation coefficients for Group 9. The dip area (bottom) shows the most probable region of origin for Group 9.

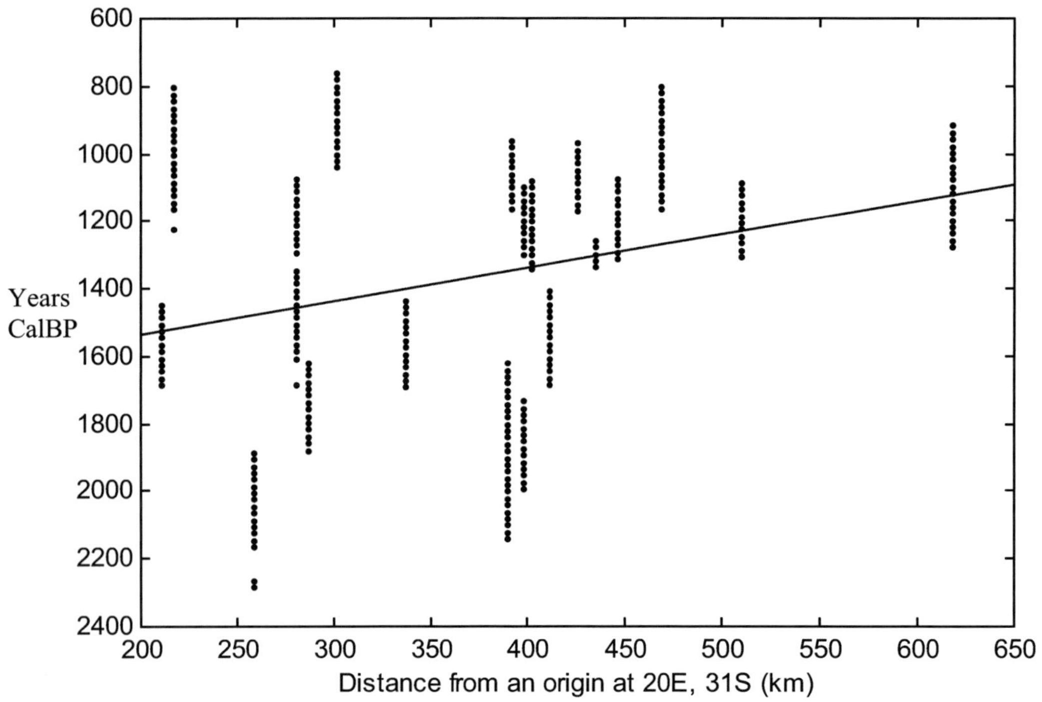

Figure 9.32 Linear regression of the weighted calibrated probability distributions of dates and distance from the origin for Group 9.

The equation of the regression line is:

$$y_{date} = (-0.98 \pm 0.19)\, x_{dist} + (1730 \pm 70)$$

This gives a predicted date of departure from the origin between 1660 and 1800 calBP. The predicted rate of spread is in the range 0.9 to 1.3 km/year.

9.4 Summary of results for Analysis 1.

The results show an unexpectedly consistent point of origin, both herder sites (Group 6) and non-herder sites both point to the same origin area between 19° E to 21°E and 30°S to 32°S when Groups 1 and 2 are excluded (Table 9.1). Even the Group 1 and 2 origins, once revised, do not contradict the general area of origin seen for all other groups. When the data used in Groups 1 and 2 is combined with other sites (Group 3 for example) then the origin point is in agreement with the general trend. These results are interesting when seen in the light of date selection. When sites have both a conventional date and an AMS date then the AMS has precedence over the conventional date (e.g. Group 3). Interestingly when conventional dates are analysed separately they show the same area of origin as that for the combined group (AMS and conventional), although the predicted date of departure is slightly older.

Group 6 (nine herder sites with reliable dates) point to an origin at 19°E, 31°S (Figure 9.21). This is considered the most reliable result as it is secure in both dating method and site classification. Groups 3 to 9 similarly all point to an origin area within a one-degree boundary of this result (Table 9.1). The direct dates on sheep bone, both herder and non-herder (Group 1 and 2) fit a *revised* origin in the far north-west corner of present day South Africa. However caution is needed as the sample size is small (four sites for Group 1 and six sites for Group 2). The results for Group 1 and 2 are inconclusive (Figure 9.6 and Figure 9.10) as they showed that any of a great number of possible origin points produced strong correlation coefficients to the sites. However when this group is combined with sites that are herder with reliable conventional dates then the origin shifts southwards (to 19°E, 31°S), there is a much narrower range of possible origin points, and the predicted departure date is more plausible archaeologically.

Group number	Group content	Predicted origin (decimal degrees)	Predicted departure date (calBP)	Predicted rate of spread (km/year)	Correlation coefficient	Revised origin	Revised departure date (calBP)	Revised spread rate (km/year)	Correlation coefficient for revised origin point
1	Herder (AMS)	25ºE, 28ºS	9000	None	-0.99	17ºE, 28ºS	2400	0.6 – 0.7	-0.97
2	Herder/non-herder (AMS)	21ºE, 31ºS	4000	None	-0.74	17ºE, 28ºS	2110– 2340	0.8 – 1.1	-0.59
3	Herder (AMS & conventional, grade 2)	19ºE, 31ºS	1810 – 1960	0.6 – 0.7	-0.63	None	None	None	None
4	Herder (conventional, grade 2)	20ºE, 30ºS	2600 – 2830	0.2 – 0.3	-0.82	None	None	None	None
5	Herder/non herder (AMS & conventional, grade 2)	20ºE, 31ºS	1760 – 1980	0.6 - 0.9	-0.35	None	None	None	None
6	Herder (AMS & conventional grade 2 and 3)	19ºE, 31ºS	1790 – 1950	0.6 – 0.7	-0.59	None	None	None	None
7	Non-herder (AMS & conventional grade 2 and 3)	21ºE, 32ºS	2330 – 2690	0.2 – 0.3	-0.46	None	None	None	None
8	Group 6 & Group 7	20ºE, 31ºS	1910 – 2070	0.5 – 0.7	-0.47	None	None	None	None
9	Group 8 & sites unclassified as herder or non-herder	20ºE, 31ºS	1660 – 1800	0.9 – 1.3	-0.29	None	None	None	None

Table 9.1 The predicted origins and departure dates for sites that contain the first evidence for sheep in the western half of southern Africa.

Chapter 9: PART TWO - Spatial and statistical analysis of the southern African sheep database

9.5 Analysis 2: to test the models of Stow (1905), Cooke (1965), Elphick (1977) and Bousman (1998) for the route of the first introduction of sheep into southern Africa by the spatial analysis of the radiocarbon dates for this process.

9.5.1 The test of Stow & Cooke and Bousman's model

Both Stow and Cooke and Bousman support a route for the introduction of sheep that moves down the western edge of southern Africa from present day Namibia southwards to the Cape. Radiocarbon determinations along this route should show a pattern of decreasing age with movement in the direction of the spread if this route is correct (Figure 9.33).

To test this model, sites were grouped, and the best-fit origin for each group was sought. To fulfil the model's predictions Group A should have an origin in the north-west of present day South Africa and Group B should have an origin in the south-west corner at the southern tip of Africa.

Group A area and sites:

Lat. 28° S to 33°S, long. 16° E to 19°E
- Jakkalsberg A/B (herder, grade 2)
- Spoegrivier (herder, grade 1)
- Kasteelberg A (herder, grade 1)
- Kasteelberg B (herder, grade 2)

Group B area and sites:

Lat. 33°S to the southern coast, long. 19°E to 27°E
- Boomplaas (herder, conventional grade 2)
- Nelson bay Cave (non-herder, grade 1)
- Blombos (non-herder, grade 1)
- De Kelders (herder, grade 1)
- Byneskranskop (herder, grade 1)
- Smitswinkelbaai (herder, grade 3)
- Scott's Cave (non-herder, grade 3)

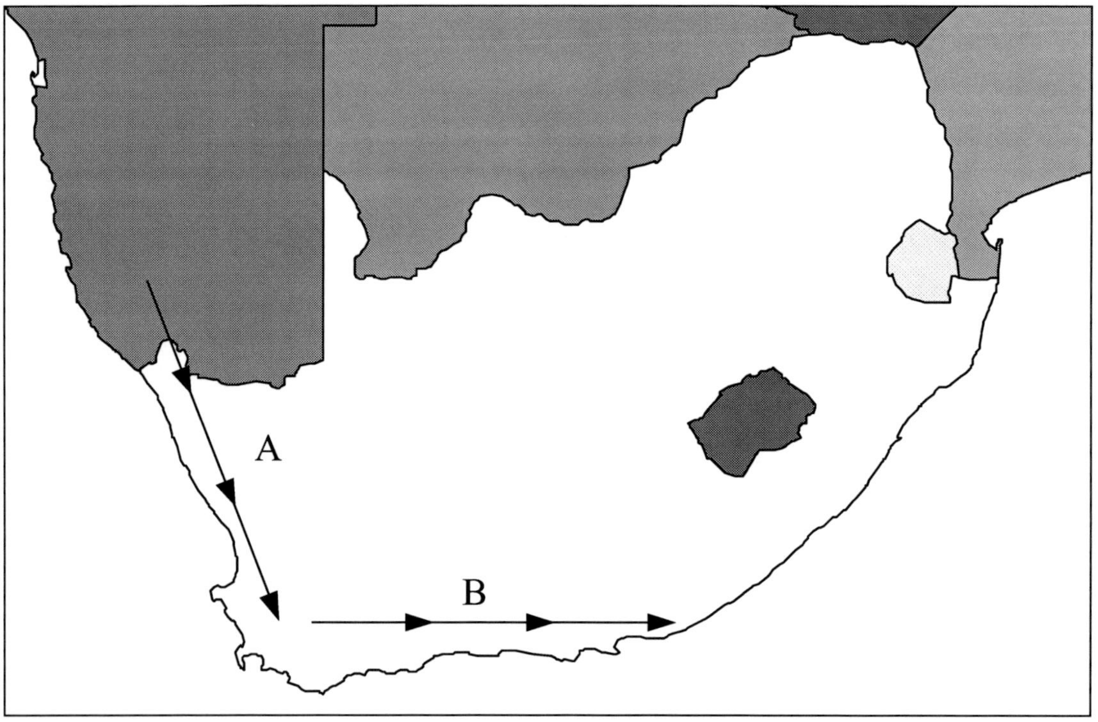

Figure 9.33 Stow and Cooke and Bousman's route for the introduction of the first sheep. Radiocarbon dates for the first sheep along A should show a trend of decreasing antiquity with movement south. Along B, sites should show a similar trend of decreasing antiquity eastward. A-sites should be older than B-sites.

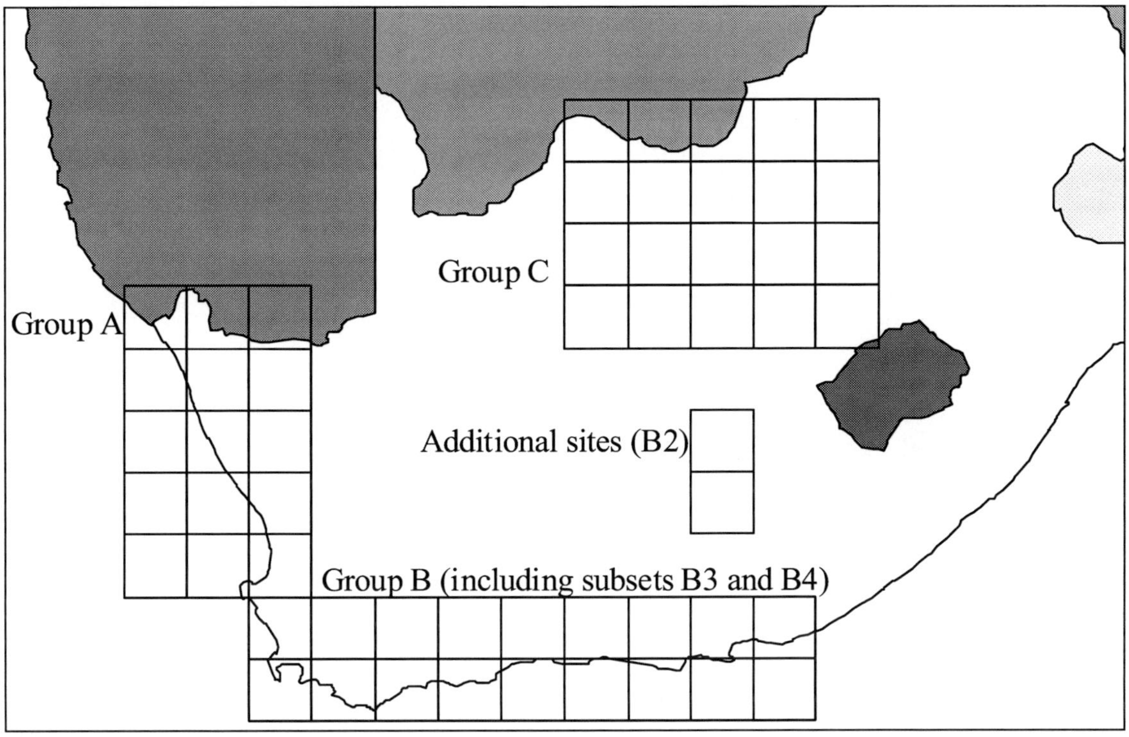

Figure 9.34 The divisions made in order to test the western route of introduction of the first sheep proposed by Stow & Cooke and Bousman. Group B3 and B4 fall within the area of Group B (they are simply a reduced subset of B to test the effect of including/excluding site types and grade 3 determinations). Group B2 is an extension of Group B.

Four alternative ways of combining Group B sites were tested, labelled B2, B3 and B4 (Figure 9.34): B2 has an extended range to include Blydefontein and Lame Sheep Shelter, sites that were out of the Stow and Cooke's model range. B3 and B4 are subsets of B1 to test the effect on the Group B results of excluding grade 3 determinations, at Smitswinkelbaai (B3) and Scott's Cave (B3), and non-herder sites, Scott's Cave (B4). Additionally there is Group C; this was added to include sites in the central part of South Africa and outside of the range of the Stow and Cooke's model. This group of sites is very recent; their relationship to the first introduction of sheep is unclear.

Group B2 area and sites:

> As Group B but with the area extended to include Blydefontein (unsure of site type, grade 3) and Lame Sheep Shelter (unsure of site type, grade 3).

Group B3 area and sites:

> As Group B but excluding the sites with grade 3 determinations (i.e. Smitswinkelbaai and Scott's Cave).

Group B4 area and sites:

> As Group B but excluding Scott's cave (non-herder grade 3).

Group C area and sites:

> Latitude 25°S to 29°S, longitude 23°E to 28°E.
> All group A sites with the addition of the four listed:
> Blinkklipkop (herder, grade 2)
> Wonderwerk (non-herder, grade 1)
> Jubilee Shelter (non-herder, grade 1)
> Dikbosch (non-herder, grade 3).

9.5.1.1 Results for Group A:

The correlation coefficients calculated in the search of 200 potential origin points to find the most likely origin to sites in Group A are shown Figure 9.35. There are four sites in this group.

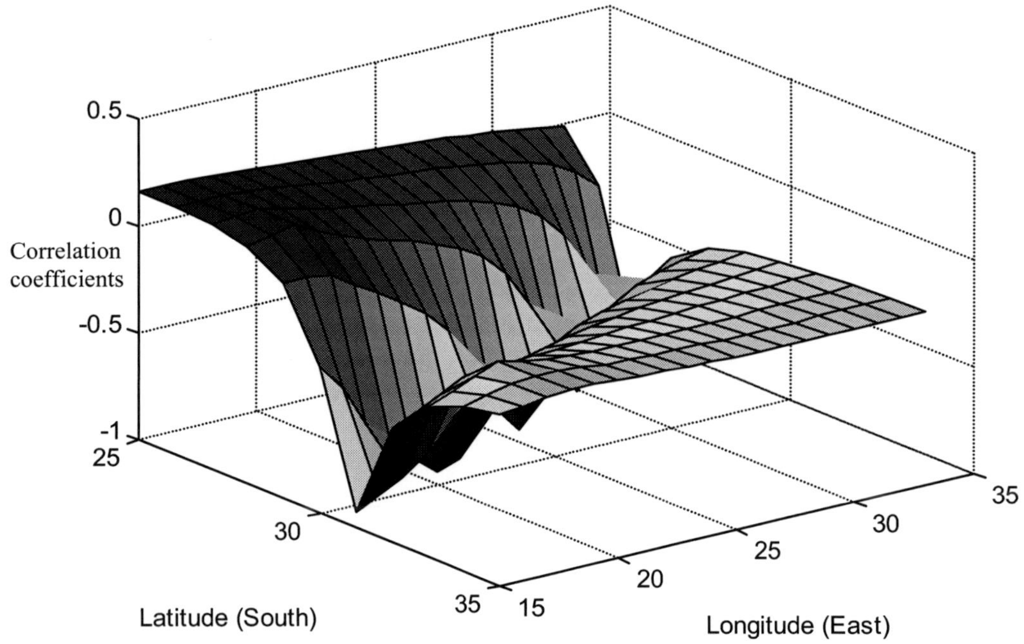

Figure 9.35 Correlation coefficients to Group A. There are four minima (dips), these have strong negative correlations (calculated from the linear regression of date and distance of each site in Group A to each potential origin point). They are the most probable origins to sites in Group A.

All of the four minima shown in Figure 9.35 were explored. They all produce predicted departure dates that are older than the archaeological evidence would suggest (e.g. an at origin at 15°E, 31°S produced departure date ±3000 calBP, origin at 23°E, 30°S gave a departure date at ±4000 calBP; departure dates become older with movement east). The origin that gave an archaeologically plausible date was at 18°E, 31°S (correlation coefficient, r = -0.72). The regression plot of date and distance from this origin point is shown in Figure 9.36 .

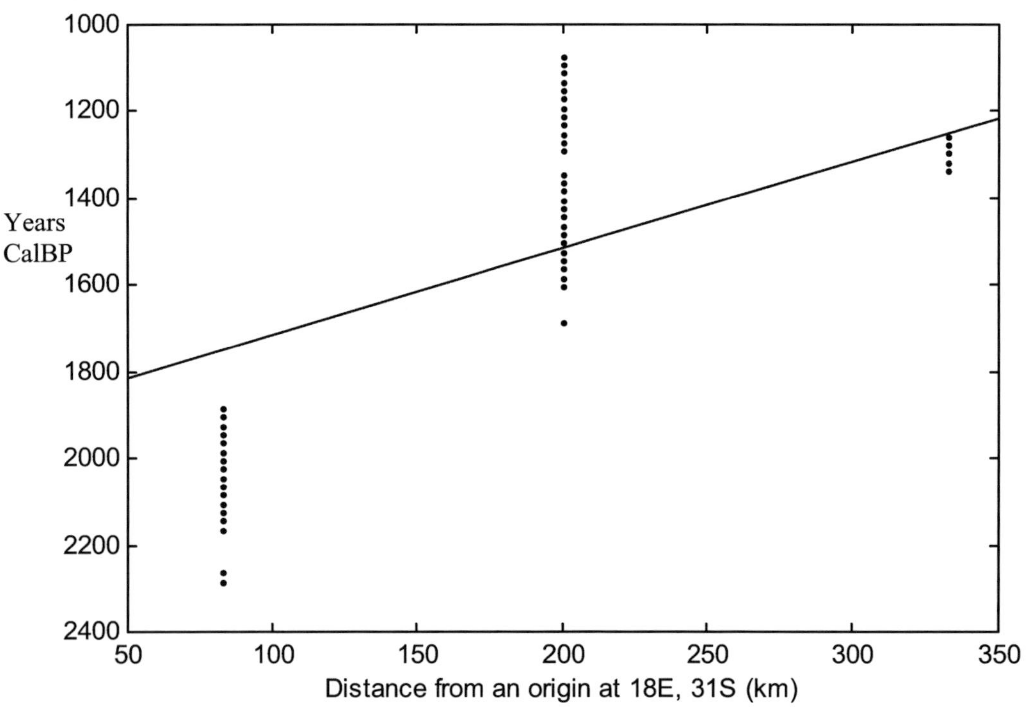

Figure 9.36 Linear regression of the weighted calibrated probability distributions of Group A dates and their distance from an origin at 18°E, 31°S.

The equation of the regression line is:

$y_{date} = (-1.99 \pm 0.28)\ x_{dist} + (1910 \pm 80)$

From the equation, the predicted departure time is in the time range 1830 to 1990 calBP and the predicted rate of spread is 0.4 to 0.6 km/year.

As a test of the model, the age of sites as predicted by the equation was compared to the calibrated date at the site. The equation gives an oldest possible date at the site of Spoegrivier (82 km from the origin) of 1850 calBP. The calibrated 2-sigma range of the oldest AMS date at Spoegrivier is 1870 – 2310 calBP. Although the predicted date and the AMS date do not match, this is a problem that can be explained, in part, by the assumption that the model makes: that spread from the origin is at a constant rate to all points in the group. A constant rate of spread is a good approximation to keep the model simple. If the rates of spread are allowed to vary slightly then the date at Spoegrivier can be accommodated by the model.

9.5.1.2 Results for Group B:

The correlation coefficients calculated at 200 potential origin points across the landscape to search for the most likely origin to Group B are shown in Figure 9.33. There are seven sites in this group.
All the correlation coefficients in the blue area are small, so the linear trend between date and distance is poor from all of the potential origin points to sites in this group.
An origin point from the blue area (Figure 9.37) with coordinates 25°E, 32°S (correlation coefficient, r = -0.42) was used to draw the regression diagram (Figure 9.38).

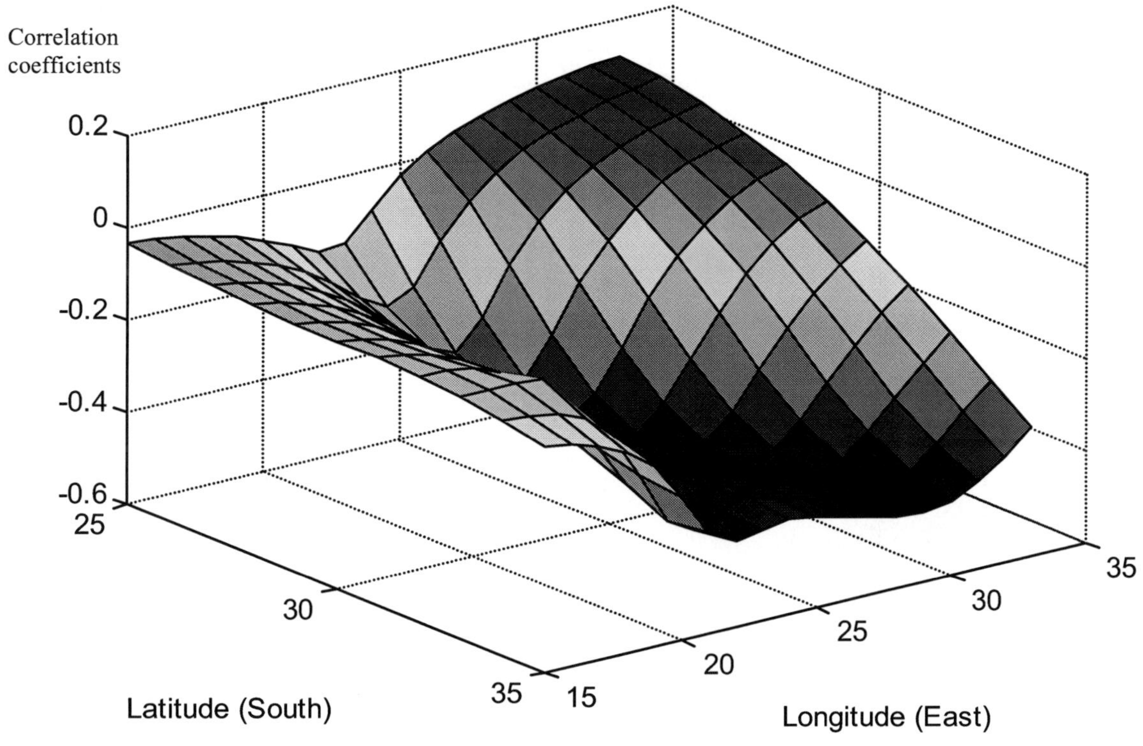

Figure 9.37 Correlation coefficients for Group B. The dip area shows the region of highest negative correlation. These are the points on the landscape that are most likely to be the origin for this group based on the linear regression of date and distance from each site to each potential origin.

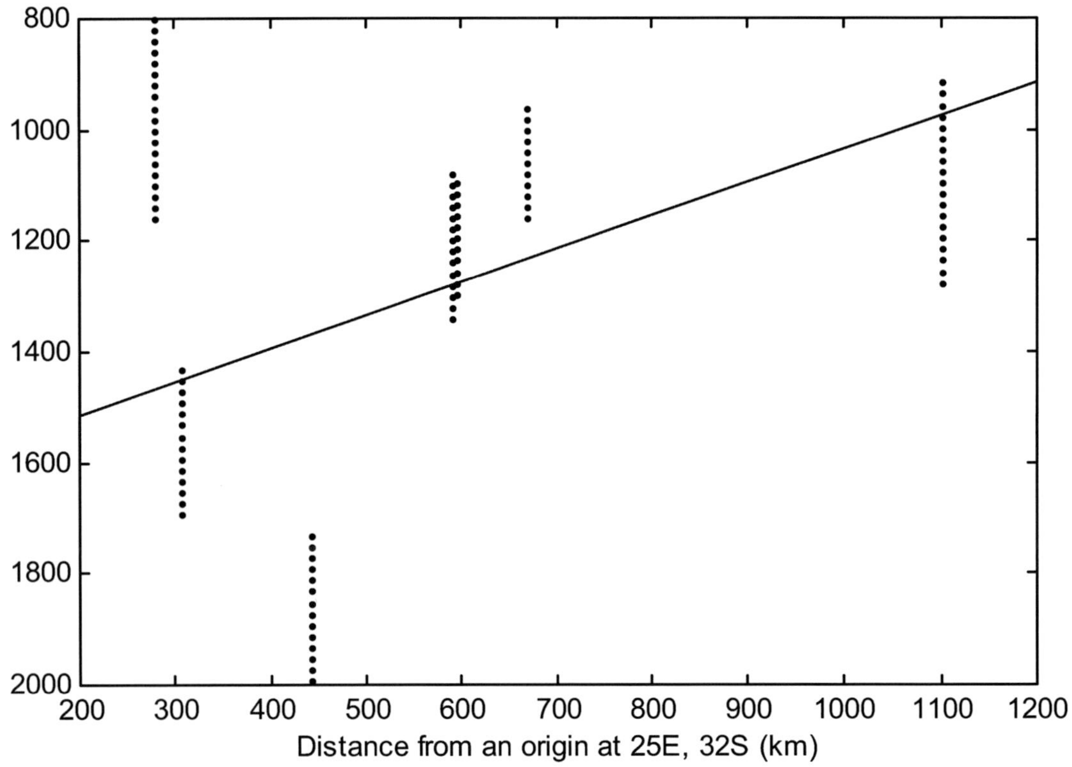

Figure 9.38 Linear regression of the weighted calibrated probability distributions of Group B1 dates and their distance from an origin at 25°E, 32°S. The vertical axis is calibrated years BP.

The equation of the regression line is:

$y_{date} = (-0.60 \pm 0.13)\ x_{dist} + (1640 \pm 80)$

This equation gives a predicted date of departure from the origin in the time range 1560 to 1720 calBP and a predicted rate of spread of 1.4 to 2.1 km/year. The correlation coefficient for this model is low. As a test for the validity of the model the predicted date for the site of Blombos (the site with the earliest date for sheep in this group) was calculated using the equation. The geodesic distance to Blombos from the origin at 25°E, 32°S is 440 km. Using the equation, the oldest predicted date for Blombos is ±1500 calBP. The 2-sigma range for a calibrated AMS date on sheep bone at Blombos is 1700 – 1930 calBP. This date is only met if the rate of spread to Blombos was 10 km/year. The correlation coefficient for this origin was low
(r = -0.42). This reflects the non-linearity of the date age and distance from an origin in this group and suggests that varying rates of spread to sites might explain the shortfalls in the model.

9.5.1.3 Results for Group B2:

This was the extended set of determinations, including two sites to the north-east of Group B. There are nine sites in this group. The correlation coefficients to search for the most likely origin to Group B2 are shown in Figure9.39.

The correlation coefficients in the blue area are all low, so the linear trend between dates and distance from an origin is poor. An origin in the blue region (Figure 9.39) at 22°E, 34°S with a correlation coefficient of -0.44 was used to draw the linear regression plot (Figure 9.40).

The equation of the regression line is:

$y_{date} = (-0.37 \pm 0.07)\ x_{dist} + (1410 + 30)$

Using this equation, the predicted time of departure from the origin is in the range 1380 to 1440 calBP and the predicted rate of spread is 2.3 to 3.3 km/year.

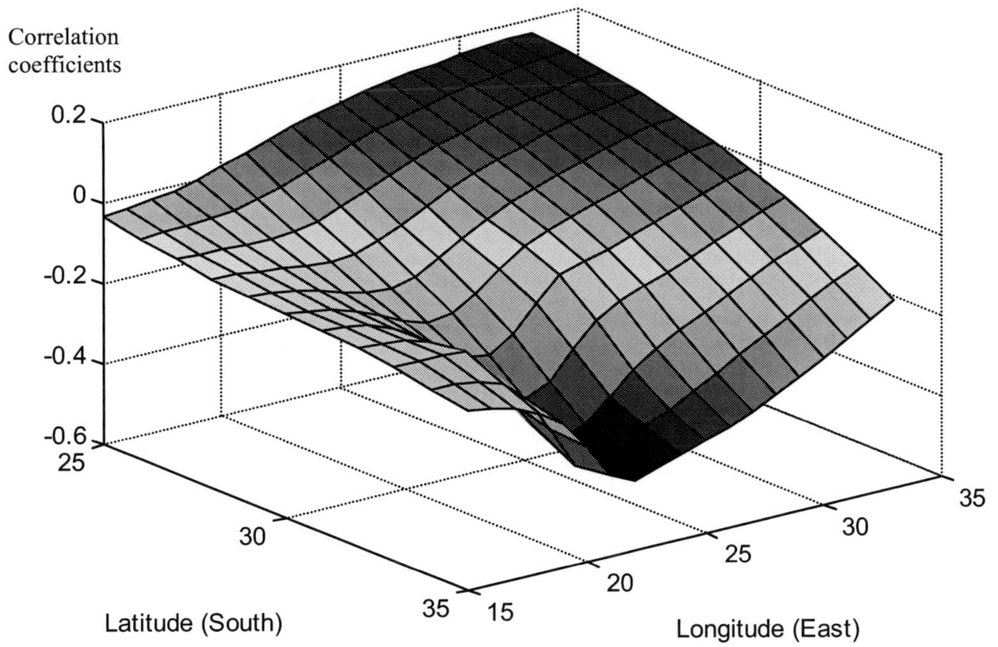

Figure 9.39 Correlation coefficients to find the most likely origin to Group B2. The dip is the region that contains strong negative correlation coefficients and it is, therefore, the most likely origin area for this group.

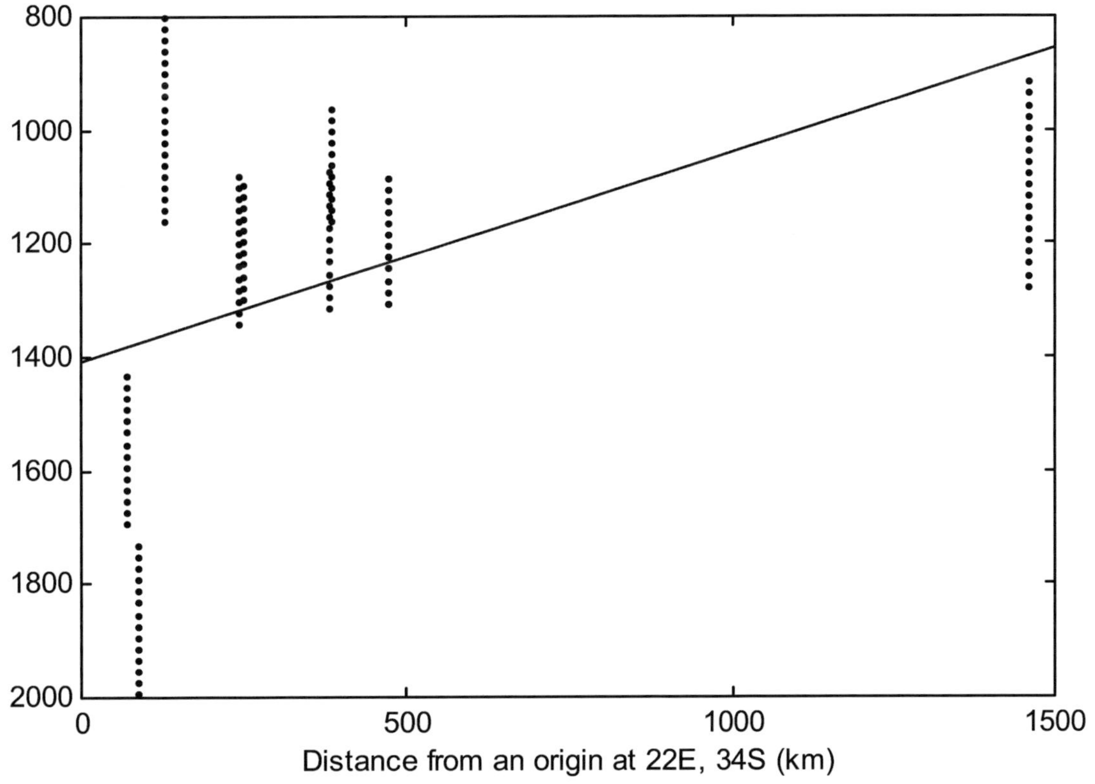

Figure 9.40 Linear regression of the weighted calibrated probability distributions of Group B2 dates and their distance from an origin at 22°E, 34°S. The vertical axis is calibrated years BP.

9.5.1.4 Results for Group B3.

This subset consisted only of sites that are herder with reliable radiocarbon dates. There are five dates in this group. The correlation coefficients for the whole of the surface that was searched to find the most likely origin to Group B3 are shown in Figure 9.41.

The origin range for this Group is very narrowly defined (Figure 9.41). The correlation coefficients at the origin point are large, so there is a linear trend between dates and distance from the origin point. An origin point at 21°E, 34°S with a correlation coefficient of –0.97 was used in the linear regression of dates in Group B3 with their distance from this origin. The results are shown in Figure 9.42.

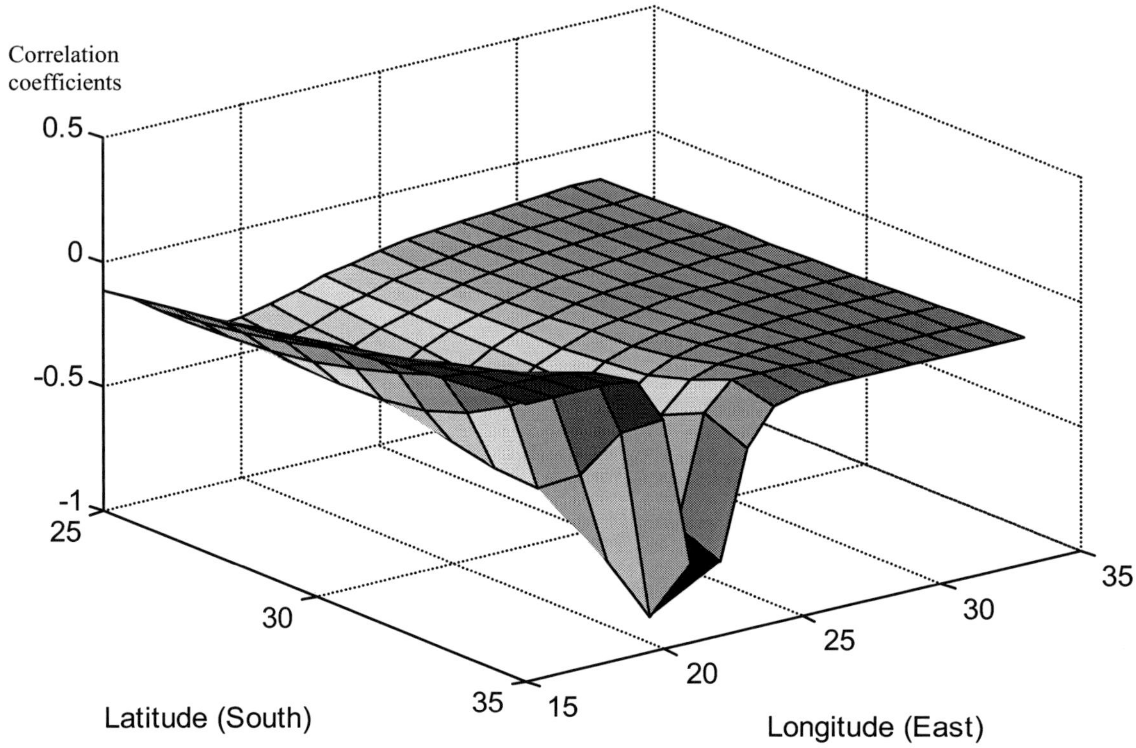

Figure 9.41 Correlation coefficients for Group B3. The origin points with the strongest negative correlation coefficients, and thus the most probable origins for this group, are in the dip area. The vertical axis is correlation coefficient.

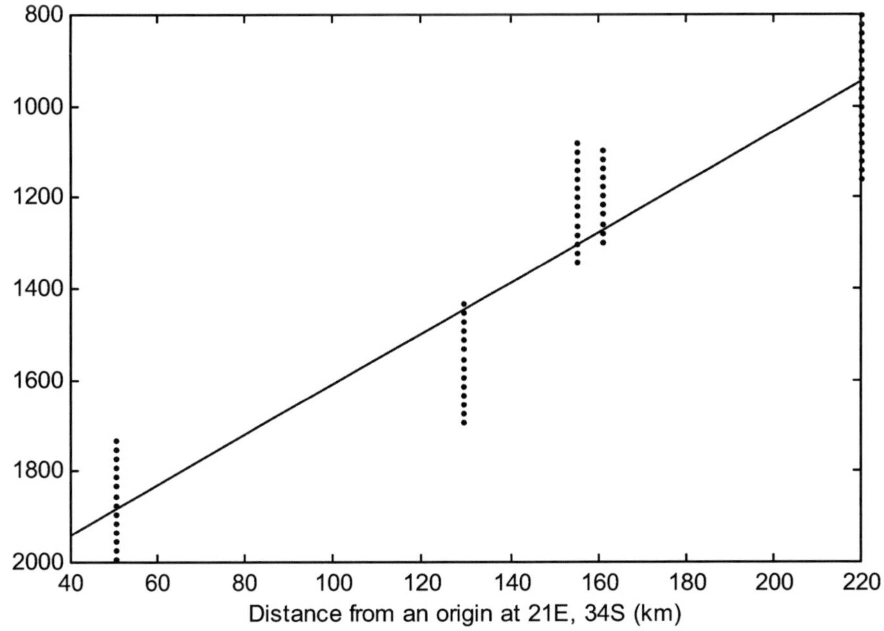

Figure 9.42 Linear regression of the weighted calibrated probability distributions of Group B3 dates and their distance from an origin at 21°E, 34°S. The vertical axis is calibrated years BP.

The equation for the regression line is:

$y_{date} = (-5.52 \pm 0.18)\ x_{dist} + (2160 \pm 30)$

From the equation, the predicted departure date from the origin is in the range 2130 to 2190 calBP and the predicted rate of spread is ± 0.2 km/year.

9.5.1.5 Results for Group B4

This subset excluded the non-herder site at Scott's cave. There are six sites in this group. The correlation coefficients calculated in the search of 200 potential origin points to find the most likely origin to sites in Group B4 are shown Figure9.43.

An origin point from the dip area (Figure 9.43) with coordinates 21°E, 34°S (correlation coefficient, r = -0.89) was used to in the linear regression diagram (Figure 9.44).

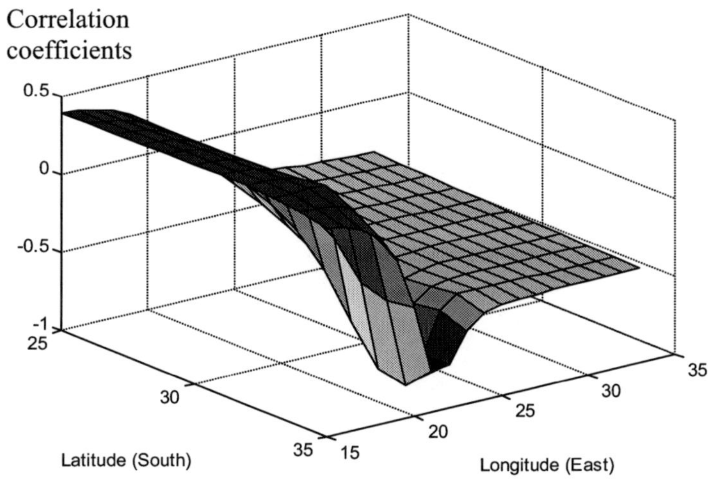

Figure 9.43 Correlation coefficients to find the most likely origin to Group B4. The dip shows the region with large negative correlation coefficients, these are the most likely origins to this group.

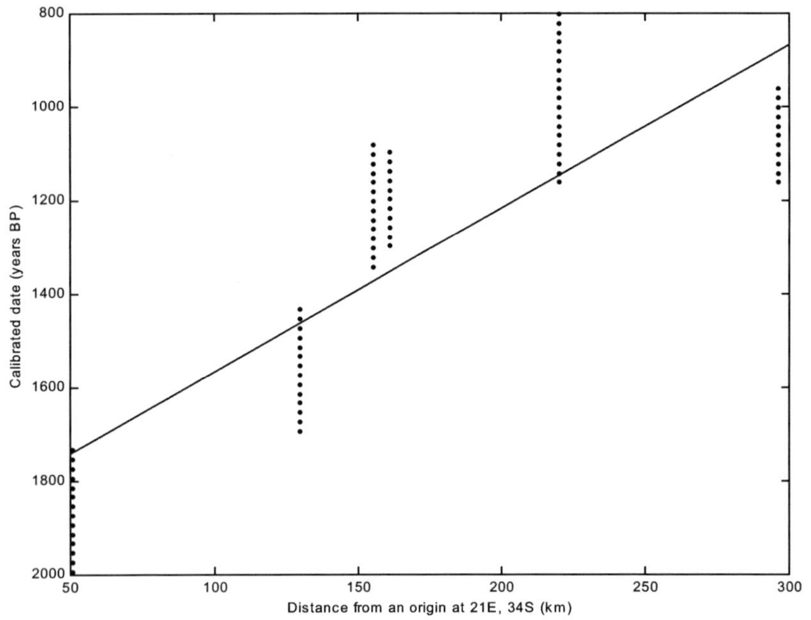

Figure 9.44 Linear regression of the weighted calibrated probability distributions of Group B4 dates and their distance from an origin at 21°E, 34°S.

The equation of the regression line is:

$$y_{date} = (-3.50 \pm 0.20)\ x_{dist} + (1920 \pm 40)$$

This equation gives a predicted date of departure from the origin in the time range 1880 to 1960 calBP and a predicted rate of spread ±0.3 km/year.
The high correlation coefficient –0.9 confirms the strength of relationship between site age and site distance from this origin point.

9.5.1.6 Results for Group A plus Group C

This group consisted of the western coastal sites plus four more central sites (see Figure 9.34). There are eight sites in this group. The correlation coefficients for the whole of the area searched to find the most likely origin for Group A plus Group C are shown Figure 9.45.

The blue area is much broader (Figure 9.45). There are a higher number of possible origin points that fit a linear trend of decreasing date with increasing distance from them. An origin with coordinates 17°E, 31°S (correlation coefficient, r = -0.9) was used to create the linear regression plot in Figure 9.46.

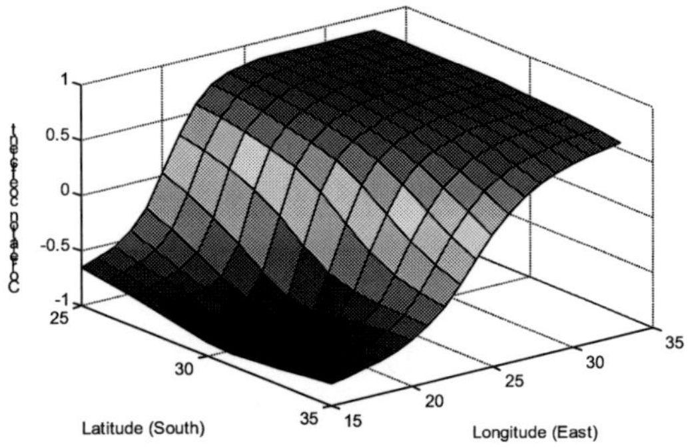

Figure 9.45 Correlation coefficients for Group A plus Group C. The region shows correlation coefficients that are strong and negative and are the most likely origin point for this group. The vertical axis shows correlation coefficients.

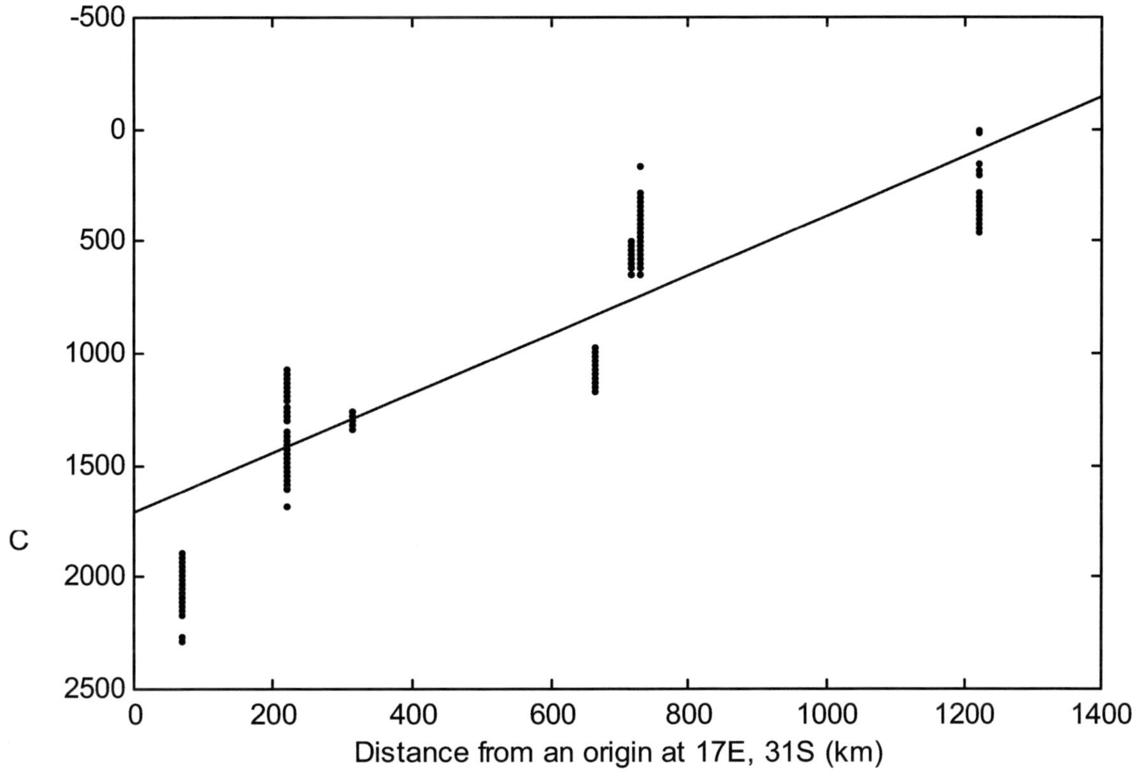

Figure 9.46 Linear regression of the weighted calibrated probability distributions of dates and distance from an origin for Group A plus Group C. The vertical axis is calibrated years BP.

The equation of the regression line is:

$y_{date} = (-1.33 \pm 0.07)\ x_{dist} + (1710 \pm 40)$

From the equation the predicted date of departure from the origin is in the time range 1670 to 1750 calBP and the predicted rate of spread is in the range 0.7 to 0.8 km/year. The problem with this result is that is not reconcilable with the AMS date at Spoegrivier of ±2000 calBP. A number of other origin points were tested and none gave a departure date that was old enough to include the Spoegrivier. This suggests that the two Groups of data combined here cannot in fact be explained by a common origin. The dates in the central region of present day South Africa are all very recent if only AMS dates are taken into account, as is the case here. These site do have older dates associated with sheep but they are single conventional dates on charcoal and they are not considered reliable in the light of the more recent AMS dates. Because of this, until the archaeological evidence for these sites has been re-examined, the decision was made to treat these dates as though they are a later event, post-dating the first introduction of farming.

9.5.2 Discussion of the results

For the radiocarbon dates to support the Stow, Cooke and Bousman western route of introduction there would be a trend for dates to get progressively more recent from the north west corner of present day South Africa southwards to the Cape (Group A). From this southern point a further trend should be discernible with dates getting progressively more recent eastwards along the coast (Group B, B2, B3, B4).

9.5.2.1 The origin point and trend in Group A data:

A search for the most likely origin to Group A by the linear regression of dates and distance from each of the 200 potential origin points did not point to an origin in the north-west corner of South Africa as predicted by the western route of introduction. The position of the most likely origin to Group A data is shown in Figure 9.47. Of the origins predicted by the correlation coefficients only one was plausible when compared to the archaeological evidence for the spread of sheep (Figure 9.47).

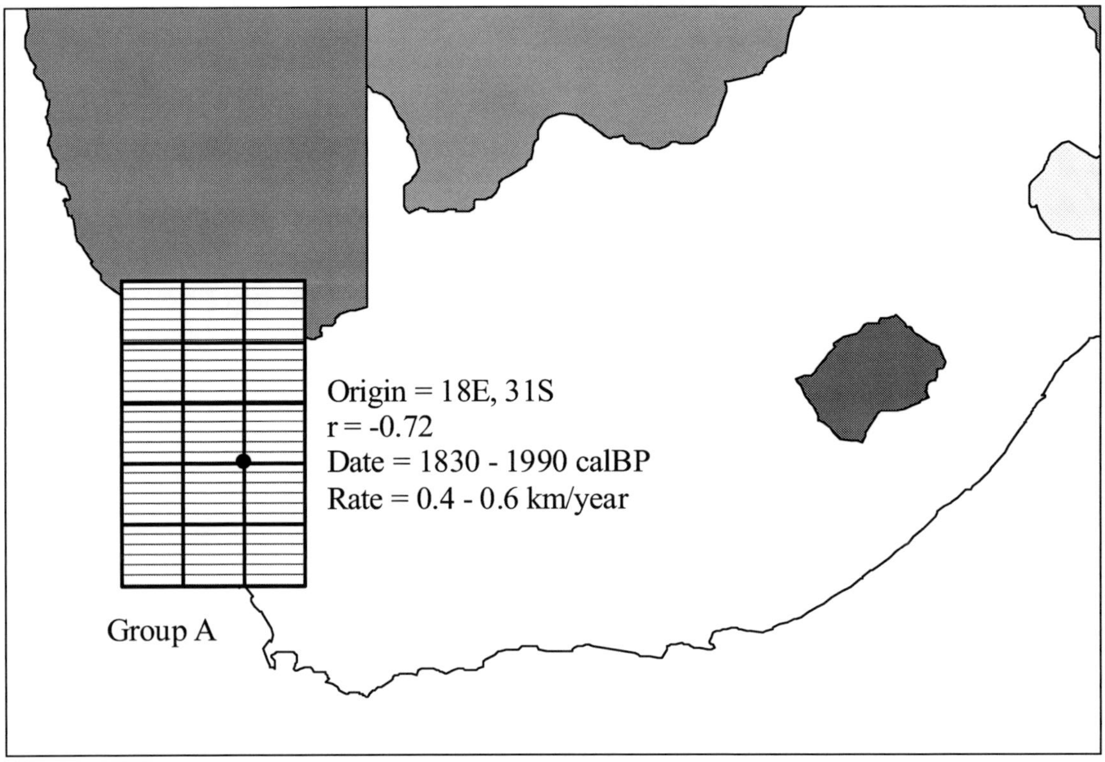

Figure 9.47 The origin point and its correlation coefficient (r), the predicted date of departure from the origin and the predicted rate of spread from the origin to other sites in Group A.

As discussed in section 9.5.1.1 the set of predictions for the rate of spread and time of departure from this origin does not agree with the AMS date for the earliest occurrence of sheep in the group at the site of Spoegrivier. Other possible origin points for Group A gave a predicted date of departure at 3000 calBP or older. This is at odds with the archaeological time frame for the first occurrence of sheep. The model assumes a constant rate of spread. This is most certainly not the way migration/diffusion occurs. By slightly varying the rate of spread the date at Spoegrivier can be accommodated by the model.

Although the predicted origin point is further south than would be expected for a north-south migration, it is not so far south to rule out a north-west origin. However the surface trend shows that from the suggested origin point movement is in two directions north and south (Figure 9.48).

In conclusion the radiocarbon dates in Group A do not show a trend of decreasing age with distance from an origin in the north west corner of present day South Africa. These results do not confirm the western route of introduction of sheep predicted by Stow, Cooke and Bousman.

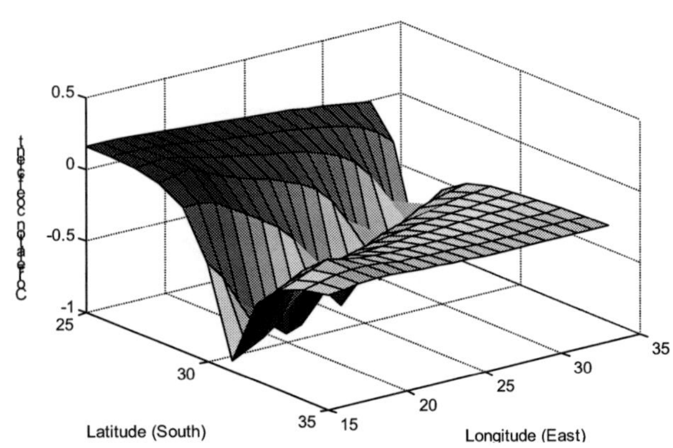

Figure 9.48 This shows that movement from the origin (falling between longitude 15° and 20°) occurs in two directions north and south (full-scale figure is in section 9.5.1.1). The vertical axis is correlation coefficient.

9.5.2.2 The origin point and trend in the Group B data:

To confirm the second leg of the Stow, Cooke and Bousman western route for the introduction of the first sheep, radiocarbon dates should show a trend of a progressive decrease in age eastwards from an origin point in the south-western corner of southern Africa. To test this four variations on the data within this region were analysed, namely Group B, Group B2, Group B3 and Group B4. The routine to search 200 potential origin points across South Africa was applied to these four groups and the results are shown in Figure 9.49 to Figure 9.52 below.

At first glance the origin points in Figure 9.51 and Figure9.52 seem to fit with an origin in the south- west as predicted by the second branch of Stow and Cooke's model. However the trend in the correlation coefficient data seen for both Groups B3 and B4 show that at the origin point movement to other sites occurs in two directions west and east (the figures from previous sections are reproduced below to show the general trend, for detail see the original figure in the relevant section).

For group B4 (Figure 9.53 top) the trend from the origin surface shows that movement away from origin is favoured in a westward direction (in direct contradiction to the eastern direction of movement that the model predicts for this leg of the spread). For Group B3 (Figure 9.53 bottom) the trend in correlation coefficients suggests that from the origin point movement to other sites is in both directions east and west. This does not fit the western migration route hypothesis either.

For Group B and Group B2 the origin points in Figure 9.49 and Figure 9.50 are clearly not in agreement with the second leg of the western route. The correlation coefficient surfaces in Figure9.52 show that whilst the trend (movement both east and west from the origin) seen for Groups B3 and B4 is apparent, it is not as definitely defined. This is because the linear trend of dates and distance from the origin is not strong. The correlation coefficients for both Groups B and B2 are below –0.5. Group B and B2 contain the sites that are either not as securely dated (namely, Smitswinkelbaai) or not clearly defined as either herder or non-herder (namely Scott's Cave, Blydefontein, Lame Sheep Shelter, and Smitswinkelbaai).

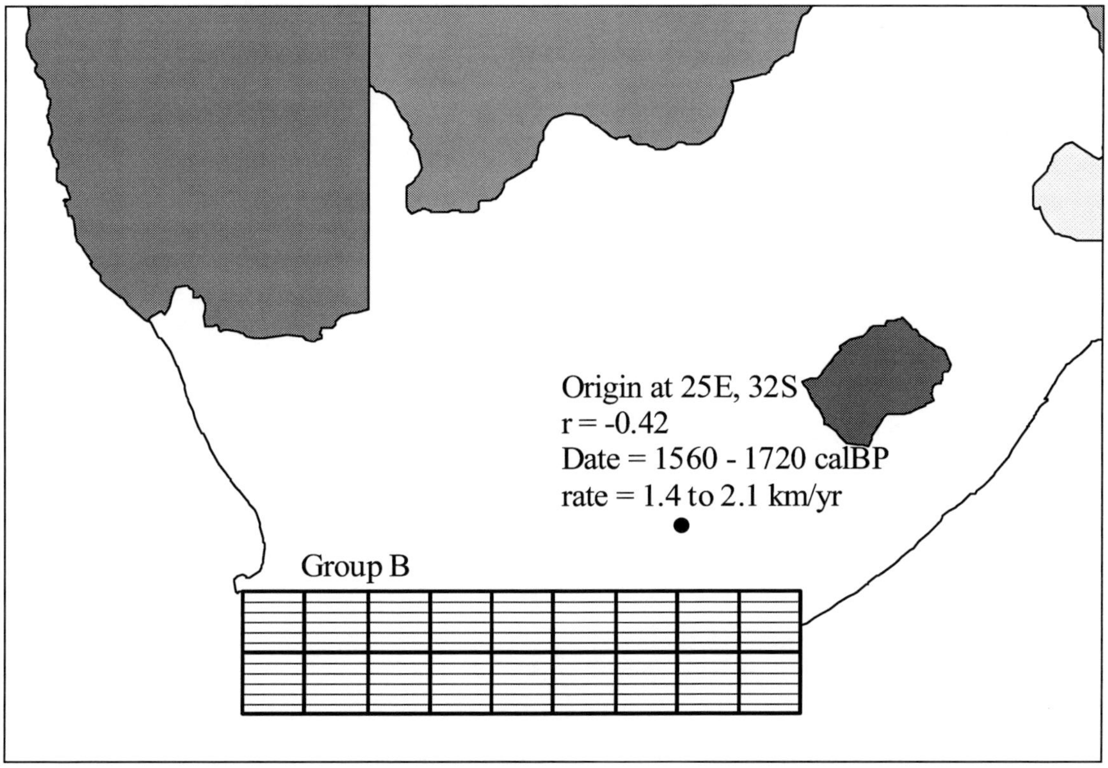

Figure 9.49 The origin and its correlation coefficient (r), the predicted date of departure from the origin and the predicted rate of spread from the origin to all sites in Group B.

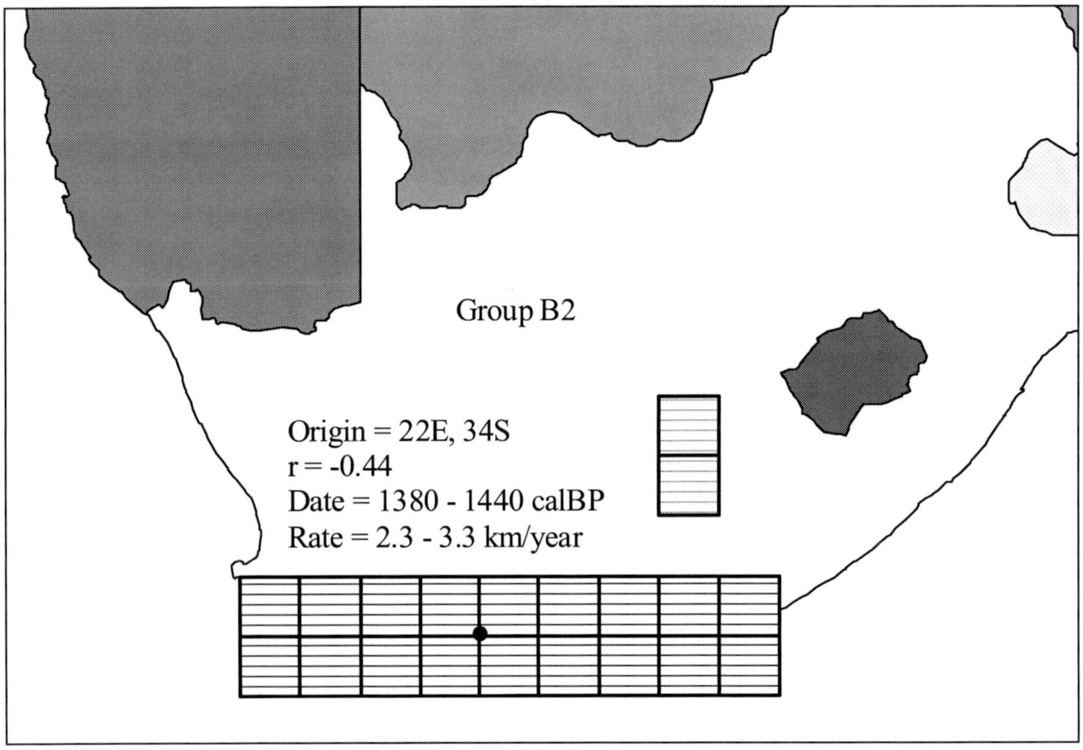

Figure 9.50 The origin point and its correlation coefficient (r), the predicted departure time from the origin and the predicted rate of spread from the origin to all other sites in Group B2.

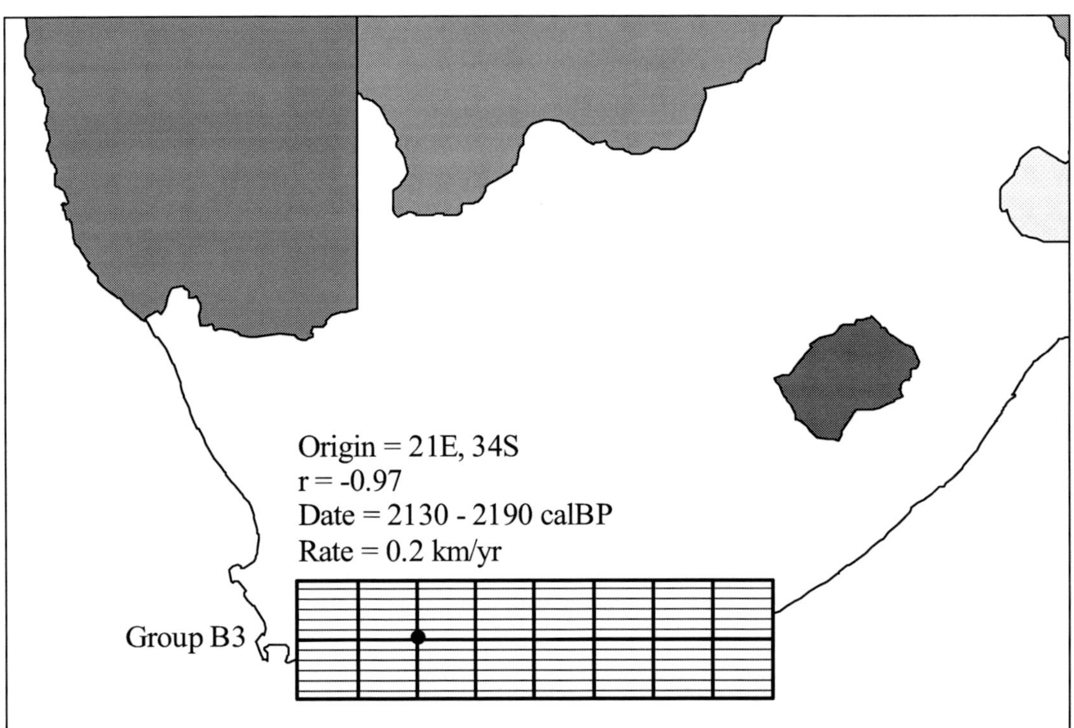

Figure 9.51 The most probable origin and its correlation coefficient (r), the date of departure from the origin and the consequent rate of spread from the origin to all sites in Group B3.

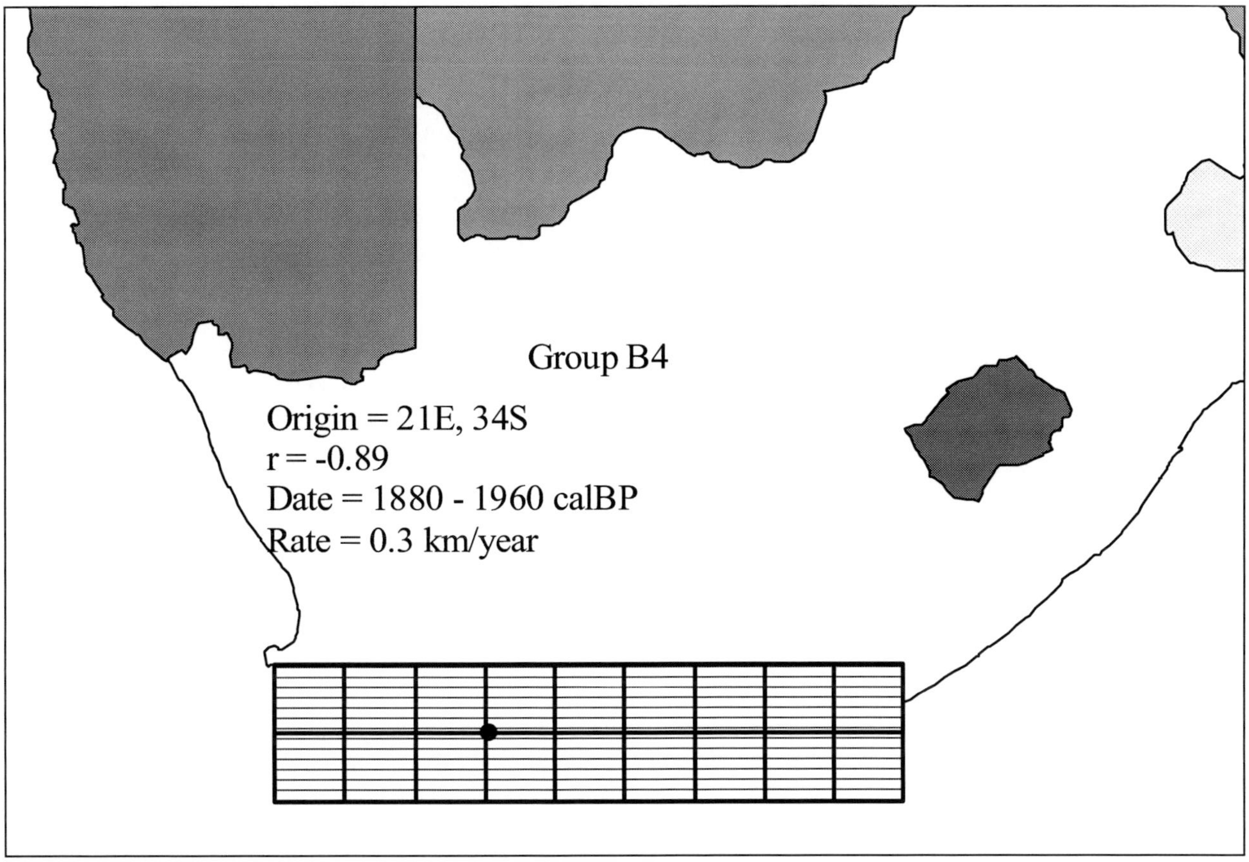

Figure 9.52 The predicted rate of spread and the predicted date of departure from the origin point that showed the best correlation coefficient (r) for the linear regression of date and distance for Group B4.

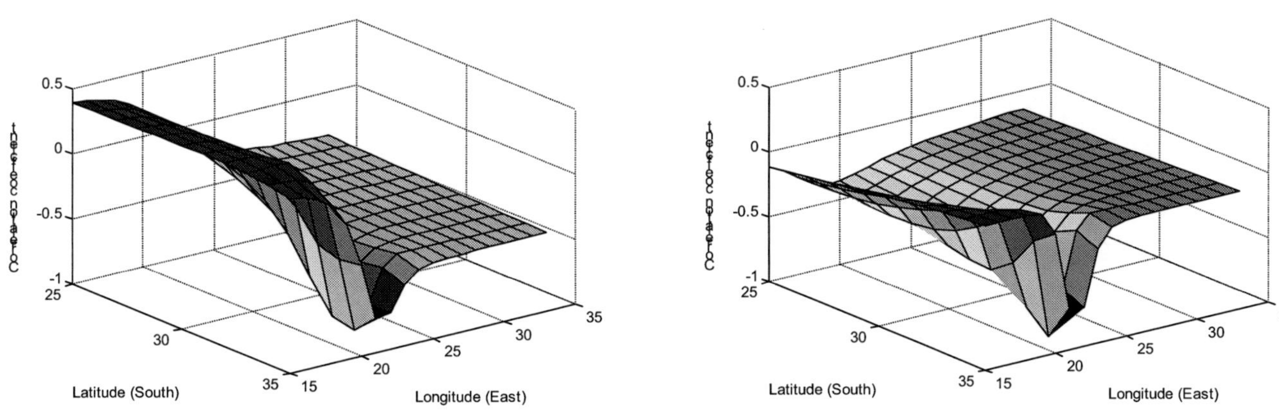

Figure 9.53 The correlation coefficient surfaces produced for Group B4 (left) and Group B3 (right) showing the trend for spread to occur westwards (left) and in both directions (right). The full-scale figures are in sections 9.5.1.3 and 9.5.1.1 respectively.

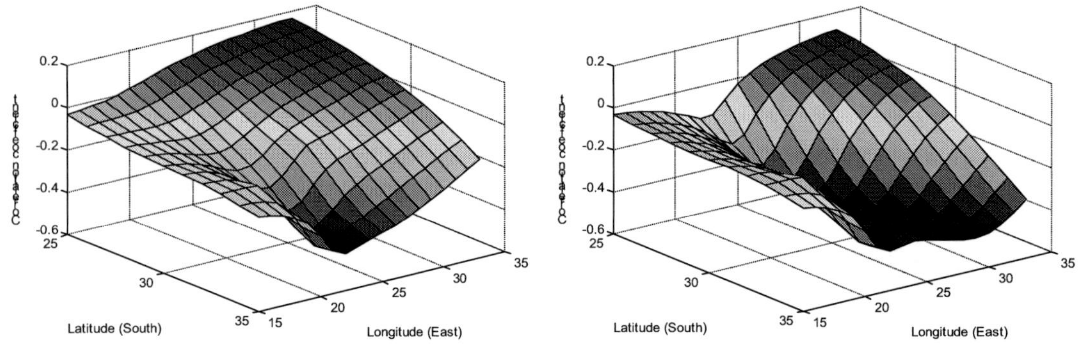

Figure 9.54 Correlation coefficient surfaces for Group B2 (left) and Group B (right) (the full-scale figures are in sections 9.5.1.3 and 9.5.1.2 respectively) to show the trend of spread in two directions from the origin area.

Lame Sheep Shelter and Blydefontein are the two sites found outside of the coastal route (Group B2, Figure 9.34) and just as with the sites in the central area of South Africa (Group C) it is not clear if these sites are linked to the first introduction of sheep in South Africa. Nevertheless, the inclusion of these sites does not change the overall trend in the data. In conclusion the radiocarbon dates in Group B data, in all four variations, do not support the western route for the introduction of the first sheep.

9.5.3 The test of Elphick's route

Elphick's route is shown above (Figure9.55). In an attempt to test Elphick's route I divided it into eight sections (labelled A to H in Figure9.56). If Elphick's route is correct then the radiocarbon dates along each of these eight sections will show decreasing antiquity in the direction of the spread. The radiocarbon dates were allocated to these sections of Elphick's routes (Figure 9.56).

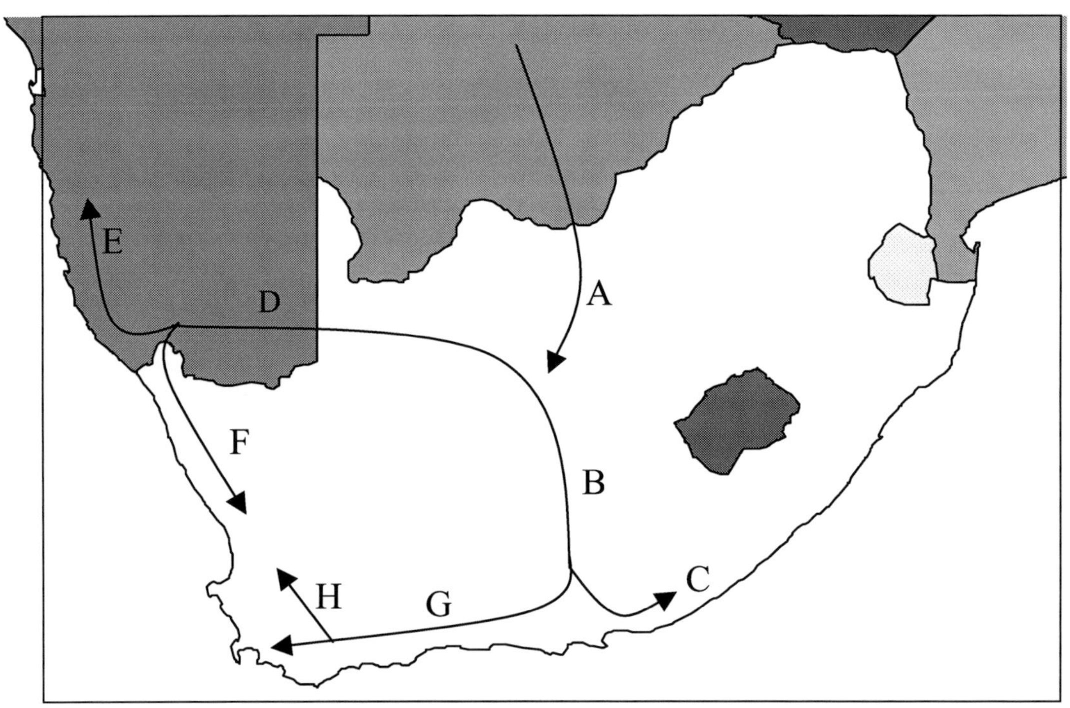

Figure 9.55 Elphick's route for the introduction of the first sheep. The letters A to H and the directional arrows show the sections of the route along which we should expect dates to decrease in age.

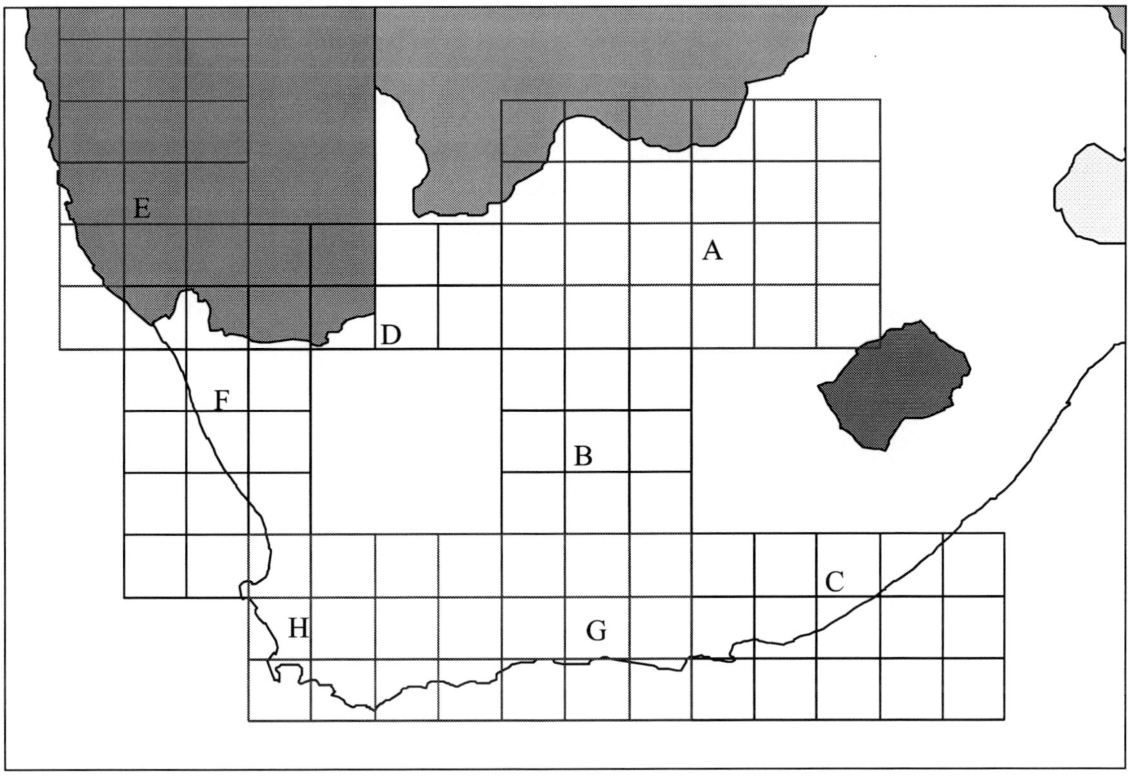

Figure 9.56 Allocation of the radiocarbon dates associated with the spread of the first sheep in order to test the route of spread suggested by Elphick.

A. 22° E to 28°E and 25°S to 29°S: 4 sites (Jubilee shelter (grade 1, non-herder), Wonderwerk (grade 1, non-herder), Blinkklipkop (grade 2, herder) and Dikbosch (grade 3, non-herder)).

B. 22° E to 25°E, 29°S to 32°S: No sites available.

C. 25°E to east coast: 3 sites (Scott's cave (grade 3, non-herder), Lame Sheep Shelter (grade 3, unsure), Blydefontein (grade 3, unsure).

D. 18°E to 22°E, 27°S to 29°S: No sites available.

E. Namibia, outside of study area.

F. 16°E to 19°E, 28°S to 33°S: 4 sites (Spoegrivier (grade 1, herder), Kasteelberg A (grade1, AMS), Kasteelberg B (grade 2, herder), Jakkalsberg A/B (grade 2, herder).

G. & H.[1] 18°E to 25°E, 32°S to 35°S: 6 sites (Boomplaas (grade 2, herder), Nelson Bay Cave (grade 1, Non-herder), Blombos (grade 1, non-herder), De Kelders (grade 1, herder),

H. Byneskranskop (grade 1, herder), Smitswinkelbaai (grade 3, herder)).

9.5.3.1 Results for Group A:

The correlation coefficients to search for the best-fit origin to Group A are shown in Figure 9.57.

There is a large area of possible origin points to this group (Figure 9.57). An origin at 21°E, 25°S with a correlation coefficient of –0.64 was used to create the linear regression plot of date and distance from this origin (Figure 9.58).

The equation of the regression line is:

$$y_{date} = (-1.70 \pm 0.28)\, x_{dist} + (1485 \pm 145)$$

This gives a predicted departure date in the range 1340 to 1630 calBP. The predicted rate of spread is 0.5 to 0.7 km/year.

[1] G and H are combined because there is only one site in Group H. They can be combined because they both would follow the same directional trend according to Elphick's model.

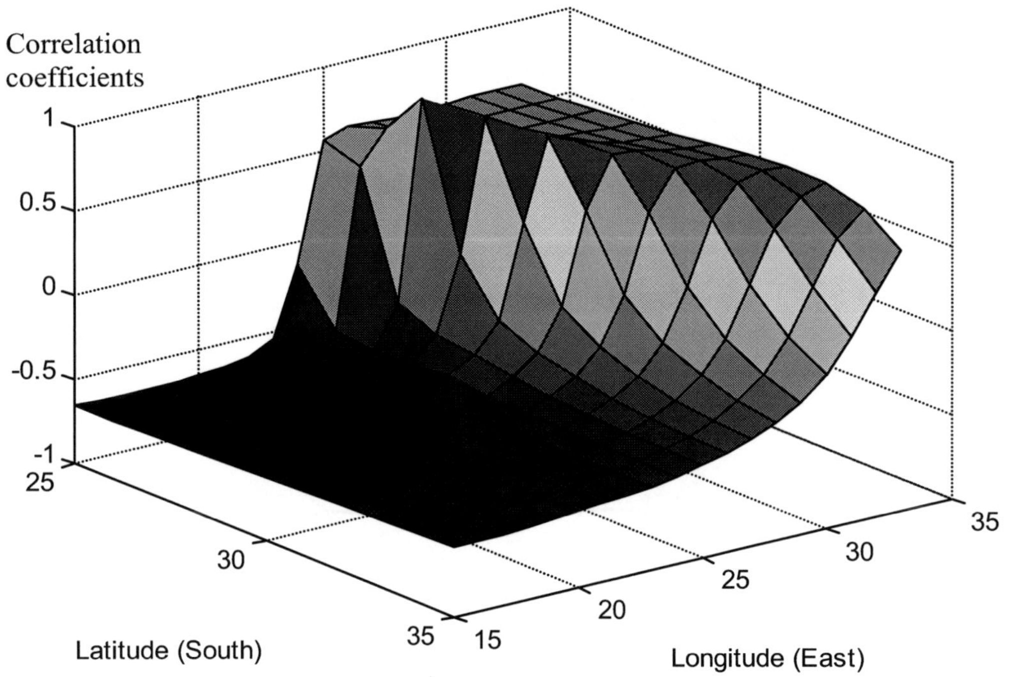

Figure 9.57 Correlation coefficients with the best fit to Group A. The flat dark area shows possible origin points for this group (i.e. the largest negative correlation coefficients). Correlation coefficients are shown on the vertical axis.

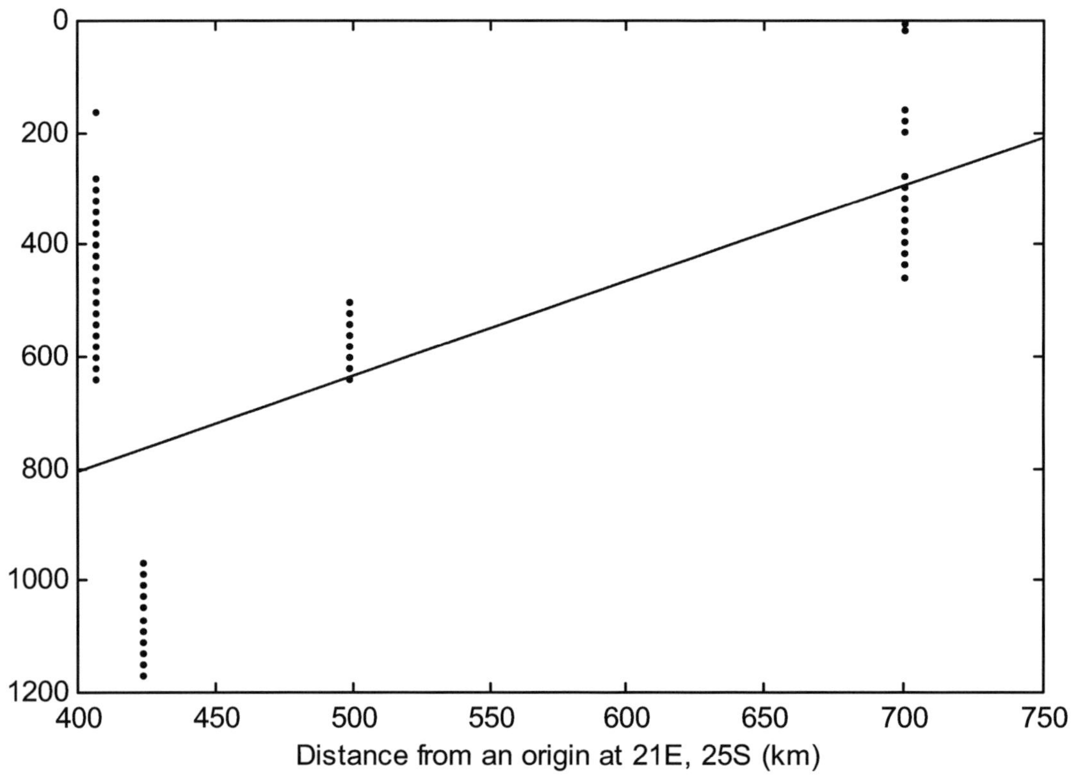

Figure 9.58 Linear regression of the weighted calibrated probability distributions of Group A dates and their distance from an origin at 21°E, 25°S. The vertical axis is calibrated years BP.

This group is equivalent to Group C (to test Stow, Cooke and Bousman's route) in the previous section (section 9.5.1.6), where it was analysed only in combination with Group A. As discussed there, this group is not definitely associated with the first introduction of sheep in to southern Africa. The dating of this group is also not secure as the AMS dates from these sites (used here) are very much younger than their conventional radiocarbon dates.

9.5.3.2 Results for Group C:

The correlation coefficients produced in the search for the most likely origin to Group C are shown in Figure 9.59.

This group only contains 3 sites so the origin area is very broad.

Any of a wide number of origin points has good linear relationship to the three dates (Figure 9.33). Due to the large number of possible origins (with large negative correlation coefficients) to choose for further analysis, the decision was made to take an origin point that would coincide with Elphick's route. An origin at 25ºE, 32ºS with a correlation coefficient of -0.80 was used to draw the regression plot (Figure 9.60).

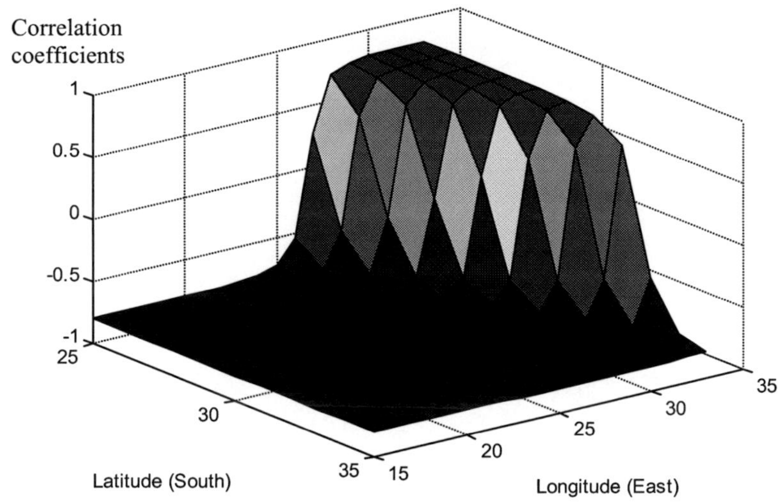

Figure 9.59 Correlation coefficients for Group C. The flat dark area shows the possible origins for Group C based on the strength of the linear relationship between date and distance from each origin point to all the sites in the group.

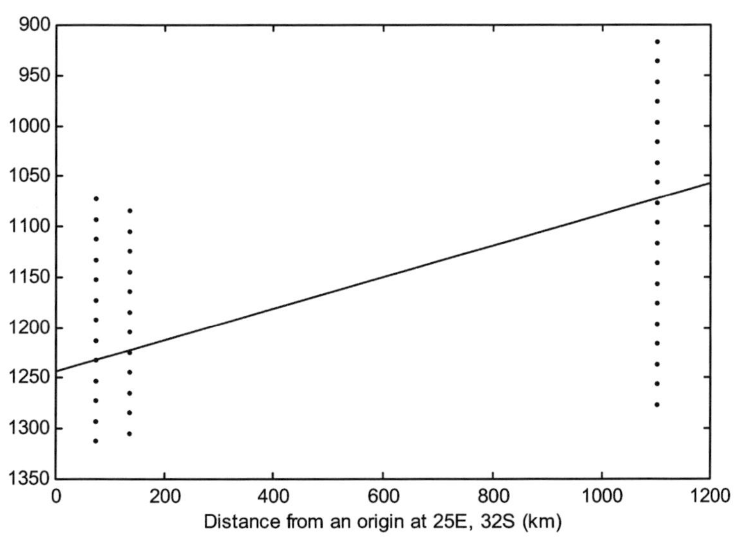

Figure 9.60 Linear regression of the weighted calibrated probability distributions of Group C dates and their distance from an origin at 25ºE, 32ºS. The vertical axis is calibrated years BP.

The equation of the regression line is:

$y_{date} = (-0.15 \pm 0.02)\ x_{dist} + (1240 \pm 10)$

The predicted date of departure from the origin is 1230 to 1250 calBP. The predicted rate of spread from the equation is 5.9 to 7.7 km/year. Two of the sites in this group, Lame Sheep Shelter and Blydefontein are not securely associated with the spread of the first sheep into South Africa.

9.5.3.3 Results for Group F:

The correlation coefficients calculated in the search for the most likely origin to Group F sites are shown in Figure 9.61. This group is identical to Group A in the section to test the western route of introduction (section 9.5.1.1). So the results are identical.

All of the four minima shown in Figure 9.61 were explored. As was discussed in the section 9.5.1.1 to test Stow, Cooke and Bousman's western route, all but one of the origins produced a predicted departure dates that was older than the archaeological evidence would suggest. For example, the origin point at 15°E, 31°S with a correlation coefficient of 0.92 was analysed; the regression plot is shown in Figure 9.62.

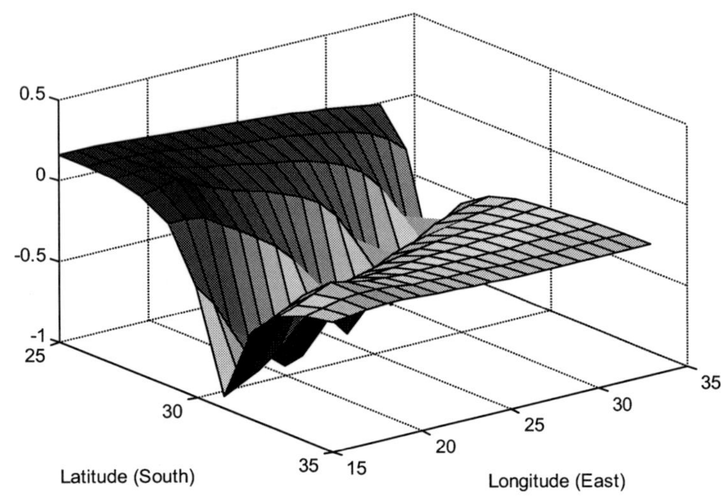

Figure 9.61 There are a number of minima (the dips), these are associated with large negative correlation coefficients and as thus the most probable origin points to this Group. The vertical axis shows correlation coefficients.

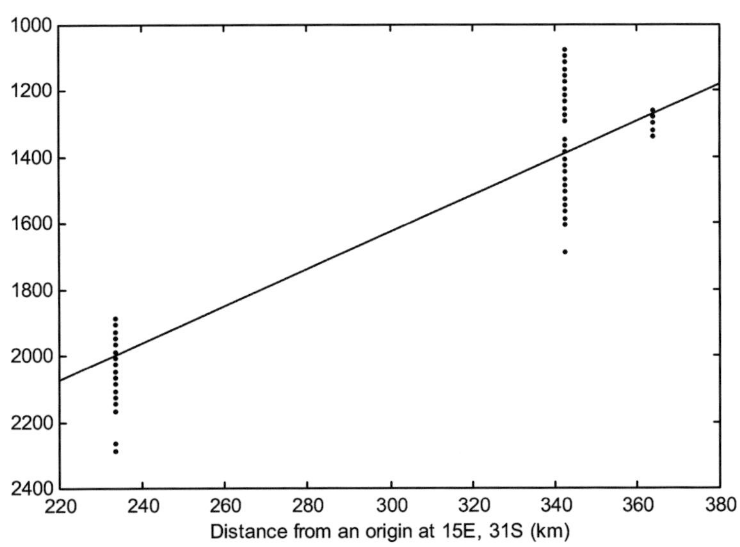

Figure 9.62 Linear regression of the weighted calibrated probability distributions of Group F dates and their distance from an origin at 15°E, 31°S. The vertical axis shows calibrated years BP.

The equation of the regression line is:

$y_{date} = (-5.6 \pm 0.34)\, x_{dist} + (3300 \pm 120)$

The predicted departure time is greater than 3000 calBP. This result does not fit in with the archaeological evidence for this region.
The origin at 18°E, 31°S (correlation coefficient r = 0.72) produces a result that is more plausible archaeologically (see section 9.5.1.1). The equation of this regression line is:

$y_{date} = (-1.99 \pm 0.28)\, x_{dist} + (1910 \pm 80)$

The predicted departure time is 1830 to 1990 calBP and the predicted rate of spread is 0.4 to 0.6 km/year. The graph for this regression is shown in section 9.5.1.1.

9.5.3.4 Results for Group G + Group H.

Group G and H are analysed together as there is only one site in the region of Group H. This combination of these two groups is not at odds with the direction of spread of the routes in Elphick's model. This group is identical to Group B4 in section 9.5.1.5. The correlation coefficients to search for the most likely origin to Group G + H are shown 9.63. An origin point from the blue area (Figure 9.63) with coordinates 21°E, 34°S (correlation coefficient, r = -0.89) was used to draw the linear regression diagram (Figure 9.64).
The equation of the regression line is:

$y_{date} = (-3.50 \pm 0.20)\, x_{dist} + (1920 \pm 40)$

This equation gives a predicted date of departure from the origin in the time range 1880 to 1960 calBP and a predicted rate of spread ±0.3 km/year.
The high correlation coefficient –0.89 confirms the strength of relationship between the site's ages and their distance from this origin point.

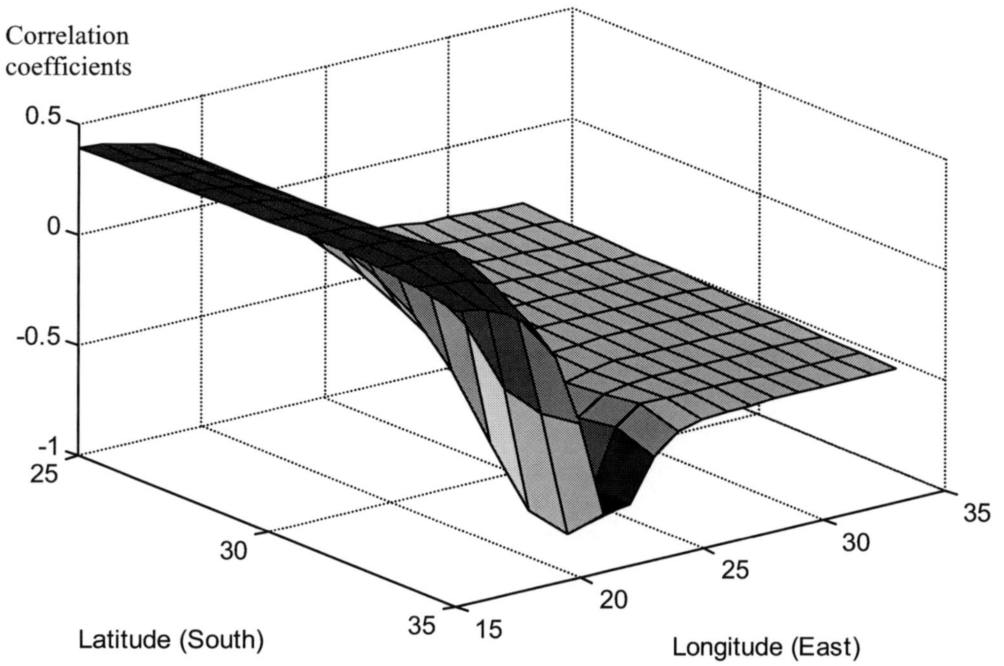

Figure 9.63 Correlation coefficients to find the most likely origin to Group G + H. The minima (dip and area to the right of it) show the region that contains the strongest negative coefficients. This is the most likely origin area for Group G + H.

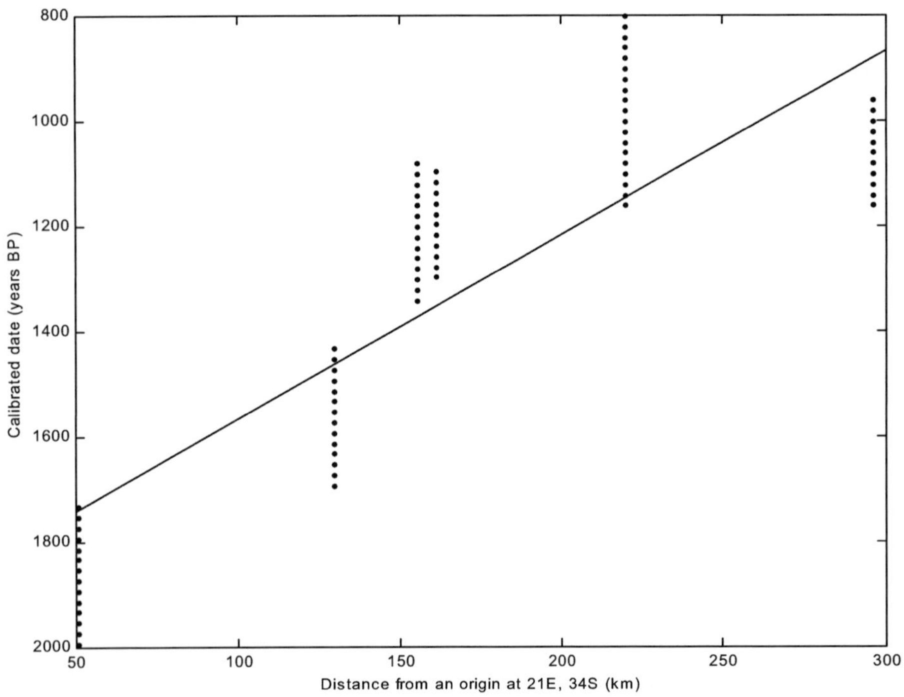

Figure 9.64 Linear regression of the weighted calibrated probability distributions of group G + H dates and their distance from an origin at 21°E, 34°S.

9.5.3.5 Results for Group G + Group H + Group C:

As there are only 3 sites in Group C the consequence was that the results for this group were very broad. Because of this the decision was made to combine Group C with Groups G + H. This was also a test to see whether the east – west split that Elphick predicts on the southern coast shows in the results. This group is identical to Group B2 in section 9.5.1.3. The correlation coefficients from the search for the most probable origin for Group G + H + C are shown in Figure 9.65.

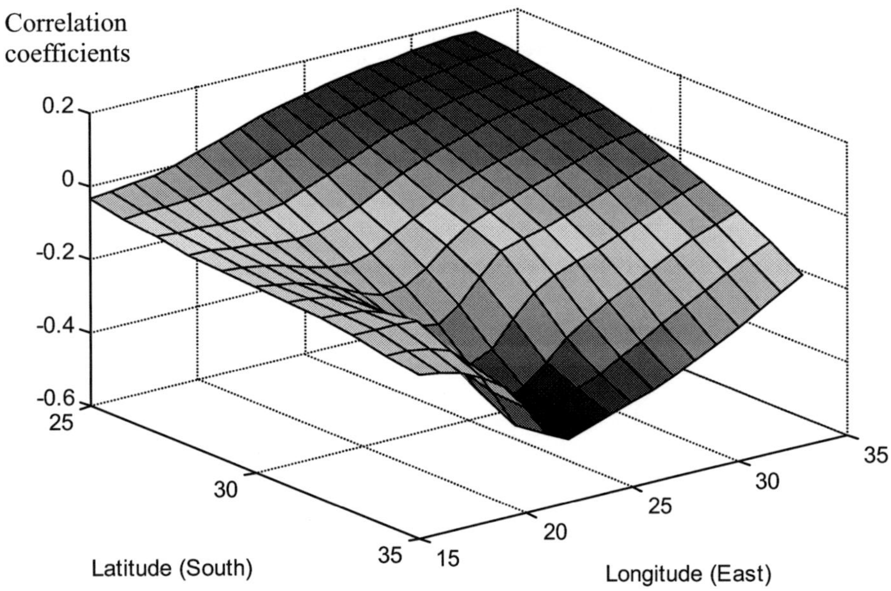

Figure 9.65 Correlation coefficients for the combined group G + H + C. The dip region shows the common origin point for these groups.

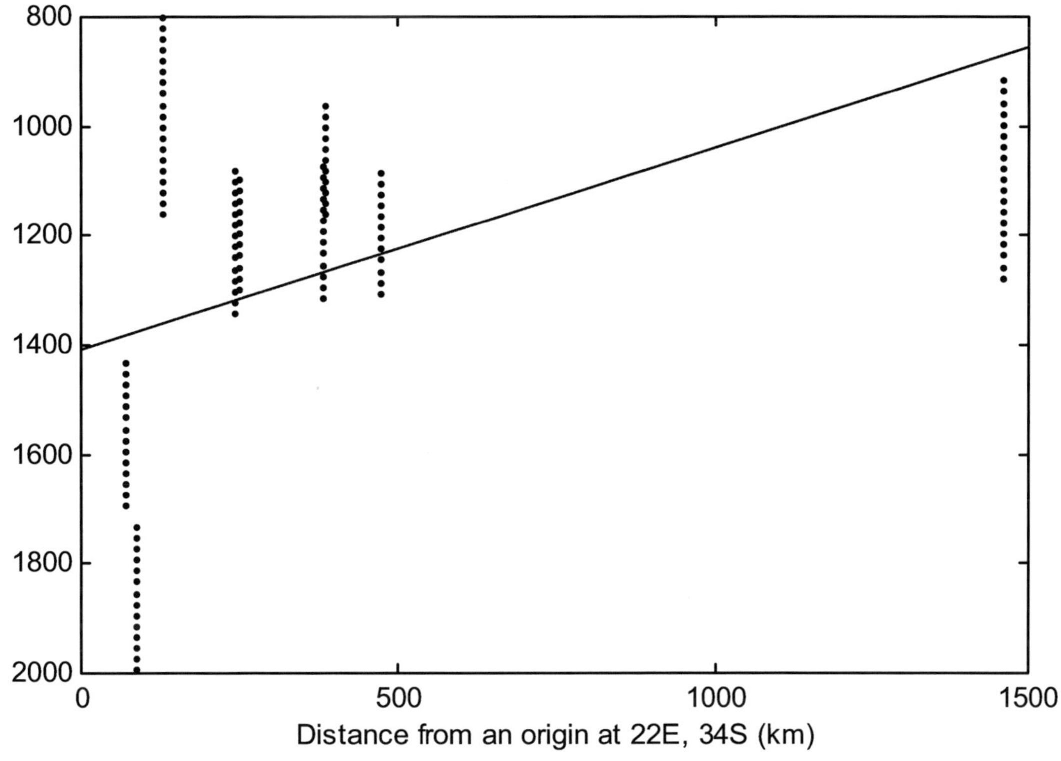

Figure 9.66 Linear regression of the weighted calibrated probability distribution of Group G + H + C dates and their distance from an origin at 22°E, 34°S. The vertical axis shows calibrated years BP.

The east-west split is clear in this diagram. However the correlation coefficients for the blue area (the most likely area of origin) are low (± -0.44), so the linear relationship between the age of sites and their distance from the origin is not strong for this group. An origin point at 22°E, 34°S (Figure 9.65) was used to create the regression plot for this combined group (Figure9.66).

The equation of the regression line is:

$y_{date} = (-0.37 \pm 0.07)\ x_{dist} + (1410 + 30)$

Using this equation, the predicted time of departure from the origin is in the range 1380 to 1440 calBP and the predicted rate of spread is 2.3 to 3.3 km/year. This predicted date of departure from the origin point does not fit with the age of the oldest sites in this group. This, coupled with the low correlation coefficient, suggests that the rate of spread in this group was too variable to yield a linear trend. The combination of these sites into one group may be misleading. As already stated the place of Blydefontein and Lame Sheep Shelter in the first spread of sheep is not clear; these two sites may represent later spread into the interior.

9.5.4 *Discussion of the results*

The object of this exercise was to try to test whether Elphick's route for the first occurrence of sheep in southern Africa is confirmed or refuted by examining the radiocarbon dates. Due to the paucity of available data the strategy proposed in section 9.5.3 (Figure 9.55) was only partly achieved. There are no sites with the evidence for the early appearance of sheep in areas B and D in Elphick's model (Figure 9.55). Area E (Namibia) does contain sites, but this area fell outside the study area. The only areas with relevant data are areas A, C, F, G and H. However the archaeological evidence in area A and C are not reliably dated nor reliably associated with the spread of the *first* sheep. This leaves the results from areas F, G and H to consider (Figure 9.67).

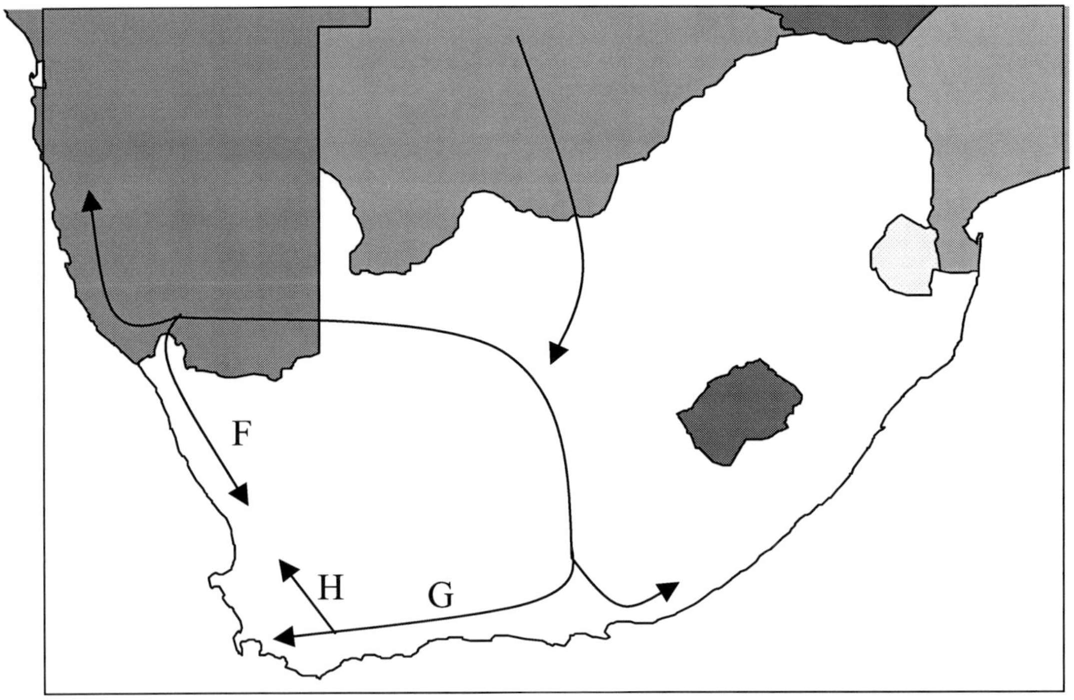

Figure 9.67 Elphick's routes for which there is reliable archaeological evidence for the spread of the first sheep.

9.5.4.1 The trend in Group F data

If Elphick's route were to be confirmed then dates in Group F should be old in the north west of southern Africa becoming progressively more recent with movement southwards to a point half way down the west coast of present day South Africa (Figure 9.67). The group F data is identical to that used to test the Stow and Cooke model (Group A, section 9.5.1.1). Results do confirm this trend.

9.5.4.2 The trend in the Group G and H data

As there is only one site in Group H these groups were analysed together. The results once again support Elphick's route. The west to east split on the south coast is confirmed (Figure 9.68, Figure 9.69, Figure 9.70).

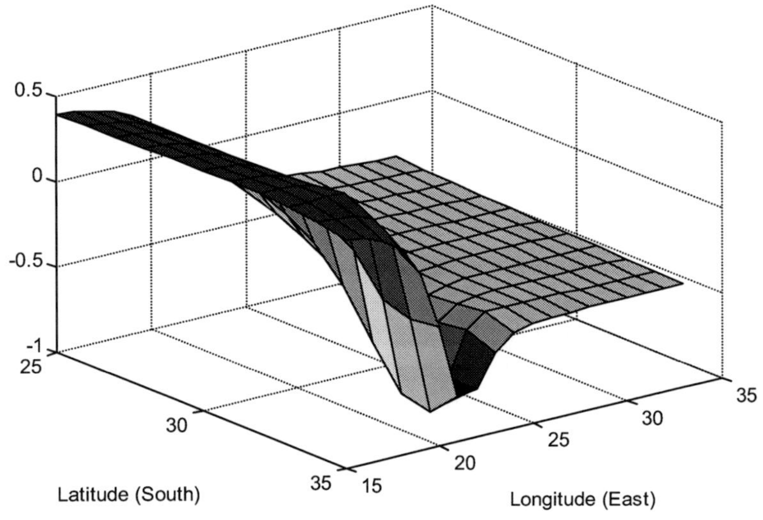

Figure 9.68 Correlation coefficient surface for Group G + H to show the east-west split around the origin point. Here the trend away from the origin in a westwards direction is predominant. Correlation coefficients are shown on the vertical axis.

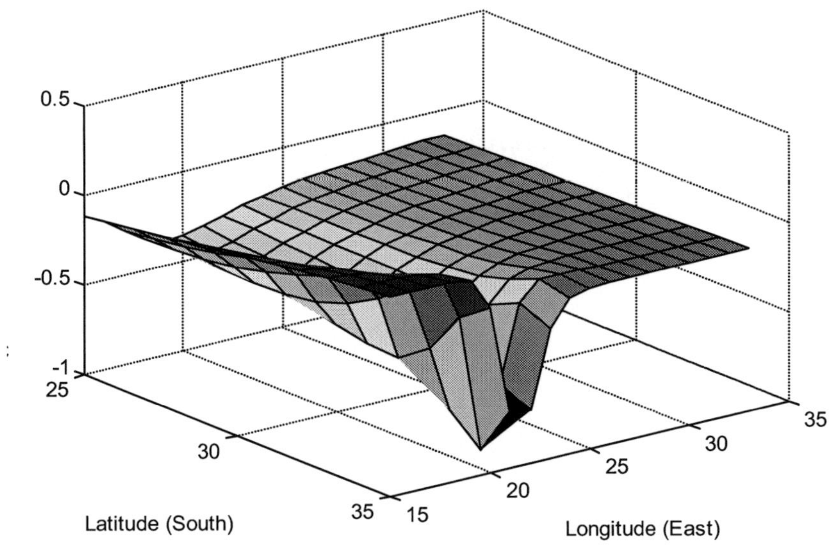

Figure 9.69 Correlation coefficient surface for Group B3 (section 9.5.1.1). This is a variation on the Group H + C data that was tested for the Stow and Cooke model, but contains just the five most securely dated sheep sites for this area. The trend for a movement in both directions, east and west from the origin point is clear. The vertical axis shows correlation coefficient.

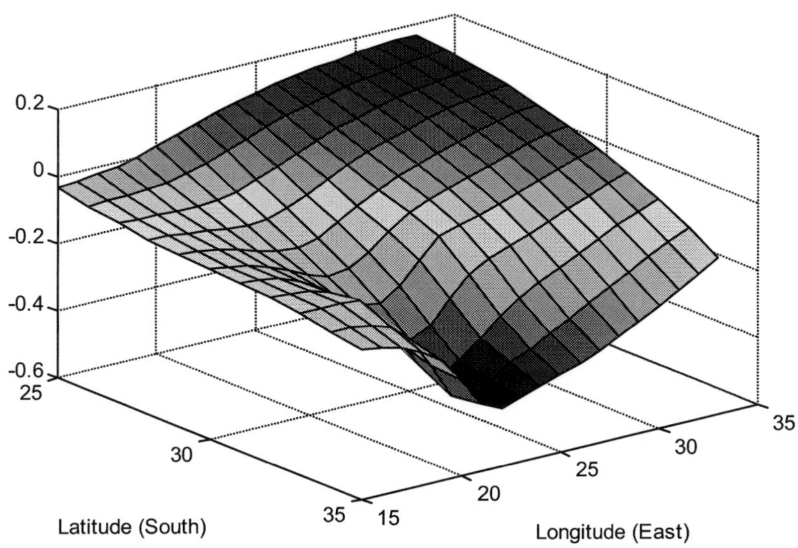

Figure 9.70 Correlation coefficient surface for Group G + H + C. Once again the movement in two directions, east and west from the origin point is evident. The vertical axis shows correlation coefficient.

From the same group of dates different combinations were drawn to see the effect of excluding and including less securely known sites. Therefore the results for Group B in the Stow and Cooke section and those considered here (Group H + C) are all variations on the same set of dates. They all showed the same trend in support of Elphick's route. However, whilst all supporting the pattern, the predicted dates of departure from the origins for each of the combinations of dates differs (Table 9.3).

Those that give high correlation coefficients fit the archaeological evidence more securely than those with poor correlation coefficients.

A summary of the results from the test of the Elphick model is shown in Figure 9.71.

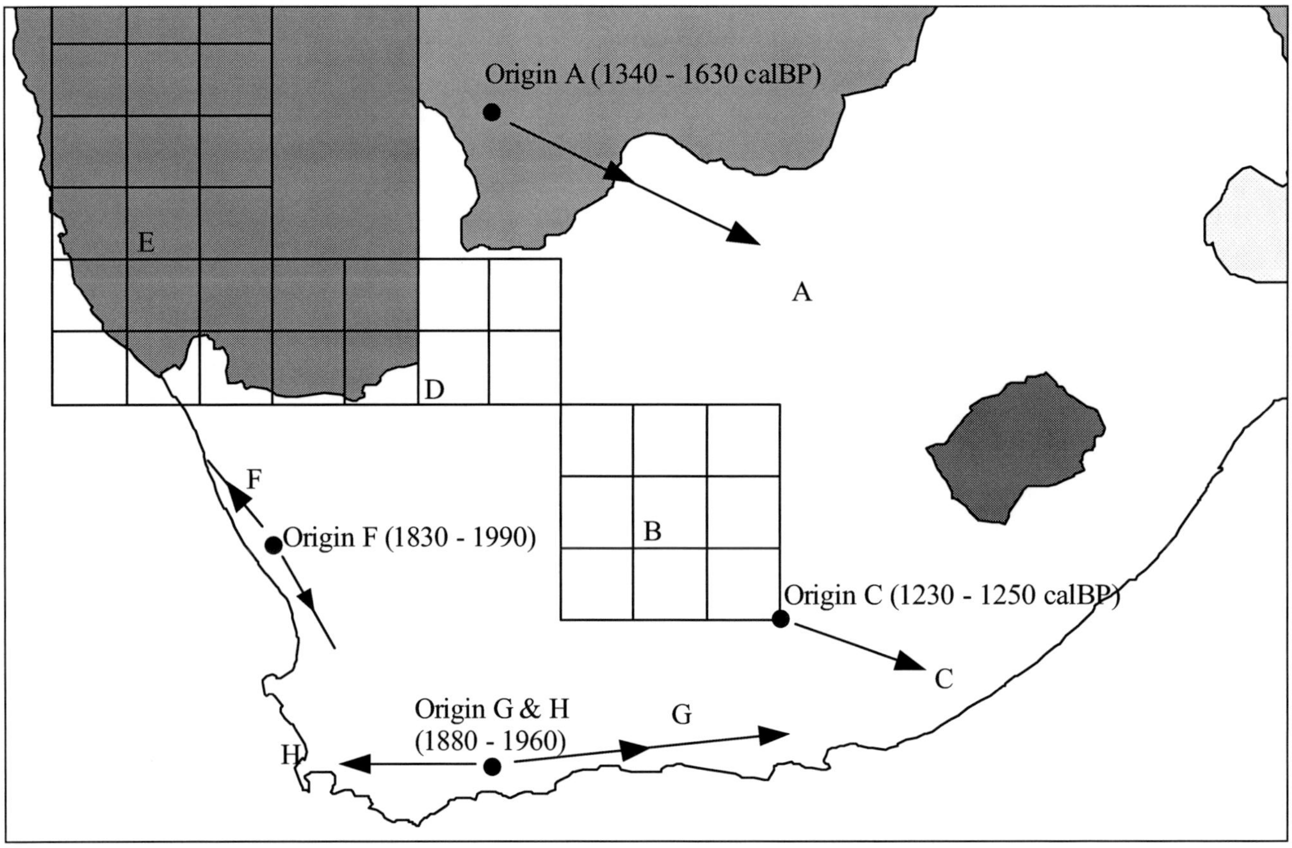

Figure 9.71 The results for the test of Elphick's route of spread. The movement in two directions, east and west from an origin point on the south coast is seen in the radiocarbon dates. There is no data for sheep in B and D. Area E has evidence for sheep but it lies outside the study area. Areas A and C contain sites with very recent dates that contradict the conventional dates or insecurely dated associations with sheep, so the way in which these two groups fit in with other earlier evidence for sheep in the rest of South Africa is unclear. The origin for Group F data is not in the far north-west corner as expected but further down the coast.

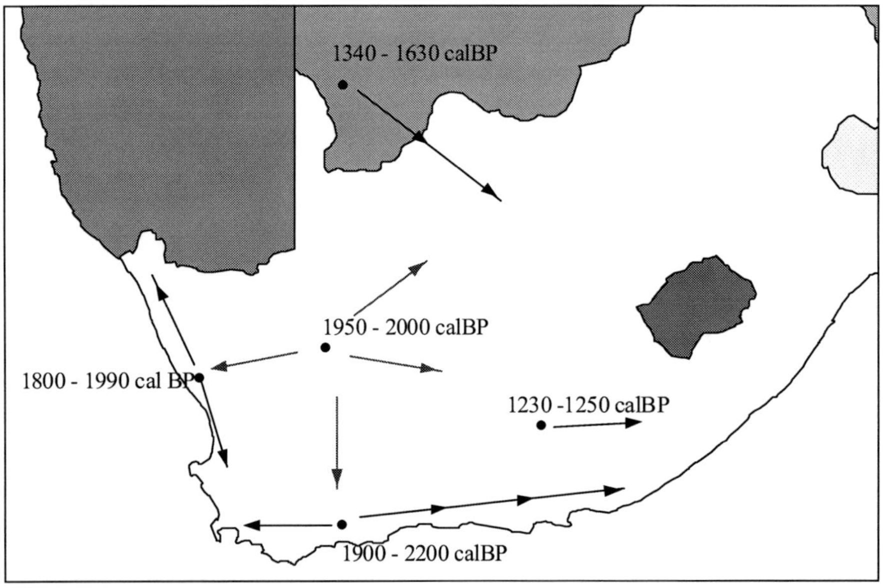

Figure 9.72 The main trends of movement of the first sheep into southern Africa suggested by the spatial analysis of radiocarbon dates. The dots are origins of dispersal and their accompanying predicted date. The arrows leave what I consider to be an area where stock keepers travelling from further north fragmented, with groups moving in different directions.

Group	Sites	Origin	Correlation coefficient	Time at origin (calBP)	
Stow & Cooke & Bousman Group B1	Smitswinkelbaai Boomplaas Nelson Bay Cave Blombos Die Kelders Byneskranskop Scott's cave	25ºE, 32ºS	-0.42	1560 -1720	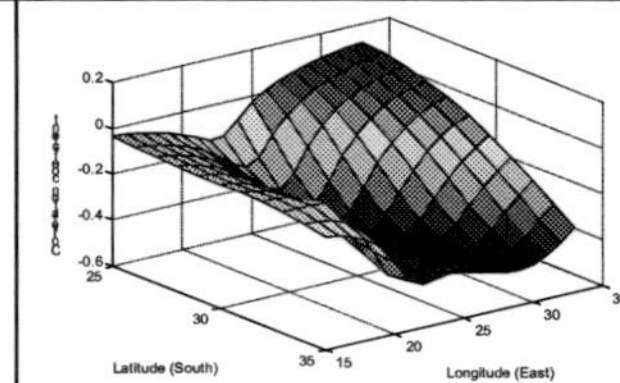
Stow & Cooke & Bousman Group B2	As Group B1 + Lame Sheep Shelter Blydefontein	22ºE, 34ºS	-0.44	1380 - 1440	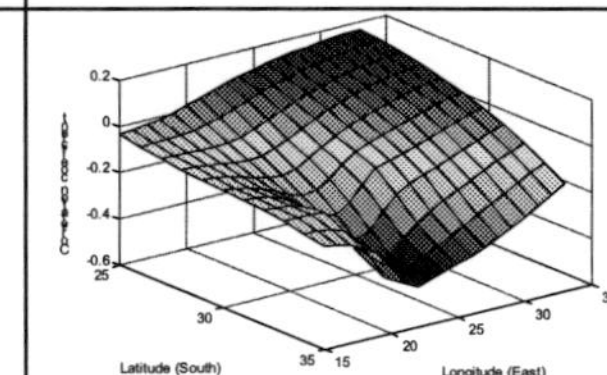
Stow & Cooke & Bousman Group B3	Boomplaas Nelson Bay Cave Blombos Die Kelders Byneskranskop	21ºE, 34ºS	-0.97	2130 - 2190	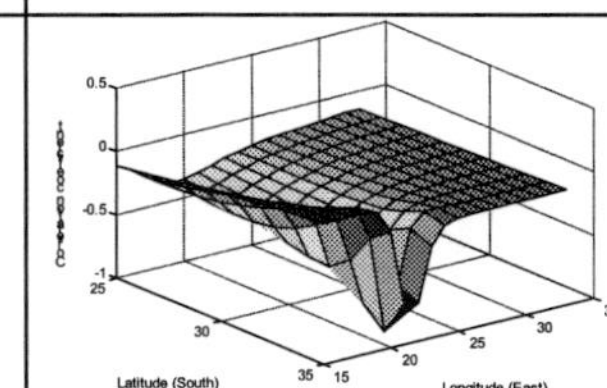
Stow & Cooke & Bousman Group B4	As Group B1 but without Scott's Cave	21ºE, 34ºS	-0.89	1880 - 1960	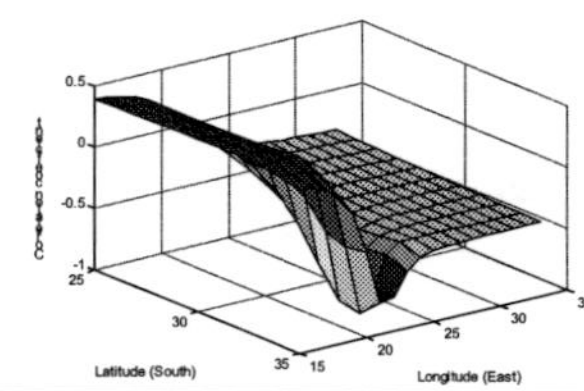
Elphick Group G + H	As Group B4	As Group B4	As GroupB4	As Group B4	
Elphick Group G+H+C	As Group B2	As Group B2	As Group B2	As Group B2	
Elphick Group C	Scott's Cave Blydefontein Lame Sheep Shelter	25ºE, 32ºS	-0.80	1230 - 1250	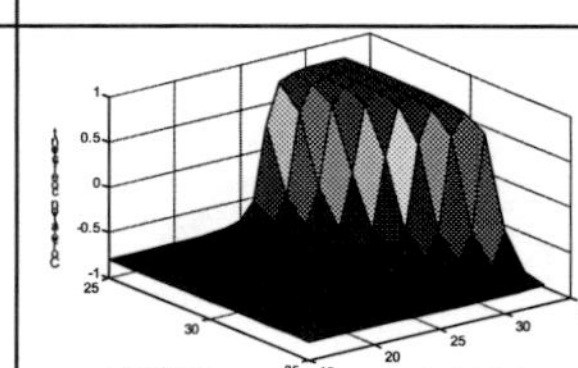

Table 9.2 Comparison of the results from different combinations of dates selected from the same pool (used to analyse the second leg of Stow, Cooke and Bousman's route and the southern coast leg of Elphick's route).

9.5.5 Conclusions on the most plausible route for the introduction of sheep based on the radiocarbon dates for their first occurrence

Of the two routes tested, the radiocarbon dates lend support to Elphick's route rather than the western route proposed by Stow and Cooke, and more recently, by Bousman. The radiocarbon analysis does not however confirm that either route is correct. When the results from both analyses are combined, the discernible trends are best illustrated diagrammatically (Figure 9.72). The two younger origin points (i.e. 1340 – 1630 calBP, far north in present day Botswana and 1230 – 1250 calBP, in the south-east) are possibly later expansions of sheep and not directly related to the first spread. The other three origin points are based on the analysis of secure data relating to the first spread of sheep. The predicted departure dates at the three origin points are very similar, so they can only be accommodated as a group if the spread to each one is very rapid.

9.6 Summary:

The aim was to see whether spatial analysis of radiocarbon data associated with the spread of the first sheep into South Africa could be used in support of either of the two routes that have been proposed for their introduction.

Radiocarbon dates have recently been used to support the western route of spread as first proposed by Stow and Cooke (Bousman 1998). The results presented here show that the radiocarbon data lends support to Elphick's route rather than that proposed by Stow and Cooke.
This analysis suffered from the small sample size, a biased distribution and the poor quality of radiocarbon data at certain sites. South Africa is a large country and much of the interior remains un-surveyed. To quote Dillehay & Meltzer (1991), "absence of evidence is not evidence of absence". Until such time as these gaps in the archaeological record are filled, firm confirmation of either route is impossible. The area for analysis must also be extended to encompass the relevant Namibian sites[2].

---

[2] Early evidence of sheep in Namibia that might support the western route of spread has not been found, see Bousman's map (Figure 8.7, Chapter 8). Sites in Namibia predating those in South Africa have not yet been found.

Chapter 10: Conclusions

The questions posed at the start of the research asked whether the new and larger body of geographically referenced radiocarbon data for Europe might throw new light on the spread of the first farming, and whether new methods of manipulating this data might further our knowledge of the spread of farming in Europe and of domestic sheep in South Africa.

The emphasis throughout the research has been on the methodological treatment of radiocarbon dates. Questions posed about the Neolithic dispersion across Europe will not be answered satisfactorily until the deficiencies with the Gkiasta database (as noted in Chapter 2) have been addressed. The quality of the underlying data has reduced the reliability of any conclusions drawn about the spread of the first farmers into Europe, and it is with this caution to the reader that tentative findings regarding the spread of the first farmers to Europe are presented below. As stated in Chapter 2[1] it is essential that the stratigraphic sequence at each site be well understood: the association between what is being dated and the event that is under analysis must be clearly defined before the methodology that has been developed in this research can be used to answer questions that are specific to the nature of the Neolithic dispersion. A few examples of the sort of problems inherent in the database are given by way of example[2].

In the case of Tell Azmak and Tell Karanovo in Bulgaria (Volume II), the dataset includes radiocarbon determinations from different levels of complex 'tell' sites which belong to different cultural entities: Karanovo I for older radiocarbon determinations and Karanovo II for the younger radiocarbon determinations. The second example is the site of "Trifauckij" in Moldavia[3]. The dates given for this site are in fact from two different sites: Soroca 2, layer 2 for the older radiocarbon determinations, and Soroca 5, for the more recent radiocarbon determinations.

Examples such as these from an archaeologist who specialises in a particular area confirm that the Gkiasta database needs to be checked and expanded before it is used to make any firm conclusions about the nature of the spread of the first farming in Europe.

The case is different for the spread of the first sheep into modern day South Africa where the limitation is in the size of the available body of data rather than on the quality of the database, a limitation that shall only be overcome by further archaeological research in the region.

Despite this obvious limitations imposed on this work by the deficiencies in the Gkiasta database, the findings remain interesting. One of the advantages of working with such a large sample of data is that much of the noise created around errors in the data is filtered out by the sheer number of samples; robust patterns in the data are still to be seen. The main finding is that the continent-wide trends noted by Clark (1965) and Ammerman and Cavalli-Sforza (1971, 1984) are little altered by the increase in data. The spatial analysis of the radiocarbon database confirms that the first farming spread from an origin in the south-east to the rest of Europe with an overall rate of spread of ± 1 km/year.

To overcome some of the criticisms commonly levelled against the practice of using just one radiocarbon determination or simply the oldest radiocarbon determination in analysis, the analysis was undertaken using seven iterations of the Gkiasta database (Chapter 2, pg 27) each with a different estimate of the oldest or initial date for the appearance of farming. Interestingly the different selections of radiocarbon determinations, namely choosing the oldest, youngest or an average of radiocarbon determinations at a site made very little real difference to the results (section 5.2.3, pg 94 – 98 and Table 5.7 – 5.12, pg 108 – 114).

Importantly the average rate for the spread of farming in Europe at a continent-wide scale hides the variation that is seen when the Gkiasta database is broken down into smaller analytical units. Rates calculated at a regional level differed from those calculated by Ammerman and Cavalli-Sforza working with a small sample of dates in 1971 (see Table 6.3 and 6.4, pg 157-159). The correlation coefficients they calculate for the linear regression of regional sets of dates from an origin in Jericho (Table 6.4, pg 159) are deceptively high. For example, r = 0.915 for the western Mediterranean is calculated using a very small subsample of determinations, the total number of determinations for their whole European database is only 51 determinations. There are now enough dates to pose questions within smaller regions without sacrificing the weight of proof that large sample sizes give to results. It was at this level of analysis that the potential of the larger dataset has been demonstrated.

Rates of spread provide clues about the mechanisms of the movement. In this research only one line of enquiry

[1] since confirmed by Dolukhanov (pers. comm.)

[2] Dolukhanov (pers. Comm.)

[3] The correct name is 'Soroca' (Dolukhanov pers. comm.).

was followed: to see if the rates of spread in areas attributed by Zvelebil and Lillie (2000) to either cultural or demic diffusion differed. The hypothesis was that people and ideas move at different rates. The results, showing that the spread of farming was marginally faster in demic areas compared to cultural diffusion areas, are suggestive rather than conclusive. Although the difference between the rate of demic spread, 2.0 to 2.7 km/yr, and the rate of cultural diffusion, 1.4 – 1.6 km/yr, is small, the difference is potentially significant: we need more radiocarbon dates and a more standardized methodology if we are to understand the expansion. Many factors doubtless affect the rate of spread besides simply its diffusion mechanism. Nevertheless this analysis, by measuring rates of spread, has provided new data for use in comparative studies of other similar processes. For example estimations of the spread of the first Bantu Iron Age farmers south of the equator in Africa, a demic diffusion process, have ranged from 5km/year (Hall 1987) to 22km/year (Vansina 1996). Such striking differences between the African and European demic diffusion rates invite further research and consideration.

The choice was made to use Zvelebil and Lillie's surface map as the basis for this analysis in this research. Theirs is of course just one of many different interpretations of the breakdown of Europe into cultural and demic areas. Table 6.1 (page 119) shows that there is by no means a consensus that Zvelebil and Lillie's map is correct. Resolution of this issue, demic versus cultural diffusion, clearly depends on the integration of evidence from many different sources. In this research the possibilities of approaching the problem through radiocarbon evidence were explored. The emphasis was methodological rather than interpretational.

However there are other kinds of evidence that we can turn to in our attempt to separate cultural and demic diffusion. The demic diffusion hypothesis clearly has a very obvious genetic prediction: if people rather than ideas are moving then their genes are moving too. Patterns in the genes of modern day Europeans offer clues to past genetic movements. There is currently considerable debate and research into this line of evidence (see for example Cavalli-Sforza & Minch 1997, Lell & Wallace 2000, Richards *et al.* 2000, Simoni *et al.* 2000 and Barbujani & Bertorelle 2001). Richards *et al.* (1996) have argued that the mitochondrial dna (mtDNA) signatures of modern day European populations are the result of much earlier population movements and that the spread of farming to Europe was a later indigenous development with very little demic diffusion. The analysis presented in this research is based on the early work of Ammerman and Cavalli-Sforza (1971, 1984) who used the similarities in the patterns of the radiocarbon data and the clines in the genetic signatures of modern Europeans to argue that some population movement (i.e. demic diffusion), must have occurred with the spread of farming to Europe (Fix 1999). I mention the genetic evidence in passing as it has the potential to offer a further line of evidence that might lead to a convergence of results and a final resolution.

In South Africa there is a parallel debate, demic versus cultural diffusion, over the spread of the first sheep. Because the dataset is small it is hard to justify the calculation of rates of spread. Instead the research considered which of two competing hypotheses about the path of diffusion - the more popular western coastal route (Stow (1905), Cooke (1965), Bousman (1998)) or the central interior route (Elphick (1977), Ehret (1982), Hall (1987)) - was lent more support by the radiocarbon evidence. Unexpectedly, although neither routes were conclusively confirmed, the radiocarbon data lent the most support to the central interior route, currently the less favoured route amongst South African archaeologists. But the sample is small and further dates for the first sheep are needed.

There are a number of limitations to the research:

1. *The assumption of a linear relationship*
   The analysis assumed a linear relationship in all calculations of rate and origin of spread. The idea that there is a constant rate of spread with linear distance from an origin is an oversimplification and, in some cases, a limiting assumption. This has been highlighted by recent work on the peopling of the Americas (Gibbs 1997, Steele *et al.* 1998, Glass *et al.* 1999) where it has been shown that the dispersal of humans was influenced by habitat variation. Cost-distance modelling, which effectively stretches the distance travelled in areas where slow dispersal may be favoured and shrinks it in areas where dispersal may have been fast, has shown that the pattern in a group of dates that does not fit a simple linear trend of decreasing age with distance from an origin point when the landscape is treated as homogenous, can be explained when habitats are weighted in this way. This is likely to be of particular relevance to Europe. Europe's landforms are unique compared to the other continents; the ratio of shoreline to landmass is extremely high: the 37 000 km of shoreline is almost exactly the length of the equator (Davies 1997). The regions are connected by a network of natural paths; the Seine, the Rhine, the Elbe, Oder, Vistula, Niemen, Ebro, Dnieper, Loire, Danube, Dniester and the Thames rivers to name just a few (*ibid.*). There are the mountain chains that act as barriers and the gaps between them act as natural gateways. Important too is the consideration of the vegetation; densely wooded forests and marshland and the open plains must have also affected movement, as too would the

presence of different groups of people on the landscape[4].

When the database was compiled, spatially referenced databases on topography, hydrology, and vegetation were not freely available. They now are. The predilection of the first farmers, in parts of Europe, for fertile active floodplains and wet bottomland environments, should be incorporated into models for diffusion.

For South Africa the case is similar. Work by Hart (1987), Avery (1992, 1995), Smith (1984) suggests that the movement of stock-keepers would have been constrained by environmental factors. The next step would be to include these variables in analysis (e.g. see Wansleeben and Verhart 1995).

2. *Unprojected geographic coordinates*
The isochron and chronological maps use unprojected geographic coordinates. This is a problem that does not apply to any of the statistical analyses using geographic coordinates since all these work with projected coordinates.

3. *Uncertainty in the data*
In a criticism of Ammerman and Cavalli-Sforza's 1984 work, Zvelebil (1986) raises the point that some of the sites used to measure the rate of spread of farming in Europe have no positive evidence of cereals and often only a few bones of domestic animals in an otherwise wild faunal assemblage. The assumption that such sites are true farming sites rests on slender evidence. The problem of classifying and labelling sites, in particular of distinguishing those of foragers from herders, was highlighted in the work on the South African database. There is a real difficulty in distinguishing between the two site types. For the Gkiasta database the classification of a site as Neolithic or Mesolithic was taken as given in the site reports. This work needs to be refined. Research by Sue Colledge and James Conolly of University College London to trace the spread of domestic plants, in a project which is an extension of that which gave rise to the Gkiasta database, is an example of the kind of refinement that is required if this type of work is to withstand criticism such as that levelled by Zvelebil.

4. *The uncertain nature of radiocarbon dates*
Consideration must be given to the resolution of radiocarbon dates in relation to the temporal scale of the processes that are being measured. The error on radiocarbon dates is frequently in the order of hundreds of years. This represents several human generations (using a generation span of 25 years after Ammerman and Cavalli-Sforza 1984). This means that in cases of rapid spread the resolution of the dates might hide the finer details of the process (Barnett 2000).

The method presented in this research, using the whole of the calibrated date range broken down into smaller time ranges (20 years) that are then weighted by the probability that they reflect the true age of the sample, goes some way towards addressing this problem.

The main contribution of this research is methodological. Two innovative approaches are suggested.

a) A technique to use the calibrated probability range of radiocarbon dates in both visualisation and statistical analyses is presented (Chapter 7, pg 180). There has been a call for archaeologists working in time periods for which the calibration curve is good, to use calibrated dates in their analyses (Buck *et al.* 1994). Early and recent work on the Neolithic had not done so, and one of the aims of this research was to develop ways of displaying and analysing calibrated dates. A technique was developed where the whole of the calibrated date can be used in both the visualisation of data and in its manipulation.

b) A technique to search for the most probable origin to a set of spatially referenced radiocarbon dates is presented. and used to test the frontier map proposed by Zvelebil and Lillie (2000) and also resolve the debate surrounding the route of introduction of the first sheep in modern day South Africa. In the case of the European Neolithic, the most interesting result was that the origin points and direction of dispersal can be used to suggest which frontiers might have been most influential to areas of cultural diffusion. In southern Africa the technique was used to show that the interior route is as likely a route for the spread of the first sheep in southern Africa as the currently more favoured coastal route.

The findings can be summarized as follows:

The spread of the first farmers to Europe:

a) Patterns in the radiocarbon data at a continent-wide scale analysis, first traced 37 years ago, are confirmed with a larger dataset. There is an overall trend for the dates associated with the first farming to be old in the south-east of Europe and

[4] Recent models that consider the effect of indigenous people on dispersal are given by Aoki and Shida 1996, Young 2002.

to become gradually more recent with spread in a north-westerly direction. Using a technique to search for the most probable origin to a set of radiocarbon dates, the assumed origin at Jericho (Ammerman & Cavalli-Sforza 1971) was shown to fall within the most likely area of origin for European farming. Linear regression of radiocarbon determinations with distance from their most likely origin point produced a rate of spread for farming at a continent-wide scale in the range of 1.1 – 1.3 km/yr depending on which of the seven iterations of the database were used in the rate calculation. This result is similar to that of ±1 km/yr as calculated by Ammerman and Cavalli-Sforza (1971) using just 51 radiocarbon determinations.

b) Patterns in the radiocarbon data at a regional level are different from those measured by Ammerman and Cavalli-Sforza (1971). There are some striking differences here. For example they calculate a rate of spread of 5.59 km/year for the spread of the Bandkeramik culture. The rate for the spread of the first farming in central Europe using the Gkiasta database is 2.1 – 2.9 km/year. Likewise there is a considerable difference between the Ammerman & Cavalli-Sforza rate of spread to the Mediterranean (1.52 km/year) and that calculated using the Gkiasta database (1.4 – 8.3 km/year). This regional breakdown deserves a great deal more research.

The spread of the first sheep to South Africa:

Spatial analysis of the radiocarbon data for the spread of the first sheep lends more support to Elphick's (1977) interior route of introduction than to Stow (1905) and Cooke's (1965) western coastal route of introduction. This result illustrates how archaeologists' opinions have been skewed by the finding of a couple of early dates for sheep on the western coast of southern Africa just south of the current Namibian border. This analysis has clearly demonstrated the need for a less subjective approach to the analysis of the evidence.

This research has shown the potential of a spatial-temporal analysis as one of many avenues that may be followed in an attempt to better understand very complex processes.

References

Aitchison, T. C., L. Morven, D. J. Michczynska, W. G. Mook, R. L. Otlet, B. S. Ottaway, M. F. Pazdur, J. Van der Plicht, P. J. Reimer, S. W. Robinson, E. M. Scott, M. Stuiver and B. Weninger. 1989. A comparison of methods used for the calibration of radiocarbon dates. *Radiocarbon* 31 (3): 846 – 864.

Aitchison, T. C., E. M. Scott, D. D. Harkness, M. S. Baxter and Cook, G. T. 1990. Report on stage 3 of the international collaborative program. *Radiocarbon* 32 (3): 271 – 278.

Aitken, M. J. 1990. *Science based dating in archaeology*. Longman: New York.

Ahlfont, K., M. Guinard, E. Gustafsson, C. Olson and S. Welinder. 1995. Patterns of Neolithic farming in Sweden. *Tor* 27: 133-184.

Alexander, J.
1977. The 'frontier' concept in prehistory: the end of the moving frontier. In Megaw, J. V. (ed.) *Hunter, gatherers and first farmers beyond Europe.* Leicester: Leicester University Press.
1984. Early frontiers in southern Africa. In Hall, M. , G. Avery, D. M. Avery, M. L. Wilson, A. J. B. Humphreys (eds) *Frontiers: southern African archaeology today.* 12-23. Oxford: British Archaeological Reports International Series 207 (10).

Ammerman, A. J. 1989. On the Neolithic transition in Europe: a comment on Zvelebil and Zvelebil (1988). *Antiquity* 63: 162-165.

Ammerman, A. J. & Cavalli-Sforza, L, L.
1971. Measuring the rate of spread of early farming in Europe. *Man* 6: 674-688.
1984. *The Neolithic transition and the genetics of populations in Europe.* Princeton: Princeton University Press.

Anati, E. 1993. *World rock art: the primordial language*. Edizioni del Centro: Italy.

Anthony, D. W.
1990. Migration in Archaeology: the baby and the bathwater. *American Anthropologist* 92: 895-914.
1997. Prehistoric migration as social process. In Chapman, J. & Hamerow, H. (eds) *Migrations and invasions in archaeological explanation.* 21-32. British Archaeological Reports International Series 664.

Aoki, K & Shida, M. 1996. Travelling wave solutions for the spread of farmers into a region occupied by hunter-gatherers. *Theoretical Population Biology* 50: 1-17.

Avery, D. M.
1992. Micromammals and the environment of early pastoralists at Spoeg Rivier, Western Cape Province, South Africa. *South African Archaeological Bulletin* 47: 116-121.
1995. Physical environment and site choice in South Africa. *Journal of Archaeological Science* 22: 343-353.

Avery, G.
1974. Open station shell midden sites and associated features from the Pearly Beach Area, south-western Cape. *South African Archaeological Bulletin* 29: 104-114.
1975. Discussion on the age and use of tidal fish traps (visvywers). *South African Archaeological Bulletin* 30: 105-113.
1994. Faunal remains from Klein Kliphuis Shelter, Clanwillian district, South Africa. South African Archaeological Bulletin 47: 40-43.

Barker, G. 1985. *Prehistoric farming in Europe.* Cambridge University Press: Cambridge.

Barnett, W. K. 2000. Cardial pottery and the agricultural transition in Mediterranean Europe. In Price, T. D. (ed.) *Europe's first farmers.* 93-116. Cambridge: Cambridge University Press.

Beaton, J. M. 1991. Colonizing continents: some problems from Australia and the Americas. In Dillehay, T. D. & Meltzer, D. J. (eds) *The first Americas: search and research.* 209–230. Boca Raton, Florida : CRC Press.

Beaumont, P. B. & Vogel, J. C. 1972. On a new radiocarbon chronology for Africa south of the Equator. *African Studies* 31 (2): 79-89 & Part 2, *African Studies* 31 (3): 161–182.

Binder, D. 2000. Mesolithic and Neolithic interaction in southern France and northern Italy: new data and current hypotheses. In Price, T. D. (ed.) *Europe's first farmers.* 177-143. Cambridge: Cambridge University Press.

Blockley, S.P.E., R.E. Donahue and A.M. Pollard. 2000. Radiocarbon calibration and the Late Glacial occupation of northwest Europe. *Antiquity* 74: 112-121.

Bocquet-Appel, J. P. & Yves-Demars, P. 2000. Neanderthal contraction and modern human colonization of Europe. *Antiquity* 74: 544-552.

Bogucki, P.
1979. Tactical and strategic settlements in the early Neolithic of Lowland Poland. Journal of Anthropological Research 35(2): 238-246.
1982. Early Neolithic subsistence and settlement in the Polish Lowlands. Oxford: BAR International Series 150.
1986. Sustainable and Unsustainable adaptions by early farming communities in Northern Poland. Journal of Anthropological Research 15: 289-311.
1988. *Forest farmers and stockherders: early agriculture and its consequences in north-central Europe.* Cambridge: Cambridge University Press.
1995. How agriculture came to Central Europe. Http: //www. princeton. edu/~bogucki/saa1995. html
1996. The spread of early farming in Europe. *American Scientist* 84: 242-253.
1997. The Neolithic diaspora in Europe. Http: //www. princeton. edu/~bogucki/saa1997. html
2000. How agriculture came to north-central Europe. In Douglas Price, T. (ed.) Europe's first farmers. pp197-218. Cambridge: Cambridge University Press.

Bousman, C. B. 1998. The chronological evidence for the introduction of domestic stock into southern Africa. *African Archaeological Review* 15 (2): 133-150.

Bowman, S. 1994. Using radiocarbon: an update. *Antiquity* 68: 838–843.

Bowman, S. 1995. *Interpreting the past: radiocarbon dating.* London: British Museum Publications.

Bowman, S. & Balaam, N. 1990. Using radiocarbon. *Antiquity* 64: 315 – 318.

Bradley, R. 1984. *The social foundations of prehistoric Britain.* London: Longman.

Brink, J. & Webley, L. 1996. Faunal evidence for pastoralist settlement at Jakkalsberg, Richtersveld, Northern Cape Province. *South African Field Archaeology* 5: 70-78.

Bronk Ramsey C.
1995. Radiocarbon Calibration and Analysis of Stratigraphy: The OxCal Program. *Radiocarbon* 37 (2): 425–430.
1998. Probability and dating. In Mook, W. G. & Van der Plicht, J. Proceedings of the 16th International ^{14}C Conference. *Radiocarbon* 40 (1): 461-474.
2000. OxCal Program v3.5 University of Oxford Radiocarbon Accelerator Unit. (http: //www. rlaha. ox. ac. uk/orau/index. htm).

Broodbank, C. & Strasser, T. F. 1991. Migrant farmers and the Neolithic colonization of Crete. *Antiquity* 65: 233-245.

Buck, C. E., W. G. Cavanagh & C. D. Litton. 1996. *The Bayesian approach to interpreting archaeological data.* Chichester: Wiley.

Buck C. E., C. D. Litton and E. M. Scott. 1994. Making the most of radiocarbon dating: some statistical considerations *Antiquity* 68: 252-263.

Chapman, J.
1993. Social power in the Iron Gates Mesolithic. In Chapman, J. & Dolukhanov, P. (eds) *Cultural transformations and interactions in Eastern Europe.* 71-122. Aldershot: Avebury Press.
1994. The origins of farming in southeast Europe. *Préhistoire Européene* 6: 133-156.

Cherry, J. F. 1990. The first colonization of the Mediterranean islands: a review of recent research. *Journal of Mediterranean Archaeology* 3: 145-221.

Childe, V. G.
1957. *The dawn of European civilization.* London: Routledge and Kegan Paul.
1958. *The prehistory of European society.* Cassell & Company: London.
1969. *Prehistoric migrations in Europe.* Anthropological Publications: The Netherlands.

Clark, J. G. D.
1965a. Radiocarbon dating and the spread of the farming economy. *Antiquity* 31: 57-73.
1965b. Radiocarbon dating and the expansion of farming culture from the Near East over Europe. *Proceedings of the Prehistoric Society* 31: 57 – 73.

Clark, R. M. 1975. A calibration curve for radiocarbon dates. *Antiquity* 49: 251–266.

Clutton-Brock, J. 1996. The legacy of Iron Age dogs and livestock in southern Africa. 161-167. In Sutton, J. E. G. (ed.) *The growth of farming communities in Africa from the equator southwards.* Azania special volume XXIX – XXX.

Cooney, G & Grogan, E. 1994. *Irish prehistory: a social perspective.* Dublin: Wordwell.

Cooke, C. K. 1965. Evidence of human migrations from the rock art of southern Rhodesia. *Africa* 35: 263-285.
Davies, N. 1997. *Europe: a history.* London: Pimlico.

Deacon, H. J.
1967. Two radiocarbon dates from Scott's Cave, Gamtoos Valley. *South African Archaeological Bulletin* 22: 51-52.
1976. *Where hunters gathered: a study of Holocene Stone Age people in the eastern Cape.* Claremont: South African Archaeological Society.

Deacon, H. J. , J. Deacon, M. Brooker & M. L. Wilson. 1978. The evidence for herding at Boomplaas Cave in the southern Cape, South Africa. *South African Archaeological Bulletin* 33: 39-65.

Deacon, J. 1978. Changing patterns in the Late Pleistocene/Early Holocene prehistory of southern Africa as seen from the Nelson Bay Cave stone artifact sequence. *Quaternary Research* 10: 84-111.

Dean, J. S. 1978. Independent dating in archaeological analysis. In Schiffer, M. B. (ed.) *Advances in archaeological method and theory, vol. 1.* 223-255. New York: Academic Press.

Dehling H. & Van der Plicht, J. 1993. Statistical problems in calibrating radiocarbon dates. *Radiocarbon* 35 (1): 239-244.

Demoule, J-P. & Perlès, C. 1993. The Greek Neolithic: a new review. *Journal of World Prehistory* 7: 355-416.

Deng, F. M. 1973. *The Dinka and their songs.* Oxford: Oxford University Press.

Dennell, R.
1983. *European economic prehistory: a new approach.* Academic Press: London.
1985. The hunter-gatherer: agricultural frontier in prehistoric temperate Europe. In Green, S. & Perlman, S. (eds) *The archaeology of frontiers and boundaries.* 113-139. London: Academic Press.
1992. The origins of crop agriculture in Europe. In Cowan, C. W. & Watson, P. J. (eds) *The origins of agriculture: an international perspective.* 71-100. Washington: Smithsonian Institution Press.

De Vries, H. L. 1958. Variation in concentration of radiocarbon with time and location on Earth. *Proc. Koninkl. Nederl. Akad. Wetenschappen* 61: 94 – 102.

Dillehay, T. D. & Meltzer, D. J. Finale: Processes and Prospects. In Dillehay, T. D. & Meltzer, D. J. (eds) The first Americans: search and research. 287-293. Boca Raton: CRC Press.

Dolukhanov, P., A. Shukurov and D. Sokoloff. 1998. Improved radiocarbon chronology and the Neolithisation of Europe. Actes du Colloque C14 Archéologie: 133 – 136.

Ehret, C. 1982. The first spread of food production to southern Africa. In Ehret, C. & Posnansky, M. (eds.) *The archaeological and linguistic reconstruction of African history.* 158 – 182. Berkeley: University of California Press.

Elphick, R.
1977. *Kraal and castle.* New Haven: Yale University Press.
1985. *Khoikhoi and the founding of white South Africa.* Johannesburg: Raven Press.

Evans-Pritchard, E. E. 1980. *The Nuer.* Oxford: Oxford University Press.

Gibbs, K. 1997. *Models of early human migration into South America.* Unpublished MSc thesis, University of Southampton.

Gimbutas, M. 1991. *Civilization of the Goddess.* San Francisco: Harper and Row.

Glass, C. Steele, J. & Wheatley, D. 1999. Modelling human range expansion across a heterogeneous cost surface. *Proceedings of the 1997 computer applications in archaeology conference,* Birmingham (UK). BAR International Series 750:67-72.

Gob, A. 1990. *Chronologie du Mésolithique en Europe.* Atlas des dates 14C. Liège: C.I.P.L.

Gowlett, J. A. J. *et al.* 1987. Radiocarbon dates from the Oxford AMS system: archaeometry datalist. *Archaeometry* 5 (29): 125-155.

Green, S. W. & Perlman, S. M. (eds) 1985. *The archaeology of frontiers and boundaries.* London: Academic Press.

Hall, M. & Vogel, J. C. Some recent radiocarbon dates from southern Africa. *Journal of African History* 21: 431-455.

Halstead, P. 1996. The development of agriculture and pastoralism in Greece: when, how, who and what? In Harris, D. H. (ed.) *The origins and spread of agriculture and pastoralism in Eurasia.* 296-309. London: London University Press.

Hall, M. 1987. *The changing past: farmers, kings and traders in southern Africa, 200 – 1860.* Cape Town: David Philip.

Hall, M. & Smith, A. B. 1986. Introduction. In Hall, M. & Smith, A. B. (eds) Prehistoric pastoralism in southern Africa. *South African Archaeological Society Goodwin Series* 5: 3-4.

Hall, M. , G. Avery, D. M. Avery, M. L. Wilson, A. J. B. Humphreys (eds) 1994. *Frontiers: southern African archaeology today.* 96-110. Oxford: British Archaeological Reports International Series 207 (10).

Hart, T. 1987. Porterville survey. In Parkington, J. & Hall, M. (eds.) *Papers in prehistory of the western Cape, South Africa.* BAR International Series 332 (ii): 403-423.

Harke, H. 1998. Archaeologists and migrations: a problem of attitude? *Current Anthropology* 39 (1): 19 – 45.

Hayden, B. 1990. Nimrods, Piscators, pluckers and planters: the emergence of food production. *Journal of Anthropological Archaeology* 9: 31-69.

Helmer, D. 1984. Les parcages des moutons et des chèvres au Néolithique ancien et moyen dans le sud-est de la France. In Clutton-Brock, J. & Grigson, C. (eds) *Animals and archaeology, 3: early herders and their flocks.* 39-45. Oxford: British Archaeological Reports International Series 202.

Henshilwood, C.
1995. Holocene archaeology of the coastal Garcia State forest, southern Cape, South Africa. Unpublished PhD thesis. University of Cambridge.
1996. A revised chronology for pastoralism in the southernmost Africa: new evidence of sheep at *c.*2000 b.p. from Blombos Cave, South Africa. *Antiquity* 70: 945-949.

Herskovits, M. J. 1926. The cattle complex in East Africa. *American Anthropologist* 28: 230 – 554.

Holl, A. F. C. 1998. The dawn of African pastoralists: an introductory note. *Journal of Anthropological Archaeology* 17 (2): 81-96.

Housley, R. A. , C. S. Gamble, M. Street and P. Pettitt. 1997. Radiocarbon evidence for the Lateglacial human recolonisation of northern Europe. *Proceedings of the Prehistoric Society* 63: 25-54.

Huffman, T. N.
1993. Broederstroom and the central cattle pattern. South African Journal of Science 89: 220-226.
1996. Archaeological evidence for climatic change during the last 2000 years in southern Africa. *Quaternary International* 33: 55-66.

Huffman, T. N. & Herbert, R. K. 1996. New perspectives on Eastern Bantu. In Sutton, J. E. G. (ed.) *The growth of farming communities in Africa from the equator southwards.* Azania special volume XXIX –XXX: 27-36.

Humphreys, A. J. B. 1974. A preliminary report on test excavations at Dikbosch Shelter 1, Herbert district, Northern Cape. *South African Archaeological Bulletin* 29: 115-119.

Inskeep, R. R.
1972. Nelson Bay Cave, Robberg Peninsula, Plettenberg Bay. *Palaeoecology of Africa* 6: 247-248.
1987. Nelson Bay Cave, Cape Province, South Africa. BAR 357 (I).

Inskeep, R. R. & Vogel, J. C. 1985. Radiocarbon dates from the Holocene levels at Nelson Bay cave, and an interim report on their associations. *South African Archaeological Bulletin.* 40: 103-105.

Ingold, T. 1980. *Hunters, pastoralists & ranchers.* Cambridge: Cambridge University Press.

Jackes, M. D., D. Lubell and C. Meiklejohn. 1997. On physical anthropological aspects of the Mesolithic-Neolithic transition in the Iberian Peninsula. *Current Anthropology* 38: 839-846.

Jacobsen, T. W. 1981. Franchthi Cave and the beginning of settled village life in Greece. *Hesperia* 50: 303-319.

Jacobson, L. 1987. The size variability of ostrich eggshell beads from central Namibia and its relevance as a stylistic and temporal marker. *South African Archaeological Bulletin* 42: 55-58.

Jarman, M. 1996. Human influence in the development of the Cretan fauna. In Reese, D. (ed.) *The Pleistocene and Holocene fauna of Crete and its first settlers.* 211-230. Madison, WI: Prehistory Press.

Jennbert, K. 1985. Neolithisation: a Scanian perspective. *Journal of Danish Archaeology* 4: 196-197.

Jerardino, A. 1999. A first account of fat-tailed sheep in the rock paintings of the Western Cape coast. *South African Archaeological Bulletin* 54: 64-66.

Jerardino, A & Yates, R. 1996. Preliminary results from excavations at Steenbokfontein Cave: implications for past and future research. *South African Archaeological Bulletin* 51: 7-16.

Jochim, M. 2000. The origins of agriculture in south-central Europe. In Price, T. D. (ed.) *Europe's first farmers.* 183-196. Cambridge: Cambridge University Press.

Kaplan, J. 1987. Settlement and subsistence at Renbaan Cave. In Parkington, J. & Hall, M. (eds.) *Papers in the prehistory of the western Cape, South Africa.* BAR 332 (ii): 350-376.

Keeley, L. 1992. The introduction of agriculture to the western North European Plain. In Gebauer, A. B. & Price, T. D. (eds) *Transitions to agriculture in prehistory.* 81-95. Madison, WI: Prehistory Press.

Kinahan, J.
1991. *Pastoral nomads of the central Namib desert.* Windhoek: New Namibia books for the Namibia Archaeological Trust.
1996. A new archaeological perspective on nomadic pastoralist expansion in south-western Africa. 211-226. In Sutton, J. E. G. (ed.) *The growth of farming communities in Africa from the equator southwards.* Azania special volume XXIX – XXX.

Klein, J. C. Lerman, P. E. Damon and E. K. Ralph. 1982. Calibration of radiocarbon dates: tables based on the consensus data of the workshop on calibrating the radiocarbon timescale. *Radiocarbon* 24: 103 – 150.

Klein, R. G.
1972. Preliminary report on the July through September 1970 excavations at Nelson bay Cave, Plettenberg Bay (Cape Province, South Africa). *Paleoecology of Africa* 6: 177-208.
1978. A preliminary report on the larger mammals from the Boomplaas Stone Age cave site, Cango Valley, Oudtshoorn District, South Africa. *South African Archaeological Bulletin* 33: 66-75.
1984. Southern African prehistory and paleoenvironments. Balkema: Rotterdam.
1986. The prehistory of Stone Age herders in the Cape Province of South Africa. *South African archaeological Society Goodwin series* 5: 5-12.

Klein, R. G. & Cruz-Uribe, K.
1984. The analysis of animal bones from archaeological sites. Chicago: University of Chicago Press.
1989. Faunal evidence for prehistoric herder-forager activities at Kasteelberg, Western Cape Province, South Africa. *South African Archaeological Bulletin* 44: 82-97.

Klein, R. G. & Scott, K. 1974. The fauna of Scott's Cave, Gamtoos Valley, South-eastern Cape Province. *South African Journal of Science* 70: 186-187.

Layton, R. H. 2001. Hunter-gatherers, their neighbours and the nation state. In Panter-Brick, C. , Layton, R. H. & Rowley-Conwy, P. (eds) *Hunter-gatherers: an interdisciplinary perspective.* 292-321. Cambridge: Cambridge University Press.

Lewthwaite, J. 1986. The transition to food production: a Mediterranean perspective. In Zvelebil, M. (ed.) *Hunters in transition.* 53-66. Cambridge: Cambridge University Press.

Levin, I. , B. Kromer, H. Schoch-Fischer, M. Bruns, M. Munnich, D. Berdau, J. C. Vogel and K. O. Munnich. 1985. 25 years of tropospheric 14C observations in central Europe. *Radiocarbon* 27 (1): 1– 19.

Levine, M. A. 1990. Accommodating age: radiocarbon results and fluted point sites in northeastern North America. *Archaeology of Eastern North America* 18: 33–63.

Libby, W. F. 1952. *Radiocarbon dating.* Chicago: University of Chicago Press.

Long, A. & Rippeteau, B. 1974. Testing contemporaneity and averaging radiocarbon dates. *American Antiquity* 39 (2): 205 – 215.

Longley, P. A. , M. F. Goodchild, D. J. Maguire, D. W. Rhind. 2001. *Geographic Information systems and science.* Chichester: John Wiley.

Maggs, T. 1996. The Early Iron Age in the extreme south: some patterns and problems. 171-178. In Sutton, J. E. G. (ed.) *The growth of farming communities in Africa from the equator southwards.* Azania special volume XXIX – XXX.

Maggs, T. & Whitelaw, G. 1991. A review of recent archaeological research on food-producing communities in southern Africa. *Journal of African History* 32: 3-24.

Manhire, A. H. , J. E. Parkington, A. D. Mazel and T. M. Maggs. 1986. Cattle, sheep and horses: A review of domestic animals in the rock art of southern Africa. *South African society Goodwin Series* 5: 22-30.

Marean, C. W. 1985. The faunal remains from Smitswinkelbaai, Cape Peninsula. *South African Archaeological Bulletin* 40: 100-102.

Mazel, A. D. 1989. People making history: the last ten thousand years of hunter-gather communities in the Thukela Basin. *Natal Museum Journal of Humanities* 1: 1-168.

McCormac, F. G. , M. G. L. Baillie, J. R. Pilcher, R. M. Kalin. 1995. Location-dependent differences in the ^{14}C content of wood. In Cook, G. T. , D. D. Harkness, B. F. Miller and E. M. Scott

(eds) Proceedings of the 15th International ^{14}C conference. *Radiocarbon* 37 (2): 395-407.
Megaw, J. V. S. (ed.) 1977. *Hunters, gatherers and first farmers beyond Europe.* Leicester: Leicester University Press.

Mercer, R. 1986. The Neolithic in Cornwall. *Cornish Archaeology* 25: 35-80.

Mook, W. G. 1983. ^{14}C calibration curves depending on sample time width. In Mook, W. G. and Waterbolk, H. T. (eds) ^{14}C and Archaeology. *Proceedings of the 1st International Symposium (=PACT 8).* Strasbourg: 517 – 525.
1986. Business meeting, 12th International Radiocarbon Conference. *Radiocarbon* 28: 799.

Moore, J. A. 1985. Forager/farmer interactions: information, social organisation and the frontier. In Green, S. W. & Perleman, S. M. (eds) *The archaeology of frontiers and boundaries.* 93-112. London: Academic Press.

Nash, S. E. & Dean, J. S. 2000. The surprising deficient history of archaeochronology. In Nash, S. E. (ed.) *It's about time: A history of archaeological dating in North America.* 2-13. Salt Lake City: University of Utah Press.

Papathanassopoulos, G. A. (ed.) 1996. *Neolithic culture in Greece.* Athens: N. P. Goulandris Foundation.

Parkington, J.
1972. Seasonal mobility in the Late Stone Age. *African Studies* 31: 223-243.
1976. *Follow the San.* Unpublished PhD thesis. University of Cambridge.
1984. Changing views of the Later Stone Age of South Africa. *Advances in World Archaeology* 3: 89-142.
1992. Making sense of sequence at the Elands Bay Cave, western Cape, South Africa. 6-12. In Smith, A. B. & Mutti, B. (eds) *Guide to the Archaeological sites in the south-western Cape.* Unpublished guide for the South African Association of Archaeologists Conference July 1992.

Parkington, J. & Hall, M. 1989. Patterning in recent radiocarbon dates from southern Africa as a reflection of prehistoric settlement and interaction. *Journal of African History* 28: 1-25.

Parkington, J. & Poggenpoel, C. 1987. Diepkloof rock shelter. In Parkington, J. & Hall, M. *Papers in the prehistory of the western Cape*, South Africa. BAR 332 (ii): 269-293.

Pearson, G. W. 1987. How to cope with calibrations. *Antiquity* 61: 98-103.

Pearson, G. W. & Stuiver, M.
1986. High-precision calibration of the radiocarbon timescale, 500 – 2500 BC. In Stuiver, M. and Kra, R. S. (eds) Proceedings of the 12th International Radiocarbon Conference. *Radiocarbon* 28(2B): 839-862.
1993. High-precision bidecadel calibration of the radiocarbon time scale, 500 – 2500 BC. In Stuiver, M. , A.Long and R.S. Kra (eds) Calibration 1993. *Radicarbon* 35 (1): 25 – 33.

Pearson, G. W., B. Becker and F. Qua. 1993. High-precision ^{14}C measurement of German and Irish oaks to show the natural ^{14}C variations from 7890 to 5000 BC. In Stuiver, M. *et al.* (eds) Calibration 1993. *Radiocarbon* 35 (1): 93–104.

Peltzer, E. 2000. History of Model II regression analysis. http: //www. mbari. org/~etp3/regress/history. html

Perlès, C. 1993. Les début du Néolithique en Grèce. *La Recherche* 266: 642-649.

Piggott, S. 1965. *Ancient Europe.* Edinburgh: Edinburgh University Press.

Pinhasi, R, Foley, R.A. & Lahr, M. 2000. Spatial and temporal patterns in the Mesolithic-Neolithic Archaeological record of Europe. In Renfrew, C. & Boyle, K. *Archaeogenetics: DNA and the population prehistory of Europe.* 45–56. Cambridge: McDonald Institute for Archaeological research.

Pluccienik, M.
1994. Holocene hunter-gatherers in Italy. Http://www.archaeology.lamp.ac. uk/amarkp/PUBLIST/C14TEXT. HTM
1998. Radiocarbon determinations and the Mesolithic-Neolithic transition in southern Italy. *Journal of Mediterranean archaeology* 10: 115-150.

Plug, I. , C.A.Bollong, T. J. G. Hart and C. G. Sampson. 1994. Context and direct dating of pre-european livestock in the Upper Seacow river valley. *Annals of the South African Museum* 104 (3): 31-48.

Poggenpoel, C. A. & Robertshaw, P. T. 1981. The excavation of Smitswinkelbaai Cave, Cape Peninsula. *South African Archaeological Bulletin* 36: 29-35.

Price, T. D.
1991. The Mesolithic of northern Europe. *Annual Review of Anthropology* 20: 211-233.
2000. Europe's first farmers: an introduction. In Price, T. D. (ed.) *Europe's first farmers.* 1–18. Cambridge: Cambridge University Press.

Price, T. D., A. B. Gebauer and L.H. Keeley. 1995. The spread of farming into Europe north of the Alps. In Price, T. D. & Gebauer, A. B. (eds) *Last hunters-first farmers: new perspectives on the prehistoric transition to agriculture.* 95-126. Santa Fe, NM: School for American Research Press.

Ravenhart, R. 1967. *Before Van Riebeek.* Struick: Cape Town.

Renfrew, C.
1987. *Archaeology and language: the puzzle of Indo-European origins.* London: Cape.
1989. The origins of Indo-European languages. *Scientific American* 261 (4): 106-114.

Robertshaw, P. T. 1978. The origin of pastoralism in the Cape. *South African Historical Journal* 10: 117-133.

Robey, T. S.
1984. Burrow & Bedding: site taphonomy & spatial archaeology at Tortoise Cave. MA thesis, University of Cape Town.
1987. The stratigraphic and cultural sequence at Tortoise Cave, Verlorenvlei. In Parkington, J. & Hall, M. (eds.) *Papers in the prehistory of the western Cape, South Africa.* Bar 332 (ii): 294-325.

Runnels, C. 1995. Review of Aegean Prehistory IV: the Stone Age of Greece from the Palaeolithic to the advent of the Neolithic. *American Journal of Archaeology* 99: 699-728.

Russell, T. 2000. The application of the Harris Matrix to San rock art at Main Caves North. *South African Archaeological Bulletin* 55: 60-70.

Russell, T. 2002. The spatial analysis of radiocarbon databases: the spread of the first farmers in Europe and the fat-tailed sheep in southern Africa. Unpublished PhD thesis. University of Southampton.

Russell, T & Steele, J. (forthcoming) *Visualising the Neolithic transition in Europe.* Presentation at the 14C and Archaeology Fourth Symposium, Oxford April 2002.

Sadr, K. 1998. The first herders at the Cape of Good Hope. *African Archaeological Review* 15 (2): 101-132.

Sampson, C. G. 1984. A prehistoric pastoralist frontier in the upper Zeekoe Valley, South Africa. In Hall, M., G. Avery, D.M. Avery, M.L. Wilson, A.J.B. Humphreys (eds) *Frontiers: southern African archaeology today.* 96-110. Oxford: British Archaeological Reports International Series 207 (10).

Sampson, C. G. & Vogel, J. C. 1995. Radiocarbon chronology of Later Stone Age pottery and decoration in the Upper Seacow Valley. *South African Journal of Field Archaeology* 4: 84-94.

Schweitzer, F. R.
1974. Archaeological evidence for sheep at the Cape. *South African Archaeological Bulletin* 29: 75-82.
1979. Excavations at Die Kelders, Cape Province. *Annals of the South African Museum* 78: 101-232.

Schweitzer, F. R. & Scott, K. J. 1973. Early occurrence of domestic sheep in sub-Saharan Africa. *Nature* 24: 547.

Schweitzer, F. R. & Wilson.A. 1982. Byneskrankop I. *Annals of the South African Museum.*

Scott, E. M., T. C. Aitchison, D.D. Harkness, G.T. Cook and M.S. Baxter. 1990. An overview of all three stages of the international radiocarbon intercomparison. *Radiocarbon* 32 (3): 309 – 319.

Scott, L.
Palynology of hyrax middens: 2000 years of paleoenvironmental history in Namibia. Quaternary International 33: 73-79.

Sealy, J. & Yates, R.
1994. The chronology of the introduction of pastoralism to the Cape, South Africa. *Antiquity* 68: 58-67.
1996. Direct radiocarbon dating of early sheep bones: two further results. *South African Archaeological Bulletin* 51: 109-110.

Séfériadès, M. L. 1993. The European Neolithisation process. *Porocilo* 21: 137-162.

Smith, A. B.
1984. Environmental limitations on prehistoric pastoralism in Africa. *The African Archaeological Review* 2: 99-111.
1987. Seasonal exploitation of resources on the Vredenberg Peninsula after 2000 B. P. In Parkington, J. & Hall, M (eds) *Papers in prehistory of the western Cape*, South Africa. BAR International Series 332 (ii): 393-402.
1990. On becoming herders: Khoikhoi and San ethnicity in southern Africa. *African Studies* 49 (2): 51-73.
1992a. Kasteelberg. In Smith, A. B. & Mutti, B. (eds) *Guide to the Archaeological sites in the south-western Cape.* 28-30. Unpublished guide for the South African Association of Archaeologists Conference July 1992.
1992b. *Pastoralism in Africa.* Ohio: Ohio University Press.

1997. Southern African pastoralists. 210-214. In Vogel, J. O. (ed.) *Encyclopedia of precolonial Africa*. Altamira press: California.
1998a. Early domestic stock in southern Africa: a commentary. *African Archaeological Review* 15 (2): 151-156.
1998b. Keeping people on the periphery: the ideology of social hierarchies between hunters and herders. *Journal of Anthropological Archaeology* 17: 201-215.

Smith, A. B. & Jacobson, L. 1995. Excavations at Geduld and the appearance of early domestic stock in Namibia. *South African Archaeological Bulletin* 50: 3-14.

Smith, A. B., K. Sadr, J. Gribble, R.Yates.
1991. Excavations in the south-western Cape, South Africa, and the archaeological identity of prehistoric hunter-gatherers within the last 2000 years. *South African Archaeological Bulletin* 46: 71-91.
1992.Witklip and Posberg Reserve. Guide to the archaeological sites in the south-western Cape. 31-40. *South African Association of Archaeologists Conference,* Unpublished conference excursion guide.

Solberg, B. 1989. The Neolithic transition in southern Scandinavia: internal development or migration. Oxford Journal of Archaeology 8: 261-296.

Sondaar, P. Y. 1971. Palaeozoogeography of the Pleistocene mammals from the Aegean. *Opera Botanica* 30: 65-70.

Soodyall, H. & Jenkins, T. 1997. *Khoisan prehistory: the evidence of the genes.* Unpublished paper delivered at the conference on Khoisan identities and cultural heritage at the South African Museum, Cape Town, South Africa, 1997.

Spaulding, A. C. 1958. The significance of differences in Carbon-14 dates. *American Antiquity* 23: 309–311.

Staüble, H. 1995. Radiocarbon dates of the earliest Neolithic in central Europe. *Radiocarbon* 37 (2): 227-237.

Steele, J. Gamble, C. & Sluckin, T. (In press 2000) Estimating the rate of Paleo-Indian expansion into South America. 125-133. In Nicholson, R. & O'Connor, T. (eds) People as agents of environmental change. Oxford: Oxbow Books

Steele, J. Adams, J. & Sluckin, T. 1998. Modelling Paleo-Indian dispersals. *World Archaeology.*
Stow, G. W. 1905. *The native races of South Africa.* London: Swan Sonnenschein.

Stuiver, M. & Pearson, G. W.
1986. High-precision calibration of the radiocarbon time scale, AD 1950-500 BC. In Stuiver, M. and Kra, R. S. (eds) Proceedings of the 12th International Radiocarbon Conference. *Radiocarbon* 28(2B): 805-838.
1993. High-precision bidecadel calibration of the radiocarbon time scale, AD 1950 – 500 BC and 2500 – 6000 BC. In Stuiver, M., A.Long and R. S. Kra (eds) Calibration 1993. *Radicarbon* 35 (1): 1 – 23.

Stuiver M. and Reimer, P. J. 1993. Extended 14C data base and revised CALIB 3. 0 14C Age calibration program. *Radiocarbon* 35 (1): 215-230.

Stuiver M. , P. J. Reimer, E. Bard, J. W. Beck, G. S. Burr, K. A. Hughen, B. Kromer, G. McCormac, J. van der Plicht and M. Spurk. 1998. INTCAL98 Radiocarbon Age Calibration, 24000-0 cal BP. *Radiocarbon* 40 (3): 1041-1083.

Stuiver, M. 1961. Variations in radiocarbon concentration and sunspot activity. *Journal of Geophysical research* 66 (1): 273–276.

Stuiver, M. 1965. Carbon-14 content of 18th and 19th century wood: variations correlated with sunspot activity. *Science* 149: 533–535.

Stuiver, M. & Polach, H. A. 1977. Discussion: reporting of ^{14}C data. *Radiocarbon* 19 (3): 355–363.

Stuiver, M. & Suess, H. E. 1966. On the relationship between radiocarbon dates and true sample ages. *Radiocarbon* 8: 534 – 540.

Stuiver M. , P. J. Reimer, E. Bard, J. W. Beck, G. S. Burr, K. A. Hughen, B. Kromer, G. McCormac, J. van der Plicht and M. Spurk. 1998. INTCAL98 Radiocarbon Age Calibration, 24000-0 cal BP. *Radiocarbon* 40 (3): 1041-1083.

Suess, H. E. 1965. Secular variations in the cosmic-ray-produced carbon-14 in the atmosphere and their interpretations. *Journal of Geophysical Research* 70: 5937–5951.

Suess, H. E. 1967. Bristlecone pine calibration of the radiocarbon time-scale from 4100 BC to 1500 BC. In *Radioactive dating and methods of low level-counting.* 143 – 151. I. A. E. A, Vienna.

Suess, H. E. 1970. The three causes of secular C14 fluctuations, their amplitudes and time constraints. In Olsson, I. U. (ed.) *Radiocarbon variations and absolute chronology.* Nobel Symposium 12. 595 – 605. Stockholm: Almqvist & Wiksell.

Taylor, R. E. 2000. The introduction of radiocarbon dating. In Nash, S. E. (ed.) *It's about time: a history of archaeological dating in North America*. 84-104. Salt Lake City: University of Utah Press.

Taylor, R. E. 2001. Radiocarbon dating. In Brothwell, D. R. & Pollard, A. M. (eds) *Handbook of the archaeological sciences*. 23-34. Chichester: John Wiley

Taylor, R. E. & Aitken, M. J. 1997. *Chronometric dating in archaeology*. London: Plenum Press.

Thackeray, A. I. & Humphreys, A. J. B. 1983. Ghaap and Gariep: Later Stone Age studies in the northern Cape. *South African Archaeological Society Monograph Series 2*.

Thackeray, A. I. Thackeray, J. F. & Beaumont, P. B. 1983. Excavations at the Blinklipkop specularite mine near Postmasburg, Northern Cape. *South African Archaeological Bulletin* 38: 17-25.

Theocaris, D. 1973. *Neolithic Greece*. Athens: National Bank of Greece.

Thorp, C. R.
1996. A preliminary report on evidence of interaction between hunter-gatherers and farmers along a hypothesised frontier in the eastern Free State. *South African Archaeological Bulletin* 51(164): 57-63.
2000. Hunter-gatherers and farmers: an enduring frontier in the Caledon Valley, South Africa. *Bar International Series* 860.

Tringham, R. 2000. Southeastern Europe in the transition to agriculture in Europe: bridge, buffer or mosaic. In Price, T. D. (ed.) *Europe's first farmers*. 19-56. Cambridge: Cambridge University Press.

Turner, G. 1986. Faunal remains form Jubilee Shelter, Transvaal. *South African Archaeological Bulletin* 41: 63-68.

Tyson, P. D. & Lindesay, J. A. 1992. The climate of the last 2000 years in southern Africa. *The Holocene* 2 (3): 271-278.

Van Andel, T. H. & Runnels, C. N. The earliest farmers in Europe. *Antiquity* 69: 481-500.

Van der Plicht, J. 1993. The Groningen radiocarbon calibration program. *Radiocarbon* 35 (1): 231–237.

Van der Plicht, J. & Mook, W. G. 1987. Automatic radiocarbon calibration: illustrative examples. *Palaeohistoria* 29: 173 – 182.

Van Rijssen, W. J. 1994. The late Holocene deposits of Klein Kliphuis shelter, Cedarberg, western Cape province. *South African Archaeological Bulletin* 47: 34-39.

Vansina, J. 1996. A slow revolution: farming in subequatorial Africa. In Sutton, J. E. G. (ed.). *The growth of farming communities in Africa from the Equator southwards*. The British Institute in Eastern Africa: Kenya 15-26.

Vencl, S. 1986. The role of hunter-gathering populations in the transition to farming: a central-European perspective. In Zvelebil, M. (ed.) *Hunters in transition: Mesolithic societies of Temperate Eurasia and their transition to farming*. 43-51. Cambridge: Cambridge University Press.

Vogel, J. C.
1965. Carbon-14 content of wood from different locations: International C^{14} and H^3 conference, Pullman, Washington, Proceedings.
2000. Radiocarbon dating of the Iron Age sequence in the Limpopo valley. *South African Archaeological Society Goodwin Series* 8: 51 – 57.

Vogel, J. C. & Fuls, A. 1999. The spatial distribution of radiocarbon dates for the Iron Age in southern Africa. *South African Archaeological Bulletin* 54: 97-101.

Vogel, J. C. , A. Fuls and E. Visser. 1993. Pretoria calibration curve for short-lived samples, 1930 – 3350 BC. *Radiocarbon* 35 (1): 73 – 85.

Vogel, J. C. , A. Fuls, E. Visser and B. Becker. 1986. Radiocarbon fluctuations during the third millennium BC. In Stuiver, M. and Kra, R. S. (eds) Proceedings of the 12th International ^{14}C Conference. *Radiocarbon* 28 (2B): 935-938.

Vogel *et al*. 1997. New dates for the introduction of sheep into South Africa: the evidence from Spoeg Rivier Cave. *South African Journal of Science* 89: 220-226.

Von den Driesch, A & Deacon, H. J. 1985. Sheep remains from Boomplaas Cave, South Africa. *South African Archaeological Bulletin* 40: 39-44.

Wadley, L. 1996. Changes in the social relations of precolonial hunter-gatherers after agropastoralist contact: An example from the Magaliesberg, South Africa. *Journal of Anthropological Archaeology* 15: 205-217.

Walker, N. J. 1983. The significance of an early date for pottery and sheep in Zimbabwe. *South African Archaeological Bulletin* 38: 88-92.

Ward, G. K. & Wilson, S. R. 1978. Procedures for comparing and combining radiocarbon age determinations: a critique. *Archaeometry* 20 (1): 19-31.

Watkins, T. 1975. *Radiocarbon: calibration and prehistory*. Edinburgh: Edinburgh University Press.

Webley, L.
1992a. Unpublished PhD thesis. University of Cape Town.
1992b. Tracing the emergence of a pastoralist identity in Namaqualand. Unpublished Khoisan Conference paper, Cape Town 1992.
1997a. Tracing the emergence of a pastoralist identity in Namaqualand. Unpublished paper given at the Khoisan Conference. Cape Town, July 1997.
1997b. Jakkalsberg A and B: The cultural material from the agropastoralist sites in the Richtersveld, Northern Cape. *South African Field Archaeology* 6: 3-19.

Webley, L & Brink 1996
Faunal evidence for pastoralist settlement at Jakkalsberg, Richterveld, Northern Cape province. South African Field Archaeology 5: 70-78.

Whittle, A.
1985. *Neolithic Europe: a survey*. Cambridge: Cambridge University Press.
1996. *Europe in the Neolithic*. Cambridge: Cambridge University Press.

Wilkie, N. C. & Savina, M. E. 1997. The earliest farmers in Macedonia. *Antiquity* 71: 201-207.

Wilson, M. L. 1996. The late Holocene occupants of Die Kelders: hunter-gatherers or herders? *South African Archaeological bulletin* 5: 79-83.

Woodman, P. C. 1985. Prehistoric settlement and environment. In Edwards, K. J. & Warren, W. P. (eds) *The Quaternary history of Ireland*. 251-278. London: Academic Press.

Young, D. A. 2002. A new space-time computer simulation method for human migration. *American Anthropologist* 104 (1): 138-158.

Zilhao, J.
1993. The spread of agro-pastoral economies across Mediterranean Europe. *Journal of Mediterranean Archaeology* 6: 5-63.
2000. From the Mesolithic to the Neolithic in the Iberian peninsula. In Price, T. D. (ed.) *Europe's first farmers.* 144-182. Cambridge: Cambridge University Press.
Zvelebil, M.
1986a. Mesolithic prelude and Neolithic revolution. In Zvelebil, M. (ed.) *Hunters in transition: Mesolithic societies of Temperate Eurasia and their transition to farming.* 5-15. Cambridge: Cambridge University Press.
1986b. Mesolithic societies and the transition to farming: problems of time, scale and organisation. In Zvelebil, M. (ed.) *Hunters in transition: Mesolithic societies of Temperate Eurasia and their transition to farming.* 167-188. Cambridge: Cambridge University Press.
2000. The social context of the agricultural transition in Europe. In Renfrew, C. & Boyle, K (eds) *Archaeogenetics: DNA and the population prehistory of Europe.* 57-80. Cambridge: MacDonald Institute for Archaeological Research.

Zvelebil, M. & Dolukhanov, P. 1991. Transition to farming in eastern and northern Europe. *Journal of World Prehistory* 5: 233-278.

Zvelebil, M & Lillie, M. 2000. Transition to agriculture in eastern Europe. In Price, T. D. (ed.) *Europe's first farmers.* 57–92. Cambridge: Cambridge University Press.

Zvelebil, M. & Rowley-Conwy, P.
1984. Transition to farming in Northern Europe: A hunter-gatherer perspective. *Norwegian Archaeological Review* 17(2): 104-128.
1986. Foragers and farmers in Atlantic Europe. In Zvelebil, M. (ed.) *Hunters in transition: Mesolithic societies of Temperate Eurasia and their transition to farming.* 67 – 93. Cambridge: Cambridge University Press.

Appendix 1: The Gkiasta European Database: Datasets A to G.
Dataset A (weighted mean or oldest radiocarbon determination at 509 Neolithic sites).

SITE NAME	Country	Latitude (decimal degrees)	Longitude (decimal degrees)	Date (uncal. bp)	Standard deviation
ABINGDON	Great Britain	51.68055556	-1.260833333	4883	57
ABRI DE LA COMA FRANCEZE	France	42.83333333	2.916666667	5180	60
ABRI DE LA GRANDE RIVOIRE	France	45.21666667	5.65	5135	35
ABRI DE LA POUJADE	France	44.11666667	3.166666667	6990	120
ABRI DE PENDIMOUN	France	44.88333333	5.651	6100	120
ABRI DES PRES MOUREY	France	47.05	4.75	5081	52
ABRI DES ROCS	France	46.73333333	0.533333333	5300	50
ABRI DES USCLADES	France	43.98333333	12.86666667	6080	50
ABRI JEAN CROS	France	43.08333333	2.466666667	6600	130
ACHILLEION	Greece	39.28333333	22.38333333	7550	60
AGEROD	Sweden	56.93333333	13.41666667	6770	49
ALDWINCLE	Great Britain	52.41	-0.54	4560	70
ALFORT	France	45.68333333	5.35	6000	60
ANDERBY CREEK	Great Britain	53.27	0.32	4625	55
ANTONNAIRE	France	44.68333333	5.383333333	5570	130
ANTRE CORYCIEN	Greece	38.48333333	22.5	6315	64
ANZA	Macedonia	41.7	22	7210	50
APATIN	Yugoslavia	45.66666667	18.98333333	7130	60
ARAGUINA SENNOLA	France	41.38333333	9.166666667	6540	99
ARDCRONY	Ireland	52.88	-8.15	4675	35
ARGISSA MAGOULA	Greece	39.63333333	22.46666667	8130	100
ARMEAU	France	48.13333333	3.366666667	6260	65
ASCOTT-UNDER-WYCHWOOD	Great Britain	51.85472222	-1.565833333	4881	38
ASHCOTT HEATH	Great Britain	51.16	-2.78	4625	120
AUVERNIER-PORT	Switzerland	46.96666667	6.866666667	5075	57
AWANS	Belgium	50.66666667	5.2	6295	69
AZMAK	Bulgaria	42.45	25.76666667	6483	100
BADSHOT	Great Britain	51.23	-0.79	4746	26
BAKER PLATFORM	Great Britain	51.183	-2.82	4950	80
BAKER TRACK	Great Britain	51.18	-2.88	4760	80
BALBRIDIE	Great Britain	57.05	-2.441	4919	45
BALBRIDIE TIMBER HAL	Great Britain	57.05	-2.44	5150	60
BALLINTRUER MORE	Ireland	52.99	-6.68	4800	70
BALLOY LES REAUDINS	France	44.4	3.166666667	6180	90
BALLYGLASS (MA.13)	Ireland	54.3	-9.4	4590	48
BALLYMACDERMOT	Ireland	54.17	-6.38	4830	95
BALLYNAGILLY	Ireland	54.71	-6.85	5745	90
BALMA MARGINEDA	Andora	42.41666667	1.583333333	6670	85
BARNENEZ	France	48.66666667	-3.85	5345	45
BARNHOUSE	Great Britain	59	-3.22	4558	42
BASAROGA	Hungary	46.25	21	6580	60
BAUME BOURBON	France	44.9	4.633333333	6068	77
BAUME DE FONTBREGOUA	France	43.55	6.233333333	6700	100
BAUME DE MONTCLUS	France	44.26666667	4.433333333	6334	70
BAUME D'OULLINS	France	44.31666667	4.466666667	6630	110
BAVANS	France	47.46666667	6.7	7130	70
BEESTON CASTLE	Great Britain	53.13	-2.7	5140	90
BEKESSZENTANDRAS	Hungary	47.83333333	20.5	6080	60
BERCY QUARTIER SUD	France	48.86666667	2.333333333	5420	60
BETHENAS	France	45.71666667	5.25	5060	110
BEVILACQUA	Italy	38.92	16.61	6430	90
BILBO	France	49.06666667	-0.05	5560	140
BILGROIX	France	47.55	-2.916666667	5420	60
BLACKPATCH	Great Britain	50.86861111	-0.445	5090	130
BOGHEAD MOUND, FOCHABERS	Great Britain	57.62	-3.07	6006	60
BOYNTON, HIGH EASTON	Great Britain	54.12	-0.23	4840	80
BRADWELL-ON-SEA SITE 8	Great Britain	51.75	0.88	4690	70
BRIAR HILL	Great Britain	52.2275	-0.923888889	5440	110
BRIDGEMERE	Great Britain	51.21	-2.41	4630	50
BROKEN CAVE	Great Britain	50.53	-3.69	4949	50
BROOME HEATH	Great Britain	52.46777778	1.451388889	5424	117
BROVST	Denmark	57.1	9.5	5464	37
BUDDELIN	Germany	54.48333333	13.53333333	5815	100
BUHOLOUP	France	43.15	1.183333333	5490	120
BULLTOFTAGARDEN	Sweden	55.6	13.06666667	6660	80
BURNTWOOD FARM (SITE R6)	Great Britain	51.12	-1.29	4750	140
BURY HILL	Great Britain	50.92	-1.37	4750	50
BYLANY	Czech Rep.	50.96666667	15.3	6270	65
CALLIS WOLD, BARROW 275	Great Britain	53.99	-0.73	4875	48
CAMP DEL GINEBRE	France	42.75	2.566666667	5060	45
CAMSTER LONG CAIRN, WICK PARISH	Great Britain	58.38	-3.27	4927	45
CANNON HILL, BRAY	Great Britain	51.5	-0.72	5270	110
CAPDENAC LE HAUT	France	44.58333333	2.066666667	5148	73
CAPEL EITHIN	Great Britain	53.22	-4.29	5520	80
CARN	France	48.56666667	-4.683333333	5230	75
CARN BREA	Great Britain	50.22027778	-5.246944444	4999	64
CARNANBANE	Ireland	54.75	-6.82	4978	61
CARREG COETAN	Great Britain	52.02	-4.82	4697	46
CARROWMORE	Ireland	54.27	-8.53	5750	85
CARROWMORE - CULLEENAMORE	Ireland	54.23	-8.62	4710	100
CASABIANCA OR CASTERRAGIO	France	42.08333333	9.5	6670	130
CASHELKEELTY I	Ireland	51.72	-9.82	5845	100
CAUCADE	France	43.7	7.35	6800	130
CAVERNA DELLE ARENE CANDIDE	Italy	44.5	22.41666667	7082	42
CERON	France	44	5.05	5990	75
CHABAROVICE	Czech Rep.	50.66666667	5.45	6070	90
CHAMP DE BATAILLE	France	50.03333333	2.65	5390	100
CHAMP DE LA GRANGE	France	46.76666667	2.433333333	5247	61

Appendix 1: The Gkiasta European Database: Datasets A to G.
Dataset A (weighted mean or oldest radiocarbon determination at 509 Neolithic sites).

SITE NAME	Country	Latitude (decimal degrees)	Longitude (decimal degrees)	Date (uncal. bp)	Standard deviation
CHANTIER DE BERCY	France	48.86666667	2.333333333	5820	60
CHASSEY	France	46.88333333	4.766666667	5234	49
CHATEAU PERCIN	France	43.7	0.35	5250	100
CHATEAUNEUF LES MARTIGUES	France	43.38333333	5.166666667	6900	100
CHATELLIERS DU VIEL	France	46.43333333	-0.866666667	5200	110
CHATTON SANDYFORD	Great Britain	55.29416667	-1.983888889	4840	90
CHAZELLES	France	44.33333333	4.333333333	5660	110
CHEMIN DE NAVIOT	France	48.4	6.166666667	6330	140
CHERHILL	Great Britain	51.42944444	-1.955277778	4715	90
CHILTON 1-2	Great Britain	51.18	-1.81	4500	120
CHILTON NO. 4 TRACK	Great Britain	51.15	-2.9	4760	65
COJOUX	France	47.76666667	-1.95	5577	69
COMA FRANCEZE	France	42.83333333	2.916666667	5200	70
COMBE HILL	Great Britain	50.79	0.21	4590	110
COMBOIRE	France	45.13333333	5.683333333	5360	70
CONEYBURY HENGE	Great Britain	51.17	-1.81	5050	100
COPPA NEVIGATA	Italy	41.5	15.66666667	6520	70
CORHAMPTON HOUSE, CORHAMPTON	Great Britain	50.98	-1.15	4790	70
CORON MATON	Belgium	50.55	5.883333333	6420	120
COUEDAN	France	47.98333333	-2.916666667	5260	70
COUFIN	France	45.06666667	5.4	5260	120
COUTURE DE LA CHAUSSE	Belgium	50.58333333	3.683333333	6485	75
COVETA DE L'OR	Spain	38.7	-0.466666667	6265	75
COYGAN CAMP	Great Britain	51.75388889	-4.486388889	5000	95
CREUX-ROUGE	France	45.78333333	3.083333333	6070	140
CRISNEE	Belgium	50.68333333	5.4	6320	75
CROIX SAINT PIERRE	France	47.76666667	-2	6070	80
CUEVA CHICA DE SANTIAGO	Spain	38.95	-6.916666667	6160	100
CUEVA DE CHAVES	Spain	42.36666667	0.216666667	6770	70
CUEVA DE LA DEHESILLA	Spain	36.75	-5.8	6260	100
CUEVA DEL NACIMIENTO	Spain	38.08333333	-2.3	6780	130
CUGNAUX	France	43.61666667	1.433333333	5460	70
CUL FROID	France	49.23333333	1.05	5180	100
CURNIC	France	48.58333333	-4.416666667	5340	60
DALLADIES FARM, FETTERCAIRN	Great Britain	56.79583333	-2.610833333	4603	37
DARION	Belgium	50.65	5.183333333	6180	43
DERRIERE LE VILLAGE	France	49.41666667	4.016666667	6067	95
DERRIERE LES PRES	France	49.06666667	-0.05	5110	70
DEVIL'S KNEADING TROUGH	Great Britain	51.16777778	0.971388889	4540	105
DIVOSTIN	Yugoslavia	44.03333333	21.83333333	7010	38
DOLMEN DE CHATEAU BU	France	47.81666667	1.95	5230	50
DOMICA CAVE	Slovakia	48.48333333	20.48333333	6080	75
DOOEY'S CAIRN, BALLYMACALDRACK	Ireland	55.01	-6.41	5150	90
DORCHESTER SITE VIII	Great Britain	51.66	-1.17	4510	100
DORSET CURSUS (DOWN FARM)	Great Britain	50.93	-2	6800	100
DRAYTON	Great Britain	51.66	-1.31	4990	100
DRESDEN-NICKERN	Germany	51.03333333	12.11666667	5932	100
DUDERSTADT	Germany	51.51666667	10.28333333	6309	48
DUNDRUM NATURE RESERVE	Ireland	54.25	-5.85	4666	97
EAST HESLERTON WOLD	Great Britain	54.16	-0.57	5020	110
EASTON DOWN	Great Britain	51.39	-1.91	4624	36
EATON HEATH	Great Britain	52.60638889	1.262777778	5095	49
EGOLZWIL 4	Switzerland	47.18333333	8.016666667	5087	50
EILSLEBEN	Germany	51.48333333	11.51666667	6114	47
ELATEIA	Greece	38.61666667	22.73333333	7435	55
ELLERBEK	Germany	54.38333333	10.18333333	5170	110
ELSLOO	Netherlands	51.93333333	5.766666667	6320	90
ENDROD	Hungary	47.91666667	21.83333333	6945	50
ENVEIG	France	42.51666667	1.866666667	5290	60
ERTEBOLLE	Denmark	56.8	9.166666667	5624	45
ESCANIN	France	43.75	6.25	5630	99
ETANG DAVID	France	47.38333333	4.516666667	5600	120
ETON WICK	Great Britain	51.5	-0.66	4697	40
FELDBACH	Switzerland	47.23333333	8.783333333	5170	70
FELSDACH LAUTERACH	Germany	48.33333333	2.45	5190	130
FENDMEILEN-VORDERELD	Switzerland	47.28333333	8.633333333	5415	60
FENGATE	Great Britain	52.57388889	-0.209722222	4960	64
FERRITERS COVE	Ireland	52.16666667	10.5	5190	110
FLAGSTONES HOUSE, DORCHESTER	Great Britain	50.71	-2.43	4960	80
FLYNDERHAGE	Denmark	56.01666667	10.23333333	5705	85
FOCHABERS - BOGHEAD MOUND	Great Britain	57.61861111	-3.071388889	4895	37
FONT JUVENAL	France	43.28333333	2.333333333	6235	85
FRANCHTHI	Greece	37.43333333	23.13333333	7980	110
FRANKENAU	Austria	47.5	16.5	5660	100
FRIEDBERG	Germany	50.33333333	8.75	6120	100
FRIGOURAS	France	44.13333333	5.95	5450	100
GELEEN	Netherlands	51.96666667	5.75	6370	60
GIANTS' HILLS	Great Britain	53.22	0.13	5450	80
GIRIBALDI CIMIEZ CARABACEL	France	43.7	7.233333333	5750	60
GODDELAU	Germany	61.01666667	8.116666667	6300	90
GORNJA TUZLA	Bosnia-Herz.	44.45	18.76666667	6640	75
GRAND LOUVRE	France	48.86666667	2.333333333	5620	70
GRANDE BARME SAVIGNY	France	45.73333333	5.916666667	5010	140
GREAT LANGDALE	Great Britain	54.47	-3.14	4790	80
GREIFENSEE-STOREN/WILDSBERG	Switzerland	47.36666667	8.683333333	5140	49
GRINDALE, BARROW 1	Great Britain	54.12	-0.24	4910	120
GROTTA ARMA DELLO STEFANIN	Italy	44.1	8.016666667	6610	60
GROTTA CORBEDDU	Italy	40.25	9.483333333	6602	60

Appendix 1: The Gkiasta European Database: Datasets A to G.
Dataset A (weighted mean or oldest radiocarbon determination at 509 Neolithic sites).

SITE NAME	Country	Latitude (decimal degrees)	Longitude (decimal degrees)	Date (uncal. bp)	Standard deviation
GROTTA DEI PICCIONI	Italy	42.21666667	14.96666667	6247	130
GROTTA DEL SANT. DELLA MADONNA	Italy	40.9	15.78333333	5555	75
GROTTA DEL'ORSO	Italy	43.98333333	3.233333333	5550	70
GROTTA DI PORTO BADISCO	Italy	40.08333333	18.48333333	5850	55
GROTTA DI ROMITO	Italy	40.86666667	16.9	6420	70
GROTTA FILIESTRU	Italy	40.5	8.666666667	6710	75
GROTTA PODOSOJNA	Croatia	45.25	14.23333333	6470	95
GROTTE DE CAMPRAFAUD	France	43.43333333	3.9	6480	130
GROTTE DE GAZEL	France	43.31666667	2.416666667	6905	90
GROTTE DE LA VIEILLE EGLISE	France	45.91666667	6.283333333	5295	52
GROTTE DE L'ABBE PRIALET	France	43.88333333	3.733333333	6070	100
GROTTE DE L'ABEURADOR	France	43.35	2.7	6260	90
GROTTE DE L'AIGLE	France	44.06666667	4.633333333	6139	77
GROTTE DES ESCABASSES	France	44.71666667	1.8	5160	35
GROTTE DES PERRATS	France	45.78333333	0.35	5450	40
GROTTE DES TEPPES	France	46.43333333	5.333333333	5390	50
GROTTE DESPERRATS	France	45.78333333	3.083333333	5330	120
GROTTE DU GARDON	France	45.95	5.333333333	6325	40
GROTTE DU SANGLIER	France	44.68333333	5.333333333	5440	130
GROTTE DU VIEUX MOUNOI	France	43.3	5.866666667	5625	85
GROTTE LOMBARD	France	43.68333333	7.866666667	6150	85
GROTTE TOURNIE	France	43.35	2.8	5900	100
GUENNOC	France	48.58333333	-4.583333333	5075	140
GWERNVALE	Great Britain	51.87	-3.15	5050	75
HALDRUP STRAND	Denmark	56.86666667	10.96666667	5630	120
HALLERTAU	Germany	48.55	11.75	5990	90
HAMANGIA-BAIA 1	Romania	44.7	28.66666667	5880	70
HAMBLEDON HILL	Great Britain	50.90944444	-2.209166667	4665	26
HANGING GRIMSTONE	Great Britain	54.04	-0.77	4710	90
HAZLETON NORTH	Great Britain	51.87	-1.9	4875	15
HIENHEIM	Germany	48.88333333	11.75	6235	45
HOLMEGAARD	Denmark	56.26666667	10.76666667	6039	48
HONEYGORE (W) TRACK, WESTHAY	Great Britain	51.18	-2.84	4760	30
HONEYGORE TRACK	Great Britain	51.18	-2.82	4590	40
HORNE LEFANTOVCE	Slovakia	48.41666667	18.16666667	5775	140
HUDE 1	Germany	53.16666667	8.433333333	6060	115
INCHTUTHIL	Great Britain	56.54	-3.44	5100	41
IRCHONWELZ	Belgium	50.61666667	3.75	6030	60
JONES'S TRACK, ASHCOTT / W. HEATH	Great Britain	51.14	-2.78	4590	70
KANNE	Belgium	50.8	5.666666667	6260	75
KATALSZEG	Hungary	46.21666667	20.08333333	6370	100
KERGOGLE	France	48	-4.483333333	5190	80
KILHAM	Great Britain	54.09055556	-0.386666667	4830	125
KING BARROW RIDGE	Great Britain	51.18	-2.83	5640	120
KINLOCH FARM, COLLESSIE	Great Britain	56.29	-3.16	4725	70
KNAP HILL	Great Britain	51.37083333	-1.826111111	4710	115
KNAP OF HOWAR, PAPA WESTRAY	Great Britain	59.35	-2.92	5706	85
KNOSSOS	Greece	35.51666667	25.16666667	7910	140
KOLN-MENGENICH	Germany	50.91666667	7	6320	70
KORLAT	Hungary	48.36666667	21.25	6440	100
LA BAUME DE RONZE	France	44.31666667	4.416666667	5265	66
LA CARRIERE MERQUE	France	50.01666667	1.983333333	5250	120
LA CAUNE	France	42.71666667	2.616666667	5640	120
LA CAUNE OUEST	France	42.91666667	1.933333333	5164	125
LA CHAUVE SOURIS	France	44.46666667	4.683333333	6790	70
LA CROIX MAIGRET	France	49.4	4.883333333	6030	130
LA FOSSE TOUMISE	France	49.38333333	3.4	5870	105
LA GRANDE RIVOIRE	France	45.2	5.633333333	6000	70
LA LEDE DU GURP	France	45.45	-1.15	6101	65
LA MOLLE PIERRE	France	47.05	6.583333333	5340	70
LA MOTTE DES JUSTICES	France	47	-0.2	5340	100
LA PLAISANCE	France	50	2.133333333	5255	80
LA POSTE VIEILLE	France	43.25	2.216666667	5190	75
LA ROCHE AUX FEES	France	45.98333333	0.983333333	5525	85
LA VESIGNIE	France	44.25	4.616666667	5815	75
LAGNANO DA PIEDE	Italy	38	12.33333333	6700	100
LAMERSDORF	Germany	50.8	6.466666667	6410	45
LANGDALE-PIKE OF STICKLE/ T. KNOTT	Great Britain	54.45472222	-3.123055556	4680	135
LANGWEILER	Germany	50.81666667	6.433333333	7090	135
LARCUSTE	France	47.85	-2.783333333	5490	120
LARZICOURT	France	48.63333333	4.716666667	5800	100
LE BARATIN	France	44.08333333	5.883333333	6600	140
LE BOISANNE	France	48.53333333	-2	5610	70
LE CAMP	France	48.15	2.733333333	5830	45
LE CAMP DE BIERRE	France	48.85	0.533333333	5470	140
LE CAMP DE CESAR	France	49.36666667	2.516666667	5280	140
LE CAMP DE LAURE	France	43.55	5.216666667	5020	100
LE CHEVAL	France	43.93333333	4.35	5475	110
LE COQ GALLEUX	France	49.4	2.733333333	5300	140
LE FOSSE DE TRAVERS	France	48.85	2.533333333	5410	60
LE FOURNET	France	44.68333333	1.716666667	6915	70
LE GARDON	France	45.95	5.033333333	5815	40
LE GUERZIT	France	48.71666667	-3.833333333	5950	110
LE MONT D'HUETTE	France	49.4	2.966666667	5890	120
LE MOULIN	France	44.18333333	5.5	5695	92
LE PETIT BOIS	France	46.71666667	4.933333333	5816	45
LE ROC DE DOURGNE	France	42.61666667	2.15	6850	100
LE ROUICK	France	47.56666667	-2.95	5293	40
LE SANGLIER	France	44.68333333	1.716666667	6095	100

Appendix 1: The Gkiasta European Database: Datasets A to G.
Dataset A (weighted mean or oldest radiocarbon determination at 509 Neolithic sites).

SITE NAME	Country	Latitude (decimal degrees)	Longitude (decimal degrees)	Date (uncal. bp)	Standard deviation
LE TROU QUI FUME	France	47	4.716666667	5480	130
LE VERGER	France	46.98333333	4.716666667	5710	140
L'EGLISE	France	43.75	4.8	5111	81
LES BOURIES	France	46.43333333	1.666666667	5210	70
LES BROUILLARDS	France	49.3	2.75	5050	70
LES CHAMBONS	France	45.95	3.083333333	5130	140
LES COUDOUMINES	France	42.75	2.566666667	5135	36
LES DEPENDANCES DE DIGNY	France	48.16666667	-3.9	5304	67
LES ERVES	France	48.11666667	-0.333333333	5580	140
LES FIEFS	France	48.23333333	2.45	5611	35
LES FONTINETTES	France	49.38333333	3.75	5942	62
LES GOUILLAUDS	France	45.5	-0.6	5950	120
LES GRANDS TRAQUIERS	France	49.01666667	4.266666667	6200	140
LES LONGUES RAIES	France	48.91666667	2.766666667	5510	140
LES MOULINS	France	44.35	4.766666667	5262	63
LES OBEAUX	France	49.4	2.8	6080	110
LES PICHELOTS	France	47.31666667	-1.5	5530	70
LES PLANCHES	France	46.91666667	5.766666667	5520	130
LES PLOTS	France	43.21666667	2.35	5190	41
LES REAUDINS	France	48.4	3.166666667	5865	65
LES RESSOURS	France	52.03333333	1.3	5040	70
LES RIVAUX	France	45.05	3.85	6240	110
LES SABLES DE LOI	France	46.43333333	0.883333333	5110	60
LES SABLINS	France	50.51666667	1.65	5675	85
LINLITHGOW	Great Britain	55.97	-3.6	5265	55
LISCUIS EN LANISCAT	France	48.21666667	-3.133333333	5140	100
LISMORE FIELDS	Great Britain	53.26	-1.94	5270	100
LITTLE WALTHAM	Great Britain	51.81	0.45	5120	130
LO PLA DEL BACH	France	42.45	2.066666667	5700	80
LOCH CALDER - AN T'SIONNAICH	Great Britain	58.54	-3.6	4685	60
LOCH CALDER - TULLOCH OF ASSERY B	Great Britain	58.53	-3.6	4965	60
LOCHHILL	Great Britain	54.96972222	-3.610833333	5070	105
LOCMARIAQUER ER GRAH	France	47.56666667	-2.966666667	5300	60
LONG DOWN	Great Britain	50.88	-2.17	4675	44
L'OUCHE DE BEAUCE	France	48.38333333	2.383333333	5570	70
MACHRIE MOOR SITE	Great Britain	55.54	-5.32	5500	70
MAD MAN'S WINDOW, GLENARM	Ireland	54.98	-5.93	5095	120
MADDALENA DI MUCCIA	Italy	43.06666667	13.06666667	6580	75
MAES HOWE	Great Britain	58.99722222	-3.187222222	5094	60
MAIDEN CASTLE	Great Britain	50.7	-2.48	5030	80
MALVOISIN	France	43.8	5.016666667	5270	140
MANNLEFELSEN	France	47.45	7.233333333	5140	140
MAR DYKE	Great Britain	51.5	0.27	5740	80
MARKLAND GRIPS	Great Britain	53.27	-1.21	5209	48
MEARE HEATH	Great Britain	51.163	-2.78	4583	45
MEARE HEATH FIELD	Great Britain	51.17	-2.79	5180	70
MEARE LAKE VILLAGE WEST	Great Britain	51.17	-2.8	5210	80
MELDON BRIDGE, NEAR PEEBLES	Great Britain	55.65055556	-3.263611111	4686	90
MIDGEHOLME MOSS	Great Britain	54.98	-2.61	5270	90
MILETO	Italy	43.79	11.21	6140	57
MILLBARROW	Great Britain	51.45	-1.87	4708	52
MIOUVIN	France	43.51666667	3.283333333	6450	100
MISBOURNE VIADUCT	Great Britain	51.58	-0.55	6094	58
MOHELNICE	Czech Rep.	49.78333333	16.91666667	6290	34
MOLINO CASAROTO	Italy	45.46666667	11.6	6450	110
MONAMORE, ARRAN	Great Britain	55.5125	-5.141111111	5110	110
MONTE LEONE	France	41.4	9.166666667	5855	95
MONTE MAULO	Italy	41.57	14.64	6245	49
MONTE REVINCU	France	42.6	9.266666667	5405	55
MOULIN DE LA GARONNE	France	43.5	1.333333333	5120	120
MUDDERSHEIM	Germany	50.7	-2.15	5210	70
MYARD	France	47.38333333	4.516666667	5175	135
NEA NIKOMEDEIA	Greece	40.6	22.25	7557	91
NECROPOLE DE LA BOIXE DOLMEN C	France	45.8	0.133333333	5380	50
NETHER KINNEIL	Great Britain	56	-3.67	5060	50
NETHERAVON BAKE	Great Britain	51.22	-1.85	4760	90
NEUVY EN DUNOIS	France	48.2	1.55	5250	140
NEWGRANGE	Ireland	53.71	-6.49	4505	41
NEWTON	Great Britain	55.79	-6.26	4923	42
NIEDERMERZ	Germany	50.86666667	6.333333333	6800	80
NORMANTON DOWN, AMESBURY	Great Britain	51.16861111	-1.835555556	4510	103
NORROIS	France	48.66666667	4.55	6020	110
NORSLUND	Denmark	56.01666667	10.95	5530	61
NORTH MARDEN	Great Britain	50.94	-0.87	4710	110
NORTH STOKE	Great Britain	51.57	-1.13	4672	49
OBRE	Bosnia-Herz.	44.18333333	18.1	7240	60
OFFHAM HILL	Great Britain	50.89	-0.02	4807	48
OLAJKUT	Hungary	46.21666667	20.25	6443	46
OLBY LYNG	Denmark	55.5	12.21666667	5265	92
OLEYE	Belgium	50.68333333	5.216666667	6281	38
OLSZANICA	Poland	50.1	19.83333333	6430	75
ORSETT	Great Britain	50.85277778	0.348611111	4636	31
OSTOROS	Hungary	48.86666667	20.43333333	6180	100
OSZENTIVAN	Hungary	46.2	20.18333333	6460	80
PADINA	Yugoslavia	44.5	8.5	6980	115
PARC DU BRINBORION	France	48.33333333	9.566666667	6140	45
PAS DE LA CHARMATE	France	45.06666667	5.383333333	5651	84
PASSY	France	48.1	3.266666667	6065	55

Appendix 1: The Gkiasta European Database: Datasets A to G.
Dataset A (weighted mean or oldest radiocarbon determination at 509 Neolithic sites).

SITE NAME	Country	Latitude (decimal degrees)	Longitude (decimal degrees)	Date (uncal. bp)	Standard deviation
PENDIMOUM	France	43.81666667	7.5	6463	42
PENTE DE CROUPETON	France	48.88333333	22.6	6136	100
PEN-Y-WYRLOD	Great Britain	51.97583333	-3.236111111	4970	80
PETIT MONT	France	47.53333333	-2.9	5650	70
PHYN	Switzerland	47.58333333	8.933333333	4993	28
PIECE DE PINCEVENT	France	48.36666667	2.883333333	5630	120
PITNACREE	Great Britain	56.65944444	-3.748888889	4810	90
PLACE SAINT LAMBERT	Belgium	50.65	5.566666667	6316	27
PLANTA	Switzerland	46.23333333	7.366666667	6500	80
PLATIA MAGOULA ZAKROU	Greece	36.95	22.76666667	7083	44
POIGEN	Austria	48.66666667	38.66666667	5935	90
PONTCHARAUD	France	45.78333333	3.083333333	7095	50
PORT CHARLOTTE	Great Britain	55.73	-6.39	5020	90
PORT LEUCATE	France	42.83333333	3	5900	140
PORT ST.MARY RUSHEN	Great Britain	54.08333333	-4.75	4970	80
PRISTINA	Yugoslavia	42.65	21.15	6280	80
PULKAU	Austria	48.7	15.86666667	6215	100
QUANTERNESS	Great Britain	58.99972222	-3.015	4574	62
QUEBARN	France	47.83333333	-4.233333333	5760	120
RAFFEGUES	France	43.41666667	3.583333333	5180	71
RAIGMORE	Great Britain	57.48	-4.19	4813	74
RECTORY FARM	Great Britain	52.32	-0.16	4950	36
REDLANDS FARM	Great Britain	52.33	-0.59	4823	50
REHMSDORF	Germany	51.03333333	13.53333333	5880	71
REICHSTETT	France	48.63333333	7.75	5940	140
RENAGHJU	France	41.51666667	8.916666667	6525	60
RENDINA	Italy	41	16	6891	81
REVELLATA	France	42.56666667	8.733333333	6280	75
RHUM (KINLOCH)	Great Britain	57.04	-6.27	4725	140
RICOH	France	48.06666667	7.3	6400	70
RINGKLOSTER	Denmark	56.01666667	10.23333333	5230	100
RIPABIANCA DI MONTERADO	Italy	43.71666667	13.1	6260	85
RIPARO ARMA DI NASINO	Italy	44.11666667	8.05	6470	120
RIPARO DI ROMAGNANO III	Italy	46	11.1	6480	50
RIPARO GABAN	Italy	46.07	11.1	6030	45
ROBIN HOOD'S BALL	Great Britain	51.21	-1.86	4613	67
ROC TROUE	France	43.95	3.183333333	5850	80
ROCHER MOSS 2	Great Britain	53.61666667	-2.033333333	5830	100
RODINGEN	Germany	50.96666667	6.55	6250	75
ROQUEMISSOU	France	44.41666667	2.733333333	5785	105
ROSDORF	Germany	51.51666667	9.933333333	6350	70
ROTTEN BOTTOM, GAMES HOPE	Great Britain	55.42	-3.36	5040	100
ROUCADOUR	France	44.83333333	1.5	5940	140
ROUTE DE ROUFFACH	France	48.08333333	7.333333333	6062	81
RUDE 2	Germany	54.61666667	9.583333333	5644	41
RUFACHER HUBEN	France	48.03333333	7.333333333	5740	130
RUNNYMEDE	Great Britain	51.43	-0.56	5780	85
SAINT JUST	France	47.76666667	-1.95	5770	80
SAINT MARCEL D'ARDECHE	France	44.35	4.466666667	6330	90
SAINT MICHEL-DE-TOUCH	France	43.61666667	1.433333333	5600	90
SAINT MITRE:ABRI 2	France	44.88333333	5.65	6512	79
SAINT MITRE:ABRI 3	France	44.88333333	7.483333333	6431	50
SAINT PRIEST	France	45.7	13.81666667	6300	60
SAINTE LUCE	France	44.3	5.333333333	5760	100
SALIAGOS	Greece	37.05	25.08333333	6172	74
SALPETERMOSE	Denmark	56.91666667	12.3	6020	100
SAMMARDENCHIA, UDINE	Italy	45.72	13.76	6120	60
SAN MARCO, GUBBIO	Italy	43.13	12.4	6276	41
SANDORRESZE	Hungary	47.65	20.15	6141	62
SANDUN	France	47.33333333	2.566666667	5660	100
SANTA TECCHIA	Italy	41.5	15.68333333	7000	100
SARN-Y-BRYN-CALOD	Great Britain	52.64	-3.16	4960	70
SCAFFA PIANA	France	42.68333333	9.2	5340	71
SCAMUSO	Italy	41.1	16.89	7290	110
SCHENKON	Switzerland	47.16666667	8.133333333	4980	120
SCORD OF BROUSTER	Great Britain	50.93	-1.66	5050	90
SEAMER MOOR, EAST AYTON	Great Britain	54.26333333	-0.464444444	5030	90
SELEDIN	France	48.21666667	-3.05	5149	63
SELEVAC	Yugoslavia	44.5	21.88333333	6192	51
SENTE SAILLANCOURT	France	49.08333333	2	5220	110
SESKLO	Greece	39.38333333	22.81666667	7755	97
SHANBALLYEDMOND	Ireland	52.69	-8.25	4930	60
SHAPWICK HEATH	Great Britain	51.16	-2.81	4540	130
SHIPPEA HILL, PEACOCKS FARM	Great Britain	52.42888889	0.4125	5465	120
SHURTON HILL	Great Britain	50.92	-0.58	4625	57
SIDARI	Greece	39.78333333	19.73333333	7670	120
SIDLINGS COPSE	Great Britain	51.79	-1.22	5250	75
SIERENTZ	France	47.56666667	7.45	5570	100
SILBURY HILL, AVEBURY	Great Britain	51.415	-1.856111111	4593	65
SION-PETIT CHASSEUR	Switzerland	46.23333333	7.35	5110	57
SITAGROI	Greece	41.15	24.2	6425	100
SKATEHOLM	Sweden	55.38333333	13.48333333	7030	70
SKOTINI CAVE	Greece	38.51666667	23.98333333	6247	49
SO'N MATGE	Spain	39.33333333	3	6680	120
SOURCE DE RESELAUZE	France	43.51666667	4.983333333	5380	110
SOUTH CADBURY CASTLE	Great Britain	51.0225	-2.541944444	4705	115
SOUTH STREET, AVEBURY	Great Britain	51.42111111	-1.870555556	4641	64
STANWICK SOUTH	Great Britain	52.32	-0.59	4764	55
STARCEVO	Yugoslavia	44.81666667	20.7	6975	60

Appendix 1: The Gkiasta European Database: Datasets A to G.
Dataset A (weighted mean or oldest radiocarbon determination at 509 Neolithic sites).

SITE NAME	Country	Latitude (decimal degrees)	Longitude (decimal degrees)	Date (uncal. bp)	Standard deviation
STEPLETON	Great Britain	50.88	-0.7	4975	71
STREET HOUSE FARM, LOFTUS	Great Britain	54.56	-0.83	4978	31
STRETTO, PARTANNA	Italy	38.12	13.36	6630	120
STRZELCE	Poland	53.31666667	18.11666667	6260	60
STUMBLE	Great Britain	51.73	0.73	4728	49
STUROVO	Slovakia	47.8	18.73333333	6215	71
SUDENSEE DAMM	Germany	54.63333333	9.583333333	5954	40
SWEET TRACK, SHAPWICK	Great Britain	51.163	-2.83	5224	75
SWIFTERBANT	Netherlands	52.56666667	5.6	6288	27
SZAMOSSALYI	Hungary	48.88333333	2.716666667	5910	80
SZARVAS	Hungary	47.83333333	20.58333333	6620	60
TANKARDSTOWN	Ireland	52.42	-8.62	5105	45
TARNABOD	Hungary	47.68333333	20.21666667	6280	100
TELL AZMAK	Bulgaria	42.45	25.76666667	6758	33
TELL KARANOVO	Bulgaria	42.5	25.93333333	6690	71
TERRES DE PREPOUX	France	48.33333333	3.066666667	6730	110
THE ORD NORTH, LAIRG	Great Britain	58.01	-4.43	4614	57
THEOPETRA CAVE	Greece	38.46666667	23.23333333	7036	44
THREE HOLES CAVE	Great Britain	50.49	-3.68	5010	70
TISZAVASVARI-KERESZTFAL	Hungary	48.96666667	21.38333333	6226	69
TOMASZOW	Poland	51.36666667	21.26666667	6145	70
TREMONTEIX	France	45.78333333	0.35	5880	70
TRIFAUCKIJ	Moldavia	48.18333333	28.31666667	7420	80
TROU AL'WESSE	Belgium	50.46666667	5.25	5950	70
TROU ARNAUD	France	44.55	5.316666667	5343	50
TULLY, ALDERGROVE AIRPORT	Ireland	54.46	-7.82	4960	85
TUMBA OR VELUSKA	Macedonia	41.9	21.36666667	6874	55
TWANN	Switzerland	47.1	7.15	5019	34
TY FLOCH	France	48.16666667	2.416666667	5990	50
UNANG	France	44.05	5.15	5865	65
UNTERPULLENDORF	Austria	47.5	16.5	6130	140
VAENGE SO	Denmark	56.13333333	10.51666667	5474	39
VALLON-DES-VAUX	Switzerland	46.78333333	6.783333333	4985	35
VEJLEBRO	Denmark	55.68333333	9.583333333	5362	54
VIEUX MANOIR TALANGE	France	49.25	6.15	5310	80
VILAGGIO A, SCARAMELLA	Italy	41.5	15.75	6863	60
VILLAGGIO LEOPARDI	Italy	42.46666667	14.93333333	6578	135
VILLENEUVE TOLOSANE	France	43.51666667	1.3	5400	90
VLASAC	Yugoslavia	44.53333333	22.05	6335	92
VRSNIK-TARINCI	Macedonia	41.75	22.15	6892	62
WANGE	Belgium	50.8	5.05	6310	75
WAREMME	Belgium	50.66666667	13.95	6400	120
WAYLAND'S SMITHY, ASHBURY	Great Britain	51.56638889	-1.594444444	4770	130
WEST KENNET	Great Britain	51.41	-1.85	4767	44
WESTEREGELN	Germany	52.96666667	11.4	6093	71
WESTWARD HO!	Great Britain	51.03333333	-4.25	5740	100
WHITWELL LONG CAIRN	Great Britain	53.27	-1.24	4750	64
WINDEN AM SEE	Austria	47.95	16.83333333	5880	71
WINDMILL HILL	Great Britain	50.92	-1.88	4715	21
WINNALL DOWN	Great Britain	51.08	-1.32	4725	53
WINTERBOURNE STOKE	Great Britain	51.2	-1.85	4603	77
WOLLA BANK	Great Britain	53.25	0.33	4865	65
WORGRET, WAREHAM	Great Britain	50.7	6.55	6227	38
ZALAVAR	Hungary	46.66666667	17.16666667	6180	100
ZILGENDORF	Germany	50.16666667	10.96666667	6233	37
ZILGENDORF I	Germany	50.16666667	10.96666667	6310	80
ZOPY	Czech Rep.	49.33333333	17.58333333	6430	100
ZORLENTU MARE	Romania	45.31	21.81	6070	40
ZURICH	Switzerland	47.36666667	8.583333333	5145	70
ZURICH-BAUSCHANZE	Switzerland	47.41	8.52	5320	60
ZURICH-WOLLISHOFEN	Switzerland	47.41	8.52	4993	46
ZWENKAU	Germany	51.23333333	12.35	6160	70

Appendix 1: The Gkiasta European database: Datasets A to G.
Dataset B (weighted mean or most recent radiocarbon determinations at 509 Neolithic sites).

SITE NAME	Country	Latitude (decimal degrees)	Longitude (decimal degrees)	Date (uncal. bp)	Standard deviation
KNOSSOS	Greece	35.51666667	25.16666667	5967	41
CUEVA DE LA DEHESILLA	Spain	36.75	-5.8	6260	100
PLATIA MAGOULA ZAKROU	Greece	36.95	22.76666667	6913	40
SALIAGOS	Greece	37.05	25.08333333	5716	85
FRANCHTHI	Greece	37.43333333	23.13333333	6031	105
LAGNANO DA PIEDE	Italy	38	12.33333333	6700	100
CUEVA DEL NACIMIENTO	Spain	38.08333333	-2.3	5480	120
STRETTO, PARTANNA	Italy	38.12	13.36	5690	100
THEOPETRA CAVE	Greece	38.46666667	23.23333333	5485	102
ANTRE CORYCIEN	Greece	38.48333333	22.5	6315	64
SKOTINI CAVE	Greece	38.51666667	23.98333333	4812	42
ELATEIA	Greece	38.61666667	22.73333333	7435	55
COVETA DE L'OR	Spain	38.7	-0.466666667	6265	75
BEVILACQUA	Italy	38.92	16.61	6430	90
CUEVA CHICA DE SANTIAGO	Spain	38.95	-6.916666667	6160	100
ACHILLEION	Greece	39.28333333	22.38333333	7210	90
SO'N MATGE	Spain	39.33333333	3	6680	120
SESKLO	Greece	39.38333333	22.81666667	7110	70
ARGISSA MAGOULA	Greece	39.63333333	22.46666667	6700	130
SIDARI	Greece	39.78333333	19.73333333	7670	120
GROTTA DI PORTO BADISCO	Italy	40.08333333	18.48333333	5850	55
GROTTA CORBEDDU	Italy	40.25	9.483333333	6602	60
GROTTA FILIESTRU	Italy	40.5	8.666666667	5625	65
NEA NIKOMEDEIA	Greece	40.6	22.25	7281	74
GROTTA DI ROMITO	Italy	40.86666667	16.9	6420	70
GROTTA DEL SANT. DELLA MADONNA	Italy	40.9	15.78333333	5555	75
RENDINA	Italy	41	16	6891	81
SCAMUSO	Italy	41.1	16.89	5870	70
SITAGROI	Greece	41.15	24.2	6425	100
ARAGUINA SENNOLA	France	41.38333333	9.166666667	6540	99
MONTE LEONE	France	41.4	9.166666667	5855	95
COPPA NEVIGATA	Italy	41.5	15.66666667	6520	70
SANTA TECCHIA	Italy	41.5	15.68333333	6540	65
VILAGGIO A, SCARAMELLA	Italy	41.5	15.75	6863	60
RENAGHJU	France	41.51666667	8.916666667	6095	45
MONTE MAULO	Italy	41.57	14.64	6245	49
ANZA	Macedonia	41.7	22	6230	60
VRSNIK-TARINCI	Macedonia	41.75	22.15	6892	62
TUMBA OR VELUSKA	Macedonia	41.9	21.36666667	6874	55
CASABIANCA OR CASTERRAGIO	France	42.08333333	9.5	6670	130
GROTTA DEI PICCIONI	Italy	42.21666667	14.96666667	6247	130
CUEVA DE CHAVES	Spain	42.36666667	0.216666667	6120	70
BALMA MARGINEDA	Andora	42.41666667	1.583333333	6670	85
AZMAK	Bulgaria	42.45	25.76666667	6483	100
LO PLA DEL BACH	France	42.45	2.066666667	5300	65
TELL AZMAK	Bulgaria	42.45	25.76666667	6758	33
VILLAGGIO LEOPARDI	Italy	42.46666667	14.93333333	6578	135
TELL KARANOVO	Bulgaria	42.5	25.93333333	6690	71
ENVEIG	France	42.51666667	1.866666667	5290	60
REVELLATA	France	42.56666667	8.733333333	6280	75
MONTE REVINCU	France	42.6	9.266666667	5175	45
LE ROC DE DOURGNE	France	42.61666667	2.15	4250	120
PRISTINA	Yugoslavia	42.65	21.15	6280	80
SCAFFA PIANA	France	42.68333333	9.2	5340	71
LA CAUNE	France	42.71666667	2.616666667	5640	120
CAMP DEL GINEBRE	France	42.75	2.566666667	5060	45
LES COUDOUMINES	France	42.75	2.566666667	5135	36
ABRI DE LA COMA FRANCEZE	France	42.83333333	2.916666667	5180	60
COMA FRANCEZE	France	42.83333333	2.916666667	5200	70
PORT LEUCATE	France	42.83333333	3	5410	140
LA CAUNE OUEST	France	42.91666667	1.933333333	5164	125
MADDALENA DI MUCCIA	Italy	43.06666667	13.06666667	6580	75
ABRI JEAN CROS	France	43.08333333	2.466666667	4450	100
SAN MARCO, GUBBIO	Italy	43.13	12.4	6276	41
BUHOLOUP	France	43.15	1.183333333	5490	120
LES PLOTS	France	43.21666667	2.35	5190	41
LA POSTE VIEILLE	France	43.25	2.216666667	5190	75
FONT JUVENAL	France	43.28333333	2.333333333	5070	110
GROTTE DU VIEUX MOUNOI	France	43.3	5.866666667	5250	60
GROTTE DE GAZEL	France	43.31666667	2.416666667	6090	65
GROTTE DE L'ABEURADOR	France	43.35	2.7	6260	90
GROTTE TOURNIE	France	43.35	2.8	5050	120
CHATEAUNEUF LES MARTIGUES	France	43.38333333	5.166666667	5460	130
RAFFEGUES	France	43.41666667	3.583333333	5180	71
GROTTE DE CAMPRAFAUD	France	43.43333333	3.9	5100	130
MOULIN DE LA GARONNE	France	43.5	1.333333333	5120	120
VILLENEUVE TOLOSANE	France	43.5	1.333333333	5060	45
MIOUVIN	France	43.51666667	3.283333333	5865	100
SOURCE DE RESELAUZE	France	43.51666667	4.983333333	5380	110
LE CAMP DE LAURE	France	43.55	5.216666667	5020	100
BAUME DE FONTBREGOUA	France	43.55	6.233333333	5330	120
CUGNAUX	France	43.61666667	1.433333333	5020	50
SAINT MICHEL-DE-TOUCH	France	43.61666667	1.433333333	5130	100
GROTTE LOMBARD	France	43.68333333	7.866666667	6150	85
CAUCADE	France	43.7	7.35	6800	130
CHATEAU PERCIN	France	43.7	0.35	5250	100
GIRIBALDI CIMIEZ CARABACEL	France	43.7	7.233333333	5175	85
RIPABIANCA DI MONTERADO	Italy	43.71666667	13.1	6260	85
ESCANIN	France	43.75	6.25	5630	99

Appendix 1: The Gkiasta European database: Datasets A to G.
Dataset B (weighted mean or most recent radiocarbon determinations at 509 Neolithic sites).

SITE NAME	Country	Latitude (decimal degrees)	Longitude (decimal degrees)	Date (uncal. bp)	Standard deviation
MILETO	Italy	43.79	11.21	6140	57
MALVOISIN	France	43.8	5.016666667	5270	140
PENDIMOUM	France	43.81666667	7.5	6463	42
GROTTE DE L'ABBE PRIALET	France	43.88333333	3.733333333	5270	100
LE CHEVAL	France	43.93333333	4.35	5475	110
ROC TROUE	France	43.95	3.183333333	5850	80
ABRI DES USCLADES	France	43.98333333	12.86666667	6080	50
GROTTA DEL'ORSO	Italy	43.98333333	3.233333333	5550	70
CERON	France	44	5.05	5990	75
DIVOSTIN	Yugoslavia	44.03333333	21.83333333	7010	38
UNANG	France	44.05	5.15	5865	65
GROTTE DE L'AIGLE	France	44.06666667	4.633333333	6139	77
LE BARATIN	France	44.08333333	5.883333333	6600	140
GROTTA ARMA DELLO STEFANIN	Italy	44.1	8.016666667	5600	80
ABRI DE LA POUJADE	France	44.11666667	3.166666667	5130	90
RIPARO ARMA DI NASINO	Italy	44.11666667	8.05	5955	65
FRIGOURAS	France	44.13333333	5.95	5450	100
LE MOULIN	France	44.18333333	5.5	5695	92
OBRE	Bosnia-Herz.	44.18333333	18.1	6710	60
LA VESIGNIE	France	44.25	4.616666667	5815	75
BAUME DE MONTCLUS	France	44.26666667	4.433333333	6334	70
SAINTE LUCE	France	44.3	5.333333333	5760	100
BAUME D'OULLINS	France	44.31666667	4.466666667	5590	100
LA BAUME DE RONZE	France	44.31666667	4.416666667	5265	66
CHAZELLES	France	44.33333333	4.333333333	5660	110
LES MOULINS	France	44.35	4.466666667	5220	120
SAINT MARCEL D'ARDECHE	France	44.35	4.766666667	5262	63
BALLOY LES REAUDINS	France	44.4	3.166666667	6180	90
ROQUEMISSOU	France	44.41666667	2.733333333	5785	105
GORNJA TUZLA	Bosnia-Herz.	44.45	18.76666667	6640	75
LA CHAUVE SOURIS	France	44.46666667	4.683333333	6790	70
SELEVAC	Yugoslavia	44.5	22.41666667	7082	42
CAVERNA DELLE ARENE CANDIDE	Italy	44.5	21.88333333	6192	51
PADINA	Yugoslavia	44.5	8.5	5700	90
VLASAC	Yugoslavia	44.53333333	22.05	6335	92
TROU ARNAUD	France	44.55	5.316666667	5343	50
CAPDENAC LE HAUT	France	44.58333333	2.066666667	5148	73
ANTONNAIRE	France	44.68333333	5.383333333	5570	130
GROTTE DU SANGLIER	France	44.68333333	5.333333333	5440	130
LE FOURNET	France	44.68333333	1.716666667	6915	70
LE SANGLIER	France	44.68333333	1.716666667	6095	100
HAMANGIA-BAIA 1	Romania	44.7	28.66666667	5880	70
GROTTE DES ESCABASSES	France	44.71666667	1.8	5160	35
STARCEVO	Yugoslavia	44.81666667	20.7	6475	45
ROUCADOUR	France	44.83333333	1.5	5025	50
ABRI DE PENDIMOUN	France	44.88333333	5.651	6100	120
SAINT MITRE:ABRI 2	France	44.88333333	5.65	6512	79
SAINT MITRE:ABRI 3	France	44.88333333	7.483333333	6431	50
BAUME BOURBON	France	44.9	4.633333333	6068	77
LES RIVAUX	France	45.05	3.85	5030	110
PAS DE LA CHARMATE	France	45.06666667	5.383333333	5651	84
COUFIN	France	45.06666667	5.4	5260	120
COMBOIRE	France	45.13333333	5.683333333	5360	70
LA GRANDE RIVOIRE	France	45.2	5.633333333	5030	70
ABRI DE LA GRANDE RIVOIRE	France	45.21666667	5.65	5135	35
GROTTA PODOSOJNA	Croatia	45.25	14.23333333	6470	95
ZORLENTU MARE	Romania	45.31	21.81	6070	40
LA LEDE DU GURP	France	45.45	-1.15	5305	45
MOLINO CASAROTO	Italy	45.46666667	11.6	5555	130
LES GOUILLAUDS	France	45.5	-0.6	5950	120
APATIN	Yugoslavia	45.66666667	18.98333333	6910	45
ALFORT	France	45.68333333	5.35	5755	65
GROTTA DEI CICLAMI	Italy	45.7	4.916666667	5445	60
SAINT PRIEST	France	45.7	13.81666667	6300	60
BETHENAS	France	45.71666667	5.25	5060	110
SAMMARDENCHIA, UDINE	Italy	45.72	13.76	6120	60
GRANDE BARME SAVIGNY	France	45.73333333	5.916666667	5010	140
CREUX-ROUGE	France	45.78333333	3.083333333	6070	140
GROTTE DES PERRATS	France	45.78333333	0.35	5450	40
GROTTE DESPERRATS	France	45.78333333	3.083333333	5330	120
TREMONTEIX	France	45.78333333	0.35	5400	50
PONTCHARAUD	France	45.78333333	3.083333333	5110	130
NECROPOLE DE LA BOIXE DOLMEN C	France	45.8	0.133333333	5380	50
GROTTE DE LA VIEILLE EGLISE	France	45.91666667	6.283333333	5295	52
LES CHAMBONS	France	45.95	3.083333333	5130	140
GROTTE DU GARDON	France	45.95	5.333333333	5060	40
LE GARDON	France	45.95	5.033333333	5095	70
LA ROCHE AUX FEES	France	45.98333333	0.983333333	5525	85
RIPARO DI ROMAGNANO III	Italy	46	11.1	5560	50
RIPARO GABAN	Italy	46.07	11.1	5750	60
OSZENTIVAN	Hungary	46.2	20.18333333	6050	100
OLAJKUT	Hungary	46.21666667	20.25	6443	46
KATALSZEG	Hungary	46.21666667	20.08333333	6370	100
SION-PETIT CHASSEUR	Switzerland	46.23333333	7.35	5110	57
PLANTA	Switzerland	46.23333333	7.366666667	5900	80
BASAROGA	Hungary	46.25	21	6580	60
CHATELLIERS DU VIEL	France	46.43333333	-0.866666667	5200	110
GROTTE DES TEPPES	France	46.43333333	5.333333333	5390	50
LES BOURIES	France	46.43333333	1.666666667	5210	70

Appendix 1: The Gkiasta European database: Datasets A to G.
Dataset B (weighted mean or most recent radiocarbon determinations at 509 Neolithic sites).

SITE NAME	Country	Latitude (decimal degrees)	Longitude (decimal degrees)	Date (uncal. bp)	Standard deviation
ZALAVAR	Hungary	46.66666667	17.16666667	6180	100
LE PETIT BOIS	France	46.71666667	4.933333333	5816	45
ABRI DES ROCS	France	46.73333333	0.533333333	5300	50
CHAMP DE LA GRANGE	France	46.76666667	2.433333333	5247	61
VALLON-DES-VAUX	Switzerland	46.78333333	6.783333333	4985	35
CHASSEY	France	46.88333333	4.766666667	5234	49
LES PLANCHES	France	46.91666667	5.766666667	5520	130
AUVERNIER-PORT	Switzerland	46.96666667	6.866666667	5075	57
LE VERGER	France	46.98333333	4.716666667	5710	140
LA MOTTE DES JUSTICES	France	47	-0.2	5340	100
LE TROU QUI FUME	France	47	4.716666667	5480	130
ABRI DES PRES MOUREY	France	47.05	4.75	5081	52
LA MOLLE PIERRE	France	47.05	6.583333333	5040	60
TWANN	Switzerland	47.1	7.15	5019	34
SCHENKON	Switzerland	47.16666667	8.133333333	4980	120
EGOLZWIL 4	Switzerland	47.18333333	8.016666667	5087	50
FELDBACH	Switzerland	47.23333333	8.783333333	5170	70
FENDMEILEN-VORDERELD	Switzerland	47.28333333	8.633333333	4940	70
LE TROU DU DIABLE	France	47.31666667	4.783333333	5105	55
LES PICHELOTS	France	47.31666667	-1.5	5530	70
SANDUN	France	47.33333333	2.566666667	5120	70
GREIFENSEE-STOREN/WILDSBERG	Switzerland	47.36666667	8.683333333	5140	49
ZURICH	Switzerland	47.36666667	8.583333333	4640	70
ETANG DAVID	France	47.38333333	4.516666667	5600	120
MYARD	France	47.38333333	4.516666667	5175	135
ZURICH-BAUSCHANZE	Switzerland	47.41	8.52	5320	60
ZURICH-WOLLISHOFEN	Switzerland	47.41	8.52	4993	46
MANNLEFELSEN	France	47.45	7.233333333	5140	140
BAVANS	France	47.46666667	6.7	6410	95
FRANKENAU	Austria	47.5	16.5	5660	100
UNTERPULLENDORF	Austria	47.5	16.5	6130	140
PETIT MONT	France	47.53333333	-2.9	5020	70
BILGROIX	France	47.55	-2.916666667	5420	60
LE ROUICK	France	47.56666667	-2.95	5293	40
LOCMARIAQUER ER GRAH	France	47.56666667	-2.966666667	5300	60
SIERENTZ	France	47.56666667	7.45	5570	100
PHYN	Switzerland	47.58333333	8.933333333	4993	28
SANDORRESZE	Hungary	47.65	20.15	6141	62
TARNABOD	Hungary	47.68333333	20.21666667	6280	100
COJOUX	France	47.76666667	-1.95	5577	69
CROIX SAINT PIERRE	France	47.76666667	-2	6070	80
SAINT JUST	France	47.76666667	-1.95	5770	80
STUROVO	Slovakia	47.8	18.73333333	6215	71
DOLMEN DE CHATEAU BU	France	47.81666667	1.95	5230	50
BEKESSZENTANDRAS	Hungary	47.83333333	20.5	6080	60
QUEBARN	France	47.83333333	-4.233333333	5760	120
SZARVAS	Hungary	47.83333333	20.58333333	6190	140
LARCUSTE	France	47.85	-2.783333333	5490	120
ENDROD	Hungary	47.91666667	21.83333333	6090	60
WINDEN AM SEE	Austria	47.95	16.83333333	5880	71
COUEDAN	France	47.98333333	-2.916666667	5260	70
KERGOGLE	France	48	-4.483333333	5190	80
RUFACHER HUBEN	France	48.03333333	7.333333333	5740	130
RICOH	France	48.06666667	7.3	5150	100
ROUTE DE ROUFFACH	France	48.08333333	7.333333333	6062	81
PASSY	France	48.1	3.266666667	5545	50
LES ERVES	France	48.11666667	-0.333333333	5580	140
ARMEAU	France	48.13333333	3.366666667	6260	65
LE CAMP	France	48.15	2.733333333	5830	45
LES DEPENDANCES DE DIGNY	France	48.16666667	-3.9	5304	67
TY FLOCH	France	48.16666667	2.416666667	5560	115
TRIFAUCKIJ	Moldavia	48.18333333	28.31666667	6495	100
NEUVY EN DUNOIS	France	48.2	1.55	5250	140
SELEDIN	France	48.21666667	-3.05	5149	63
LISCUIS EN LANISCAT	France	48.21666667	-3.133333333	5140	100
LES FIEFS	France	48.23333333	2.45	5611	35
FELSDACH LAUTERACH	Germany	48.33333333	2.45	5190	130
PARC DU BRINBORION	France	48.33333333	9.566666667	6140	45
TERRES DE PREPOUX	France	48.33333333	3.066666667	5980	120
KORLAT	Hungary	48.36666667	21.25	6440	100
PIECE DE PINCEVENT	France	48.36666667	2.883333333	5630	120
L'OUCHE DE BEAUCE	France	48.38333333	2.383333333	5570	70
CHEMIN DE NAVIOT	France	48.4	6.166666667	6000	80
LES REAUDINS	France	48.4	3.166666667	5160	85
HORNE LEFANTOVCE	Slovakia	48.41666667	18.16666667	5775	140
DOMICA CAVE	Slovakia	48.48333333	20.48333333	6080	75
LE BOISANNE	France	48.53333333	-2	5610	70
HALLERTAU	Germany	48.55	11.75	5990	90
CARN	France	48.56666667	-4.683333333	5230	75
CHANGE	France	48.56666667	1.583333333	6180	120
CURNIC	France	48.58333333	-4.416666667	5340	60
GUENNOC	France	48.58333333	-4.583333333	5075	140
LARZICOURT	France	48.63333333	4.716666667	5800	100
REICHSTETT	France	48.63333333	7.75	5940	140
BARNENEZ	France	48.66666667	-3.85	5345	45
NORROIS	France	48.66666667	4.55	6020	110
POIGEN	Austria	48.66666667	38.66666667	5935	90
PULKAU	Austria	48.7	15.86666667	6215	100
LE GUERZIT	France	48.71666667	-3.833333333	5950	110

Appendix 1: The Gkiasta European database: Datasets A to G.
Dataset B (weighted mean or most recent radiocarbon determinations at 509 Neolithic sites).

SITE NAME	Country	Latitude (decimal degrees)	Longitude (decimal degrees)	Date (uncal. bp)	Standard deviation
LE FOSSE DE TRAVERS	France	48.85	2.533333333	5410	60
BERCY QUARTIER SUD	France	48.86666667	2.333333333	5200	70
CHANTIER DE BERCY	France	48.86666667	2.333333333	5820	60
GRAND LOUVRE	France	48.86666667	2.333333333	5620	70
OSTOROS	Hungary	48.86666667	20.43333333	6180	100
PENTE DE CROUPETON	France	48.88333333	22.6	6136	100
SZAMOSSALYI	Hungary	48.88333333	2.716666667	5630	60
HIENHEIM	Germany	48.88333333	11.75	5910	50
LES LONGUES RAIES	France	48.91666667	2.766666667	5510	140
TISZAVASVARI-KERESZTFAL	Hungary	48.96666667	21.38333333	6226	69
LES GRANDS TRAQUIERS	France	49.01666667	4.266666667	5480	90
BILBO	France	49.06666667	-0.05	5150	50
DERRIERE LES PRES	France	49.06666667	-0.05	5110	70
SENTE SAILLANCOURT	France	49.08333333	2	5220	110
CUL FROID	France	49.23333333	1.05	5180	100
VIEUX MANOIR TALANGE	France	49.25	6.15	5310	80
LES BROUILLARDS	France	49.3	2.75	5050	70
ZOPY	Czech Rep.	49.33333333	17.58333333	6430	100
LE CAMP DE CESAR	France	49.36666667	2.516666667	5280	140
LA FOSSE TOUMISE	France	49.38333333	3.4	5280	75
LES FONTINETTES	France	49.38333333	3.75	5942	62
LA CROIX MAIGRET	France	49.4	4.883333333	6030	130
LE COQ GALLEUX	France	49.4	2.733333333	5300	140
LE MONT D'HUETTE	France	49.4	2.966666667	5890	120
LES OBEAUX	France	49.4	2.8	5330	110
DERRIERE LE VILLAGE	France	49.41666667	4.016666667	6067	95
MOHELNICE	Czech Rep.	49.78333333	16.91666667	6290	34
LA PLAISANCE	France	50	2.133333333	5255	80
LA CARRIERE MERQUE	France	50.01666667	1.983333333	5250	120
CHAMP DE BATAILLE	France	50.03333333	2.65	5390	100
OLSZANICA	Poland	50.1	19.83333333	6430	75
ZILGENDORF	Germany	50.16666667	10.96666667	6233	37
ZILGENDORF I	Germany	50.16666667	10.96666667	6310	80
CARN BREA	Great Britain	50.22027778	-5.246944444	4561	47
FRIEDBERG	Germany	50.33333333	8.75	6120	100
TROU AL'WESSE	Belgium	50.46666667	5.25	5950	70
THREE HOLES CAVE	Great Britain	50.49	-3.68	4650	70
LES SABLINS	France	50.51666667	1.65	5675	85
BROKEN CAVE	Great Britain	50.53	-3.69	4949	50
CORON MATON	Belgium	50.55	5.883333333	6420	120
COUTURE DE LA CHAUSSE	Belgium	50.58333333	3.683333333	6020	55
IRCHONWELZ	Belgium	50.61666667	3.75	6030	60
DARION	Belgium	50.65	5.183333333	6180	43
PLACE SAINT LAMBERT	Belgium	50.65	5.566666667	6316	27
AWANS	Belgium	50.66666667	5.2	6295	69
CHABAROVICE	Czech Rep.	50.66666667	5.45	6070	90
WAREMME	Belgium	50.66666667	13.95	6400	120
CRISNEE	Belgium	50.68333333	5.4	6320	75
OLEYE	Belgium	50.68333333	5.216666667	6281	38
MUDDERSHEIM	Germany	50.7	-2.15	5210	70
WORGRET, WAREHAM	Great Britain	50.7	6.55	6227	38
MAIDEN CASTLE	Great Britain	50.7	-2.48	4570	80
FLAGSTONES HOUSE, DORCHESTER	Great Britain	50.71	-2.43	4960	80
COMBE HILL	Great Britain	50.79	0.21	4590	110
KANNE	Belgium	50.8	5.666666667	6260	75
LAMERSDORF	Germany	50.8	6.466666667	6410	45
WANGE	Belgium	50.8	5.05	6310	75
LANGWEILER	Germany	50.81666667	6.433333333	5930	145
ORSETT	Great Britain	50.85277778	0.348611111	4636	31
NIEDERMERZ	Germany	50.86666667	6.333333333	5830	130
BLACKPATCH	Great Britain	50.86861111	-0.445	5090	130
LONG DOWN	Great Britain	50.88	-0.7	4975	71
STEPLETON	Great Britain	50.88	-2.17	4675	44
OFFHAM HILL	Great Britain	50.89	-0.02	4807	48
HAMBLEDON HILL	Great Britain	50.90944444	-2.209166667	4665	26
KOLN-MENGENICH	Germany	50.91666667	7	6320	70
BURY HILL	Great Britain	50.92	-1.37	4750	50
SHURTON HILL	Great Britain	50.92	-0.58	4625	57
WINDMILL HILL	Great Britain	50.92	-1.88	4715	21
DORSET CURSUS (DOWN FARM)	Great Britain	50.93	-2	4770	120
SCORD OF BROUSTER	Great Britain	50.93	-1.66	4500	80
NORTH MARDEN	Great Britain	50.94	-0.87	4710	110
BYLANY	Czech Rep.	50.96666667	15.3	5635	65
RODINGEN	Germany	50.96666667	6.55	6250	75
CORHAMPTON HOUSE, CORHAMPTON	Great Britain	50.98	-1.15	4790	70
SOUTH CADBURY CASTLE	Great Britain	51.0225	-2.541944444	4705	115
DRESDEN-NICKERN	Germany	51.03333333	12.11666667	5932	100
REHMSDORF	Germany	51.03333333	13.53333333	5880	71
WESTWARD HO!	Great Britain	51.03333333	-4.25	4840	70
WINNALL DOWN	Great Britain	51.08	-1.32	4725	53
BURNTWOOD FARM (SITE R6)	Great Britain	51.12	-1.29	4750	140
JONES'S TRACK, ASHCOTT / W. HEATH	Great Britain	51.14	-2.78	4590	70
CHILTON NO. 4 TRACK	Great Britain	51.15	-2.9	4760	65
ASHCOTT HEATH	Great Britain	51.16	-2.78	4625	120
SHAPWICK HEATH	Great Britain	51.16	-2.81	4540	130
MEARE HEATH	Great Britain	51.163	-2.78	4583	45
SWEET TRACK, SHAPWICK	Great Britain	51.163	-2.83	4550	70
DEVIL'S KNEADING TROUGH	Great Britain	51.16777778	0.971388889	4540	105
NORMANTON DOWN, AMESBURY	Great Britain	51.16861111	-1.835555556	4510	103

Appendix 1: The Gkiasta European database: Datasets A to G.
Dataset B (weighted mean or most recent radiocarbon determinations at 509 Neolithic sites).

CONEYBURY HENGE	Great Britain	51.17	-1.81	5050	100
MEARE HEATH FIELD	Great Britain	51.17	-2.79	5180	70
MEARE LAKE VILLAGE WEST	Great Britain	51.17	-2.8	5210	80
BAKER TRACK	Great Britain	51.18	-2.88	4760	80
CHILTON 1-2	Great Britain	51.18	-1.81	4500	120
HONEYGORE (W) TRACK, WESTHAY	Great Britain	51.18	-2.84	4760	30
HONEYGORE TRACK	Great Britain	51.18	-2.82	4590	40
KING BARROW RIDGE	Great Britain	51.18	-2.83	4750	130
BAKER PLATFORM	Great Britain	51.183	-2.82	4520	90
WINTERBOURNE STOKE	Great Britain	51.2	-1.85	4603	77
BRIDGEMERE	Great Britain	51.21	-2.41	4630	50
ROBIN HOOD'S BALL	Great Britain	51.21	-1.86	4613	67
NETHERAVON BAKE	Great Britain	51.22	-1.85	4760	90
BADSHOT	Great Britain	51.23	-0.79	4746	26
ZWENKAU	Germany	51.23333333	12.35	5890	100
TOMASZOW	Poland	51.36666667	21.26666667	5700	70
KNAP HILL	Great Britain	51.37083333	-1.826111111	4710	115
EASTON DOWN	Great Britain	51.39	-1.91	4624	36
WEST KENNET	Great Britain	51.41	-1.85	4767	44
SILBURY HILL, AVEBURY	Great Britain	51.415	-1.856111111	4593	65
SOUTH STREET, AVEBURY	Great Britain	51.42111111	-1.870555556	4641	64
CHERHILL	Great Britain	51.42944444	-1.955277778	4715	90
RUNNYMEDE	Great Britain	51.43	-0.56	4600	90
MILLBARROW	Great Britain	51.45	-1.87	4708	52
EILSLEBEN	Germany	51.48333333	11.51666667	6114	47
CANNON HILL, BRAY	Great Britain	51.5	-0.72	5270	110
ETON WICK	Great Britain	51.5	0.27	4650	90
MAR DYKE	Great Britain	51.5	-0.66	4697	40
DUDERSTADT	Germany	51.51666667	10.28333333	6309	48
ROSDORF	Germany	51.51666667	9.933333333	5685	75
WAYLAND'S SMITHY, ASHBURY	Great Britain	51.56638889	-1.594444444	4770	130
NORTH STOKE	Great Britain	51.57	-1.13	4672	49
MISBOURNE VIADUCT	Great Britain	51.58	-0.55	6094	58
DORCHESTER SITE VIII	Great Britain	51.66	-1.17	4510	100
DRAYTON	Great Britain	51.66	-1.31	4620	80
ABINGDON	Great Britain	51.68055556	-1.260833333	4883	57
CASHELKEELTY I	Ireland	51.72	-9.82	5845	100
STUMBLE	Great Britain	51.73	0.73	4728	49
BRADWELL-ON-SEA SITE 8	Great Britain	51.75	0.88	4690	70
COYGAN CAMP	Great Britain	51.75388889	-4.486388889	5000	95
SIDLINGS COPSE	Great Britain	51.79	-1.22	5250	75
LITTLE WALTHAM	Great Britain	51.81	0.45	5120	130
PEAK CAMP	Great Britain	51.83	-2.15	4747	35
ASCOTT-UNDER-WYCHWOOD	Great Britain	51.85472222	-1.565833333	4881	38
GWERNVALE	Great Britain	51.87	-3.15	4590	75
HAZLETON NORTH	Great Britain	51.87	-1.9	4875	15
ELSLOO	Netherlands	51.93333333	5.766666667	5080	70
GELEEN	Netherlands	51.96666667	5.75	6175	60
PEN-Y-WYRLOD	Great Britain	51.97583333	-3.236111111	4970	80
CARREG COETAN	Great Britain	52.02	-4.82	4697	46
LES RESSOURS	France	52.03333333	1.3	5040	70
FERRITERS COVE	Ireland	52.16666667	10.5	5190	110
BRIAR HILL	Great Britain	52.2275	-0.923888889	4600	90
STANWICK SOUTH	Great Britain	52.32	-0.59	4764	55
RECTORY FARM	Great Britain	52.32	-0.16	4950	36
REDLANDS FARM	Great Britain	52.33	-0.59	4823	50
ALDWINCLE	Great Britain	52.41	-0.54	4560	70
TANKARDSTOWN	Ireland	52.42	-8.62	4840	80
SHIPPEA HILL, PEACOCKS FARM	Great Britain	52.42888889	0.4125	4870	120
BROOME HEATH	Great Britain	52.46777778	1.451388889	4523	67
SWIFTERBANT	Netherlands	52.56666667	5.6	6288	27
FENGATE	Great Britain	52.57388889	-0.209722222	4960	64
EATON HEATH	Great Britain	52.60638889	1.262777778	4903	56
SARN-Y-BRYN-CALOD	Great Britain	52.64	-3.16	4960	70
SHANBALLYEDMOND	Ireland	52.69	-8.25	4930	60
ARDCRONY	Ireland	52.88	-8.15	4675	35
WESTEREGELN	Germany	52.96666667	11.4	6093	71
BALLINTRUER MORE	Ireland	52.99	-6.68	4800	70
BEESTON CASTLE	Great Britain	53.13	-2.7	5140	90
HUDE 1	Germany	53.16666667	8.433333333	6060	115
CAPEL EITHIN	Great Britain	53.22	-4.29	4740	80
GIANTS' HILLS	Great Britain	53.22	0.13	4650	80
WOLLA BANK	Great Britain	53.25	0.33	4865	65
LISMORE FIELDS	Great Britain	53.26	-1.94	4680	70
ANDERBY CREEK	Great Britain	53.27	0.32	4625	55
MARKLAND GRIPS	Great Britain	53.27	-1.24	4750	64
WHITWELL LONG CAIRN	Great Britain	53.27	-1.21	5209	48
STRZELCE	Poland	53.31666667	18.11666667	6260	60
ROCHER MOSS 2	Great Britain	53.61666667	-2.033333333	5830	100
NEWGRANGE	Ireland	53.71	-6.49	4505	41
CALLIS WOLD, BARROW 275	Great Britain	53.99	-0.73	4875	48
HANGING GRIMSTONE	Great Britain	54.04	-0.77	4710	90
PORT ST.MARY RUSHEN	Great Britain	54.08333333	-4.75	4970	80
KILHAM	Great Britain	54.09055556	-0.386666667	4830	125
BOYNTON, HIGH EASTON	Great Britain	54.12	-0.23	4840	80
GRINDALE, BARROW 1	Great Britain	54.12	-0.24	4510	90
EAST HESLERTON WOLD	Great Britain	54.16	-0.57	4640	70
BALLYMACDERMOT	Ireland	54.17	-6.38	4830	95
CARROWMORE - CULLEENAMORE	Ireland	54.23	-8.62	4710	100
DUNDRUM NATURE RESERVE	Ireland	54.25	-5.85	4666	97

Appendix 1: The Gkiasta European database: Datasets A to G.
Dataset B (weighted mean or most recent radiocarbon determinations at 509 Neolithic sites).

SITE NAME	Country	Latitude (decimal degrees)	Longitude (decimal degrees)	Date (uncal. bp)	Standard deviation
CARROWMORE	Ireland	54.27	-8.53	4940	85
BALLYGLASS (MA.13)	Ireland	54.3	-9.4	4590	48
ELLERBEK	Germany	54.38333333	10.18333333	5170	110
LANGDALE-PIKE OF STICKLE/ T. KNOTT	Great Britain	54.45472222	-3.123055556	4680	135
TULLY, ALDERGROVE AIRPORT	Ireland	54.46	-7.82	4575	50
GREAT LANGDALE	Great Britain	54.47	-3.14	4790	80
BUDDELIN	Germany	54.48333333	13.53333333	5815	100
STREET HOUSE FARM, LOFTUS	Great Britain	54.56	-0.83	4978	31
RUDE 2	Germany	54.61666667	9.583333333	5644	41
SUDENSEE DAMM	Germany	54.63333333	9.583333333	5954	40
BALLYNAGILLY	Ireland	54.71	-6.85	4835	55
CARNANBANE	Ireland	54.75	-6.82	4978	61
LOCHHILL	Great Britain	54.96972222	-3.610833333	5070	105
MAD MAN'S WINDOW, GLENARM	Ireland	54.98	-5.93	5095	120
MIDGEHOLME MOSS	Great Britain	54.98	-2.61	5270	90
DOOEY'S CAIRN, BALLYMACALDRACK	Ireland	55.01	-6.41	4630	130
CHATTON SANDYFORD	Great Britain	55.29416667	-1.983888889	4840	90
SKATEHOLM	Sweden	55.38333333	13.48333333	5640	60
ROTTEN BOTTOM, GAMES HOPE	Great Britain	55.42	-3.36	5040	100
OLBY LYNG	Denmark	55.5	12.21666667	5265	92
MONAMORE, ARRAN	Great Britain	55.5125	-5.141111111	5110	110
MACHRIE MOOR SITE	Great Britain	55.54	-5.32	4820	50
BULLTOFTAGARDEN	Sweden	55.6	13.06666667	6660	80
MELDON BRIDGE, NEAR PEEBLES	Great Britain	55.65055556	-3.263611111	4686	90
VEJLEBRO	Denmark	55.68333333	9.583333333	5362	54
PORT CHARLOTTE	Great Britain	55.73	-6.39	4540	70
STENBOCKSVALLAR	Sweden	55.76666667	13.91666667	5390	110
NEWTON	Great Britain	55.79	-6.26	4923	42
LINLITHGOW	Great Britain	55.97	-3.6	5265	55
NETHER KINNEIL	Great Britain	56	-3.67	4630	50
FLYNDERHAGE	Denmark	56.01666667	10.23333333	5705	85
NORSLUND	Denmark	56.01666667	10.95	5530	61
RINGKLOSTER	Denmark	56.01666667	10.23333333	5230	100
VAENGE SO	Denmark	56.13333333	10.51666667	5474	39
HOLMEGAARD	Denmark	56.26666667	10.76666667	6039	48
KINLOCH FARM, COLLESSIE	Great Britain	56.29	-3.16	4725	70
INCHTUTHIL	Great Britain	56.54	-3.44	5100	41
PITNACREE	Great Britain	56.65944444	-3.748888889	4810	90
DALLADIES FARM, FETTERCAIRN	Great Britain	56.79583333	-2.610833333	4603	37
ERTEBOLLE	Denmark	56.8	9.166666667	5624	45
HALDRUP STRAND	Denmark	56.86666667	10.96666667	5630	120
SALPETERMOSE	Denmark	56.91666667	12.3	5410	120
AGEROD	Sweden	56.93333333	13.41666667	6770	49
RHUM (KINLOCH)	Great Britain	57.04	-6.27	4725	140
BALBRIDIE	Great Britain	57.05	-2.441	4919	45
BALBRIDIE TIMBER HAL	Great Britain	57.05	-2.44	4740	130
BROVST	Denmark	57.1	9.5	5464	37
RAIGMORE	Great Britain	57.48	-4.19	4813	74
FOCHABERS - BOGHEAD MOUND	Great Britain	57.61861111	-3.071388889	4895	37
BOGHEAD MOUND, FOCHABERS	Great Britain	57.62	-3.07	6006	60
THE ORD NORTH, LAIRG	Great Britain	58.01	-4.43	4614	57
CAMSTER LONG CAIRN, WICK PARISH	Great Britain	58.38	-3.27	4927	45
LOCH CALDER - TULLOCH OF ASSERY B	Great Britain	58.53	-3.6	4655	60
LOCH CALDER - AN T'SIONNAICH	Great Britain	58.54	-3.6	4685	60
MAES HOWE	Great Britain	58.99722222	-3.187222222	5094	60
QUANTERNESS	Great Britain	58.99972222	-3.015	4574	62
BARNHOUSE	Great Britain	59	-3.22	4558	42
KNAP OF HOWAR, PAPA WESTRAY	Great Britain	59.35	-2.92	4532	70
GODDELAU	Germany	61.01666667	8.116666667	6300	90

Appendix 1: The Gkiasta European database: datasets A to G.
Dataset C (oldest radiocarbon determination at 509 Neolithic sites).

Site Name	Country	Latitude (decimal degrees)	Longitude (decimal degrees)	Sample ID	Date (uncal. bp)	SD
Knossos	Greece	35.51666667	25.16666667	BM-278	7910	140
Cueva de la Dehesilla	Spain	36.75	-5.8	UGRA-261	6260	100
Platia Magoula Zakrou	Greece	36.95	22.76666667	DEM-163	7083	44
Saliagos	Greece	37.05	25.08333333	P-1311	6172	74
Franchthi	Greece	37.43333333	23.13333333	P-2095	7980	110
Lagnano da Piede	Italy	38	12.33333333	UCLA-2148	6700	100
Cueva del Nacimiento	Spain	38.08333333	-2.3	Gif-2368	6780	130
Stretto, Partanna	Italy	38.12	13.36	Rome-291	6630	120
Theopetra Cave	Greece	38.46666667	23.23333333	DEM-575	7036	44
Antre Corycien	Greece	38.48333333	22.5	Gif-2124	6380	90
Skotini Cave	Greece	38.51666667	23.98333333	DEM-137	6247	49
Elateia	Greece	38.61666667	22.73333333	GrN-2973	7480	70
Coveta de l'Or	Spain	38.7	-0.46666667	K-1754	6265	75
Bevilacqua	Italy	38.92	16.61	P-3047	6430	90
Cueva Chica de Santiago	Spain	38.95	-6.91666667	UGRA-254	6160	100
Achilleion	Greece	39.28333333	22.38333333	LJ-3180	7550	60
So'n Matge	Spain	39.33333333	3	QL-29	6680	120
Sesklo	Greece	39.38333333	22.81666667	P-1681	7755	97
Argissa Magoula	Greece	39.63333333	22.46666667	UCLA-1657A	8130	100
Sidari	Greece	39.78333333	19.73333333	GXO-771	7670	120
Grotta di Porto Badisco	Italy	40.08333333	18.48333333	R-1225	5850	55
Grotta Corbeddu	Italy	40.25	9.483333333	UtC-1251	6690	80
Grotta Filiestru	Italy	40.5	8.666666667	Q-3020	6710	75
Nea Nikomedeia	Greece	40.6	22.25	P-1202	7557	91
Grotta di Romito	Italy	40.86666667	16.9	R-223	6420	70
Grotta del Sant. della Madonna	Italy	40.9	15.78333333	R-284	5555	75
Rendina	Italy	41	16	LJ-4548	7110	140
Scamuso	Italy	41.1	16.89	Gif-6339	7290	110
Sitagroi	Greece	41.15	24.2	Bln-778	6425	100
Araguina Sennola	France	41.38333333	9.166666667	Gif-2325	6650	140
MONTE LEONE	France	41.4	9.166666667	Ly-6099	5855	95
Coppa Nevigata	Italy	41.5	15.75	OxA-1475	6880	90
Santa Tecchia	Italy	41.5	15.66666667	BM-2414	6520	70
Vilaggio A, Scaramella	Italy	41.5	15.68333333	R-350	7000	100
RENAGHJU	France	41.51666667	8.916666667	Ly-8327	6525	60
Monte Maulo	Italy	41.57	14.64	OxA-652	6280	70
ANZA	Macedonia	41.7	22	LJ-3032	7210	50
Vrsnik-Tarinci	Macedonia	41.75	22.15	Bln-339	6950	100
Tumba or Veluska	Macedonia	41.9	21.36666667	Tx-1785	6950	120
Casabianca or Casterragio	France	42.08333333	9.5	MC-2243	6670	130
Grotta dei Piccioni	Italy	42.21666667	14.96666667	PI-46	6247	130
Cueva de Chaves	Spain	42.36666667	0.216666667	GrN-12685	6770	70
Balma Margineda	Andora	42.41666667	1.583333333	Ly-2439	6670	120
Azmak	Bulgaria	42.45	25.76666667	Bln-301	6483	100
LO PLA DEL BACH	France	42.45	2.066666667	Gif-6749	5700	80
Tell Azmak	Bulgaria	42.45	25.76666667	Bln-203	6880	100
Villaggio Leopardi	Italy	42.46666667	14.93333333	PI-101	6578	135
Tell Karanovo	Bulgaria	42.5	25.93333333	Bln-152	6807	100
ENVEIG	France	42.51666667	1.866666667	Ly-7541	5290	60
REVELLATA	France	42.56666667	8.733333333	Ly-6085	6280	75
MONTE REVINCU	France	42.6	9.266666667	Ly-8396	5405	55
Le Roc de Dourgne	France	42.61666667	2.15	MC-1107	6850	100
Pristina	Yugoslavia	42.65	21.15	Bln-435	6280	80
SCAFFA PIANA	France	42.68333333	9.2	MC-2057	5360	100
LA CAUNE	France	42.71666667	2.616666667	Ly-3302	5640	120
CAMP DEL GINEBRE	France	42.75	2.566666667	Ly-226	5355	45
LES COUDOUMINES	France	42.75	2.566666667	Ly-225	5060	45
Abri de la Coma Franceze	France	42.83333333	2.916666667	Gif-9080	5180	60
COMA FRANCEZE	France	42.83333333	2.916666667	Gif-7292	5200	70
PORT LEUCATE	France	42.83333333	3	Gif-2749	5900	140
LA CAUNE OUEST	France	42.91666667	1.933333333	Ly-5104	5164	125
Maddalena di Muccia	Italy	43.06666667	13.06666667	R-643a	6580	75
Abri Jean Cros	France	43.08333333	2.466666667	Gif-3575	6600	130
San Marco, Gubbio	Italy	43.13	12.4	OxA-1853	6430	80
BUHOLOUP	France	43.15	1.183333333	Gif-5497	5490	120
LES PLOTS	France	43.21666667	2.35	Gif-9337	5230	70
LA POSTE VIEILLE	France	43.25	2.216666667	Ly-6241	5190	75
Font Juvenal	France	43.28333333	2.333333333	Ly-4770	6235	85
Grotte du Vieux Mounoi	France	43.3	5.866666667	Ly-5362	5625	85
Grotte de Gazel	France	43.31666667	2.416666667	GrN-6702	6905	90
Grotte de l'Abeurador	France	43.35	2.7	MC-2145	6260	90
Grotte Tournie	France	43.35	2.8	MC-797	5900	100
Chateauneuf Les Martigues	France	43.38333333	5.166666667	MC-2515	6900	100
RAFFEGUES	France	43.41666667	3.583333333	Gif-7349	5200	100
Grotte de Camprafaud	France	43.43333333	3.9	Gif-3078	6480	130
MOULIN DE LA GARONNE	France	43.5	1.333333333	Gif-5845	5120	120
VILLENEUVE TOLOSANE	France	43.51666667	1.3	Gif-6313	5400	90
MIOUVIN	France	43.51666667	4.983333333	MC-1224	5380	110
Source de Reselauze	France	43.51666667	3.283333333	MC-1227	6450	100
LE CAMP DE LAURE	France	43.55	5.216666667	MC-1381	5020	100
Baume de Fontbregoua	France	43.55	6.233333333	Gif-2990	6700	100
CUGNAUX	France	43.61666667	1.433333333	Gif-9026	5460	70
Saint Michel-de-Touch	France	43.61666667	1.433333333	MC-847	5600	90
Grotte Lombard	France	43.68333333	7.866666667	Ly-3331	6190	120
Caucade	France	43.7	7.35	MC-2342	6800	130
CHATEAU PERCIN	France	43.7	0.35	MC-947	5250	100
Giribaldi cimiez Carabacel	France	43.7	7.233333333	Ly-6945	5750	60
Ripabianca di Monterado	Italy	43.71666667	13.1	R-599	6260	85
ESCANIN	France	43.75	4.8	Gif-1995	5160	120

Appendix 1: The Gkiasta European database: datasets A to G.
Dataset C (oldest radiocarbon determination at 509 Neolithic sites).

Site Name	Country	Latitude (decimal degrees)	Longitude (decimal degrees)	Sample ID	Date (uncal. bp)	SD
L'EGLISE	France	43.75	6.25	Gif-1334	5760	140
Mileto	Italy	43.79	11.21	Beta-44114	6180	80
MALVOISIN	France	43.8	5.016666667	Ly-3020	5270	140
PENDIMOUM	France	43.81666667	7.5	Ly-6495	6565	75
Grotte de l'Abbe Prialet	France	43.88333333	3.733333333	MC-1290	6070	100
LE CHEVAL	France	43.93333333	4.35	MC-943	5475	110
ROC TROUE	France	43.95	3.183333333	Gif-7688	5850	80
ABRI DES USCLADES	France	43.98333333	3.233333333	Gif-9143	5550	70
Grotta del'Orso	Italy	43.98333333	12.86666667	R-676	6080	50
CERON	France	44	5.05	Gif-9997	5990	75
Divostin	Yugoslavia	44.03333333	21.83333333	Bln-899	7200	100
UNANG	France	44.05	5.15	Gif-1796	5950	130
Grotte de l'Aigle	France	44.06666667	4.633333333	MC-563	6200	100
Le Baratin	France	44.08333333	5.883333333	Gif-1855	6600	140
Grotta Arma dello Stefanin	Italy	44.1	8.016666667	Bln-3276	6610	60
Abri de la Poujade	France	44.11666667	3.166666667	MC-1367	6990	120
Riparo Arma di Nasino	Italy	44.11666667	8.05	R-267	6470	120
FRIGOURAS	France	44.13333333	5.95	Gif-8479	5450	100
LE MOULIN	France	44.18333333	5.5	Ly-3804	5840	130
Obre	Bosnia-Herz.	44.18333333	18.1	UCLA-1605.I	7240	60
LA VESIGNIE	France	44.25	4.616666667	Ly-4745	5815	75
Baume de Montclus	France	44.26666667	4.433333333	MC-695	6450	100
SAINTE LUCE	France	44.3	5.333333333	Ly-4486	5760	100
Baume d'Oullins	France	44.31666667	4.466666667	MC-2034	6630	110
LA BAUME DE RONZE	France	44.31666667	4.416666667	Ly-2747	5400	130
CHAZELLES	France	44.33333333	4.333333333	Lv-365	5660	110
LES MOULINS	France	44.35	4.766666667	Ly-3801	5430	120
Saint Marcel d'Ardeche	France	44.35	4.466666667	MC-2376	6330	90
BALLOY LES REAUDINS	France	44.4	3.166666667	Ly-12	6180	90
ROQUEMISSOU	France	44.41666667	2.733333333	Ly-4687	5785	105
Gornja Tuzla	Bosnia-Herz.	44.45	18.76666667	GrN-2059	6640	75
LA CHAUVE SOURIS	France	44.46666667	4.683333333	Ly-4755	6790	70
Selevac	Yugoslavia	44.5	21.88333333	Z-233	6366	100
Caverna delle Arene Candide	Italy	44.5	8.5	UB-2423	6980	115
Padina	Yugoslavia	44.5	22.41666667	GrN-8230	7100	80
Vlasac	Yugoslavia	44.53333333	22.05	Z-264	6335	92
TROU ARNAUD	France	44.55	5.316666667	Ly-4696	5535	100
CAPDENAC LE HAUT	France	44.58333333	2.066666667	Gif-3714	5190	120
ANTONNAIRE	France	44.68333333	5.383333333	Ly-4080	5570	130
GROTTE DU SANGLIER	France	44.68333333	1.716666667	Ly-7791	6095	100
LE FOURNET	France	44.68333333	5.333333333	Ly-2434	5440	130
LE SANGLIER	France	44.68333333	1.716666667	Ly-33	6915	70
Hamangia-Baia 1	Romania	44.7	28.66666667	GrN-1986	5880	70
GROTTE DES ESCABASSES	France	44.71666667	1.8	Ly-9197	5160	35
Starcevo	Yugoslavia	44.81666667	20.7	OxA-8561	6975	60
ROUCADOUR	France	44.83333333	1.5	Gif-36C	5940	140
Abri de Pendimoun	France	44.88333333	7.483333333	Ly-5692	6615	105
Saint Mitre:abri 2	France	44.88333333	5.65	MC-264	6700	130
Saint Mitre:abri 3	France	44.88333333	5.651	MC-265	6100	120
Baume Bourbon	France	44.9	4.633333333	MC-794	6080	100
LES RIVAUX	France	45.05	3.85	Ly-3710	6240	110
PAS DE LA CHARMATE	France	45.06666667	5.383333333	Ly-3785	5680	130
COUFIN	France	45.06666667	5.4	Ly-3321	5260	120
COMBOIRE	France	45.13333333	5.683333333	Ly-4668	5360	70
La Grande Rivoire	France	45.2	5.633333333	Ly-176	6000	70
Abri de la Grande Rivoire	France	45.21666667	5.65	Ly-6096	5135	35
Grotta Podosojna	Croatia	45.25	14.23333333	Z-198	6470	95
Zorlentu Mare	Romania	45.31	21.81	GrN-10316	6070	40
LA LEDE DU GURP	France	45.45	-1.15	Ly-5318	6101	65
Molino Casaroto	Italy	45.46666667	11.6	Birm-175	6450	110
LES GOUILLAUDS	France	45.5	-0.6	Gif-4878	5950	120
Apatin	Yugoslavia	45.66666667	18.98333333	GrN-15973	7130	60
Alfort	France	45.68333333	5.35	Ly-386	6000	60
Grotta dei Ciclami	Italy	45.7	13.81666667	R-1040	6300	60
SAINT PRIEST	France	45.7	4.916666667	Ly-8685	5445	60
BETHENAS	France	45.71666667	5.25	Ly-3444	5060	110
Sammardenchia, Udine	Italy	45.72	13.76	Bln-3373	6120	60
GRANDE BARME SAVIGNY	France	45.73333333	5.916666667	Ly-2077	5010	140
Creux-Rouge	France	45.78333333	3.083333333	Ly-1944	6070	140
GROTTE DES PERRATS	France	45.78333333	0.35	Gif-9535	5880	70
GROTTE DESPERRATS	France	45.78333333	0.35	Gif-9536	5450	40
TREMONTEIX	France	45.78333333	3.083333333	Ly-3880	5330	120
PONTCHARAUD	France	45.78333333	3.083333333	Ly-7625	7095	50
Necropole de la Boixe Dolmen C	France	45.8	0.133333333	Gif-8909	5380	50
Grotte de la Vieille Eglise	France	45.91666667	6.283333333	CRG-411	5384	128
LES CHAMBONS	France	45.95	3.083333333	Ly-3406	5130	140
GROTTE DU GARDON	France	45.95	5.333333333	Ly-8422	6325	40
LE GARDON	France	45.95	5.033333333	Ly-7011	5815	40
LA ROCHE AUX FEES	France	45.98333333	0.983333333	Ly-4536A	5530	120
Riparo di Romagnano III	Italy	46	11.1	R-1136	6480	50
Riparo Gaban	Italy	46.07	11.1	Bln-1777	6030	45
Oszentivan	Hungary	46.2	20.18333333	Bln-479	6460	80
Olajkut	Hungary	46.21666667	20.25	Bln-581	6605	100
Katalszeg	Hungary	46.21666667	20.08333333	Bln-86	6370	100
Sion-Petit Chasseur	Switzerland	46.23333333	7.35	B-2110	5130	100
Planta	Switzerland	46.23333333	7.366666667	CRG-280	6500	80
Basaroga	Hungary	46.25	21	BM-1862	6580	60
CHATELLIERS DU VIEL	France	46.43333333	-0.86666667	Gif-5717	5200	110
GROTTE DES TEPPES	France	46.43333333	5.333333333	Gif-8999	5390	50

Appendix 1: The Gkiasta European database: datasets A to G.
Dataset C (oldest radiocarbon determination at 509 Neolithic sites).

Site Name	Country	Latitude (decimal degrees)	Longitude (decimal degrees)	Sample ID	Date (uncal. bp)	SD
LES BOURIES	France	46.43333333	1.666666667	Ly-721	5210	70
LES SABLES DE LOI	France	46.43333333	0.883333333	Ly-7275	5110	60
Zalavar	Hungary	46.66666667	17.16666667	Bln-87	6180	100
LE PETIT BOIS	France	46.71666667	4.933333333	Ly-753	5840	62
ABRI DES ROCS	France	46.73333333	0.533333333	Gif-9956	5300	50
CHAMP DE LA GRANGE	France	46.76666667	2.433333333	Ly-4439	5310	110
Vallon-des-Vaux	Switzerland	46.78333333	6.783333333	B-659	5180	120
CHASSEY	France	46.88333333	4.766666667	Ly-3474	5400	100
LES PLANCHES	France	46.91666667	5.766666667	Gif-3826	5520	130
Auvernier-Port	Switzerland	46.96666667	6.866666667	B-2559	5130	120
LE VERGER	France	46.98333333	4.716666667	Ly-2971	5710	140
LA MOTTE DES JUSTICES	France	47	-0.2	Gif-8292	5340	100
LE TROU QUI FUME	France	47	4.716666667	Ly-2613	5480	130
ABRI DES PRES MOUREY	France	47.05	6.583333333	Gif-9096	5340	70
LA MOLLE PIERRE	France	47.05	4.75	Ly-5408	5238	120
Twann	Switzerland	47.1	7.15	B-2698	5200	90
Schenkon	Switzerland	47.16666667	8.133333333	B-881	4980	120
Egolzwil 4	Switzerland	47.18333333	8.016666667	H-228/276	5150	100
Feldbach	Switzerland	47.23333333	8.783333333	UCLA-1809A	5170	70
Fendmeilen-Vordereld	Switzerland	47.28333333	8.633333333	UCLA-1691 F	5415	60
LE TROU DU DIABLE	France	47.31666667	4.783333333	Ly-6506	5105	55
LES PICHELOTS	France	47.31666667	-1.5	Gif-6316	5530	70
SANDUN	France	47.33333333	2.566666667	Gif-7703	5660	100
Greifensee-Storen/Wildsberg	Switzerland	47.36666667	8.683333333	UCLA-1824	5180	70
Zurich	Switzerland	47.36666667	8.583333333	UCLA-1772 B	5145	70
Etang David	France	47.38333333	4.516666667	Gif-3354	5600	120
MYARD	France	47.38333333	4.516666667	Gif-2342	5175	135
Zurich-Bauschanze	Switzerland	47.41	8.52	UCLA-1765 B	5040	60
Zurich-Wollishofen	Switzerland	47.41	8.52	UCLA-1785	5320	60
Mannlefelsen	France	47.45	7.233333333	Gif-2634	5140	140
Bavans	France	47.46666667	6.7	Lv-1415	7130	70
Frankenau	Austria	47.5	16.5	VRI-207	5660	100
Unterpullendorf	Austria	47.5	16.5	VRI-42	6130	140
PETIT MONT	France	47.53333333	-2.9	Gif-6844	5650	70
BILGROIX	France	47.55	-2.91666667	Gif-8961	5420	60
LE ROUICK	France	47.56666667	-2.96666667	Gif-8702	5300	60
LOCMARIAQUER ER GRAH	France	47.56666667	-2.95	Gif-7693	5370	70
SIERENTZ	France	47.56666667	7.45	Ly-4498	5570	100
Phyn	Switzerland	47.58333333	8.933333333	GrN-5957	5020	40
Sandorresze	Hungary	47.65	20.15	Bln-676	6155	80
Tarnabod	Hungary	47.68333333	20.21666667	Bln-123	6280	100
COJOUX	France	47.76666667	-1.95	Gif-5458	5600	120
CROIX SAINT PIERRE	France	47.76666667	-2	Gif-8840	6070	80
SAINT JUST	France	47.76666667	-1.95	Gif-5763	5770	80
Sturovo	Slovakia	47.8	18.73333333	Bln-559	6260	100
DOLMEN DE CHATEAU BU	France	47.81666667	1.95	Gif-9239	5230	50
Bekesszentandras	Hungary	47.83333333	20.5	BM-1860	6080	60
Quebarn	France	47.83333333	-4.23333333	Gif-5061	5760	120
Szarvas	Hungary	47.83333333	20.58333333	BM-1866	6620	60
LARCUSTE	France	47.85	-2.78333333	Gif-2826	5490	120
Endrod	Hungary	47.91666667	21.83333333	GrN-10319	6945	50
Winden am See	Austria	47.95	16.83333333	Bln-55	5940	100
COUEDAN	France	47.98333333	-2.91666667	Gif-9638	5260	70
KERGOGLE	France	48	-4.48333333	Gif-4689	5190	80
Rufacher Huben	France	48.03333333	7.333333333	Ly-2733	5740	130
RICOH	France	48.06666667	7.3	Ly-4582	6400	70
ROUTE DE ROUFFACH	France	48.08333333	7.333333333	Ly-3811	6100	120
PASSY	France	48.1	3.266666667	Ly-8823	6065	55
LES ERVES	France	48.11666667	-0.33333333	Ly-3099	5580	140
Armeau	France	48.13333333	3.366666667	GrN-6781	6260	65
LE CAMP	France	48.15	2.733333333	Ly-8139	5830	45
LES DEPENDANCES DE DIGNY	France	48.16666667	2.416666667	Ly-5668	5990	50
TY FLOCH	France	48.16666667	-3.9	Gif-5234	5580	120
Trifauckij	Moldavia	48.18333333	28.31666667	Bln-587	7420	80
NEUVY EN DUNOIS	France	48.2	1.55	Gif-785	5250	140
SELEDIN	France	48.21666667	-3.05	Gif-1877	5270	140
LISCUIS EN LANISCAT	France	48.21666667	-3.13333333	Gif-3099	5140	100
LES FIEFS	France	48.23333333	2.45	Ly-6481	5650	55
Felsdach Lauterach	Germany	48.33333333	9.566666667	GrN-4750	6140	45
PARC DU BRINBORION	France	48.33333333	2.45	Gif-2721	5190	130
TERRES DE PREPOUX	France	48.33333333	3.066666667	Ly-4503	6730	110
Korlat	Hungary	48.36666667	21.25	Bln-119	6440	100
PIECE DE PINCEVENT	France	48.36666667	2.883333333	Gif-5005	5630	120
L'OUCHE DE BEAUCE	France	48.38333333	2.383333333	Gif-7705	5570	70
CHEMIN DE NAVIOT	France	48.4	6.166666667	Ly-4479	6330	140
Les Reaudins	France	48.4	3.166666667	Ly-5540	5160	85
Horne Lefantovce	Slovakia	48.41666667	18.16666667	Bln-304	5775	140
Domica Cave	Slovakia	48.48333333	20.48333333	GrN-2435	6080	75
LE BOISANNE	France	48.53333333	-2	Gif-7713	5610	70
Hallertau	Germany	48.55	11.75	HAM-197	5990	90
Carn	France	48.56666667	-4.68333333	GrN-1968	5230	75
CHANGE	France	48.56666667	1.583333333	Gif-9014	6180	120
Curnic	France	48.58333333	-4.41666667	GrN-1966	5340	60
GUENNOC	France	48.58333333	-4.58333333	Gif-1870	5075	140
LARZICOURT	France	48.63333333	4.716666667	Ly-3907	5800	100
Reichstett	France	48.63333333	7.75	Ly-865	5940	140
BARNENEZ	France	48.66666667	-3.85	Gif-1556	5550	140
NORROIS	France	48.66666667	4.55	Ly-3891	6020	110
Poigen	Austria	48.66666667	38.66666667	H-2121/1552	5935	90

Appendix 1: The Gkiasta European database: datasets A to G.
Dataset C (oldest radiocarbon determination at 509 Neolithic sites).

Site Name	Country	Latitude (decimal degrees)	Longitude (decimal degrees)	Sample ID	Date (uncal. bp)	SD
Pulkau	Austria	48.7	15.86666667	Bln-83	6215	100
LE GUERZIT	France	48.71666667	-3.83333333	Gif-7614	5950	110
LE CAMP DE BIERRE	France	48.85	0.533333333	Ly-3724	5470	140
LE FOSSE DE TRAVERS	France	48.85	2.533333333	Gif-7710	5410	60
BERCY QUARTIER SUD	France	48.86666667	2.333333333	Gif-9810	5420	60
CHANTIER DE BERCY	France	48.86666667	2.333333333	Gif-9224	5820	60
GRAND LOUVRE	France	48.86666667	2.333333333	Gif-7618	5620	70
Ostoros	Hungary	48.86666667	20.43333333	Bln-549	6180	100
PENTE DE CROUPETON	France	48.88333333	2.716666667	Gif-8358	5910	80
Szamossalyi	Hungary	48.88333333	22.6	Bln-404	6136	100
Hienheim	Germany	48.88333333	11.75	GrN-7557	6235	45
LES LONGUES RAIES	France	48.91666667	2.766666667	Gif-5002	5510	140
Tiszavasvari-Keresztfal	Hungary	48.96666667	21.38333333	Bln-505	6305	100
LES GRANDS TRAQUIERS	France	49.01666667	4.266666667	Ly-4453	6200	140
BILBO	France	49.06666667	-0.05	Gif-8797	5110	70
DERRIERE LES PRES	France	49.06666667	-0.05	Gif-89160	5560	140
SENTE SAILLANCOURT	France	49.08333333	2	Gif-5840	5220	110
CUL FROID	France	49.23333333	1.05	Ly-2711	5180	100
VIEUX MANOIR TALANGE	France	49.25	6.15	Gif-9327	5310	80
LES BROUILLARDS	France	49.3	2.75	Gif-8244	5050	70
Zopy	Czech Rep.	49.33333333	17.58333333	Bln-57	6430	100
LE CAMP DE CESAR	France	49.36666667	2.516666667	Ly-2966	5280	140
LA FOSSE TOUMISE	France	49.38333333	3.4	Ly-6478	5870	105
Les Fontinettes	France	49.38333333	3.75	Ly-2331	6000	120
La Croix Maigret	France	49.4	4.883333333	Ly-2327	6030	130
Le Coq Galleux	France	49.4	2.8	Ly-2716	6080	110
LE MONT D'HUETTE	France	49.4	2.733333333	Ly-2970	5300	140
LES OBEAUX	France	49.4	2.966666667	Ly-3460	5890	120
Derriere le Village	France	49.41666667	4.016666667	Ly-2324	6110	140
Mohelnice	Czech Rep.	49.78333333	16.91666667	Bln-102a	6405	100
LA PLAISANCE	France	50	2.133333333	OxA-460	5255	80
LA CARRIERE MERQUE	France	50.01666667	1.983333333	Gif-4356	5250	120
CHAMP DE BATAILLE	France	50.03333333	2.65	Ly-2679	5390	100
Olszanica	Poland	50.1	19.83333333	GrN-5384	6430	75
Zilgendorf	Germany	50.16666667	10.96666667	GrN-7011	6250	65
Zilgendorf I	Germany	50.16666667	10.96666667	GrN-4245	6310	80
Carn Brea	Great Britain	50.22027778	-5.24694444	BM-825	4999	64
Friedberg	Germany	50.33333333	8.75	Bln-56	6120	100
Trou al'Wesse	Belgium	50.46666667	5.25	Lv-1752	5950	70
Three Holes Cave	Great Britain	50.49	-3.68	OxA-4495	5010	70
LES SABLINS	France	50.51666667	1.65	Gif-4024	5690	120
Broken Cave	Great Britain	50.53	-3.69	OxA-3207	5015	80
Coron Maton	Belgium	50.55	5.883333333	Gif-5257	6420	120
Couture de la Chausse	Belgium	50.58333333	3.683333333	Hv-8404	6485	75
Irchonwelz	Belgium	50.61666667	3.75	Bln-2531	6030	60
Darion	Belgium	50.65	5.183333333	Lv-1579	6240	100
Place Saint Lambert	Belgium	50.65	5.566666667	Lv-1340	6460	60
Awans	Belgium	50.66666667	5.45	Lv-1160	6070	90
Chabarovice	Czech Rep.	50.66666667	13.95	Bln-438	6400	120
Waremme	Belgium	50.66666667	5.2	Lv-1693	6300	95
Crisnee	Belgium	50.68333333	5.4	Lv-1580	6320	75
Oleye	Belgium	50.68333333	5.216666667	Lv-1685	6370	95
Muddersheim	Germany	50.7	6.55	H-924b/1121	6330	80
Worgret, Wareham	Great Britain	50.7	-2.15	HAR-9595	5210	70
Maiden Castle	Great Britain	50.7	-2.48	OxA-1337	5030	80
Flagstones House, Dorchester	Great Britain	50.71	-2.43	HAR-9161	4960	80
Combe Hill	Great Britain	50.79	0.21	I-11613	4590	110
Kanne	Belgium	50.8	5.666666667	Lv-1025	6260	75
Lamersdorf	Germany	50.8	6.466666667	KN-I.367	6410	45
Wange	Belgium	50.8	5.05	Lv-1116	6310	75
Langweiler	Germany	50.81666667	6.433333333	KN-2695	7090	135
Orsett	Great Britain	50.85277778	0.348611111	BM-1213	4741	113
Niedermerz	Germany	50.86666667	6.333333333	KN-2289	6800	80
Blackpatch	Great Britain	50.86861111	-0.445	BM-290	5090	130
Long Down	Great Britain	50.88	-0.7	OxA-1152	5050	100
Stepleton	Great Britain	50.88	-2.17	OxA-932	4690	70
Offham Hill	Great Britain	50.89	-0.02	BM-1414	4925	80
Hambledon Hill	Great Britain	50.90944444	-2.20916667	HAR-3062	4850	100
Koln-Mengenich	Germany	50.91666667	7	KN-I.369	6320	70
Bury Hill	Great Britain	50.92	-0.58	HAR-3596	4680	80
Shurton Hill	Great Britain	50.92	-1.37	UB-2122	4750	50
Windmill Hill	Great Britain	50.92	-1.88	OxA-2395	4730	80
Dorset Cursus (Down Farm)	Great Britain	50.93	-2	OxA-627	6800	100
Scord of Brouster	Great Britain	50.93	-1.66	CAR-253	5050	90
North Marden	Great Britain	50.94	-0.87	HAR-5544	4710	110
Bylany	Czech Rep.	50.96666667	15.3	GrN-4754	6270	65
Rodingen	Germany	50.96666667	6.55	H-2117/1536	6250	75
Corhampton House, Corhampton	Great Britain	50.98	-1.15	BM-1889	4790	70
South Cadbury Castle	Great Britain	51.0225	-2.54194444	I-5972	4705	115
Dresden-Nickern	Germany	51.03333333	13.53333333	Bln-73	5945	100
Rehmsdorf	Germany	51.03333333	12.11666667	Bln-176	5932	100
Westward Ho!	Great Britain	51.03333333	-4.25	HAR-5641	5740	100
Winnall Down	Great Britain	51.08	-1.32	HAR-2196	4800	80
Burntwood Farm (Site R6)	Great Britain	51.12	-1.29	OxA-1384	4750	140
Jones's Track, Ashcott /W.Heath	Great Britain	51.14	-2.78	HAR-3078	4590	70
Chilton No. 4 Track	Great Britain	51.15	-2.9	Lu-327	4760	65
Ashcott Heath	Great Britain	51.16	-2.78	Q-598	4625	120
Shapwick Heath	Great Britain	51.16	-2.81	Q-430	4540	130
Meare Heath	Great Britain	51.163	-2.78	Q-646	4650	120

Appendix 1: The Gkiasta European database: datasets A to G.
Dataset C (oldest radiocarbon determination at 509 Neolithic sites).

Site Name	Country	Latitude (decimal degrees)	Longitude (decimal degrees)	Sample ID	Date (uncal. bp)	SD
Sweet Track, Shapwick	Great Britain	51.163	-2.83	Q-968	5224	75
Devil's Kneading Trough	Great Britain	51.16777778	0.971388889	BM-254	4540	105
Normanton Down, Amesbury	Great Britain	51.16861111	-1.83555556	BM-505	4510	103
Coneybury Henge	Great Britain	51.17	-1.81	OxA-1402	5050	100
Meare Heath Field	Great Britain	51.17	-2.79	HAR-3195	5180	70
Meare Lake Village West	Great Britain	51.17	-2.8	HAR-2616	5210	80
Baker Track	Great Britain	51.18	-2.82	GaK-1600	4840	100
Chilton 1-2	Great Britain	51.18	-2.88	HAR-649	4760	80
Honeygore (W) Track, Westhay	Great Britain	51.18	-2.84	Q-1028	4780	50
Honeygore Track	Great Britain	51.18	-2.83	GaK-1939	5640	120
King Barrow Ridge	Great Britain	51.18	-1.81	OxA-1397	4500	120
Baker Platform	Great Britain	51.183	-2.82	HAR-2845	4950	80
Winterbourne stoke	Great Britain	51.2	-1.85	OxA-1405	4640	100
Bridgemere	Great Britain	51.21	-2.41	BM-2565	4630	50
Robin Hood's Ball	Great Britain	51.21	-1.86	OxA-1400	4740	100
Netheravon Bake	Great Britain	51.22	-1.85	OxA-1407	4760	90
Badshot	Great Britain	51.23	-0.79	BM-2273N1	4780	40
Zwenkau	Germany	51.23333333	12.35	GrN-1581	6160	70
Tomaszow	Poland	51.36666667	21.26666667	GrN-7591	6145	70
Knap Hill	Great Britain	51.37083333	-1.82611111	BM-205	4710	115
Easton Down	Great Britain	51.39	-1.91	OxA-3760	4730	65
West Kennet	Great Britain	51.41	-1.85	OxA-449	4825	90
Silbury Hill, Avebury	Great Britain	51.415	-1.85611111	SI-910A	4675	110
South Street, Avebury	Great Britain	51.42111111	-1.87055556	BM-356	4760	130
Cherhill	Great Britain	51.42944444	-1.95527778	BM-493	4715	90
Runnymede	Great Britain	51.43	-0.56	OxA-3582	5780	85
Millbarrow	Great Britain	51.45	-1.87	OxA-3172	4900	110
Eilsleben	Germany	51.48333333	11.51666667	OxA-1627	6190	90
Cannon Hill, Bray	Great Britain	51.5	-0.72	HAR-1198	5270	110
Eton Wick	Great Britain	51.5	-0.66	BM-2533	4750	80
Mar Dyke	Great Britain	51.5	0.27	HAR-4522	5740	80
Duderstadt	Germany	51.51666667	10.28333333	H-919/889	6340	100
Rosdorf	Germany	51.51666667	9.933333333	Hv-586	6350	70
Wayland's Smithy, Ashbury	Great Britain	51.56638889	-1.59444444	I-1468	4770	130
North Stoke	Great Britain	51.57	-1.13	BM-1405	4672	49
Misbourne Viaduct	Great Britain	51.58	-0.55	OxA-601	6190	90
Dorchester Site VIII	Great Britain	51.66	-1.17	BM-2443	4510	100
Drayton	Great Britain	51.66	-1.31	HAR-6477	4990	100
Abingdon	Great Britain	51.68055556	-1.26083333	BM-351	5060	130
Cashelkeelty I	Ireland	51.72	-9.82	UB-2413	5845	100
Stumble	Great Britain	51.73	0.73	OxA-2298	4780	70
Bradwell-on-Sea Site 8	Great Britain	51.75	0.88	HAR-6617	4690	70
Coygan Camp	Great Britain	51.75388889	-4.48638889	NPL-132	5000	95
Sidlings Copse	Great Britain	51.79	-1.22	OxA-3560	5250	75
Little Waltham	Great Britain	51.81	0.45	HAR-1087	5120	130
Peak Camp	Great Britain	51.83	-2.15	OxA-1622	4865	80
Ascott-under-Wychwood	Great Britain	51.85472222	-1.56583333	BM-833	5020	92
Gwernvale	Great Britain	51.87	-3.15	CAR-113	5050	75
Hazleton North	Great Britain	51.87	-1.9	OxA-910	5000	70
Elsloo	Netherlands	51.93333333	5.766666667	GrN-2159	6320	90
Geleen	Netherlands	51.96666667	5.75	GrN-995	6370	60
Pen-y-Wyrlod	Great Britain	51.97583333	-3.23611111	HAR-674	4970	80
Carreg Coetan	Great Britain	52.02	-4.82	CAR-392	4830	80
LES RESSOURS	France	52.03333333	1.3	Ly-7484	5040	70
Ferriters Cove	Ireland	52.16666667	10.5	BM-2227A	5190	110
Briar Hill	Great Britain	52.2275	-0.92388889	HAR-2282	5440	110
Stanwick South	Great Britain	52.32	-0.59	OxA-3001	4810	80
Rectory Farm	Great Britain	52.32	-0.16	OxA-3370	5050	80
Redlands Farm	Great Britain	52.33	-0.59	OxA-5632	4825	65
Aldwincle	Great Britain	52.41	-0.54	HAR-1411	4560	70
Tankardstown	Ireland	52.42	-8.62	GrN-14713	5105	45
Shippea Hill, Peacocks Farm	Great Britain	52.42888889	0.4125	Q-584	5465	120
Broome Heath	Great Britain	52.46777778	1.451388889	BM-679	5424	117
Swifterbant	Netherlands	52.56666667	5.6	GrN-7215	6330	45
Fengate	Great Britain	52.57388889	-0.20972222	GaK-4196	4960	64
Eaton Heath	Great Britain	52.60638889	1.262777778	BM-770	5095	49
Sarn-y-Bryn-Calod	Great Britain	52.64	-3.16	OxA-3997	4960	70
Shanballyedmond	Ireland	52.69	-8.25	GrN-11431	4930	60
Ardcrony	Ireland	52.88	-8.15	GrN-9708	4675	35
Westeregeln	Germany	52.96666667	11.4	Bln-92	6140	100
Ballintruer More	Ireland	52.99	-6.68	GrN-10469	4800	70
Beeston Castle	Great Britain	53.13	-2.7	HAR-6462	5140	90
Hude 1	Germany	53.16666667	8.433333333	Hv-1220	6060	115
Capel Eithin	Great Britain	53.22	-4.29	CAR-480	5520	80
Giants' Hills	Great Britain	53.22	0.13	OxA-641	5450	80
Wolla Bank	Great Britain	53.25	0.33	OxA-5965	4865	65
Lismore Fields	Great Britain	53.26	-1.94	OxA-2433	5270	100
Anderby Creek	Great Britain	53.27	0.32	OxA-5964	4625	55
Markland Grips	Great Britain	53.27	-1.24	OxA-4447	4760	90
Whitwell Long Cairn	Great Britain	53.27	-1.21	OxA-4176	5380	90
Strzelce	Poland	53.31666667	18.11666667	GrN-5087	6260	60
Rocher Moss 2	Great Britain	53.61666667	-2.03333333	Q-1190	5830	100
Newgrange	Ireland	53.71	-6.49	UB-361	4535	105
Callis Wold, Barrow 275	Great Britain	53.99	-0.73	BM-1170	4933	64
Hanging Grimstone	Great Britain	54.04	-0.77	HAR-2160	4710	90
Port St.Mary Rushen	Great Britain	54.08333333	-4.75	OxA-2481	4970	80
Kilham	Great Britain	54.09055556	-0.38666667	BM-293	4830	125
Boynton, High Easton	Great Britain	54.12	-0.23	HAR-268	4840	80
Grindale, Barrow 1	Great Britain	54.12	-0.24	HAR-269	4910	120

Appendix 1: The Gkiasta European database: datasets A to G.
Dataset C (oldest radiocarbon determination at 509 Neolithic sites).

Site Name	Country	Latitude (decimal degrees)	Longitude (decimal degrees)	Sample ID	Date (uncal. bp)	SD
East Heslerton Wold	Great Britain	54.16	-0.57	HAR-7030	5020	70
Ballymacdermot	Ireland	54.17	-6.38	UB-694	4830	95
Carrowmore - Culleenamore	Ireland	54.23	-8.62	St-7624	4710	100
Dundrum Nature Reserve	Ireland	54.25	-5.85	UB-412	4775	140
Seamer Moor, East Ayton	Great Britain	54.26333333	-0.46444444	NPL-73	5030	90
Carrowmore	Ireland	54.27	-8.53	Lu-1840	5750	85
Ballyglass (Ma.13)	Ireland	54.3	-9.4	SI-1450	4680	95
Ellerbek	Germany	54.38333333	10.18333333	KI-152	5170	110
Langdale -Pike of Stickle/T.Knott	Great Britain	54.45472222	-3.12305556	BM-281	4680	135
Tully, Aldergrove Airport	Ireland	54.46	-7.82	UB-2115	4960	85
Great Langdale	Great Britain	54.47	-3.14	OxA-2181	4790	80
Buddelin	Germany	54.48333333	13.53333333	Bln-561	5815	100
Street House Farm, Loftus	Great Britain	54.56	-0.83	BM-2061N	5080	60
Rude 2	Germany	54.61666667	9.583333333	Y-160	5690	70
Sudensee Damm	Germany	54.63333333	9.583333333	GrN-6588	6025	90
Ballynagilly	Ireland	54.71	-6.85	UB-305	5745	90
Carnanbane	Ireland	54.75	-6.82	UB-535	5045	95
Lochhill	Great Britain	54.96972222	-3.61083333	I-6409	5070	105
Mad Man's Window, Glenarm	Ireland	54.98	-5.93	UB-205	5095	120
Midgeholme Moss	Great Britain	54.98	-2.61	OxA-2326	5270	90
Dooey's Cairn, Ballymacaldrack	Ireland	55.01	-6.41	UB-2030	5150	90
Chatton Sandyford	Great Britain	55.29416667	-1.98388889	GaK-1507	4840	90
Skateholm	Sweden	55.38333333	13.48333333	LU-2110	7030	70
Rotten Bottom, Games Hope	Great Britain	55.42	-3.36	OxA-3540	5040	100
Olby Lyng	Denmark	55.5	12.21666667	K-1231	5320	130
Monamore, Arran	Great Britain	55.5125	-5.14111111	Q-675	5110	110
Machrie Moor Site	Great Britain	55.54	-5.32	GU-2320	5500	70
Bulltoftagarden	Sweden	55.6	13.06666667	LU-1617	6660	80
Meldon Bridge, near Peebles	Great Britain	55.65055556	-3.26361111	SRR-644	4686	90
Vejlebro	Denmark	55.68333333	9.583333333	K-1802	5540	110
Port Charlotte	Great Britain	55.73	-6.39	HAR-3487	5020	90
Stenbocksvallar	Sweden	55.76666667	13.91666667	LU-2198	5390	110
Newton	Great Britain	55.79	-6.26	GU-1952	4965	60
Linlithgow	Great Britain	55.97	-3.6	GU-1875	5265	55
Nether Kinneil	Great Britain	56	-3.67	SRR-1486.01	5060	50
Flynderhage	Denmark	56.01666667	10.23333333	K-1450	5230	100
Norslund	Denmark	56.01666667	10.23333333	K-990	5730	120
Ringkloster	Denmark	56.01666667	10.95	K-1652	5610	110
Vaenge So	Denmark	56.13333333	10.51666667	K-3921	5540	65
Holmegaard	Denmark	56.26666667	10.76666667	OxA-118	6280	130
Kinloch Farm, Collessie	Great Britain	56.29	-3.16	GU-1375	4725	70
Inchtuthil	Great Britain	56.54	-3.44	GU-2760	5160	70
Pitnacree	Great Britain	56.65944444	-3.74888889	GaK-601	4810	90
Dalladies Farm, Fettercairn	Great Britain	56.79583333	-2.61083333	SRR-289	4660	50
Ertebolle	Denmark	56.8	9.166666667	K-1529	5760	100
Haldrup Strand	Denmark	56.86666667	10.96666667	K-1612	5630	120
Salpetermose	Denmark	56.91666667	12.3	K-1233	6020	100
Agerod	Sweden	56.93333333	13.41666667	LU-1623	6860	70
Rhum (Kinloch)	Great Britain	57.04	-6.27	GU-2043	4725	140
Balbridie	Great Britain	57.05	-2.44	GU-1038	5150	60
Balbridie Timber Hal	Great Britain	57.05	-2.441	OxA-1769	5010	90
Brovst	Denmark	57.1	9.5	K-1613	5610	100
Raigmore	Great Britain	57.48	-4.19	SRR-188	4983	130
Fochabers - Boghead Mound -	Great Britain	57.61861111	-3.07138889	SRR-685	5031	100
Boghead Mound, Fochabers	Great Britain	57.62	-3.07	SRR-690	6006	60
The Ord North, Lairg	Great Britain	58.01	-4.43	GU-1169	4665	70
Camster Long Cairn, Wick parish	Great Britain	58.38	-3.27	GU-1707	4950	80
Loch Calder-Tulloch of Assery B	Great Britain	58.53	-3.6	GU-1332	4965	60
Loch Calder - an t'Sionnaich	Great Britain	58.54	-3.6	GU-1334	4685	60
Maes Howe	Great Britain	58.99722222	-3.18722222	SRR-791	5094	60
Quanterness	Great Britain	58.99972222	-3.015	Q-1294	4590	75
Barnhouse	Great Britain	59	-3.22	OxA-3499	4590	75
Knap of Howar, Papa Westray	Great Britain	59.35	-2.92	SRR-347	5706	85
Goddelau	Germany	61.01666667	8.116666667	OxA-1628	6300	90

Appendix 1: The Gkiasta European database: Datasets A to G.
Dataset D (most recent radiocarbon determinations at 509 Neolithic sites).

SITE NAME	Country	Latitude (decimal degees)	Longitude (decimal degrees)	Sample ID	Date	SD
KNOSSOS	Greece	35.51666667	25.16666667	BM-719	5967	41
CUEVA DE LA DEHESILLA	Spain	36.75	-5.8	UGRA-261	6260	100
PLATIA MAGOULA ZAKROU	Greece	36.95	22.76666667	DEM-158	6913	40
SALIAGOS	Greece	37.05	25.08333333	P-3766	5716	85
FRANCHTHI	Greece	37.43333333	23.13333333	P-2093	4990	90
LAGNANO DA PIEDE	Italy	38	12.33333333	UCLA-2148	6700	100
CUEVA DEL NACIMIENTO	Spain	38.08333333	-2.3	Gif-5422	5480	120
STRETTO, PARTANNA	Italy	38.12	13.36	Rome-290	5690	100
THEOPETRA CAVE	Greece	38.46666667	23.23333333	DEM-141	5485	102
ANTRE CORYCIEN	Greece	38.48333333	22.5	Gif-2123	6250	90
SKOTINI CAVE	Greece	38.51666667	23.98333333	DEM-104	4812	42
ELATEIA	Greece	38.61666667	22.73333333	GrN-3037	7360	90
COVETA DE L'OR	Spain	38.7	-0.466666667	K-1754	6265	75
BEVILACQUA	Italy	38.92	16.61	P-3047	6430	90
CUEVA CHICA DE SANTIAGO	Spain	38.95	-6.916666667	UGRA-254	6160	100
ACHILLEION	Greece	39.28333333	22.38333333	LJ-3201	7210	90
SO'N MATGE	Spain	39.33333333	3	QL-29	6680	120
SESKLO	Greece	39.38333333	22.81666667	GrN-16843	7110	70
ARGISSA MAGOULA	Greece	39.63333333	22.46666667	UCLA-1657E	6700	130
SIDARI	Greece	39.78333333	19.73333333	GXO-771	7670	120
GROTTA DI PORTO BADISCO	Italy	40.08333333	18.48333333	R-1225	5850	55
GROTTA CORBEDDU	Italy	40.25	9.483333333	UtC-15/233	6490	90
GROTTA FILIESTRU	Italy	40.5	8.666666667	Q-3026	5625	65
NEA NIKOMEDEIA	Greece	40.6	22.25	P-1203	7281	74
GROTTA DI ROMITO	Italy	40.86666667	16.9	R-223	6420	70
GROTTA DEL SANTUARIO DELLA MADONNA	Italy	40.9	15.78333333	R-284	5555	75
RENDINA	Italy	41	16	LJ-4549	6780	100
SCAMUSO	Italy	41.1	16.89	Gif-7054	5870	70
SITAGROI	Greece	41.15	24.2	Bln-778	6425	100
ARAGUINA SENNOLA	France	41.38333333	9.166666667	Gif-2324	6430	140
MONTE LEONE	France	41.4	9.166666667	Ly-6099	5855	95
SANTA TECCHIA	Italy	41.5	15.66666667	BM-2414	6520	70
COPPA NEVIGATA	Italy	41.5	15.75	OxA-1474	6850	80
VILAGGIO A, SCARAMELLA	Italy	41.5	15.68333333	R-351	6540	65
RENAGHJU	France	41.51666667	8.916666667	Ly-8328	6095	45
MONTE MAULO	Italy	41.57	14.64	OxA-653	6210	70
ANZA	Macedonia	41.7	22	LJ-2329	6230	60
VRSNIK-TARINCI	Macedonia	41.75	22.15	Bln-339a	6855	80
TUMBA OR VELUSKA	Macedonia	41.9	21.36666667	Tx-1787	6760	110
CASABIANCA OR CASTERRAGIO	France	42.08333333	9.5	MC-2243	6670	130
GROTTA DEI PICCIONI	Italy	42.21666667	14.96666667	PI-46	6247	130
CUEVA DE CHAVES	Spain	42.36666667	0.216666667	CSIC-380	6120	70
BALMA MARGINEDA	Andora	42.41666667	1.583333333	Ly-2439	6670	120
AZMAK	Bulgaria	42.45	25.76666667	Bln-301	6483	100
LO PLA DEL BACH	France	42.45	2.066666667	Gif-7492	5300	65
TELL AZMAK	Bulgaria	42.45	25.76666667	Bln-298	6540	100
VILLAGGIO LEOPARDI	Italy	42.46666667	14.93333333	PI-101	6578	135
TELL KARANOVO	Bulgaria	42.5	25.93333333	Bln-201	6573	100
ENVEIG	France	42.51666667	1.866666667	Ly-7541	5290	60
REVELLATA	France	42.56666667	8.733333333	Ly-6085	6280	75
MONTE REVINCU	France	42.6	9.266666667	Ly-9095	5175	45
LE ROC DE DOURGNE	France	42.61666667	2.15	MC-780	4250	120
PRISTINA	Yugoslavia	42.65	21.15	Bln-435	6280	80
SCAFFA PIANA	France	42.68333333	9.2	MC-2056	5320	100
LA CAUNE	France	42.71666667	2.616666667	Ly-3302	5640	120
LES COUDOUMINES	France	42.75	2.566666667	Ly-225	5060	45
CAMP DEL GINEBRE	France	42.75	2.566666667	Ly-267	5100	60
ABRI DE LA COMA FRANCEZE	France	42.83333333	2.916666667	Gif-9080	5180	60
COMA FRANCEZE	France	42.83333333	2.916666667	Gif-7292	5200	70
PORT LEUCATE	France	42.83333333	3	Gif-2747	5410	140
LA CAUNE OUEST	France	42.91666667	1.933333333	Ly-5104	5164	125
MADDALENA DI MUCCIA	Italy	43.06666667	13.06666667	R-643a	6580	75
ABRI JEAN CROS	France	43.08333333	2.466666667	MC-786	4450	100
SAN MARCO, GUBBIO	Italy	43.13	12.4	OxA-1854	6120	90
BUHOLOUP	France	43.15	1.183333333	Gif-5497	5490	120
LES PLOTS	France	43.21666667	2.35	Gif-9636	5170	50
LA POSTE VIEILLE	France	43.25	2.216666667	Ly-6241	5190	75
FONT JUVENAL	France	43.28333333	2.333333333	Ly-5005	5070	110
GROTTE DU VIEUX MOUNOI	France	43.3	5.866666667	Ly-5361	5250	60
GROTTE DE GAZEL	France	43.31666667	2.416666667	GrN-6705	6090	65
GROTTE DE L'ABEURADOR	France	43.35	2.7	MC-2145	6260	90
GROTTE TOURNIE	France	43.35	2.8	MC-1375	5050	120
CHATEAUNEUF LES MARTIGUES	France	43.38333333	5.166666667	Ly-2824	5460	130
RAFFEGUES	France	43.41666667	3.583333333	Gif-7350	5160	100
GROTTE DE CAMPRAFAUD	France	43.43333333	3.9	Gif-1485	5100	130
MOULIN DE LA GARONNE	France	43.5	1.333333333	Gif-5845	5120	120
VILLENEUVE TOLOSANE	France	43.5	1.333333333	Ly-8253	5060	45
MIOUVIN	France	43.51666667	4.983333333	MC-1224	5380	110
SOURCE DE RESELAUZE	France	43.51666667	3.283333333	MC-1226	5865	100
LE CAMP DE LAURE	France	43.55	5.216666667	MC-1381	5020	100
BAUME DE FONTBREGOUA	France	43.55	6.233333333	Ly-4287	5330	120
CUGNAUX	France	43.61666667	1.433333333	Gif-9023	5020	50
SAINT MICHEL-DE-TOUCH	France	43.61666667	1.433333333	MC-2217	5130	100
GROTTE LOMBARD	France	43.68333333	7.866666667	Ly-3332	6110	120
CAUCADE	France	43.7	7.35	MC-2342	6800	130
CHATEAU PERCIN	France	43.7	0.35	MC-947	5250	100
GIRIBALDI CIMIEZ CARABACEL	France	43.7	7.233333333	Ly-6349	5175	85
RIPABIANCA DI MONTERADO	Italy	43.71666667	13.1	R-599	6260	85
ESCANIN	France	43.75	4.8	Gif-3919	5070	110

Appendix 1: The Gkiasta European database: Datasets A to G.
Dataset D (most recent radiocarbon determinations at 509 Neolithic sites).

SITE NAME	Country	Latitude (decimal degees)	Longitude (decimal degrees)	Sample ID	Date	SD
L'EGLISE	France	43.75	6.25	Gif-1333	5500	140
MILETO	Italy	43.79	11.21	Beta-44155	6100	80
MALVOISIN	France	43.8	5.016666667	Ly-3020	5270	140
PENDIMOUM	France	43.81666667	7.5	Ly-5339	6320	95
GROTTE DE L'ABBE PRIALET	France	43.88333333	3.733333333	MC-1288	5270	100
LE CHEVAL	France	43.93333333	4.35	MC-943	5475	110
ROC TROUE	France	43.95	3.183333333	Gif-7688	5850	80
GROTTA DEL'ORSO	Italy	43.98333333	12.86666667	R-676	6080	50
ABRI DES USCLADES	France	43.98333333	3.233333333	Gif-9143	5550	70
CERON	France	44	5.05	Gif-9997	5990	75
DIVOSTIN	Yugoslavia	44.03333333	21.83333333	Bln-827	6910	100
UNANG	France	44.05	5.15	Gif-1795	5770	130
GROTTE DE L'AIGLE	France	44.06666667	4.633333333	MC-1251	6050	120
LE BARATIN	France	44.08333333	5.883333333	Gif-1855	6600	140
GROTTA ARMA DELLO STEFANIN	Italy	44.1	8.016666667	R-124	5600	80
ABRI DE LA POUJADE	France	44.11666667	3.166666667	MC-1297	5130	90
RIPARO ARMA DI NASINO	Italy	44.11666667	8.05	R-316	5955	65
FRIGOURAS	France	44.13333333	5.95	Gif-8479	5450	100
LE MOULIN	France	44.18333333	5.5	Ly-3805	5550	130
OBRE	Bosnia-Herz.	44.18333333	18.1	UCLA-1605.G	6710	60
LA VESIGNIE	France	44.25	4.616666667	Ly-4745	5815	75
BAUME DE MONTCLUS	France	44.26666667	4.433333333	Ly-304	6140	140
SAINTE LUCE	France	44.3	5.333333333	Ly-4486	5760	100
BAUME D'OULLINS	France	44.31666667	4.466666667	MC-2033	5590	100
LA BAUME DE RONZE	France	44.31666667	4.416666667	Ly-2177	5160	100
CHAZELLES	France	44.33333333	4.333333333	Lv-365	5660	110
SAINT MARCEL D'ARDECHE	France	44.35	4.466666667	Ly-3760	5220	120
LES MOULINS	France	44.35	4.766666667	Ly-4367	5130	110
BALLOY LES REAUDINS	France	44.4	3.166666667	Ly-12	6180	90
ROQUEMISSOU	France	44.41666667	2.733333333	Ly-4687	5785	105
GORNJA TUZLA	Bosnia-Herz.	44.45	18.76666667	GrN-2059	6640	75
LA CHAUVE SOURIS	France	44.46666667	4.683333333	Ly-4755	6790	70
SELEVAC	Yugoslavia	44.5	21.88333333	Z-233A	6113	80
CAVERNA DELLE ARENE CANDIDE	Italy	44.5	8.5	LJ-4137	5700	90
PADINA	Yugoslavia	44.5	22.41666667	GrN-7981	7075	50
VLASAC	Yugoslavia	44.53333333	22.05	Z-264	6335	92
TROU ARNAUD	France	44.55	5.316666667	Ly-4698	5255	75
CAPDENAC LE HAUT	France	44.58333333	2.066666667	Gif-2632	5100	140
ANTONNAIRE	France	44.68333333	5.383333333	Ly-4080	5570	130
LE FOURNET	France	44.68333333	5.333333333	Ly-2434	5440	130
LE SANGLIER	France	44.68333333	1.716666667	Ly-33	6915	70
GROTTE DU SANGLIER	France	44.68333333	1.716666667	Ly-7791	6095	100
HAMANGIA-BAIA 1	Romania	44.7	28.66666667	GrN-1986	5880	70
GROTTE DES ESCABASSES	France	44.71666667	1.8	Ly-9197	5160	35
STARCEVO	Yugoslavia	44.81666667	20.7	GrN-9033	6475	45
ROUCADOUR	France	44.83333333	1.5	Ly-416	5025	50
SAINT MITRE:ABRI 3	France	44.88333333	5.651	MC-265	6100	120
SAINT MITRE:ABRI 2	France	44.88333333	5.65	MC-263	6400	100
ABRI DE PENDIMOUN	France	44.88333333	7.483333333	Ly-5691	6165	130
BAUME BOURBON	France	44.9	4.633333333	Ly-412	6050	120
LES RIVAUX	France	45.05	3.85	Ly-2194	5030	100
PAS DE LA CHARMATE	France	45.06666667	5.383333333	Ly-4381	5630	110
COUFIN	France	45.06666667	5.4	Ly-3321	5260	120
COMBOIRE	France	45.13333333	5.683333333	Ly-4668	5360	70
LA GRANDE RIVOIRE	France	45.2	5.633333333	Ly-137	5030	70
ABRI DE LA GRANDE RIVOIRE	France	45.21666667	5.65	Ly-6096	5135	35
GROTTA PODOSOJNA	Croatia	45.25	14.23333333	Z-198	6470	95
ZORLENTU MARE	Romania	45.31	21.81	GrN-10316	6070	40
LA LEDE DU GURP	France	45.45	-1.15	Ly-7970	5305	45
MOLINO CASAROTO	Italy	45.46666667	11.6	Birm-266	5555	130
LES GOUILLAUDS	France	45.5	-0.6	Gif-4878	5950	120
APATIN	Yugoslavia	45.66666667	18.98333333	GrN-15971	6910	45
ALFORT	France	45.68333333	5.35	Ly-385	5755	65
SAINT PRIEST	France	45.7	4.916666667	Ly-8685	5445	60
GROTTA DEI CICLAMI	Italy	45.7	13.81666667	R-1040	6300	60
BETHENAS	France	45.71666667	5.25	Ly-3444	5060	110
SAMMARDENCHIA, UDINE	Italy	45.72	13.76	Bln-3373	6120	60
GRANDE BARME SAVIGNY	France	45.73333333	5.916666667	Ly-2077	5010	140
CREUX-ROUGE	France	45.78333333	3.083333333	Ly-1944	6070	140
GROTTE DESPERRATS	France	45.78333333	0.35	Gif-9536	5450	40
TREMONTEIX	France	45.78333333	3.083333333	Ly-3880	5330	120
GROTTE DES PERRATS	France	45.78333333	0.35	Gif-9534	5400	50
PONTCHARAUD	France	45.78333333	3.083333333	Ly-3903	5110	130
NECROPOLE DE LA BOIXE DOLMEN C	France	45.8	0.133333333	Gif-8909	5380	50
GROTTE DE LA VIEILLE EGLISE	France	45.91666667	6.283333333	CRG-412	5135	105
LES CHAMBONS	France	45.95	3.083333333	Ly-3406	5130	140
LE GARDON	France	45.95	5.033333333	Ly-4836	5095	70
GROTTE DU GARDON	France	45.95	5.333333333	Ly-8421	5060	40
LA ROCHE AUX FEES	France	45.98333333	0.983333333	Ly-4536B	5520	120
RIPARO DI ROMAGNANO III	Italy	46	11.1	R-776	5560	50
RIPARO GABAN	Italy	46.07	11.1	Bln-1777a	5750	60
OSZENTIVAN	Hungary	46.2	20.18333333	Bln-480	6050	100
OLAJKUT	Hungary	46.21666667	20.25	Bln-582	6260	100
KATALSZEG	Hungary	46.21666667	20.08333333	Bln-86	6370	100
SION-PETIT CHASSEUR	Switzerland	46.23333333	7.35	B-2111	5100	70
PLANTA	Switzerland	46.23333333	7.366666667	CRG-278	5900	80
BASAROGA	Hungary	46.25	21	BM-1862	6580	60
CHATELLIERS DU VIEL	France	46.43333333	-0.866666667	Gif-5717	5200	110

Appendix 1: The Gkiasta European database: Datasets A to G.
Dataset D (most recent radiocarbon determinations at 509 Neolithic sites).

SITE NAME	Country	Latitude (decimal degees)	Longitude (decimal degrees)	Sample ID	Date	SD
LES BOURIES	France	46.43333333	1.666666667	Ly-721	5210	70
LES SABLES DE LOI	France	46.43333333	0.883333333	Ly-7275	5110	60
ZALAVAR	Hungary	46.66666667	17.16666667	Bln-87	6180	100
LE PETIT BOIS	France	46.71666667	4.933333333	Ly-752	5790	65
ABRI DES ROCS	France	46.73333333	0.533333333	Gif-9956	5300	50
CHAMP DE LA GRANGE	France	46.76666667	2.433333333	Ly-4610	5219	73
VALLON-DES-VAUX	Switzerland	46.78333333	6.783333333	GrN-5601	4930	40
CHASSEY	France	46.88333333	4.766666667	Ly-3472	5120	110
LES PLANCHES	France	46.91666667	5.766666667	Gif-3826	5520	130
AUVERNIER-PORT	Switzerland	46.96666667	6.866666667	B-2561	4980	110
LE VERGER	France	46.98333333	4.716666667	Ly-2971	5710	140
LA MOTTE DES JUSTICES	France	47	-0.2	Gif-8292	5340	100
LE TROU QUI FUME	France	47	4.716666667	Ly-2613	5480	130
LA MOLLE PIERRE	France	47.05	4.75	Ly-5407	5044	58
ABRI DES PRES MOUREY	France	47.05	6.583333333	Gif-9095	5040	60
TWANN	Switzerland	47.1	7.15	B-2965	4950	50
SCHENKON	Switzerland	47.16666667	8.133333333	B-881	4980	120
EGOLZWIL 4	Switzerland	47.18333333	8.016666667	H-227/277	5040	100
FELDBACH	Switzerland	47.23333333	8.783333333	UCLA-1809A	5170	70
FENDMEILEN-VORDERELD	Switzerland	47.28333333	8.633333333	UCLA-1691 H	4940	70
LE TROU DU DIABLE	France	47.31666667	4.783333333	Ly-6506	5105	55
LES PICHELOTS	France	47.31666667	-1.5	Gif-6316	5530	70
SANDUN	France	47.33333333	2.566666667	Gif-7702	5120	70
GREIFENSEE-STOREN/WILDSBERG	Switzerland	47.36666667	8.683333333	UCLA-1825	5100	70
ZURICH	Switzerland	47.36666667	8.583333333	UCLA-1722 C	4640	70
ETANG DAVID	France	47.38333333	4.516666667	Gif-3354	5600	120
MYARD	France	47.38333333	4.516666667	Gif-2342	5175	135
ZURICH-WOLLISHOFEN	Switzerland	47.41	8.52	UCLA-1785	5320	60
ZURICH-BAUSCHANZE	Switzerland	47.41	8.52	UCLA-1765 A	4930	70
MANNLEFELSEN	France	47.45	7.233333333	Gif-2634	5140	140
BAVANS	France	47.46666667	6.7	Lv-1590	6410	95
FRANKENAU	Austria	47.5	16.5	VRI-207	5660	100
UNTERPULLENDORF	Austria	47.5	16.5	VRI-42	6130	140
PETIT MONT	France	47.53333333	-2.9	Gif-7014	5020	70
BILGROIX	France	47.55	-2.916666667	Gif-8961	5420	60
LE ROUICK	France	47.56666667	-2.966666667	Gif-8702	5300	60
SIERENTZ	France	47.56666667	7.45	Ly-4498	5570	100
LOCMARIAQUER ER GRAH	France	47.56666667	-2.95	Gif-7691	5250	70
PHYN	Switzerland	47.58333333	8.933333333	GrN-5958	4965	40
SANDORRESZE	Hungary	47.65	20.15	Bln-506	6120	100
TARNABOD	Hungary	47.68333333	20.21666667	Bln-123	6280	100
CROIX SAINT PIERRE	France	47.76666667	-2	Gif-8840	6070	80
SAINT JUST	France	47.76666667	-1.95	Gif-5763	5770	80
COJOUX	France	47.76666667	-1.95	Gif-5457	5550	120
STUROVO	Slovakia	47.8	18.73333333	Bln-558	6170	100
DOLMEN DE CHATEAU BU	France	47.81666667	1.95	Gif-9239	5230	50
BEKESSZENTANDRAS	Hungary	47.83333333	20.5	BM-1860	6080	60
QUEBARN	France	47.83333333	-4.233333333	Gif-5061	5760	120
SZARVAS	Hungary	47.83333333	20.58333333	BM-1865	6190	140
LARCUSTE	France	47.85	-2.783333333	Gif-2826	5490	120
ENDROD	Hungary	47.91666667	21.83333333	BM-1864	6090	60
WINDEN AM SEE	Austria	47.95	16.83333333	Bln-107	5820	100
COUEDAN	France	47.98333333	-2.916666667	Gif-9638	5260	70
KERGOGLE	France	48	-4.483333333	Gif-4689	5190	80
RUFACHER HUBEN	France	48.03333333	7.333333333	Ly-2733	5740	130
RICOH	France	48.06666667	7.3	Ly-4583	5150	100
ROUTE DE ROUFFACH	France	48.08333333	7.333333333	Ly-3812	6030	110
PASSY	France	48.1	3.266666667	Ly-7808	5545	50
LES ERVES	France	48.11666667	-0.333333333	Ly-3099	5580	140
ARMEAU	France	48.13333333	3.366666667	GrN-6781	6260	65
LE CAMP	France	48.15	2.733333333	Ly-8139	5830	45
TY FLOCH	France	48.16666667	-3.9	Ly-199	5270	80
LES DEPENDANCES DE DIGNY	France	48.16666667	2.416666667	Ly-5667	5560	115
TRIFAUCKIJ	Moldavia	48.18333333	28.31666667	Bln-589	6495	100
NEUVY EN DUNOIS	France	48.2	1.55	Gif-785	5250	140
SELEDIN	France	48.21666667	-3.05	Gif-1874	5075	140
LISCUIS EN LANISCAT	France	48.21666667	-3.133333333	Gif-3099	5140	100
LES FIEFS	France	48.23333333	2.45	Ly-6482	5555	70
PARC DU BRINBORION	France	48.33333333	2.45	Gif-2721	5190	130
FELSDACH LAUTERACH	Germany	48.33333333	9.566666667	GrN-4750	6140	45
TERRES DE PREPOUX	France	48.33333333	3.066666667	Ly-4502	5980	120
KORLAT	Hungary	48.36666667	21.25	Bln-119	6440	100
PIECE DE PINCEVENT	France	48.36666667	2.883333333	Gif-5005	5630	120
L'OUCHE DE BEAUCE	France	48.38333333	2.383333333	Gif-7705	5570	70
CHEMIN DE NAVIOT	France	48.4	6.166666667	Ly-4478	6000	80
LES REAUDINS	France	48.4	3.166666667	Ly--5886	5610	45
HORNE LEFANTOVCE	Slovakia	48.41666667	18.16666667	Bln-304	5775	140
DOMICA CAVE	Slovakia	48.48333333	20.48333333	GrN-2435	6080	75
LE BOISANNE	France	48.53333333	-2	Gif-7713	5610	70
HALLERTAU	Germany	48.55	11.75	HAM-197	5990	90
CARN	France	48.56666667	-4.683333333	GrN-1968	5230	75
CHANGE	France	48.56666667	1.583333333	Gif-9014	6180	120
CURNIC	France	48.58333333	-4.416666667	GrN-1966	5340	60
GUENNOC	France	48.58333333	-4.583333333	Gif-1870	5075	140
LARZICOURT	France	48.63333333	4.716666667	Ly-3907	5800	100
REICHSTETT	France	48.63333333	7.75	Ly-865	5940	140
NORROIS	France	48.66666667	4.55	Ly-3891	6020	110
POIGEN	Austria	48.66666667	38.66666667	H-2121/1552	5935	90
BARNENEZ	France	48.66666667	-3.85	Gif-1116	5100	140

Appendix 1: The Gkiasta European database: Datasets A to G.
Dataset D (most recent radiocarbon determinations at 509 Neolithic sites).

SITE NAME	Country	Latitude (decimal degees)	Longitude (decimal degrees)	Sample ID	Date	SD
PULKAU	Austria	48.7	15.86666667	Bln-83	6215	100
LE GUERZIT	France	48.71666667	-3.833333333	Gif-7614	5950	110
LE CAMP DE BIERRE	France	48.85	0.533333333	Ly-3724	5470	140
LE FOSSE DE TRAVERS	France	48.85	2.533333333	Gif-7710	5410	60
CHANTIER DE BERCY	France	48.86666667	2.333333333	Gif-9224	5820	60
GRAND LOUVRE	France	48.86666667	2.333333333	Gif-7618	5620	70
OSTOROS	Hungary	48.86666667	20.43333333	Bln-549	6180	100
BERCY QUARTIER SUD	France	48.86666667	2.333333333	Gif-9811	5200	70
SZAMOSSALYI	Hungary	48.88333333	22.6	Bln-404	6136	100
PENTE DE CROUPETON	France	48.88333333	2.716666667	Gif-8357	5630	60
HIENHEIM	Germany	48.88333333	11.75	GrN-4830	5910	50
LES LONGUES RAIES	France	48.91666667	2.766666667	Gif-5002	5510	140
TISZAVASVARI-KERESZTFAL	Hungary	48.96666667	21.38333333	Hv-1329	6155	95
LES GRANDS TRAQUIERS	France	49.01666667	4.266666667	Ly-4452	5480	90
BILBO	France	49.06666667	-0.05	Gif-8797	5110	70
DERRIERE LES PRES	France	49.06666667	-0.05	Gif-9650	5150	50
SENTE SAILLANCOURT	France	49.08333333	2	Gif-5840	5220	110
CUL FROID	France	49.23333333	1.05	Ly-2711	5180	100
VIEUX MANOIR TALANGE	France	49.25	6.15	Gif-9327	5310	80
LES BROUILLARDS	France	49.3	2.75	Gif-8244	5050	70
ZOPY	Czech Rep.	49.33333333	17.58333333	Bln-57	6430	100
LE CAMP DE CESAR	France	49.36666667	2.516666667	Ly-2966	5280	140
LA FOSSE TOUMISE	France	49.38333333	3.4	Ly-6477	5280	75
LES FONTINETTES	France	49.38333333	3.75	Ly-2335	5840	140
LA CROIX MAIGRET	France	49.4	4.883333333	Ly-2327	6030	130
LE MONT D'HUETTE	France	49.4	2.733333333	Ly-2970	5300	140
LES OBEAUX	France	49.4	2.966666667	Ly-3460	5890	120
LE COQ GALLEUX	France	49.4	2.8	Ly-2718	5330	110
DERRIERE LE VILLAGE	France	49.41666667	4.016666667	Ly-2322	6030	130
MOHELNICE	Czech Rep.	49.78333333	16.91666667	MOC-70	6220	80
LA PLAISANCE	France	50	2.133333333	OxA-460	5255	80
LA CARRIERE MERQUE	France	50.01666667	1.983333333	Gif-4356	5250	120
CHAMP DE BATAILLE	France	50.03333333	2.65	Ly-2679	5390	100
OLSZANICA	Poland	50.1	19.83333333	GrN-5384	6430	75
ZILGENDORF I	Germany	50.16666667	10.96666667	GrN-4245	6310	80
ZILGENDORF	Germany	50.16666667	10.96666667	GrN-7010	6225	45
CARN BREA	Great Britain	50.22027778	-5.246944444	BM-823	4561	47
FRIEDBERG	Germany	50.33333333	8.75	Bln-56	6120	100
TROU AL'WESSE	Belgium	50.46666667	5.25	Lv-1752	5950	70
THREE HOLES CAVE	Great Britain	50.49	-3.68	OxA-3889	4650	70
LES SABLINS	France	50.51666667	1.65	Gif-3701	5660	120
BROKEN CAVE	Great Britain	50.53	-3.69	OxA-3206	4885	90
CORON MATON	Belgium	50.55	5.883333333	Gif-5257	6420	120
COUTURE DE LA CHAUSSE	Belgium	50.58333333	3.683333333	Hv-8409	6020	55
IRCHONWELZ	Belgium	50.61666667	3.75	Bln-2531	6030	60
DARION	Belgium	50.65	5.183333333	Hv-10161	6145	145
PLACE SAINT LAMBERT	Belgium	50.65	5.566666667	Lv-1212	6220	90
AWANS	Belgium	50.66666667	5.45	Lv-1160	6070	90
CHABAROVICE	Czech Rep.	50.66666667	13.95	Bln-438	6400	120
WAREMME	Belgium	50.66666667	5.2	Lv-1695	6290	100
CRISNEE	Belgium	50.68333333	5.4	Lv-1580	6320	75
OLEYE	Belgium	50.68333333	5.216666667	Lv-1687	6200	100
WORGRET, WAREHAM	Great Britain	50.7	-2.15	HAR-9595	5210	70
MUDDERSHEIM	Germany	50.7	6.55	H-924a/1393	6150	90
MAIDEN CASTLE	Great Britain	50.7	-2.48	OxA-1336	4570	80
FLAGSTONES HOUSE, DORCHESTER	Great Britain	50.71	-2.43	HAR-9161	4960	80
COMBE HILL	Great Britain	50.79	0.21	I-11613	4590	110
KANNE	Belgium	50.8	5.666666667	Lv-1025	6260	75
LAMERSDORF	Germany	50.8	6.466666667	KN-I.367	6410	45
WANGE	Belgium	50.8	5.05	Lv-1116	6310	75
LANGWEILER	Germany	50.81666667	6.433333333	KN-2698	5930	145
ORSETT	Great Britain	50.85277778	0.348611111	BM-1214	4533	112
NIEDERMERZ	Germany	50.86666667	6.333333333	KN-2125	5830	130
BLACKPATCH	Great Britain	50.86861111	-0.445	BM-290	5090	130
LONG DOWN	Great Britain	50.88	-0.7	OxA-1151	4900	100
STEPLETON	Great Britain	50.88	-2.17	OxA-931	4660	80
OFFHAM HILL	Great Britain	50.89	-0.02	BM-1415	4740	60
HAMBLEDON HILL	Great Britain	50.90944444	-2.209166667	HAR-2377	4510	100
KOLN-MENGENICH	Germany	50.91666667	7	KN-I.369	6320	70
SHURTON HILL	Great Britain	50.92	-1.37	UB-2122	4750	50
BURY HILL	Great Britain	50.92	-0.58	HAR-3595	4570	80
WINDMILL HILL	Great Britain	50.92	-1.88	OxA-2394	4665	70
DORSET CURSUS (DOWN FARM)	Great Britain	50.93	-2	OxA-626	4770	120
SCORD OF BROUSTER	Great Britain	50.93	-1.66	CAR-249	4500	80
NORTH MARDEN	Great Britain	50.94	-0.87	HAR-5544	4710	110
BYLANY	Czech Rep.	50.96666667	15.3	BM-568	5635	65
RODINGEN	Germany	50.96666667	6.55	H-2117/1536	6250	75
CORHAMPTON HOUSE, CORHAMPTON	Great Britain	50.98	-1.15	BM-1889	4790	70
SOUTH CADBURY CASTLE	Great Britain	51.0225	-2.541944444	I-5972	4705	115
REHMSDORF	Germany	51.03333333	12.11666667	Bln-176	5932	100
DRESDEN-NICKERN	Germany	51.03333333	13.53333333	Bln-77	5815	100
WESTWARD HO!	Great Britain	51.03333333	-4.25	HAR-5642	4840	70
WINNALL DOWN	Great Britain	51.08	-1.32	HAR-2201	4650	110
BURNTWOOD FARM (SITE R6)	Great Britain	51.12	-1.29	OxA-1384	4750	140
JONES'S TRACK, ASHCOTT /W. HEATH	Great Britain	51.14	-2.78	HAR-3078	4590	70
CHILTON NO. 4 TRACK	Great Britain	51.15	-2.9	Lu-327	4760	65
ASHCOTT HEATH	Great Britain	51.16	-2.78	Q-598	4625	120
SHAPWICK HEATH	Great Britain	51.16	-2.81	Q-430	4540	130

Appendix 1: The Gkiasta European database: Datasets A to G.
Dataset D (most recent radiocarbon determinations at 509 Neolithic sites).

SITE NAME	Country	Latitude (decimal degees)	Longitude (decimal degrees)	Sample ID	Date	SD
SWEET TRACK, SHAPWICK	Great Britain	51.163	-2.83	HAR-1379	4550	70
DEVIL'S KNEADING TROUGH	Great Britain	51.16777778	0.971388889	BM-254	4540	105
NORMANTON DOWN, AMESBURY	Great Britain	51.16861111	-1.835555556	BM-505	4510	103
CONEYBURY HENGE	Great Britain	51.17	-1.81	OxA-1402	5050	100
MEARE HEATH FIELD	Great Britain	51.17	-2.79	HAR-3195	5180	70
MEARE LAKE VILLAGE WEST	Great Britain	51.17	-2.8	HAR-2616	5210	80
CHILTON 1-2	Great Britain	51.18	-2.88	HAR-649	4760	80
KING BARROW RIDGE	Great Britain	51.18	-1.81	OxA-1397	4500	120
HONEYGORE (W) TRACK, WESTHAY	Great Britain	51.18	-2.84	Q-1027	4742	50
BAKER TRACK	Great Britain	51.18	-2.82	HAR-2920	4520	70
HONEYGORE TRACK	Great Britain	51.18	-2.83	Q-431	4750	130
BAKER PLATFORM	Great Britain	51.183	-2.82	HAR-2844	4520	90
WINTERBOURNE STOKE	Great Britain	51.2	-1.85	OxA-1404	4550	120
BRIDGEMERE	Great Britain	51.21	-2.41	BM-2565	4630	50
ROBIN HOOD'S BALL	Great Britain	51.21	-1.86	OxA-1401	4510	90
NETHERAVON BAKE	Great Britain	51.22	-1.85	OxA-1407	4760	90
BADSHOT	Great Britain	51.23	-0.79	BM-2273N2	4710	50
ZWENKAU	Germany	51.23333333	12.35	Bln-817	5890	100
TOMASZOW	Poland	51.36666667	21.26666667	GrN-7590	5700	70
KNAP HILL	Great Britain	51.37083333	-1.826111111	BM-205	4710	115
EASTON DOWN	Great Britain	51.39	-1.91	OxA-3762	4535	65
WEST KENNET	Great Britain	51.41	-1.85	OxA-450	4700	80
SILBURY HILL, AVEBURY	Great Britain	51.415	-1.856111111	SI-910D	4530	110
SOUTH STREET, AVEBURY	Great Britain	51.42111111	-1.870555556	BM-358b	4530	110
CHERHILL	Great Britain	51.42944444	-1.955277778	BM-493	4715	90
RUNNYMEDE	Great Britain	51.43	-0.56	BM-2546	4600	90
MILLBARROW	Great Britain	51.45	-1.87	OxA-3169	4620	90
EILSLEBEN	Germany	51.48333333	11.51666667	OxA-1625	6030	100
CANNON HILL, BRAY	Great Britain	51.5	-0.72	HAR-1198	5270	110
MAR DYKE	Great Britain	51.5	0.27	HAR-4523	4650	90
ETON WICK	Great Britain	51.5	-0.66	BM-2534	4680	110
DUDERSTADT	Germany	51.51666667	10.28333333	KN-I.263	6300	55
ROSDORF	Germany	51.51666667	9.933333333	GrN-5604	5685	75
WAYLAND'S SMITHY, ASHBURY	Great Britain	51.56638889	-1.594444444	I-1468	4770	130
NORTH STOKE	Great Britain	51.57	-1.13	BM-1405	4672	49
MISBOURNE VIADUCT	Great Britain	51.58	-0.55	OxA-618	5970	100
DORCHESTER SITE VIII	Great Britain	51.66	-1.17	BM-2443	4510	100
DRAYTON	Great Britain	51.66	-1.31	OxA-2074	4620	80
ABINGDON	Great Britain	51.68055556	-1.260833333	BM-352	4710	135
CASHELKEELTY I	Ireland	51.72	-9.82	UB-2413	5845	100
STUMBLE	Great Britain	51.73	0.73	OxA-2299	4675	70
BRADWELL-ON-SEA SITE 8	Great Britain	51.75	0.88	HAR-6617	4690	70
COYGAN CAMP	Great Britain	51.75388889	-4.486388889	NPL-132	5000	95
SIDLINGS COPSE	Great Britain	51.79	-1.22	OxA-3560	5250	75
LITTLE WALTHAM	Great Britain	51.81	0.45	HAR-1087	5120	130
PEAK CAMP	Great Britain	51.83	-2.15	OxA-416	4630	110
ASCOTT-UNDER-WYCHWOOD	Great Britain	51.85472222	-1.565833333	BM-492	4735	70
GWERNVALE	Great Britain	51.87	-3.15	CAR-116	4590	75
HAZLETON NORTH	Great Britain	51.87	-1.9	OxA-643	4600	120
ELSLOO	Netherlands	51.93333333	5.766666667	GrN-2310	5080	70
GELEEN	Netherlands	51.96666667	5.75	GrN-996	6175	60
PEN-Y-WYRLOD	Great Britain	51.97583333	-3.236111111	HAR-674	4970	80
CARREG COETAN	Great Britain	52.02	-4.82	CAR-391	4560	80
LES RESSOURS	France	52.03333333	1.3	Ly-7484	5040	70
FERRITERS COVE	Ireland	52.16666667	10.5	BM-2227A	5190	110
BRIAR HILL	Great Britain	52.2275	-0.923888889	HAR-3208	4600	90
STANWICK SOUTH	Great Britain	52.32	-0.59	OxA-3002	4560	140
RECTORY FARM	Great Britain	52.32	-0.16	OxA-4360	4775	100
REDLANDS FARM	Great Britain	52.33	-0.59	OxA-5633	4820	80
ALDWINCLE	Great Britain	52.41	-0.54	HAR-1411	4560	70
TANKARDSTOWN	Ireland	52.42	-8.62	OxA-1477	4840	80
SHIPPEA HILL, PEACOCKS FARM	Great Britain	52.42888889	0.4125	Q-525	4870	120
BROOME HEATH	Great Britain	52.46777778	1.451388889	BM-756	4523	67
SWIFTERBANT	Netherlands	52.56666667	5.6	GrN-8248	6240	50
FENGATE	Great Britain	52.57388889	-0.209722222	GaK-4196	4960	64
EATON HEATH	Great Britain	52.60638889	1.262777778	BM-774	4903	56
SARN-Y-BRYN-CALOD	Great Britain	52.64	-3.16	OxA-3997	4960	70
SHANBALLYEDMOND	Ireland	52.69	-8.25	GrN-11431	4930	60
ARDCRONY	Ireland	52.88	-8.15	GrN-9708	4675	35
WESTEREGELN	Germany	52.96666667	11.4	Bln-42	6045	100
BALLINTRUER MORE	Ireland	52.99	-6.68	GrN-10469	4800	70
BEESTON CASTLE	Great Britain	53.13	-2.7	HAR-6462	5140	90
HUDE 1	Germany	53.16666667	8.433333333	Hv-1220	6060	115
CAPEL EITHIN	Great Britain	53.22	-4.29	CAR-481	4740	80
GIANTS' HILLS	Great Britain	53.22	0.13	OxA-639	4650	80
WOLLA BANK	Great Britain	53.25	0.33	OxA-5965	4865	65
LISMORE FIELDS	Great Britain	53.26	-1.94	OxA-2435	4680	70
ANDERBY CREEK	Great Britain	53.27	0.32	OxA-5964	4625	55
MARKLAND GRIPS	Great Britain	53.27	-1.24	OxA-4448	4740	90
WHITWELL LONG CAIRN	Great Britain	53.27	-1.21	OxA-4326	5115	70
STRZELCE	Poland	53.31666667	18.11666667	GrN-5087	6260	60
ROCHER MOSS 2	Great Britain	53.61666667	-2.033333333	Q-1190	5830	100
NEWGRANGE	Ireland	53.71	-6.49	GrN-5462	4500	45
CALLIS WOLD, BARROW 275	Great Britain	53.99	-0.73	BM-1167	4803	71
HANGING GRIMSTONE	Great Britain	54.04	-0.77	HAR-2160	4710	90
PORT ST.MARY RUSHEN	Great Britain	54.08333333	-4.75	OxA-2481	4970	80
KILHAM	Great Britain	54.09055556	-0.386666667	BM-293	4830	125
BOYNTON, HIGH EASTON	Great Britain	54.12	-0.23	HAR-268	4840	80
GRINDALE, BARROW 1	Great Britain	54.12	-0.24	HAR-266	4510	90

Appendix 1: The Gkiasta European database: Datasets A to G.
Dataset D (most recent radiocarbon determinations at 509 Neolithic sites).

SITE NAME	Country	Latitude (decimal degees)	Longitude (decimal degrees)	Sample ID	Date	SD
EAST HESLERTON WOLD	Great Britain	54.16	-0.57	HAR-7032	4640	70
BALLYMACDERMOT	Ireland	54.17	-6.38	UB-694	4830	95
CARROWMORE - CULLEENAMORE	Ireland	54.23	-8.62	St-7624	4710	100
DUNDRUM NATURE RESERVE	Ireland	54.25	-5.85	UB-413	4565	135
SEAMER MOOR, EAST AYTON	Great Britain	54.26333333	-0.464444444	NPL-73	5030	90
CARROWMORE	Ireland	54.27	-8.53	Lu-1810	4940	85
BALLYGLASS (MA.13)	Ireland	54.3	-9.4	SI-1453	4530	95
ELLERBEK	Germany	54.38333333	10.18333333	KI-152	5170	110
LANGDALE - PIKE OF STICKLE/T. KNOTT	Great Britain	54.45472222	-3.123055556	BM-281	4680	135
TULLY, ALDERGROVE AIRPORT	Ireland	54.46	-7.82	UB-2114	4575	50
GREAT LANGDALE	Great Britain	54.47	-3.14	OxA-2181	4790	80
BUDDELIN	Germany	54.48333333	13.53333333	Bln-561	5815	100
STREET HOUSE FARM, LOFTUS	Great Britain	54.56	-0.83	BM-2013R	4840	130
RUDE 2	Germany	54.61666667	9.583333333	Y-471	5620	50
SUDENSEE DAMM	Germany	54.63333333	9.583333333	KN-139	5880	65
BALLYNAGILLY	Ireland	54.71	-6.85	UB-625	4835	55
CARNANBANE	Ireland	54.75	-6.82	UB-534	4930	80
LOCHHILL	Great Britain	54.96972222	-3.610833333	I-6409	5070	105
MAD MAN'S WINDOW, GLENARM	Ireland	54.98	-5.93	UB-205	5095	120
MIDGEHOLME MOSS	Great Britain	54.98	-2.61	OxA-2326	5270	90
DOOEY'S CAIRN, BALLYMACALDRACK	Ireland	55.01	-6.41	UB-2045	4630	130
CHATTON SANDYFORD	Great Britain	55.29416667	-1.983888889	GaK-1507	4840	90
SKATEHOLM	Sweden	55.38333333	13.48333333	LU-2229	5640	60
ROTTEN BOTTOM, GAMES HOPE	Great Britain	55.42	-3.36	OxA-3540	5040	100
OLBY LYNG	Denmark	55.5	12.21666667	K-1230	5210	130
MONAMORE, ARRAN	Great Britain	55.5125	-5.141111111	Q-675	5110	110
MACHRIE MOOR SITE	Great Britain	55.54	-5.32	GU-2321	4820	50
BULLTOFTAGARDEN	Sweden	55.6	13.06666667	LU-1617	6660	80
MELDON BRIDGE, NEAR PEEBLES	Great Britain	55.65055556	-3.263611111	SRR-644	4686	90
VEJLEBRO	Denmark	55.68333333	9.583333333	K-1801	5180	110
PORT CHARLOTTE	Great Britain	55.73	-6.39	HAR-2084	4540	70
STENBOCKSVALLAR	Sweden	55.76666667	13.91666667	LU-2198	5390	110
NEWTON	Great Britain	55.79	-6.26	GU-1951	4880	60
LINLITHGOW	Great Britain	55.97	-3.6	GU-1875	5265	55
NETHER KINNEIL	Great Britain	56	-3.67	SRR-1608.02	4630	50
FLYNDERHAGE	Denmark	56.01666667	10.23333333	K-1450	5230	100
NORSLUND	Denmark	56.01666667	10.23333333	K-991	5680	120
RINGKLOSTER	Denmark	56.01666667	10.95	K-1653	5490	100
VAENGE SO	Denmark	56.13333333	10.51666667	OxA-532	5340	80
HOLMEGAARD	Denmark	56.26666667	10.76666667	K-3099	5870	95
KINLOCH FARM, COLLESSIE	Great Britain	56.29	-3.16	GU-1375	4725	70
INCHTUTHIL	Great Britain	56.54	-3.44	GU-2761	5070	50
PITNACREE	Great Britain	56.65944444	-3.748888889	GaK-601	4810	90
DALLADIES FARM, FETTERCAIRN	Great Britain	56.79583333	-2.610833333	SRR-290	4535	55
ERTEBOLLE	Denmark	56.8	9.166666667	K-1532	5550	110
HALDRUP STRAND	Denmark	56.86666667	10.96666667	K-1612	5630	120
SALPETERMOSE	Denmark	56.91666667	12.3	K-1235	5410	120
AGEROD	Sweden	56.93333333	13.41666667	LU-1622	6680	70
RHUM (KINLOCH)	Great Britain	57.04	-6.27	GU-2043	4725	140
BALBRIDIE	Great Britain	57.05	-2.44	GU-1036	4740	130
BALBRIDIE TIMBER HAL	Great Britain	57.05	-2.441	OxA-1767	4820	80
BROVST	Denmark	57.1	9.5	K-1863	5400	110
RAIGMORE	Great Britain	57.48	-4.19	SRR-187	4732	90
FOCHABERS - BOGHEAD MOUND	Great Britain	57.61861111	-3.071388889	SRR-684	4823	60
BOGHEAD MOUND, FOCHABERS	Great Britain	57.62	-3.07	SRR-690	6006	60
THE ORD NORTH, LAIRG	Great Britain	58.01	-4.43	GU-1172	4510	100
CAMSTER LONG CAIRN, WICK PARISH	Great Britain	58.38	-3.27	GU-1708	4915	60
LOCH CALDER - TULLOCH OF ASSERY B	Great Britain	58.53	-3.6	GU-1336	4655	60
LOCH CALDER - AN T'SIONNAICH	Great Britain	58.54	-3.6	GU-1334	4685	60
MAES HOWE	Great Britain	58.99722222	-3.187222222	SRR-791	5094	60
QUANTERNESS	Great Britain	58.99972222	-3.015	Q-1363	4540	110
BARNHOUSE	Great Britain	59	-3.22	OxA-2734	4520	70
KNAP OF HOWAR, PAPA WESTRAY	Great Britain	59.35	-2.92	SRR-346	4532	70
GODDELAU	Germany	61.01666667	8.116666667	OxA-1628	6300	90

Appendix 1: The Gkiasta European database: Datasets A to G.
Dataset E (most recent radiocarbon determination at multiple dated sites for 264 Neolithic sites).

SITE NAME	Country	Latitude (decimal degrees)	Longitude (decimal degrees)	Sample ID	Date (uncal. Bp)	SD
KNOSSOS	Greece	35.51666667	25.16666667	BM-719	5967	41
PLATIA MAGOULA ZAKROU	Greece	36.95	22.76666667	DEM-158	6913	40
SALIAGOS	Greece	37.05	25.08333333	P-3766	5716	85
FRANCHTHI	Greece	37.43333333	23.13333333	P-2093	4990	90
CUEVA DEL NACIMIENTO	Spain	38.08333333	-2.3	Gif-5422	5480	120
STRETTO, PARTANNA	Italy	38.12	13.36	Rome-290	5690	100
THEOPETRA CAVE	Greece	38.46666667	23.23333333	DEM-141	5485	102
ANTRE CORYCIEN	Greece	38.48333333	22.5	Gif-2123	6250	90
SKOTINI CAVE	Greece	38.51666667	23.98333333	DEM-104	4812	42
ELATEIA	Greece	38.61666667	22.73333333	GrN-3037	7360	90
ACHILLEION	Greece	39.28333333	22.38333333	LJ-3201	7210	90
SESKLO	Greece	39.38333333	22.81666667	GrN-16843	7110	70
ARGISSA MAGOULA	Greece	39.63333333	22.46666667	UCLA-1657E	6700	130
GROTTA CORBEDDU	Italy	40.25	9.483333333	UtC-15/233	6490	90
GROTTA FILIESTRU	Italy	40.5	8.666666667	Q-3026	5625	65
NEA NIKOMEDEIA	Greece	40.6	22.25	P-1203	7281	74
RENDINA	Italy	41	16	LJ-4549	6780	100
SCAMUSO	Italy	41.1	16.89	Gif-7054	5870	70
ARAGUINA SENNOLA	France	41.38333333	9.166666667	Gif-2324	6430	140
COPPA NEVIGATA	Italy	41.5	15.75	OxA-1474	6850	80
VILAGGIO A, SCARAMELLA	Italy	41.5	15.68333333	R-351	6540	65
RENAGHJU	France	41.51666667	8.916666667	Ly-8328	6095	45
MONTE MAULO	Italy	41.57	14.64	OxA-653	6210	70
ANZA	Macedonia	41.7	22	LJ-2329	6230	60
VRSNIK-TARINCI	Macedonia	41.75	22.15	Bln-339a	6855	80
TUMBA OR VELUSKA	Macedonia	41.9	21.36666667	Tx-1787	6760	110
CUEVA DE CHAVES	Spain	42.36666667	0.216666667	CSIC-380	6120	70
BALMA MARGINEDA	Andora	42.41666667	1.583333333	Ly-2439	6670	120
LO PLA DEL BACH	France	42.45	2.066666667	Gif-7492	5300	65
TELL AZMAK	Bulgaria	42.45	25.76666667	Bln-298	6540	100
TELL KARANOVO	Bulgaria	42.5	25.93333333	Bln-201	6573	100
MONTE REVINCU	France	42.6	9.266666667	Ly-9095	5175	45
LE ROC DE DOURGNE	France	42.61666667	2.15	MC-780	4250	120
SCAFFA PIANA	France	42.68333333	9.2	MC-2056	5320	100
CAMP DEL GINEBRE	France	42.75	2.566666667	Ly-267	5100	60
PORT LEUCATE	France	42.83333333	3	Gif-2747	5410	140
ABRI JEAN CROS	France	43.08333333	2.466666667	MC-786	4450	100
SAN MARCO, GUBBIO	Italy	43.13	12.4	OxA-1854	6120	90
LES PLOTS	France	43.21666667	2.35	Gif-9636	5170	50
FONT JUVENAL	France	43.28333333	2.333333333	Ly-5005	5070	110
GROTTE DU VIEUX MOUNOI	France	43.3	5.866666667	Ly-5361	5250	60
GROTTE DE GAZEL	France	43.31666667	2.416666667	GrN-6705	6090	65
GROTTE TOURNIE	France	43.35	2.8	MC-1375	5050	120
CHATEAUNEUF LES MARTIGUES	France	43.38333333	5.166666667	Ly-2824	5460	130
RAFFEGUES	France	43.41666667	3.583333333	Gif-7350	5160	100
GROTTE DE CAMPRAFAUD	France	43.43333333	3.9	Gif-1485	5100	130
VILLENEUVE TOLOSANE	France	43.5	1.333333333	Ly-8253	5060	45
SOURCE DE RESELAUZE	France	43.51666667	3.283333333	MC-1226	5865	100
BAUME DE FONTBREGOUA	France	43.55	6.233333333	Ly-4287	5330	120
CUGNAUX	France	43.61666667	1.433333333	Gif-9023	5020	50
SAINT MICHEL-DE-TOUCH	France	43.61666667	1.433333333	MC-2217	5130	100
GROTTE LOMBARD	France	43.68333333	7.866666667	Ly-3332	6110	120
GIRIBALDI CIMIEZ CARABACEL	France	43.7	7.233333333	Ly-6349	5175	85
ESCANIN	France	43.75	4.8	Gif-3919	5070	110
L'EGLISE	France	43.75	6.25	Gif-1333	5500	140
MILETO	Italy	43.79	11.21	Beta-44155	6100	80
PENDIMOUM	France	43.81666667	7.5	Ly-5339	6320	95
GROTTE DE L'ABBE PRIALET	France	43.88333333	3.733333333	MC-1288	5270	100
DIVOSTIN	Yugoslavia	44.03333333	21.83333333	Bln-827	6910	100
UNANG	France	44.05	5.15	Gif-1795	5770	130
GROTTE DE L'AIGLE	France	44.06666667	4.633333333	MC-1251	6050	120
GROTTA ARMA DELLO STEFANIN	Italy	44.1	8.016666667	R-124	5600	80
ABRI DE LA POUJADE	France	44.11666667	3.166666667	MC-1297	5130	90
RIPARO ARMA DI NASINO	Italy	44.11666667	8.05	R-316	5955	65
LE MOULIN	France	44.18333333	5.5	Ly-3805	5550	130
OBRE	Bosnia-Herz.	44.18333333	18.1	UCLA-1605.G	6710	60
BAUME DE MONTCLUS	France	44.26666667	4.433333333	Ly-304	6140	140
BAUME D'OULLINS	France	44.31666667	4.466666667	MC-2033	5590	100
LA BAUME DE RONZE	France	44.31666667	4.416666667	Ly-2177	5160	100
SAINT MARCEL D'ARDECHE	France	44.35	4.466666667	Ly-3760	5220	120
LES MOULINS	France	44.35	4.766666667	Ly-4367	5130	110
SELEVAC	Yugoslavia	44.5	21.88333333	Z-233A	6113	80
CAVERNA DELLE ARENE CANDIDE	Italy	44.5	8.5	LJ-4137	5700	90
PADINA	Yugoslavia	44.5	22.41666667	GrN-7981	7075	50
TROU ARNAUD	France	44.55	5.316666667	Ly-4698	5255	75
CAPDENAC LE HAUT	France	44.58333333	2.066666667	Gif-2632	5100	140
STARCEVO	Yugoslavia	44.81666667	20.7	GrN-9033	6475	45
ROUCADOUR	France	44.83333333	1.5	Ly-416	5025	50
SAINT MITRE:ABRI 2	France	44.88333333	5.65	MC-263	6400	100
ABRI DE PENDIMOUN	France	44.88333333	7.483333333	Ly-5691	6165	130
BAUME BOURBON	France	44.9	4.633333333	Ly-412	6050	120
LES RIVAUX	France	45.05	3.85	Ly-2194	5030	100
PAS DE LA CHARMATE	France	45.06666667	5.383333333	Ly-4381	5630	110
LA GRANDE RIVOIRE	France	45.2	5.633333333	Ly-137	5030	70
LA LEDE DU GURP	France	45.45	-1.15	Ly-7970	5305	45
MOLINO CASAROTO	Italy	45.46666667	11.6	Birm-266	5555	130
APATIN	Yugoslavia	45.66666667	18.98333333	GrN-15971	6910	45
ALFORT	France	45.68333333	5.35	Ly-385	5755	65
GROTTE DES PERRATS	France	45.78333333	0.35	Gif-9534	5400	50

Appendix 1: The Gkiasta European database: Datasets A to G.
Dataset E (most recent radiocarbon determination at multiple dated sites for 264 Neolithic sites).

SITE NAME	Country	Latitude (decimal degrees)	Longitude (decimal degrees)	Sample ID	Date (uncal. Bp)	SD
GROTTE DE LA VIEILLE EGLISE	France	45.91666667	6.283333333	CRG-412	5135	105
LE GARDON	France	45.95	5.033333333	Ly-4836	5095	70
GROTTE DU GARDON	France	45.95	5.333333333	Ly-8421	5060	40
LA ROCHE AUX FEES	France	45.98333333	0.983333333	Ly-4536B	5520	120
RIPARO DI ROMAGNANO III	Italy	46	11.1	R-776	5560	50
RIPARO GABAN	Italy	46.07	11.1	Bln-1777a	5750	60
OSZENTIVAN	Hungary	46.2	20.18333333	Bln-480	6050	100
OLAJKUT	Hungary	46.21666667	20.25	Bln-582	6260	100
SION-PETIT CHASSEUR	Switzerland	46.23333333	7.35	B-2111	5100	70
PLANTA	Switzerland	46.23333333	7.366666667	CRG-278	5900	80
LE PETIT BOIS	France	46.71666667	4.933333333	Ly-752	5790	65
CHAMP DE LA GRANGE	France	46.76666667	2.433333333	Ly-4610	5219	73
VALLON-DES-VAUX	Switzerland	46.78333333	6.783333333	GrN-5601	4930	40
CHASSEY	France	46.88333333	4.766666667	Ly-3472	5120	110
AUVERNIER-PORT	Switzerland	46.96666667	6.866666667	B-2561	4980	110
LA MOLLE PIERRE	France	47.05	4.75	Ly-5407	5044	58
ABRI DES PRES MOUREY	France	47.05	6.583333333	Gif-9095	5040	60
TWANN	Switzerland	47.1	7.15	B-2965	4950	50
EGOLZWIL 4	Switzerland	47.18333333	8.016666667	H-227/277	5040	100
FENDMEILEN-VORDERELD	Switzerland	47.28333333	8.633333333	UCLA-1691 H	4940	70
SANDUN	France	47.33333333	2.566666667	Gif-7702	5120	70
GREIFENSEE-STOREN/WILDSBERG	Switzerland	47.36666667	8.683333333	UCLA-1825	5100	70
ZURICH	Switzerland	47.36666667	8.583333333	UCLA-1722 C	4640	70
ZURICH-BAUSCHANZE	Switzerland	47.41	8.52	UCLA-1765 A	4930	70
BAVANS	France	47.46666667	6.7	Lv-1590	6410	95
PETIT MONT	France	47.53333333	-2.9	Gif-7014	5020	70
LOCMARIAQUER ER GRAH	France	47.56666667	-2.95	Gif-7691	5250	70
PHYN	Switzerland	47.58333333	8.933333333	GrN-5958	4965	40
SANDORRESZE	Hungary	47.65	20.15	Bln-506	6120	100
COJOUX	France	47.76666667	-1.95	Gif-5457	5550	120
STUROVO	Slovakia	47.8	18.73333333	Bln-558	6170	100
SZARVAS	Hungary	47.83333333	20.58333333	BM-1865	6190	140
ENDROD	Hungary	47.91666667	21.83333333	BM-1864	6090	60
WINDEN AM SEE	Austria	47.95	16.83333333	Bln-107	5820	100
RICOH	France	48.06666667	7.3	Ly-4583	5150	100
ROUTE DE ROUFFACH	France	48.08333333	7.333333333	Ly-3812	6030	110
PASSY	France	48.1	3.266666667	Ly-7808	5545	50
TY FLOCH	France	48.16666667	-3.9	Ly-199	5270	80
LES DEPENDANCES DE DIGNY	France	48.16666667	2.416666667	Ly-5667	5560	115
TRIFAUCKIJ	Moldavia	48.18333333	28.31666667	Bln-589	6495	100
SELEDIN	France	48.21666667	-3.05	Gif-1874	5075	140
LES FIEFS	France	48.23333333	2.45	Ly-6482	5555	70
TERRES DE PREPOUX	France	48.33333333	3.066666667	Ly-4502	5980	120
CHEMIN DE NAVIOT	France	48.4	6.166666667	Ly-4478	6000	80
LES REAUDINS	France	48.4	3.166666667	Ly--5886	5610	45
LE PARC DU CHATEAU	France	48.45	1.783333333	Ly-7677	5260	65
BARNENEZ	France	48.66666667	-3.85	Gif-1116	5100	140
BERCY QUARTIER SUD	France	48.86666667	2.333333333	Gif-9811	5200	70
PENTE DE CROUPETON	France	48.88333333	2.716666667	Gif-8357	5630	60
HIENHEIM	Germany	48.88333333	11.75	GrN-4830	5910	50
TISZAVASVARI-KERESZTFAL	Hungary	48.96666667	21.38333333	Hv-1329	6155	95
LES GRANDS TRAQUIERS	France	49.01666667	4.266666667	Ly-4452	5480	90
DERRIERE LES PRES	France	49.06666667	-0.05	Gif-9650	5150	50
LA FOSSE TOUMISE	France	49.38333333	3.4	Ly-6477	5280	75
LES FONTINETTES	France	49.38333333	3.75	Ly-2335	5840	140
LE COQ GALLEUX	France	49.4	2.8	Ly-2718	5330	110
DERRIERE LE VILLAGE	France	49.41666667	4.016666667	Ly-2322	6030	130
MOHELNICE	Czech Rep.	49.78333333	16.91666667	MOC-70	6220	80
ZILGENDORF	Germany	50.16666667	10.96666667	GrN-7010	6225	45
CARN BREA	Great Britain	50.22027778	-5.246944444	BM-823	4561	47
THREE HOLES CAVE	Great Britain	50.49	-3.68	OxA-3889	4650	70
LES SABLINS	France	50.51666667	1.65	Gif-3701	5660	120
BROKEN CAVE	Great Britain	50.53	-3.69	OxA-3206	4885	90
COUTURE DE LA CHAUSSE	Belgium	50.58333333	3.683333333	Hv-8409	6020	55
DARION	Belgium	50.65	5.183333333	Hv-10161	6145	145
PLACE SAINT LAMBERT	Belgium	50.65	5.566666667	Lv-1212	6220	90
WAREMME	Belgium	50.66666667	5.2	Lv-1695	6290	100
OLEYE	Belgium	50.68333333	5.216666667	Lv-1687	6200	100
MUDDERSHEIM	Germany	50.7	6.55	H-924a/1393	6150	90
MAIDEN CASTLE	Great Britain	50.7	-2.48	OxA-1336	4570	80
LANGWEILER	Germany	50.81666667	6.433333333	KN-2698	5930	145
ORSETT	Great Britain	50.85277778	0.348611111	BM-1214	4533	112
NIEDERMERZ	Germany	50.86666667	6.333333333	KN-2125	5830	130
LONG DOWN	Great Britain	50.88	-0.7	OxA-1151	4900	100
STEPLETON	Great Britain	50.88	-2.17	OxA-931	4660	80
OFFHAM HILL	Great Britain	50.89	-0.02	BM-1415	4740	60
HAMBLEDON HILL	Great Britain	50.90944444	-2.209166667	HAR-2377	4510	100
BURY HILL	Great Britain	50.92	-0.58	HAR-3595	4570	80
WINDLMILL HILL	Great Britain	50.92	-1.88	OxA-2394	4665	70
DORSET CURSUS (DOWN FARM)	Great Britain	50.93	-2	OxA-626	4770	120
SCORD OF BROUSTER	Great Britain	50.93	-1.66	CAR-249	4500	80
BYLANY	Czech Rep.	50.96666667	15.3	BM-568	5635	65
DRESDEN-NICKERN	Germany	51.03333333	13.53333333	Bln-77	5815	100
WESTWARD HO!	Great Britain	51.03333333	-4.25	HAR-5642	4840	70
WINNALL DOWN	Great Britain	51.08	-1.32	HAR-2201	4650	110
MEARE HEATH	Great Britain	51.163	-2.78	HAR-2224	4560	80
SWEET TRACK, SHAPWICK	Great Britain	51.163	-2.83	HAR-1379	4550	70
HONEYGORE (WEST) TRACK, WESTHAY	Great Britain	51.18	-2.84	Q-1027	4742	50
BAKER TRACK	Great Britain	51.18	-2.82	HAR-2920	4520	70

Appendix 1: The Gkiasta European database: Datasets A to G.
Dataset E (most recent radiocarbon determination at multiple dated sites for 264 Neolithic sites).

SITE NAME	Country	Latitude (decimal degrees)	Longitude (decimal degrees)	Sample ID	Date (uncal. Bp)	SD
BAKER PLATFORM	Great Britain	51.183	-2.82	HAR-2844	4520	90
WINTERBOURNE STOKE	Great Britain	51.2	-1.85	OxA-1404	4550	120
ROBIN HOOD'S BALL	Great Britain	51.21	-1.86	OxA-1401	4510	90
BADSHOT	Great Britain	51.23	-0.79	BM-2273N2	4710	50
ZWENKAU	Germany	51.23333333	12.35	Bln-817	5890	100
TOMASZOW	Poland	51.36666667	21.26666667	GrN-7590	5700	70
EASTON DOWN	Great Britain	51.39	-1.91	OxA-3762	4535	65
WEST KENNET	Great Britain	51.41	-1.85	OxA-450	4700	80
SILBURY HILL, AVEBURY	Great Britain	51.415	-1.856111111	SI-910D	4530	110
SOUTH STREET, AVEBURY	Great Britain	51.42111111	-1.870555556	BM-358b	4530	110
RUNNYMEDE	Great Britain	51.43	-0.56	BM-2546	4600	90
MILLBARROW	Great Britain	51.45	-1.87	OxA-3169	4620	90
EILSLEBEN	Germany	51.48333333	11.51666667	OxA-1625	6030	100
MAR DYKE	Great Britain	51.5	0.27	HAR-4523	4650	90
ETON WICK	Great Britain	51.5	-0.66	BM-2534	4680	110
DUDERSTADT	Germany	51.51666667	10.28333333	KN-I.263	6300	55
ROSDORF	Germany	51.51666667	9.933333333	GrN-5604	5685	75
MISBOURNE VIADUCT	Great Britain	51.58	-0.55	OxA-618	5970	100
DRAYTON	Great Britain	51.66	-1.31	OxA-2074	4620	80
ABINGDON	Great Britain	51.68055556	-1.260833333	BM-352	4710	135
STUMBLE	Great Britain	51.73	0.73	OxA-2299	4675	70
PEAK CAMP	Great Britain	51.83	-2.15	OxA-416	4630	110
ASCOTT-UNDER-WYCHWOOD	Great Britain	51.85472222	-1.565833333	BM-492	4735	70
GWERNVALE	Great Britain	51.87	-3.15	CAR-116	4590	75
HAZLETON NORTH	Great Britain	51.87	-1.9	OxA-643	4600	120
ELSLOO	Netherlands	51.93333333	5.766666667	GrN-2310	5080	70
GELEEN	Netherlands	51.96666667	5.75	GrN-996	6175	60
CARREG COETAN	Great Britain	52.02	-4.82	CAR-391	4560	80
BRIAR HILL	Great Britain	52.2275	-0.923888889	HAR-3208	4600	90
STANWICK SOUTH	Great Britain	52.32	-0.59	OxA-3002	4560	140
RECTORY FARM	Great Britain	52.32	-0.16	OxA-4360	4775	100
REDLANDS FARM	Great Britain	52.33	-0.59	OxA-5633	4820	80
TANKARDSTOWN	Ireland	52.42	-8.62	OxA-1477	4840	80
SHIPPEA HILL, PEACOCKS FARM	Great Britain	52.42888889	0.4125	Q-525	4870	120
BROOME HEATH	Great Britain	52.46777778	1.451388889	BM-756	4523	67
SWIFTERBANT	Netherlands	52.56666667	5.6	GrN-8248	6240	50
EATON HEATH	Great Britain	52.60638889	1.262777778	BM-774	4903	56
WESTEREGELN	Germany	52.96666667	11.4	Bln-42	6045	100
CAPEL EITHIN	Great Britain	53.22	-4.29	CAR-481	4740	80
GIANTS' HILLS	Great Britain	53.22	0.13	OxA-639	4650	80
LISMORE FIELDS	Great Britain	53.26	-1.94	OxA-2435	4680	70
MARKLAND GRIPS	Great Britain	53.27	-1.24	OxA-4448	4740	90
WHITWELL LONG CAIRN	Great Britain	53.27	-1.21	OxA-4326	5115	70
NEWGRANGE	Ireland	53.71	-6.49	GrN-5462	4500	45
CALLIS WOLD, BARROW 275	Great Britain	53.99	-0.73	BM-1167	4803	71
GRINDALE, BARROW 1	Great Britain	54.12	-0.24	HAR-266	4510	90
EAST HESLERTON WOLD	Great Britain	54.16	-0.57	HAR-7032	4640	70
DUNDRUM NATURE RESERVE	Ireland	54.25	-5.85	UB-413	4565	135
CARROWMORE	Ireland	54.27	-8.53	Lu-1810	4940	85
BALLYGLASS (MA.13)	Ireland	54.3	-9.4	SI-1453	4530	95
TULLY, ALDERGROVE AIRPORT	Ireland	54.46	-7.82	UB-2114	4575	50
STREET HOUSE FARM, LOFTUS	Great Britain	54.56	-0.83	BM-2013R	4840	130
RUDE 2	Germany	54.61666667	9.583333333	Y-471	5620	50
SUDENSEE DAMM	Germany	54.63333333	9.583333333	KN-139	5880	65
BALLYNAGILLY	Ireland	54.71	-6.85	UB-625	4835	55
CARNANBANE	Ireland	54.75	-6.82	UB-534	4930	80
DOOEY'S CAIRN, BALLYMACALDRACK	Ireland	55.01	-6.41	UB-2045	4630	130
SKATEHOLM	Sweden	55.38333333	13.48333333	LU-2229	5640	60
OLBY LYNG	Denmark	55.5	12.21666667	K-1230	5210	130
MACHRIE MOOR SITE	Great Britain	55.54	-5.32	GU-2321	4820	50
VEJLEBRO	Denmark	55.68333333	9.583333333	K-1801	5180	110
PORT CHARLOTTE	Great Britain	55.73	-6.39	HAR-2084	4540	70
NEWTON	Great Britain	55.79	-6.26	GU-1951	4880	60
NETHER KINNEIL	Great Britain	56	-3.67	SRR-1608.02	4630	50
NORSLUND	Denmark	56.01666667	10.23333333	K-991	5680	120
RINGKLOSTER	Denmark	56.01666667	10.95	K-1653	5490	100
VAENGE SO	Denmark	56.13333333	10.51666667	OxA-532	5340	80
HOLMEGAARD	Denmark	56.26666667	10.76666667	K-3099	5870	95
INCHTUTHIL	Great Britain	56.54	-3.44	GU-2761	5070	50
DALLADIES FARM, FETTERCAIRN	Great Britain	56.79583333	-2.610833333	SRR-290	4535	55
ERTEBOLLE	Denmark	56.8	9.166666667	K-1532	5550	110
SALPETERMOSE	Denmark	56.91666667	12.3	K-1235	5410	120
AGEROD	Sweden	56.93333333	13.41666667	LU-1622	6680	70
BALBRIDIE	Great Britain	57.05	-2.44	GU-1036	4740	130
BALBRIDIE TIMBER HAL	Great Britain	57.05	-2.441	OxA-1767	4820	80
BROVST	Denmark	57.1	9.5	K-1863	5400	110
RAIGMORE	Great Britain	57.48	-4.19	SRR-187	4732	90
FOCHABERS - BOGHEAD MOUND -	Great Britain	57.61861111	-3.071388889	SRR-684	4823	60
THE ORD NORTH, LAIRG	Great Britain	58.01	-4.43	GU-1172	4510	100
CAMSTER LONG CAIRN, WICK PARISH	Great Britain	58.38	-3.27	GU-1708	4915	60
LOCH CALDER - TULLOCH OF ASSERY B	Great Britain	58.53	-3.6	GU-1336	4655	60
QUANTERNESS	Great Britain	58.99972222	-3.015	Q-1363	4540	110
BARNHOUSE	Great Britain	59	-3.22	OxA-2734	4520	70
KNAP OF HOWAR, PAPA WESTRAY	Great Britain	59.35	-2.92	SRR-346	4532	70

Appendix 1: The Gkiasta European Database (Datasets A to G).
Dataset F (oldest radiocarbon determination at multiple dated sites only for 264 Neolithic sites).

SITE NAME	Country	Latitude (decimal degrees)	Longitude (decimal degrees)	Sample ID	Date (uncal. Bp)	SD
KNOSSOS	Greece	35.51666667	25.16666667	BM-278	7910	140
PLATIA MAGOULA ZAKROU	Greece	36.95	22.76666667	DEM-163	7083	44
SALIAGOS	Greece	37.05	25.08333333	P-1311	6172	74
FRANCHTHI	Greece	37.43333333	23.13333333	P-2095	7980	110
CUEVA DEL NACIMIENTO	Spain	38.08333333	-2.3	Gif-2368	6780	130
STRETTO, PARTANNA	Italy	38.12	13.36	Rome-291	6630	120
THEOPETRA CAVE	Greece	38.46666667	23.23333333	DEM-575	7036	44
ANTRE CORYCIEN	Greece	38.48333333	22.5	Gif-2124	6380	90
SKOTINI CAVE	Greece	38.51666667	23.98333333	DEM-137	6247	49
ELATEIA	Greece	38.61666667	22.73333333	GrN-2973	7480	70
ACHILLEION	Greece	39.28333333	22.38333333	LJ-3180	7550	60
SESKLO	Greece	39.38333333	22.81666667	P-1681	7755	97
ARGISSA MAGOULA	Greece	39.63333333	22.46666667	UCLA-1657A	8130	100
GROTTA CORBEDDU	Italy	40.25	9.483333333	UtC-1251	6690	80
GROTTA FILIESTRU	Italy	40.5	8.666666667	Q-3020	6710	75
NEA NIKOMEDEIA	Greece	40.6	22.25	P-1202	7557	91
RENDINA	Italy	41	16	LJ-4548	7110	140
SCAMUSO	Italy	41.1	16.89	Gif-6339	7290	110
ARAGUINA SENNOLA	France	41.38333333	9.166666667	Gif-2325	6650	140
COPPA NEVIGATA	Italy	41.5	15.75	OxA-1475	6880	90
VILAGGIO A, SCARAMELLA	Italy	41.5	15.68333333	R-350	7000	100
RENAGHJU	France	41.51666667	8.916666667	Ly-8327	6525	60
MONTE MAULO	Italy	41.57	14.64	OxA-652	6280	70
ANZA	Macedonia	41.7	22	LJ-3032	7210	50
VRSNIK-TARINCI	Macedonia	41.75	22.15	Bln-339	6950	100
TUMBA OR VELUSKA	Macedonia	41.9	21.36666667	Tx-1785	6950	120
CUEVA DE CHAVES	Spain	42.36666667	0.216666667	GrN-12685	6770	70
BALMA MARGINEDA	Andora	42.41666667	1.583333333	Ly-2439	6670	120
LO PLA DEL BACH	France	42.45	2.066666667	Gif-6749	5700	80
TELL AZMAK	Bulgaria	42.45	25.76666667	Bln-203	6880	100
TELL KARANOVO	Bulgaria	42.5	25.93333333	Bln-152	6807	100
MONTE REVINCU	France	42.6	9.266666667	Ly-8396	5405	55
LE ROC DE DOURGNE	France	42.61666667	2.15	MC-1107	6850	100
SCAFFA PIANA	France	42.68333333	9.2	MC-2057	5360	100
CAMP DEL GINEBRE	France	42.75	2.566666667	Ly-226	5355	45
PORT LEUCATE	France	42.83333333	3	Gif-2749	5900	140
ABRI JEAN CROS	France	43.08333333	2.466666667	Gif-3575	6600	130
SAN MARCO, GUBBIO	Italy	43.13	12.4	OxA-1853	6430	80
LES PLOTS	France	43.21666667	2.35	Gif-9337	5230	70
FONT JUVENAL	France	43.28333333	2.333333333	Ly-4770	6235	85
GROTTE DU VIEUX MOUNOI	France	43.3	5.866666667	Ly-5362	5625	85
GROTTE DE GAZEL	France	43.31666667	2.416666667	GrN-6702	6905	90
GROTTE TOURNIE	France	43.35	2.8	MC-797	5900	100
CHATEAUNEUF LES MARTIGUES	France	43.38333333	5.166666667	MC-2515	6900	100
RAFFEGUES	France	43.41666667	3.583333333	Gif-7349	5200	100
GROTTE DE CAMPRAFAUD	France	43.43333333	3.9	Gif-3078	6480	130
VILLENEUVE TOLOSANE	France	43.51666667	1.3	Gif-6313	5400	90
SOURCE DE RESELAUZE	France	43.51666667	3.283333333	MC-1227	6450	100
BAUME DE FONTBREGOUA	France	43.55	6.233333333	Gif-2990	6700	100
CUGNAUX	France	43.61666667	1.433333333	Gif-9026	5460	70
SAINT MICHEL-DE-TOUCH	France	43.61666667	1.433333333	MC-847	5600	90
GROTTE LOMBARD	France	43.68333333	7.866666667	Ly-3331	6190	120
GIRIBALDI CIMIEZ CARABACEL	France	43.7	7.233333333	Ly-6945	5750	60
ESCANIN	France	43.75	4.8	Gif-1995	5160	120
L'EGLISE	France	43.75	6.25	Gif-1334	5760	140
MILETO	Italy	43.79	11.21	Beta-44114	6180	80
PENDIMOUM	France	43.81666667	7.5	Ly-6495	6565	75
GROTTE DE L'ABBE PRIALET	France	43.88333333	3.733333333	MC-1290	6070	100
DIVOSTIN	Yugoslavia	44.03333333	21.83333333	Bln-899	7200	100
UNANG	France	44.05	5.15	Gif-1796	5950	130
GROTTE DE L'AIGLE	France	44.06666667	4.633333333	MC-563	6200	100
GROTTA ARMA DELLO STEFANIN	Italy	44.1	8.016666667	Bln-3276	6610	60
ABRI DE LA POUJADE	France	44.11666667	3.166666667	MC-1367	6990	120
RIPARO ARMA DI NASINO	Italy	44.11666667	8.05	R-267	6470	120
LE MOULIN	France	44.18333333	5.5	Ly-3804	5840	130
OBRE	Bosnia-Herz.-Herzegovine	44.18333333	18.1	UCLA-1605.I	7240	60
BAUME DE MONTCLUS	France	44.26666667	4.433333333	MC-695	6450	100
BAUME D'OULLINS	France	44.31666667	4.466666667	MC-2034	6630	110
LA BAUME DE RONZE	France	44.31666667	4.416666667	Ly-2747	5400	130
SAINT MARCEL D'ARDECHE	France	44.35	4.466666667	MC-2376	6330	90
LES MOULINS	France	44.35	4.766666667	Ly-3801	5430	120
SELEVAC	Yugoslavia	44.5	21.88333333	Z-233	6366	100
CAVERNA DELLE ARENE CANDIDE	Italy	44.5	8.5	UB-2423	6980	115
PADINA	Yugoslavia	44.5	22.41666667	GrN-8230	7100	80
TROU ARNAUD	France	44.55	5.316666667	Ly-4696	5535	100
CAPDENAC LE HAUT	France	44.58333333	2.066666667	Gif-3714	5190	120
STARCEVO	Yugoslavia	44.81666667	20.7	OxA-8561	6975	60
ROUCADOUR	France	44.83333333	1.5	Gif-36C	5940	140
SAINT MITRE:ABRI 2	France	44.88333333	5.65	MC-264	6700	130
ABRI DE PENDIMOUN	France	44.88333333	7.483333333	Ly-5692	6615	105
BAUME BOURBON	France	44.9	4.633333333	MC-794	6080	100
LES RIVAUX	France	45.05	3.85	Ly-3710	6240	110
PAS DE LA CHARMATE	France	45.06666667	5.383333333	Ly-3785	5680	130
LA GRANDE RIVOIRE	France	45.2	5.633333333	Ly-176	6000	70
LA LEDE DU GURP	France	45.45	-1.15	Ly-5318	6101	65
MOLINO CASAROTO	Italy	45.46666667	11.6	Birm-175	6450	110
APATIN	Yugoslavia	45.66666667	18.98333333	GrN-15973	7130	60
ALFORT	France	45.68333333	5.35	Ly-386	6000	60
GROTTE DES PERRATS	France	45.78333333	0.35	Gif-9535	5880	70

Appendix 1: The Gkiasta European Database (Datasets A to G).
Dataset F (oldest radiocarbon determination at multiple dated sites only for 264 Neolithic sites).

SITE NAME	Country	Latitude (decimal degrees)	Longitude (decimal degrees)	Sample ID	Date (uncal. Bp)	SD
GROTTE DE LA VIEILLE EGLISE	France	45.91666667	6.283333333	CRG-411	5384	128
LE GARDON	France	45.95	5.033333333	Ly-7011	5815	40
GROTTE DU GARDON	France	45.95	5.333333333	Ly-8422	6325	40
LA ROCHE AUX FEES	France	45.98333333	0.983333333	Ly-4536A	5530	120
RIPARO DI ROMAGNANO III	Italy	46	11.1	R-1136	6480	50
RIPARO GABAN	Italy	46.07	11.1	Bln-1777	6030	45
OSZENTIVAN	Hungary	46.2	20.18333333	Bln-479	6460	80
OLAJKUT	Hungary	46.21666667	20.25	Bln-581	6605	100
SION-PETIT CHASSEUR	Switzerland	46.23333333	7.35	B-2110	5130	100
PLANTA	Switzerland	46.23333333	7.366666667	CRG-280	6500	80
LE PETIT BOIS	France	46.71666667	4.933333333	Ly-753	5840	62
CHAMP DE LA GRANGE	France	46.76666667	2.433333333	Ly-4439	5310	110
VALLON-DES-VAUX	Switzerland	46.78333333	6.783333333	B-659	5180	120
CHASSEY	France	46.88333333	4.766666667	Ly-3474	5400	100
AUVERNIER-PORT	Switzerland	46.96666667	6.866666667	B-2559	5130	120
LA MOLLE PIERRE	France	47.05	4.75	Ly-5408	5238	120
ABRI DES PRES MOUREY	France	47.05	6.583333333	Gif-9096	5340	70
TWANN	Switzerland	47.1	7.15	B-2698	5200	90
EGOLZWIL 4	Switzerland	47.18333333	8.016666667	H-228/276	5150	100
FENDMEILEN-VORDERELD	Switzerland	47.28333333	8.633333333	UCLA-1691 F	5415	60
SANDUN	France	47.33333333	2.566666667	Gif-7703	5660	100
GREIFENSEE-STOREN/WILDSBERG	Switzerland	47.36666667	8.683333333	UCLA-1824	5180	70
ZURICH	Switzerland	47.36666667	8.583333333	UCLA-1772 B	5145	70
ZURICH-BAUSCHANZE	Switzerland	47.41	8.52	UCLA-1765 B	5040	60
BAVANS	France	47.46666667	6.7	Lv-1415	7130	70
PETIT MONT	France	47.53333333	-2.9	Gif-6844	5650	70
LOCMARIAQUER ER GRAH	France	47.56666667	-2.95	Gif-7693	5370	70
PHYN	Switzerland	47.58333333	8.933333333	GrN-5957	5020	40
SANDORRESZE	Hungary	47.65	20.15	Bln-676	6155	80
COJOUX	France	47.76666667	-1.95	Gif-5458	5600	120
STUROVO	Slovakia	47.8	18.73333333	Bln-559	6260	100
SZARVAS	Hungary	47.83333333	20.58333333	BM-1866	6620	60
ENDROD	Hungary	47.91666667	21.83333333	GrN-10319	6945	50
WINDEN AM SEE	Austria	47.95	16.83333333	Bln-55	5940	100
RICOH	France	48.06666667	7.3	Ly-4582	6400	70
ROUTE DE ROUFFACH	France	48.08333333	7.333333333	Ly-3811	6100	120
PASSY	France	48.1	3.266666667	Ly-8823	6065	55
TY FLOCH	France	48.16666667	-3.9	Gif-5234	5580	120
LES DEPENDANCES DE DIGNY	France	48.16666667	2.416666667	Ly-5668	5990	50
TRIFAUCKIJ	Moldavia	48.18333333	28.31666667	Bln-587	7420	80
SELEDIN	France	48.21666667	-3.05	Gif-1877	5270	140
LES FIEFS	France	48.23333333	2.45	Ly-6481	5650	55
TERRES DE PREPOUX	France	48.33333333	3.066666667	Ly-4503	6730	110
CHEMIN DE NAVIOT	France	48.4	6.166666667	Ly-4479	6330	140
LES REAUDINS	France	48.4	3.166666667	Ly-5540	5160	85
LE PARC DU CHATEAU	France	48.45	1.783333333	Ly-4731	6655	90
BARNENEZ	France	48.66666667	-3.85	Gif-1556	5550	140
BERCY QUARTIER SUD	France	48.86666667	2.333333333	Gif-9810	5420	60
PENTE DE CROUPETON	France	48.88333333	2.716666667	Gif-8358	5910	80
HIENHEIM	Germany	48.88333333	11.75	GrN-7557	6235	45
TISZAVASVARI-KERESZTFAL	Hungary	48.96666667	21.38333333	Bln-505	6305	100
LES GRANDS TRAQUIERS	France	49.01666667	4.266666667	Ly-4453	6200	140
DERRIERE LES PRES	France	49.06666667	-0.05	Gif-89160	5560	140
LA FOSSE TOUMISE	France	49.38333333	3.4	Ly-6478	5870	105
LES FONTINETTES	France	49.38333333	3.75	Ly-2331	6000	120
LE COQ GALLEUX	France	49.4	2.8	Ly-2716	6080	110
DERRIERE LE VILLAGE	France	49.41666667	4.016666667	Ly-2324	6110	140
MOHELNICE	Czech Rep.	49.78333333	16.91666667	Bln-102a	6405	100
ZILGENDORF	Germany	50.16666667	10.96666667	GrN-7011	6250	65
CARN BREA	Great Britain	50.22027778	-5.246944444	BM-825	4999	64
THREE HOLES CAVE	Great Britain	50.49	-3.68	OxA-4495	5010	70
LES SABLINS	France	50.51666667	1.65	Gif-4024	5690	120
BROKEN CAVE	Great Britain	50.53	-3.69	OxA-3207	5015	80
COUTURE DE LA CHAUSSE	Belgium	50.58333333	3.683333333	Hv-8404	6485	75
DARION	Belgium	50.65	5.183333333	Lv-1579	6240	100
PLACE SAINT LAMBERT	Belgium	50.65	5.566666667	Lv-1340	6460	60
WAREMME	Belgium	50.66666667	5.2	Lv-1693	6300	95
OLEYE	Belgium	50.68333333	5.216666667	Lv-1685	6370	95
MUDDERSHEIM	Germany	50.7	6.55	H-924b/1121	6330	80
MAIDEN CASTLE	Great Britain	50.7	-2.48	OxA-1337	5030	80
LANGWEILER	Germany	50.81666667	6.433333333	KN-2695	7090	135
ORSETT	Great Britain	50.85277778	0.348611111	BM-1213	4741	113
NIEDERMERZ	Germany	50.86666667	6.333333333	KN-2289	6800	80
LONG DOWN	Great Britain	50.88	-0.7	OxA-1152	5050	100
STEPLETON	Great Britain	50.88	-2.17	OxA-932	4690	70
OFFHAM HILL	Great Britain	50.89	-0.02	BM-1414	4925	80
HAMBLEDON HILL	Great Britain	50.90944444	-2.209166667	HAR-3062	4850	100
BURY HILL	Great Britain	50.92	-0.58	HAR-3596	4680	80
WINDMILL HILL	Great Britain	50.92	-1.88	OxA-2395	4730	80
DORSET CURSUS (DOWN FARM)	Great Britain	50.93	-2	OxA-627	6800	100
SCORD OF BROUSTER	Great Britain	50.93	-1.66	CAR-253	5050	90
BYLANY	Czech Rep.	50.96666667	15.3	GrN-4754	6270	65
DRESDEN-NICKERN	Germany	51.03333333	13.53333333	Bln-73	5945	100
WESTWARD HO!	Great Britain	51.03333333	-4.25	HAR-5641	5740	100
WINNALL DOWN	Great Britain	51.08	-1.32	HAR-2196	4800	80
MEARE HEATH	Great Britain	51.163	-2.78	Q-646	4650	120
SWEET TRACK, SHAPWICK	Great Britain	51.163	-2.83	Q-968	5224	75
HONEYGORE (W) TRACK, WESTHAY	Great Britain	51.18	-2.84	Q-1028	4780	50
BAKER TRACK	Great Britain	51.18	-2.82	GaK-1600	4840	100

Appendix 1: The Gkiasta European Database (Datasets A to G).
Dataset F (oldest radiocarbon determination at multiple dated sites only for 264 Neolithic sites).

SITE NAME	Country	Latitude (decimal degrees)	Longitude (decimal degrees)	Sample ID	Date (uncal. Bp)	SD
BAKER PLATFORM	Great Britain	51.183	-2.82	HAR-2845	4950	80
WINTERBOURNE STOKE	Great Britain	51.2	-1.85	OxA-1405	4640	100
ROBIN HOOD'S BALL	Great Britain	51.21	-1.86	OxA-1400	4740	100
BADSHOT	Great Britain	51.23	-0.79	BM-2273N1	4780	40
ZWENKAU	Germany	51.23333333	12.35	GrN-1581	6160	70
TOMASZOW	Poland	51.36666667	21.26666667	GrN-7591	6145	70
EASTON DOWN	Great Britain	51.39	-1.91	OxA-3760	4730	65
WEST KENNET	Great Britain	51.41	-1.85	OxA-449	4825	90
SILBURY HILL, AVEBURY	Great Britain	51.415	-1.856111111	SI-910A	4675	110
SOUTH STREET, AVEBURY	Great Britain	51.42111111	-1.870555556	BM-356	4760	130
RUNNYMEDE	Great Britain	51.43	-0.56	OxA-3582	5780	85
MILLBARROW	Great Britain	51.45	-1.87	OxA-3172	4900	110
EILSLEBEN	Germany	51.48333333	11.51666667	OxA-1627	6190	90
MAR DYKE	Great Britain	51.5	0.27	HAR-4522	5740	80
ETON WICK	Great Britain	51.5	-0.66	BM-2533	4750	80
DUDERSTADT	Germany	51.51666667	10.28333333	H-919/889	6340	100
ROSDORF	Germany	51.51666667	9.933333333	Hv-586	6350	70
MISBOURNE VIADUCT	Great Britain	51.58	-0.55	OxA-601	6190	90
DRAYTON	Great Britain	51.66	-1.31	HAR-6477	4990	100
ABINGDON	Great Britain	51.68055556	-1.260833333	BM-351	5060	130
STUMBLE	Great Britain	51.73	0.73	OxA-2298	4780	70
PEAK CAMP	Great Britain	51.83	-2.15	OxA-1622	4865	80
ASCOTT-UNDER-WYCHWOOD	Great Britain	51.85472222	-1.565833333	BM-833	5020	92
GWERNVALE	Great Britain	51.87	-3.15	CAR-113	5050	75
HAZLETON NORTH	Great Britain	51.87	-1.9	OxA-910	5000	70
ELSLOO	Netherlands	51.93333333	5.766666667	GrN-2159	6320	90
GELEEN	Netherlands	51.96666667	5.75	GrN-995	6370	60
CARREG COETAN	Great Britain	52.02	-4.82	CAR-392	4830	80
BRIAR HILL	Great Britain	52.2275	-0.923888889	HAR-2282	5440	110
STANWICK SOUTH	Great Britain	52.32	-0.59	OxA-3001	4810	80
RECTORY FARM	Great Britain	52.32	-0.16	OxA-3370	5050	80
REDLANDS FARM	Great Britain	52.33	-0.59	OxA-5632	4825	65
TANKARDSTOWN	Ireland	52.42	-8.62	GrN-14713	5105	45
SHIPPEA HILL, PEACOCKS FARM	Great Britain	52.42888889	0.4125	Q-584	5465	120
BROOME HEATH	Great Britain	52.46777778	1.451388889	BM-679	5424	117
SWIFTERBANT	Netherlands	52.56666667	5.6	GrN-7215	6330	45
EATON HEATH	Great Britain	52.60638889	1.262777778	BM-770	5095	49
WESTEREGELN	Germany	52.96666667	11.4	Bln-92	6140	100
CAPEL EITHIN	Great Britain	53.22	-4.29	CAR-480	5520	80
GIANTS' HILLS	Great Britain	53.22	0.13	OxA-641	5450	80
LISMORE FIELDS	Great Britain	53.26	-1.94	OxA-2433	5270	100
MARKLAND GRIPS	Great Britain	53.27	-1.24	OxA-4447	4760	90
WHITWELL LONG CAIRN	Great Britain	53.27	-1.21	OxA-4176	5380	90
NEWGRANGE	Ireland	53.71	-6.49	UB-361	4535	105
CALLIS WOLD, BARROW 275	Great Britain	53.99	-0.73	BM-1170	4933	64
GRINDALE, BARROW 1	Great Britain	54.12	-0.24	HAR-269	4910	120
EAST HESLERTON WOLD	Great Britain	54.16	-0.57	HAR-7030	5020	70
DUNDRUM NATURE RESERVE	Ireland	54.25	-5.85	UB-412	4775	140
CARROWMORE	Ireland	54.27	-8.53	Lu-1840	5750	85
BALLYGLASS (MA.13)	Ireland	54.3	-9.4	SI-1450	4680	95
TULLY, ALDERGROVE AIRPORT	Ireland	54.46	-7.82	UB-2115	4960	85
STREET HOUSE FARM, LOFTUS	Great Britain	54.56	-0.83	BM-2061N	5080	60
RUDE 2	Germany	54.61666667	9.583333333	Y-160	5690	70
SUDENSEE DAMM	Germany	54.63333333	9.583333333	GrN-6588	6025	90
BALLYNAGILLY	Ireland	54.71	-6.85	UB-305	5745	90
CARNANBANE	Ireland	54.75	-6.82	UB-535	5045	95
DOOEY'S CAIRN, BALLYMACALDRACK	Ireland	55.01	-6.41	UB-2030	5150	90
SKATEHOLM	Sweden	55.38333333	13.48333333	LU-2110	7030	70
OLBY LYNG	Denmark	55.5	12.21666667	K-1231	5320	130
MACHRIE MOOR SITE	Great Britain	55.54	-5.32	GU-2320	5500	70
VEJLEBRO	Denmark	55.68333333	9.583333333	K-1802	5540	110
PORT CHARLOTTE	Great Britain	55.73	-6.39	HAR-3487	5020	90
NEWTON	Great Britain	55.79	-6.26	GU-1952	4965	60
NETHER KINNEIL	Great Britain	56	-3.67	SRR-1486.01	5060	50
NORSLUND	Denmark	56.01666667	10.23333333	K-990	5730	120
RINGKLOSTER	Denmark	56.01666667	10.95	K-1652	5610	110
VAENGE SO	Denmark	56.13333333	10.51666667	K-3921	5540	65
HOLMEGAARD	Denmark	56.26666667	10.76666667	OxA-118	6280	130
INCHTUTHIL	Great Britain	56.54	-3.44	GU-2760	5160	70
DALLADIES FARM, FETTERCAIRN	Great Britain	56.79583333	-2.610833333	SRR-289	4660	50
ERTEBOLLE	Denmark	56.8	9.166666667	K-1529	5760	100
SALPETERMOSE	Denmark	56.91666667	12.3	K-1233	6020	100
AGEROD	Sweden	56.93333333	13.41666667	LU-1623	6860	70
BALBRIDIE	Great Britain	57.05	-2.44	GU-1038	5150	60
BALBRIDIE TIMBER HAL	Great Britain	57.05	-2.441	OxA-1769	5010	90
BROVST	Denmark	57.1	9.5	K-1613	5610	100
RAIGMORE	Great Britain	57.48	-4.19	SRR-188	4983	130
FOCHABERS - BOGHEAD MOUND	Great Britain	57.61861111	-3.071388889	SRR-685	5031	100
THE ORD NORTH, LAIRG	Great Britain	58.01	-4.43	GU-1169	4665	70
CAMSTER LONG CAIRN, WICK PARISH	Great Britain	58.38	-3.27	GU-1707	4950	80
LOCH CALDER - TULLOCH OF ASSERY B	Great Britain	58.53	-3.6	GU-1332	4965	60
QUANTERNESS	Great Britain	58.99972222	-3.015	Q-1294	4590	75
BARNHOUSE	Great Britain	59	-3.22	OxA-3499	4590	75
KNAP OF HOWAR, PAPA WESTRAY	Great Britain	59.35	-2.92	SRR-347	5706	85

Appendix 1: The Gkiasta European database: Datasets A to G.
Dataset G (weighted mean or median radiocarbon determination at 229 Neolithic sites).

SITE NAME	Country	Latitude (decimal degrees)	Longitude (decimal degrees)	Sample ID	Date (uncal. Bp)	SD
KNOSSOS	Greece	35.51666667	25.16666667	BM-436	7740	140
PLATIA MAGOULA ZAKROU	Greece	36.95	22.76666667	DEM-161	6946	51
SALIAGOS	Greece	37.05	25.08333333	P-1368	5909	87
FRANCHTHI	Greece	37.43333333	23.13333333	P-1667	7278	89
STRETTO, PARTANNA	Italy	38.12	13.36	Rome-292	6260	110
THEOPETRA CAVE	Greece	38.46666667	23.23333333	DEM-454	6563	68
ANTRE CORYCIEN	Greece	38.48333333	22.5	Gif-2123	6315	64
SKOTINI CAVE	Greece	38.51666667	23.98333333	DEM-138	5817	37
ELATEIA	Greece	38.61666667	22.73333333	GrN-2973	7435	55
ACHILLEION	Greece	39.28333333	22.38333333	GrN-7436	7295	70
SESKLO	Greece	39.38333333	22.81666667	P-1682	7483	72
ARGISSA MAGOULA	Greece	39.63333333	22.46666667	UCLA-1657D	7990	95
GROTTA CORBEDDU	Italy	40.25	9.483333333	UtC-1251	6602	60
GROTTA FILIESTRU	Italy	40.5	8.666666667	Q-3023	6470	65
RENDINA	Italy	41	16	LJ-4548	6891	81
SCAMUSO	Italy	41.1	16.89	Gif-7345	6600	120
ARAGUINA SENNOLA	France	41.38333333	9.166666667	Gif-2324	6540	99
COPPA NEVIGATA	Italy	41.5	15.75	OxA-1474	6863	60
MONTE MAULO	Italy	41.57	14.64	OxA-652	6245	49
ANZA	Republic of Macedonia	41.7	22	LJ-2405	6940	80
VRSNIK-TARINCI	Republic of Macedonia	41.75	22.15	Bln-339	6892	62
TUMBA OR VELUSKA	Republic of Macedonia	41.9	21.36666667	Tx-1785	6874	55
CUEVA DE CHAVES	Spain	42.36666667	0.216666667	CSIC-378	6460	70
BALMA MARGINEDA	Andora	42.41666667	1.583333333	Ly-2439	6670	85
TELL AZMAK	Bulgaria	42.45	25.76666667	Bln-203	6758	33
TELL KARANOVO	Bulgaria	42.5	25.93333333	Bln-152	6690	71
MONTE REVINCU	France	42.6	9.266666667	Ly-8395	5355	55
LE ROC DE DOURGNE	France	42.61666667	2.15	MC-1105	5550	80
SCAFFA PIANA	France	42.68333333	9.2	MC-2056	5340	71
ABRI JEAN CROS	France	43.08333333	2.466666667	MC-785	4600	100
SAN MARCO, GUBBIO	Italy	43.13	12.4	OxA-1851	6276	41
LES PLOTS	France	43.21666667	2.35	Gif-9337	5190	41
FONT JUVENAL	France	43.28333333	2.333333333	MC-498	5540	100
GROTTE DU VIEUX MOUNOI	France	43.3	5.866666667	Ly-5363	5580	60
GROTTE DE GAZEL	France	43.31666667	2.416666667	GrN-6707	6350	55
GROTTE TOURNIE	France	43.35	2.8	MC-1377	5800	120
CHATEAUNEUF LES MARTIGUES	France	43.38333333	5.166666667	Ly-446	6430	140
RAFFEGUES	France	43.41666667	3.583333333	Gif-7349	5180	71
GROTTE DE CAMPRAFAUD	France	43.43333333	3.9	Gif-1488	5900	140
VILLENEUVE TOLOSANE	France	43.51666667	1.3	Gif-4848	5200	80
CUGNAUX	France	43.53333333	1.333333333	Gif-9245	5180	70
BAUME DE FONTBREGOUA	France	43.55	6.233333333	Ly-4288	5680	120
SAINT MICHEL-DE-TOUCH	France	43.61666667	1.433333333	MC-844	5350	90
GROTTE LOMBARD	France	43.68333333	7.866666667	Ly-3331	6150	85
GIRIBALDI CIMIEZ CARABACEL	France	43.7	7.233333333	Ly-6350	5495	95
L'EGLISE	France	43.75	6.25	Gif-1333	5630	99
ESCANIN	France	43.75	4.8	Gif-1995	5111	81
MILETO	Italy	43.79	11.21	Beta-44114	6140	57
PENDIMOUM	France	43.81666667	7.5	Gif-6471	6463	42
GROTTE DE L'ABBE PRIALET	France	43.88333333	3.733333333	MC-1289	5750	100
DIVOSTIN	Yugoslavia	44.03333333	21.83333333	Bln-824	7010	38
UNANG	France	44.05	5.15	Gif-1793	5865	65
GROTTE DE L'AIGLE	France	44.06666667	4.633333333	MC-1251	6139	77
ABRI DE LA POUJADE	France	44.11666667	3.166666667	MC-2187	6620	120
RIPARO ARMA DI NASINO	Italy	44.11666667	8.05	R-263	6140	110
LE MOULIN	France	44.18333333	5.5	Ly-3804	5695	92
BAUME DE MONTCLUS	France	44.26666667	4.433333333	Ly-303	6334	70
LA BAUME DE RONZE	France	44.31666667	4.416666667	Ly-2177	5265	66
LES MOULINS	France	44.35	4.766666667	Ly-3801	5262	63
PADINA	Yugoslavia	44.5	22.41666667	GrN-7981	7082	42
SELEVAC	Yugoslavia	44.5	21.88333333	Z-233	6192	51
CAVERNA DELLE ARENE CANDIDE	Italy	44.5	8.5	LJ-4140	6090	90
TROU ARNAUD	France	44.55	5.316666667	Ly-4696	5343	50
CAPDENAC LE HAUT	France	44.58333333	2.066666667	Gif-2632	5148	73
STARCEVO	Yugoslavia	44.81666667	20.7	GrN-9034	6640	45
ROUCADOUR	France	44.83333333	1.5	GSY-36B	5190	140
SAINT MITRE:ABRI 2	France	44.88333333	5.65	MC-263	6512	79
ABRI DE PENDIMOUN	France	44.88333333	7.483333333	Gif-6571	6431	50
BAUME BOURBON	France	44.9	4.633333333	Ly-412	6068	77
LES RIVAUX	France	45.05	3.85	Ly-3708	5060	90
PAS DE LA CHARMATE	France	45.06666667	5.383333333	Ly-3785	5651	84
LA GRANDE RIVOIRE	France	45.2	5.633333333	Ly-4447	5820	140
LA LEDE DU GURP	France	45.45	-1.15	Ly-6049	5790	55
MOLINO CASAROTO	Italy	45.46666667	11.6	R-748a	5570	50
APATIN	Yugoslavia	45.66666667	18.98333333	GrN-15972	7015	50
PONTCHARAUD	France	45.78333333	3.083333333	Ly-3902	5460	130
GROTTE DE LA VIEILLE EGLISE	France	45.91666667	6.283333333	CRG-302	5295	52
LE GARDON	France	45.95	5.033333333	Ly-7010	5595	60
GROTTE DU GARDON	France	45.95	5.333333333	Ly-5514	5460	50
LA ROCHE AUX FEES	France	45.98333333	0.983333333	Ly-4536A	5525	85
RIPARO DI ROMAGNANO III	Italy	46	11.1	R-781a	6060	50
RIPARO GABAN	Italy	46.07	11.1	Bln-1778	5990	45
OSZENTIVAN	Hungary	46.2	20.18333333	Bln-477	6270	80
OLAJKUT	Hungary	46.21666667	20.25	Bln-581	6443	46
SION-PETIT CHASSEUR	Switzerland	46.23333333	7.35	B-2110	5110	57
PLANTA	Switzerland	46.23333333	7.366666667	CRG-271	6400	140
LE PETIT BOIS	France	46.71666667	4.933333333	Ly-752	5816	45
CHAMP DE LA GRANGE	France	46.76666667	2.433333333	Ly-4439	5247	61
VALLON-DES-VAUX	Switzerland	46.78333333	6.783333333	B-659	4985	35

Appendix 1: The Gkiasta European database: Datasets A to G.
Dataset G (weighted mean or median radiocarbon determination at 229 Neolithic sites).

SITE NAME	Country	Latitude (decimal degrees)	Longitude (decimal degrees)	Sample ID	Date (uncal. Bp)	SD
AUVERNIER-PORT	Switzerland	46.96666667	6.866666667	B-2559	5075	57
LA MOLLE PIERRE	France	47.05	4.75	Ly-5407	5081	52
TWANN	Switzerland	47.1	7.15	B-2698	5019	34
EGOLZWIL 4	Switzerland	47.18333333	8.016666667	H-227/277	5087	50
FENDMEILEN-VORDERELD	Switzerland	47.28333333	8.633333333	UCLA-1766 A	5100	70
SANDUN		47.33333333	2.566666667	Gif-8056	5250	100
GREIFENSEE-STOREN/WILDSBERG	Switzerland	47.36666667	8.683333333	UCLA-1824	5140	49
ZURICH	Switzerland	47.36666667	8.583333333	UCLA-1654 A	4800	100
ZURICH-BAUSCHANZE	Switzerland	47.41	8.52	UCLA-1765 A	4993	46
BAVANS	France	47.46666667	6.7	Lv-1588	6500	100
PETIT MONT	France	47.53333333	-2.9	Gif-7307	5230	70
LOCMARIAQUER ER GRAH	France	47.56666667	-2.95	Gif-7691	5293	40
PHYN	Switzerland	47.58333333	8.933333333	GrN-5957	4993	28
SANDORRESZE	Hungary	47.65	20.15	Bln-506	6141	62
COJOUX	France	47.76666667	-1.95	Gif-5456	5577	69
STUROVO	Slovakia	47.8	18.73333333	Bln-558	6215	71
ENDROD	Hungary	47.91666667	21.83333333	BM-1868	6830	60
WINDEN AM SEE	Austria	47.95	16.83333333	Bln-107	5880	71
RICOH	France	48.06666667	7.3	Ly-4581	6190	80
ROUTE DE ROUFFACH	France	48.08333333	7.333333333	Ly-3811	6062	81
LES DEPENDANCES DE DIGNY	France	48.16666667	2.416666667	Ly-5669	5980	70
SELEDIN	France	48.21666667	-3.05	Gif-1872	5149	63
LES FIEFS	France	48.23333333	2.45	Ly-6481	5611	35
TERRES DE PREPOUX	France	48.33333333	3.066666667	Ly-4507	6120	110
CHEMIN DE NAVIOT	France	48.4	6.166666667	Ly-4477	6270	60
LES REAUDINS	France	48.4	3.166666667	Ly-5549	5530	50
BARNENEZ	France	48.66666667	-3.85	Gif-1116	5345	45
PENTE DE CROUPETON	France	48.88333333	2.716666667	Gif-8356	5790	70
HIENHEIM	Germany	48.88333333	11.75	GrN-5870	6125	35
TISZAVASVARI-KERESZTFAL	Hungary	48.96666667	21.38333333	Bln-505	6226	69
LA FOSSE TOUMISE	France	49.38333333	3.4	Ly-6479	5835	85
LES FONTINETTES	France	49.38333333	3.75	Ly-2330	5942	62
LE COQ GALLEUX	France	49.4	2.8	Ly-2720	5950	120
DERRIERE LE VILLAGE	France	49.41666667	4.016666667	Ly-2322	6067	95
MOHELNICE	Czech Republic	49.78333333	16.91666667	Bln-102	6290	34
ZILGENDORF	Germany	50.16666667	10.96666667	GrN-7010	6233	37
CARN BREA	Great Britain	50.22027778	-5.246944444	BM-824	4697	60
LES SABLINS	France	50.51666667	1.65	Gif-3701	5675	85
BROKEN CAVE	Great Britain	50.53	-3.69	OxA-3205	4949	50
COUTURE DE LA CHAUSSE	Belgium	50.58333333	3.683333333	Hv-9270	6305	85
DARION	Belgium	50.65	5.183333333	Hv-10161	6180	43
PLACE SAINT LAMBERT	Belgium	50.65	5.566666667	Lv-1212	6316	27
WAREMME	Belgium	50.66666667	5.2	Lv-1693	6295	69
OLEYE	Belgium	50.68333333	5.216666667	Lv-1607	6281	38
MUDDERSHEIM	Germany	50.7	6.55	H-924a/1393	6227	38
MAIDEN CASTLE	Great Britain	50.7	-2.48	OxA-1339	4740	80
LANGWEILER	Germany	50.81666667	6.433333333	KN-1885	6210	125
ORSETT	Great Britain	50.85277778	0.348611111	BM-1213	4636	31
NIEDERMERZ	Germany	50.86666667	6.333333333	KN-2371	6190	90
LONG DOWN	Great Britain	50.88	-0.7	OxA-1151	4975	71
STEPLETON	Great Britain	50.88	-2.17	OxA-931	4675	44
OFFHAM HILL	Great Britain	50.89	-0.02	BM-1414	4807	48
HAMBLEDON HILL	Great Britain	50.90944444	-2.209166667	HAR-1802	4665	26
BURY HILL	Great Britain	50.92	-0.58	HAR-3595	4625	57
WINDMILL HILL	Great Britain	50.92	-1.88	OxA-2394	4715	21
SCORD OF BROUSTER	Great Britain	50.93	-1.66	CAR-251	4540	70
BYLANY	Czech Republic	50.96666667	15.3	GrN-4752	6170	45
DRESDEN-NICKERN	Germany	51.03333333	13.53333333	Bln-73	5880	71
WESTWARD HO!	Great Britain	51.03333333	-4.25	HAR-5640	5200	120
WINNALL DOWN	Great Britain	51.08	-1.32	HAR-2196	4725	53
MEARE HEATH	Great Britain	51.163	-2.78	HAR-2224	4583	45
SWEET TRACK, SHAPWICK	Great Britain	51.163	-2.83	HAR-1472	5070	80
HONEYGORE (WEST) TRACK, WESTHAY	Great Britain	51.18	-2.84	Q-1027	4760	30
BAKER TRACK	Great Britain	51.18	-2.82	GaK-1600	4590	40
HONEYGORE TRACK	Great Britain	51.18	-2.83	Q-909	4773	80
BAKER PLATFORM	Great Britain	51.183	-2.82	HAR-2843	4720	80
WINTERBOURNE STOKE	Great Britain	51.2	-1.85	OxA-1404	4603	77
ROBIN HOOD'S BALL	Great Britain	51.21	-1.86	OxA-1400	4613	67
BADSHOT	Great Britain	51.23	-0.79	BM-2273N1	4746	26
TOMASZOW	Poland	51.36666667	21.26666667	GrN-7050	5895	40
EASTON DOWN	Great Britain	51.39	-1.91	OxA-3759	4624	36
WEST KENNET	Great Britain	51.41	-1.85	OxA-449	4767	44
SILBURY HILL, AVEBURY	Great Britain	51.415	-1.856111111	SI-910A	4593	65
SOUTH STREET, AVEBURY	Great Britain	51.42111111	-1.870555556	BM-356	4641	64
MILLBARROW	Great Britain	51.45	-1.87	OxA-3169	4708	52
EILSLEBEN	Germany	51.48333333	11.51666667	OxA-1624	6114	47
ETON WICK	Great Britain	51.5	-0.66	BM-2533	4697	40
DUDERSTADT	Germany	51.51666667	10.28333333	H-919/889	6309	48
MISBOURNE VIADUCT	Great Britain	51.58	-0.55	OxA-601	6094	58
DRAYTON	Great Britain	51.66	-1.31	OxA-2071	4810	70
ABINGDON	Great Britain	51.68055556	-1.260833333	BM-348	4883	57
STUMBLE	Great Britain	51.73	0.73	OxA-2298	4728	49
PEAK CAMP	Great Britain	51.83	-2.15	OxA-1622	4747	35
ASCOTT-UNDER-WYCHWOOD	Great Britain	51.85472222	-1.565833333	BM-491b	4881	38
HAZLETON NORTH	Great Britain	51.87	-1.9	HAR-8349	4875	15
ELSLOO	Netherlands	51.93333333	5.766666667	GrN-2160	6150	70
CARREG COETAN	Great Britain	52.02	-4.82	CAR-391	4697	46
STANWICK SOUTH	Great Britain	52.32	-0.59	OxA-3001	4764	55
RECTORY FARM	Great Britain	52.32	-0.16	OxA-3367	4950	36

Appendix 1: The Gkiasta European database: Datasets A to G.
Dataset G (weighted mean or median radiocarbon determination at 229 Neolithic sites).

Site Name	Country	Latitude (decimal degrees)	Longitude (decimal degrees)	Sample ID	Date (uncal. Bp)	SD
TANKARDSTOWN	Ireland	52.42	-8.62	GrN-16558	5070	20
SHIPPEA HILL, PEACOCKS FARM	Great Britain	52.42888889	0.4125	Q-583	5295	120
BROOME HEATH	Great Britain	52.46777778	1.451388889	BM-757	4579	65
SWIFTERBANT	Netherlands	52.56666667	5.6	GrN-7214	6288	27
WESTEREGELN	Germany	52.96666667	11.4	Bln-42	6093	71
CAPEL EITHIN	Great Britain	53.22	-4.29	CAR-618	5350	100
GIANTS' HILLS	Great Britain	53.22	0.13	CAR-822	4970	100
LISMORE FIELDS	Great Britain	53.26	-1.94	OxA-2434	4930	70
MARKLAND GRIPS	Great Britain	53.27	-1.24	OxA-4447	4750	64
WHITWELL LONG CAIRN	Great Britain	53.27	-1.21	OxA-4176	5209	48
NEWGRANGE	Ireland	53.71	-6.49	GrN-5462	4505	41
CALLIS WOLD, BARROW 275	Great Britain	53.99	-0.73	BM-1167	4875	48
EAST HESLERTON WOLD	Great Britain	54.16	-0.57	HAR-7030	5020	70
DUNDRUM NATURE RESERVE	Ireland	54.25	-5.85	UB-412	4666	97
CARROWMORE	Ireland	54.27	-8.53	LU-1698	5040	60
BALLYGLASS (MA.13)	Ireland	54.3	-9.4	SI-1450	4590	48
TULLY, ALDERGROVE AIRPORT	Ireland	54.46	-7.82	UB-2119	4890	65
STREET HOUSE FARM, LOFTUS	Great Britain	54.56	-0.83	BM-1966R	4978	31
RUDE 2	Germany	54.61666667	9.583333333	Y-160	5644	41
SUDENSEE DAMM	Germany	54.63333333	9.583333333	GrN-6588	5954	40
BALLYNAGILLY	Ireland	54.71	-6.85	UB-18	5295	90
CARNANBANE	Ireland	54.75	-6.82	UB-534	4978	61
DOOEY'S CAIRN, BALLYMACALDRACK	Ireland	55.01	-6.41	UB-2029	4940	50
SKATEHOLM	Sweden	55.38333333	13.48333333	LU-2478	6300	100
OLBY LYNG	Denmark	55.5	12.21666667	K-1230	5265	92
MACHRIE MOOR SITE	Great Britain	55.54	-5.32	GU-2315	4840	90
VEJLEBRO	Denmark	55.68333333	9.583333333	K-1799	5362	54
PORT CHARLOTTE	Great Britain	55.73	-6.39	HAR-2406	4710	70
NEWTON	Great Britain	55.79	-6.26	GU-1951	4923	42
NETHER KINNEIL	Great Britain	56	-3.67	SRR-1486.02	4940	50
NORSLUND	Denmark	56.01666667	10.23333333	K-990	5705	85
RINGKLOSTER	Denmark	56.01666667	10.95	K-1652	5530	61
VAENGE SO	Denmark	56.13333333	10.51666667	K-3920	5474	39
HOLMEGAARD	Denmark	56.26666667	10.76666667	K-3099	6039	48
INCHTUTHIL	Great Britain	56.54	-3.44	GU-2760	5100	41
DALLADIES FARM, FETTERCAIRN	Great Britain	56.79583333	-2.610833333	SRR-289	4603	37
ERTEBOLLE	Denmark	56.8	9.166666667	K-1529	5624	45
AGEROD	Sweden	56.93333333	13.41666667	LU-1622	6770	49
BALBRIDIE TIMBER HAL	Great Britain	57.05	-2.441	OxA-1767	4919	45
BALBRIDIE	Great Britain	57.05	-2.44	GU-1037	4930	70
BROVST	Denmark	57.1	9.5	K-1613	5464	37
RAIGMORE	Great Britain	57.48	-4.19	SRR-187	4813	74
FOCHABERS - BOGHEAD MOUND -	Great Britain	57.61861111	-3.071388889	SRR-684	4895	37
THE ORD NORTH, LAIRG	Great Britain	58.01	-4.43	GU-1169	4614	57
CAMSTER LONG CAIRN, WICK PARISH	Great Britain	58.38	-3.27	GU-1707	4927	45
LOCH CALDER - TULLOCH OF ASSERY B -	Great Britain	58.53	-3.6	GU-1338	4800	60
QUANTERNESS	Great Britain	58.99972222	-3.015	Q-1294	4574	62
BARNHOUSE	Great Britain	59	-3.22	OxA-2734	4558	42
KNAP OF HOWAR, PAPA WESTRAY	Great Britain	59.35	-2.92	SRR-348	4765	70

Appendix 2 The southern African database for the first occurrence of domestic stock in the western half of southern African

2.1 Introduction:

This appendix lists the southern African sites that contain the earliest evidence of sheep bones in the western half of southern Africa.

Each site is described by 'site type'. Type 1 sites are true herder sites. Type 2 sites are site that are inhabited by hunter-gatherers with access to stock. This allocation is based on the viewpoint of the archaeologist working at the sites, gleaned from both and unpublished reports, references to these publications are listed with each site. There are sites that lack sufficient evidence to be securely labelled as either herder or non-herder and these are labelled as 'unsure'.

Sites were also given a grading that relates to the reliability of the radiocarbon dating at the site. Determinations are graded on a scale 1 to 3. Grade 1, the most reliable, is given to direct AMS determinations on sheep bone. Grade 2 is given to determinations that are in good stratigraphic association with sheep bone. Grade 3, the least reliable, is a single radiocarbon determination with poor stratigraphic association with sheep bone.

2.2 Site number 1

Site name:	Die Kelders
Co-ordinates:	34°32'8"S, 19°22'33"E
Site type:	1
AMS determinations:	YES
Determination Grading:	1
Context and properties of site:	Coastal cave, shell midden
Total sheep at site:	34 MNI
Date at which sheep first occur and MNI:	Between 1960 ± 95 and 1960 ± 85, 1 MNI.
Date at which sheep peak and MNI:	Layer 2 (multiple determinations, Table 2.1), 18 MNI.
Sheep in the final occupation unit?	Yes (7 MNI)

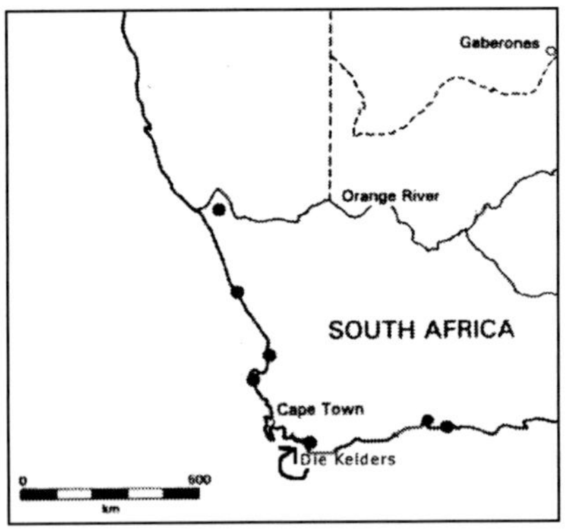

Figure 2.1 The Location of Die Kelders.

LAYER	LAB NUMBER	YEARS BP	MATERIAL
2	Gx-1685	1465 ± 100	Charcoal
	Gak-3955	1600 ± 120	Charcoal
	Gak-3877[1]	1650 ± 90	Charcoal
	Gak-3878[1]	2620 ± 100	Charcoal
3	Gak-3956	1590 ± 80	Charcoal
4	Gx-1686	2020 ± 95	Charcoal
	UW-291[1]	1480 ± 55	Charcoal
7/9	Gx-1687	1960 ± 95	Charcoal
12	Gx-1688	1960 ± 85	Charcoal

Table 2.1 Conventional radiocarbon determinations.

[1]These samples were collected from sieves and not in situ and are therefore less reliably associated with the particular layer.

LAYER	LAB NUMBER	YEARS BP	MATERIAL
7 (square B5)	OxA-3860	1325 ± 60	Sheep bone
7 (square AA8)	OxA-3961	1290 ± 60	Sheep bone

Table 2.2 AMS determinations on bone.

Layer	1	2	20	22	3	4	5	6	7	8	9	10	12
Sheep (MNI)	7	18	1?	1?	1	1	1	1?	1	1?	0	1	0
% Vol.	3	13.5			6.8	12	4.2	3	18	8.5	3.4	17.5	8

Table 2.3 Pattern of sheep occurrence and volume excavated

References: Schweitzer (1974, 1979), Schweitzer & Scott 1973, Beaumont & Vogel (1972), Wilson 1996.

2.3 Site number 2

Site name: Elands Bay Cave
Co-ordinates: 32°15'90"S, 18°15'80"E
Site type: 2
AMS determinations: none
Determination Grading: 2
Context and properties of site: Coastal cave and shell midden.
Total sheep at site: 1 MNI
Date at which sheep first occur: 1120 ± 85 bp
Date at which sheep peak: N/A (single occurrence)
Sheep in the final occupation unit? No

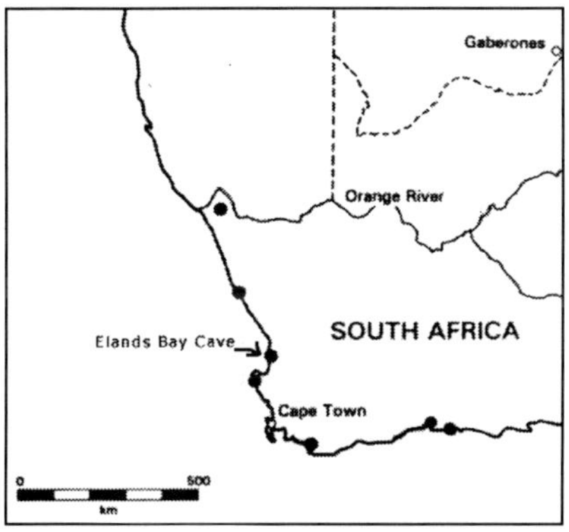

Figure 2.2 The location of Elands Bay Cave.

LAYER	LAB NUMBER	YEARS BP	MATERIAL
Hearth at surface (EDDI)	GaK-4336	20 ± 70	Charcoal from hearth
NKOM	Pta-1815	320 ± 50	Charcoal
Ash lens at top of layer 6 (ALMJ)	Gak-4335	1120 ± 85	Charcoal from hearth
JECH	Pta-5595	1310 ± 40	Charcoal
Hearth at base of layer 6 (RAYI)	GaK-4337	1520 ± 80	Charcoal form hearth
KEPL	Pta-5611	2100 ± 20	Charcoal
Bedding from base of layer 9 (JOFR)	GaK-4339	2950 ± 115	Estuarine bedding grass

Table 2.4 Conventional radiocarbon determinations

References: Parkington 1972, 1976 & 1992.

2.4 Site number 3

Site name: Boomplaas
Co-ordinates: 33°23'S, 22°11'E
Site type: 1
AMS determinations: None
Determination Grading: 3
Context and properties of site: Inland, river valley, used as a kraal during historic times.
Total sheep at site: 44 MNI/837 NISP.
Date at which sheep first occur and MNI: Before 1700 ± 55 bp (Layer BLD3/3AM), 2 MNI/26 NISP.

Date at which sheep peak and MNI: Sheep peak in DGL4 (prior to 1630 ± 50), 10 MNI/186 NISP.
Sheep in the final occupation unit? YES

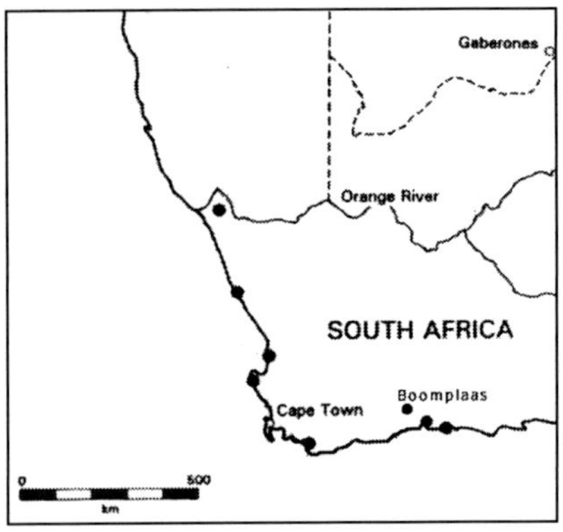

Figure 2.3 The location of Boomplaas.

LAYER	LAB NUMBER	YEARS BP	MATERIAL
Hearth at top of DGL	UW-337	1630 ± 50	Charcoal from hearth
Hearth at top of BLD2	UW-307	1510 ± 75	As above
Hearth at base of BLD2	UW-338	1700 ± 55	As above

Table 2.5 Conventional radiocarbon determinations.

References: Deacon (1976), Deacon *et al.* (1978), Deacon & Von Den Driesch (1985), Klein (1978).

2.5 Site number 4

Site name: Blombos
Co-ordinates: 34°25'S, 21°13'E
Site type: 2
AMS determinations: Yes
Determination grading: 1
Context and properties of site: Coastal cave
Total sheep at site: 3 MNI/6 NISP
Date at which sheep first occur and MNI: Layer 6, 1 MNI/1 NISP
Date at which sheep peak and MNI: N/A
Are sheep found in the final occupation unit? No

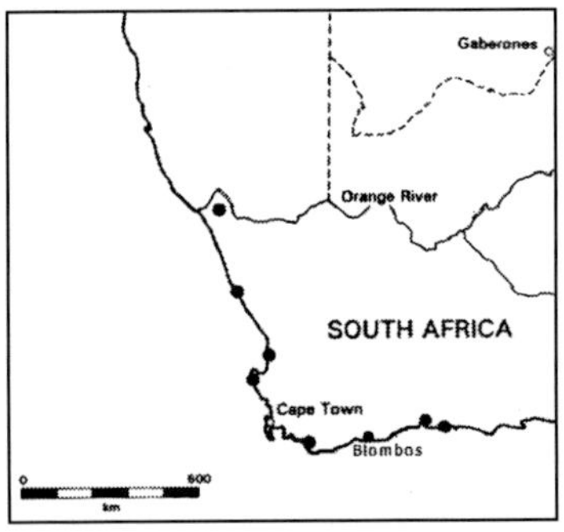

Figure 2.4 The location of Blombos

LAYER	LAB NUMBER	YEARS BP	MATERIAL
5	Pta-6185	1840 ± 50	Charcoal
	Pta-6175	1900 ± 40	Marine shell
	Pta-6246	1880 ± 50	Marine shell
	Pta-6247	1940 ± 50	Marine shell

Table 2.6 Conventional radiocarbon determinations.

LAYER	LAB NUMBER	YEARS BP	MATERIAL
5	OxA-4543	1960 ± 50	Sheep bone
6	OxA-4544	1880 ± 55	Sheep bone

Table 2.7 AMS determinations on sheep bone.

References: Henshilwood (1995, 1996)

2.6 Site number 5

Site name:	Spoegrivier
Co-ordinates:	30°29'40"S, 17°22'E
Site type:	1 (initially hunter-gatherer)
AMS determinations:	YES
Determination grading:	1
Context and properties of site:	Cave site (shell midden), coastal and riverine.
Total sheep at site:	13 MNI
Date at which sheep first occur and MNI:	Layer FBS 2105 ± 65 bp, 1 MNI/1 NISP
Date at which sheep peak and MNI:	Layer Pat (undated), 2 MNI/22 NISP
Sheep in the final occupation unit?	Yes

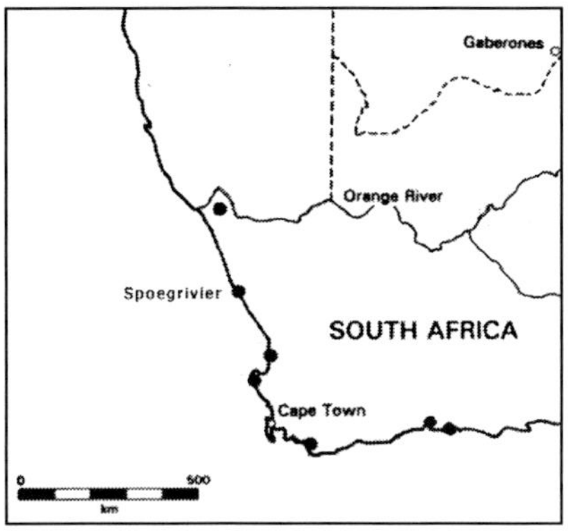

Figure 2.5 The location of Spoegrivier.

LAYER	LAB NUMBER	YEARS BP	MATERIAL
Hearth in twiggy (20cm below surface)	Pta-4753	1390 ± 50	Charcoal
Hearth 12 in Shelly patch 1 (91cm below surface)	Pta-4745	1920 ± 40	Charcoal

Table 2.8 Conventional radiocarbon determinations.

LAYER	LAB NUMBER	YEARS BP	MATERIAL
FBS C9	OxA-3862	2105 ± 65	Sheep bone

Table 2.9 AMS determinations.

References: Avery (1992), Vogel *et al.* (1997), Webley (1992).

2.7 Site number 6

Site name:	Jakkalsberg A and B
Co-ordinates:	28°10'30"S, 16°53'20"E
Site type:	1
AMS determinations:	None
Determination grading:	2
Context and properties of site:	Situated on the banks of the Orange River
Total sheep at site:	Jakkalsberg A - Sheep: 7 MNI/54 MISP - Sheep/goats: 14 MNI/446 NISP Jakkalsberg B - Sheep: 7 MNI/21 NISP - Sheep/goats: 20 MNI/253 NISP
Date at which sheep first occur and MNI:	1420 ± 25 b.p., Sheep: 1 MNI/1 NISP and sheep/goats: 2 MNI/9 NISP
Date at which sheep peak and MNI:	N/A
Sheep in the final occupation unit?	Yes, sheep are found throughout.

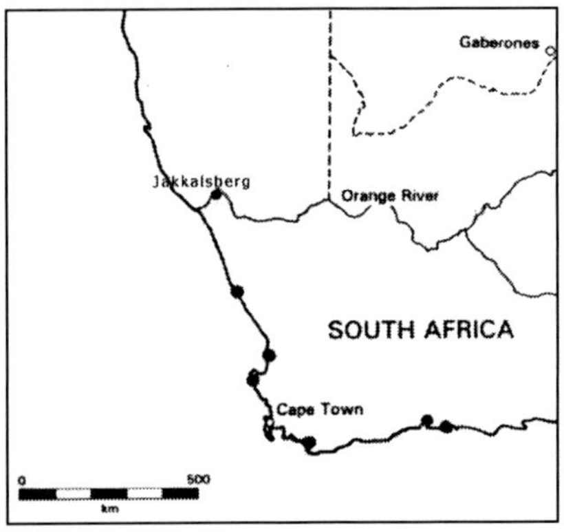

Figure 2.6 The location of Jakkalsberg.

SITE	LAYER	LAB NUMBER	YEARS BP	MATERIAL
A	Hearth 1	Pta-5958	1330 ± 60	Charcoal
	Hearth 2	Pta-6100	1300 ± 25	Charcoal
B	Hearth 1 in K11	Pta-6122	1420 ± 25	Charcoal
	Hearth 2 in MOU	Pta-6101	1380 ± 50	Charcoal

Table 2.10 Radiocarbon determinations.

References: Brink & Webley (1996), Webley (1997).

2.8 Site number 7

Site name: Kasteelberg B
Co-ordinates: 32°48'8"S, 17°56'8"E
Site type: 1
AMS determinations: Yes
Determination grading: 2
Total sheep at site: 138 MNI/6008 NISP
Date at which sheep first occur and MNI: 1300 ± 60 bp, 9 MNI/525 NISP
Date at which sheep peak and MNI: 910 ± 40 bp, 19 MNI/903 NISP
Are sheep found in the final occupation unit? Yes

LAYER	LAB NUMBER	YEARS BP	MATERIAL
2-21 cm	Pta-3737	190 ± 40	Charcoal
30-34 cm	Pta-3742	920 ± 40	Charcoal
45-50 cm	Pta-3747	880 ± 50	Charcoal
91 cm	Pta-3787	910 ± 50	Charcoal
125 cm	Pta-3994	1200 ± 45	Charcoal
	Pta-3998	1220 ± 45	Charcoal
170 cm	Pta-3995	1300 ± 60	Charcoal

Table 2.11 Conventional radiocarbon determinations.

References: Klein & Cruz-Uribe 1989, Smith 1984, 1987, & 1992a.

2.9 Site number 8

Site name: Kasteelberg A
Co-ordinates: 32°48'8"S, 17°56'8"E
Site type: 1
AMS determinations: Yes
Determination grading: 1
Context and properties of site: Coastal sites around large boulder outcrops
Total sheep at site: 34 MNI/268 NISP
Date at which sheep first occur and MNI: 1860 ± 60 b.p.
Date at which sheep peak and MNI: Unknown
Sheep in the final occupation unit? Unknown

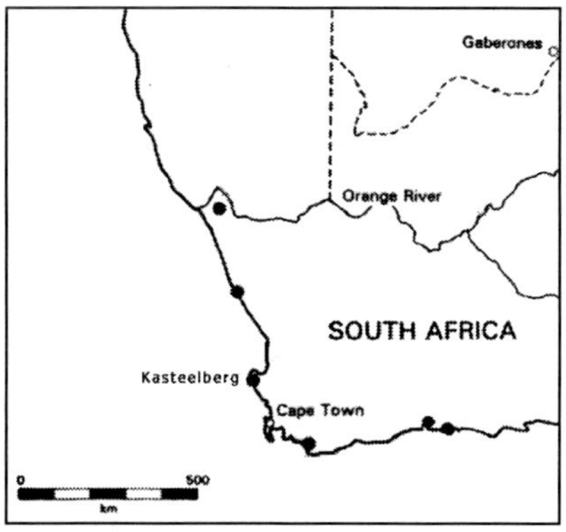

Figure 2.7 The location of Kasteelberg.

LAYER	LAB NUMBER	YEARS BP	MATERIAL
	Pta-3711	1860 ± 60	Charcoal
	Pta-3461	1790 ± 40	Charcoal

Table 2.12 Conventional radiocarbon determination

LAYER	LAB NUMBER	YEARS BP	MATERIAL
20-30cm A20	OxA-3864	1630 ± 60	Sheep bone
As above	OxA-3865	1430 ± 55	Sheep bone

Table 2.13 AMS determinations.

References: Klein & Cruz-Uribe (1989), Smith (1984, 1987, & 1992a).

2.10 Site number 9

Site name: Witklip
Co-ordinates: 32°55'1"S, 17°59E
Site type: 2
AMS determinations: None
Determination grading: 2
Context and properties of site: Inland shell midden site in small rock overhang.
Total sheep at site: Smith does not give an exact value, he combines sheep with duiker in the category small-medium bovids. This category has MNI 6.

Date at which sheep first occur and MNI: 1860 ± 50 bp.
Date at which sheep peak and MNI: Unknown
Are sheep found in the final occupation unit? No

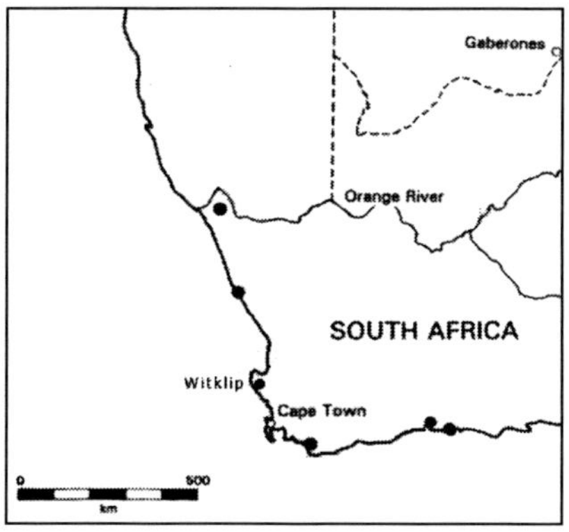

Figure 2.8 The location of Witklip.

LAYER	LAB NUMBER	YEARS BP	MATERIAL
0.12-0.15 m, sq.I1 Orange loam	Pta-5467	500 ± 50	Charcoal
0.15-0.3m, sq.I1 Hearth below orange loam	Pta-5467	330 ± 45	Charcoal
0.4 m, sq.F1 Grey ashy soil with shell	Pta-4608	1380 ± 50	Charcoal
0.8 m, sq.F1 Hearth below light brown soil	Pta-4609	1860 ± 50	Charcoal
1.3 m, sq.F1 Pale brown sand	Pta-4607	3060 ± 60	Charcoal

Table 2.14 Conventional radiocarbon determinations.

References: Smith *et al.* (1991 & 1992)

2.11 Site number 10

Site name:	Byneskranskop
Co-ordinates:	34°35'S, 19°28'E
Site type:	1 in final occupation horizon
AMS determinations:	Yes
Determination grading:	1
Context and properties of site:	Coastal cave site
Total sheep at site:	11 MNI
Date at which sheep first occur and MNI:	Layer 1, 1880 ± 50 bp, 11 MNI
Date at which sheep peak and MNI:	Sheep only occur in Layer 1
Are sheep found in the final occupation unit?	Yes

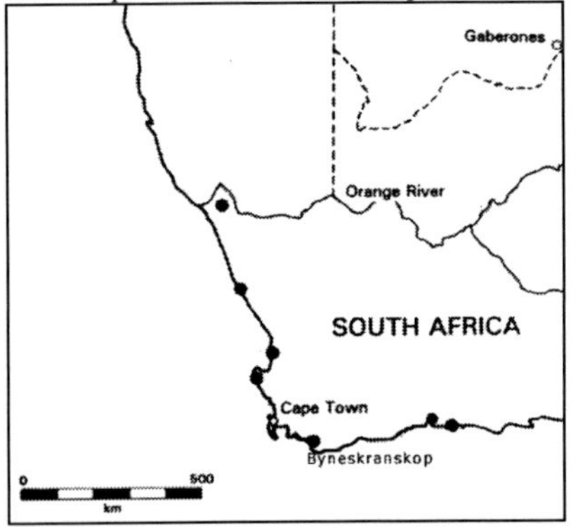

Figure 2.9 The location of Byneskranskop.

LAYER	LAB NUMBER	YEARS BP	MATERIAL
1,Upper-O31,1b	Pta-1864	255 ± 50	Charcoal
1,Upper-P32,1b	Pta-1866	535 ± 50	Charcoal
1,Middle-CLA	Pta-1865	1880 ± 50	Charcoal
Bottom-EVA	Pta-1631	3220 ± 45	Charcoal
	Pta-1567	3400 ± 55	Charcoal
	Pta-1571	3900 ± 60	Charcoal

Table 2.15 Conventional radiocarbon determinations.

LAYER	LAB NUMBER	YEARS BP	MATERIAL
1, O30	OxA-3863	1370 ± 60	Sheep bone

Table 2.16 AMS determinations.

References: Schweitzer & Wilson (1982), Sealy & Yates (1994).

2.12 Site number 11

Site name:	Nelson Bay Cave
Co-ordinates:	34°6'20"S, 23°22'50"E
Site type:	2
AMS determinations:	Yes
Determination grading:	1
Context and properties of site:	Coastal cave site
Total sheep at site:	4 MNI/14 NISP

Date at which sheep first occur and MNI:	Units 62-30, 1930 ± 60, 1 MNI/1 NISP
Date at which sheep peak and MNI:	Units 22-29, 3 MNI/13 NISP
Are sheep found in the final occupation unit?	Unclear as MNI's are given for groups of units.

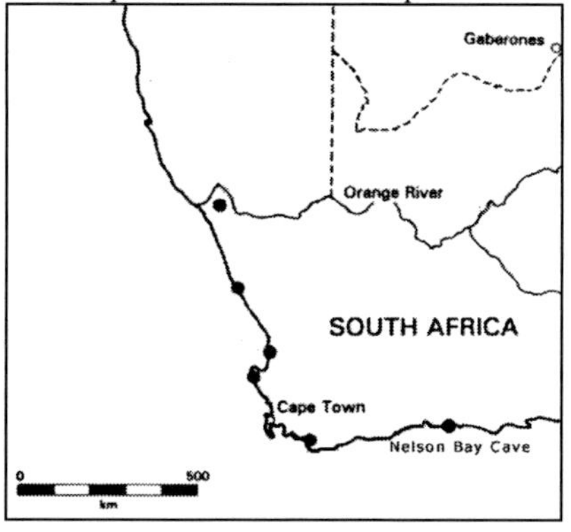

Figure 2.10 The location of Nelson Bay Cave.

LAYER	LAB NUMBER	YEARS BP	MATERIAL
Unit 22a	Pta-1361	455 ± 30	Charcoal
Unit 22b	Pta-1360	1500 ± 35	Charcoal
Unit 23	Pta-3361	850 ± 50	Charcoal
Unit 24	Pta-3362	650 ± 50	Charcoal
Unit 28	Pta-3363	2560 ± 60	Charcoal
Unit 29	GrN-5703	1930 ± 60	Charcoal
Unit 30	Pta-1363	2085 ± 35	Charcoal
Unit 6	Pta-2921	2459 ± 60	Charcoal
Unit 18	GrN-5715	2540 ± 50	Charcoal

Table 2.17 Conventional radiocarbon determinations.

LAYER	LAB NUMBER	YEARS BP	MATERIAL
Unit 29	OxA-873	1100 ± 80	Sheep bone

Table 2.18 AMS determinations.

References: Beaumont & Vogel (1972), Deacon (1978), Inskeep (1972, 1987), Inskeep & Vogel (1985), Klein (1972), Sealy & Yates (1994).

2.13 Site number 12

Site name:	Wonderwerk
Co-ordinates:	27°50'45"S, 23°33'19"E
Site type:	2
AMS determinations:	Yes
Determination grading:	3
Context and properties of site:	Large inland cave site.
Total sheep at site:	3 MNI
Date at which sheep first occur and MNI:	After 1210 ± 50 bp (Layer 2a), 1 MNI.
Date at which sheep peak and MNI:	Layer 1 (undated), 2 MNI
Sheep in the final occupation unit?	Yes

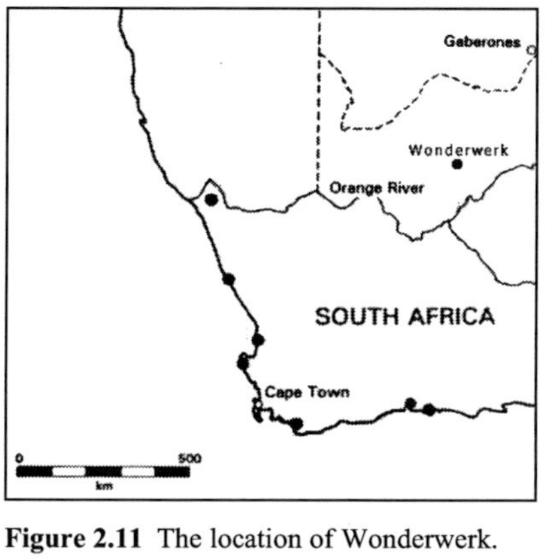

Figure 2.11 The location of Wonderwerk.

LAYER	LAB NUMBER	YEARS BP	MATERIAL
Layer 2b (10cm)	Pta-2779	1210 ± 50	Charcoal (assoc. with *Ovis* bones
Top of layer 3 (10 to 15 cm)	Pta-2542	1890 ± 50	Charcoal
Base of layer 3 (15 to 20cm)	Pta-2543	2910 ± 60	Charcoal
Layer 3a/3b interface (29 cm)	Pta-2785	3990 ± 60	Charcoal

Table 2.19 Conventional radiocarbon determination.

LAYER	LAB NUMBER	YEARS BP	MATERIAL
Top of layer 3a	OxA-622	460 ± 120	Sheep hair

Table 2.20 AMS determination.

References: Gowlett *et al.* (1987), Sealy & Yates (1994), Thackeray & Humphreys 1983, Vogel *et al.* (1986).

2.14 Site number 13

Site name: Jubilee Shelter
Co-ordinates: 25°42'S, 27°55E
Site type: Unknown
AMS determinations: Yes
Determination grading: 1
Context and properties of site: Interior cave site
Total sheep at site: 1 MNI
Date at which sheep first occur and MNI: A sheep bone occurs in B C4
Date at which sheep peak and MNI: N/A
Sheep in the final occupation unit? No

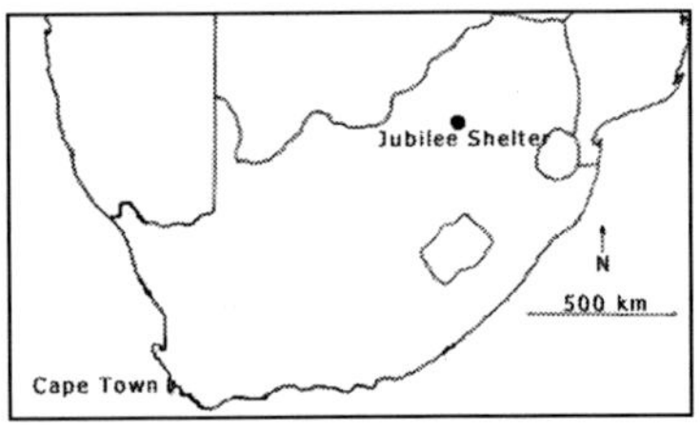

Figure 2.12 The location of Jubilee Shelter.

LAYER	LAB NUMBER	YEARS BP	MATERIAL
RR D2	WITS-1399	1350 ± 120	Charcoal
MS C2	WITS-1381	1550 ± 40	Charcoal
B C4	WITS-1398	1840 ± 50	Charcoal

Table 2.21 Conventional radiocarbon determinations.

LAYER	LAB NUMBER	YEARS BP	MATERIAL
B C4	OxA-6039	310 ± 50	Sheep bone

Table 2.22 AMS determinations.

References: Sealy & Yates (1994),Turner (1986), Vogel *et al.* (1986), Wadley (1996).

2.15 Site number 14

Site name: Blinkklipkop
Co-ordinates: 28°18'10"S, 23°07'00"E
Site type: 1
AMS determinations: None
Determination grading: 2
Context and properties of site: A prehistoric specularite mine in the interior of South Africa.
Total sheep at site: 28 MNI
Date at which sheep first occur and MNI: Between 1160 ± 50 bp and 1150 ± 40 bp (in layer 5q), 1 MNI
Date at which sheep peak and MNI: Layer 4b (undated) 6 MNI sheep/goats
Sheep in the final occupation unit? Yes

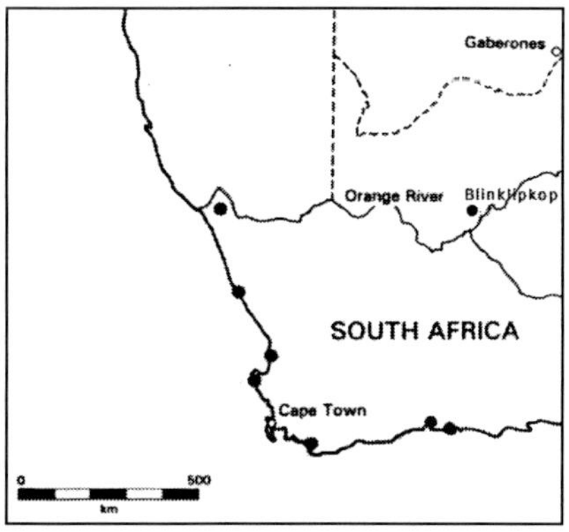

Figure 2.13 The location of Blinkklipkop.

LAYER	LAB NUMBER	YEARS BP	MATERIAL
3b	Pta-2841	80 ± 35	Grass
4a base	Pta-2833	260 ± 45	Charcoal
5c	Pta-2839	280 ± 50	Charcoal
5j	Pta-3419	830 ± 45	Charcoal
5p	Pta-2840	1160 ± 50	Charcoal
5 s/t	Pta-2835	1150 ± 40	Charcoal

Table 2.23 Conventional radiocarbon determinations.

References: Thackeray & Humphreys (1984), Thackeray, Thackeray & Beaumont (1983).

2.16 Site number 15

Site name:	/Ai Tomas
Co-ordinates:	28°35'S, 17°05'E
Site type:	1?
AMS determinations:	None (failed due to insufficient bone protein).
Determination grading:	3
Context and properties of site:	Two large boulder clusters 100 m apart from each other, 50 km east of the Atlantic ocean.
Total sheep at site:	12 (sheep/goat combined) MNI
Date at which sheep first occur and MNI:	1980 ± 120 bp, Layer L4S, 2 MNI
Date at which sheep peak:	No peak
Are sheep in the final occupation unit?	Yes

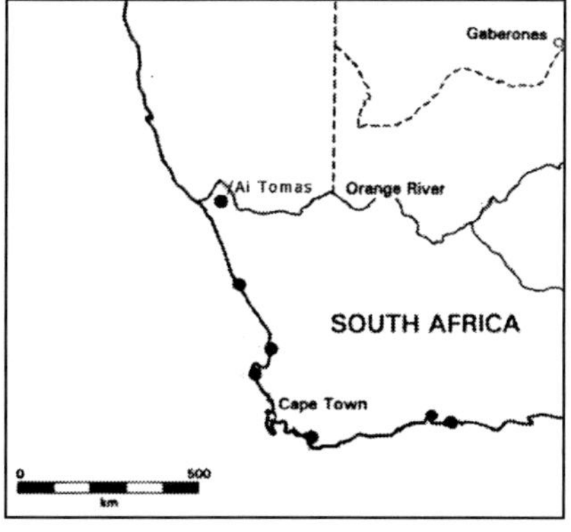

Figure 2.14 The location of /Ai Tomas.

LAYER	LAB NUMBER	YEARS BP	MATERIAL
Area 1, N33, layer BL (depth 5 cm)	Pta-5458	420 ± 50	Charcoal
Area 2,P32, layer SLS2 (depth 16 cm)	Pta-5452	330 ± 45	Charcoal
Area 3,Y21, layer LBS3 (equivalent to L4S)	Pta-5530	1980 ± 120	Charred bone

Table 2.24 Conventional radiocarbon determinations.

References: Webley (1992a, 1992b).

2.17 Site number 16

Site name:	Hawston
Co-ordinates:	34°25'S, 19°08'E
Site type:	1?
AMS determinations:	None
Determination grading:	3
Context and properties of site:	A coastal open-air shell midden
Total sheep at site:	No report given.
Date at which sheep first occur and MNI:	Sheep occur between the two dated layers (Table 2.25)
Date at which sheep peak and MNI:	Unknown (no site details given)
Are sheep found in the final occupation unit?	Unknown

Figure 2.15 The location of Hawston.

LAYER	LAB NUMBER	YEARS BP	MATERIAL
27 cm (Hawston 1A)	Pta-834	1860 ± 60	Charcoal
80 cm (Hawston 1B)	Pta-835	1900 ± 40	Charcoal

Table 2.25 Conventional radiocarbon determination.

References: Avery (1974, 1975)

2.18 Site number 17

Site name:	Scott's Cave
Co-ordinates:	25°43'E, 33°44'S
Site type:	2
AMS determinations:	None
Determination grading:	3
Context and properties of site:	Small inland rock shelter
Total sheep at site:	2 MNI
Date at which sheep first occur:	Spit 5, (earliest occupation level) 1190 ± 100 bp, 1 MNI
Date at which sheep peak and MNI:	N/A
Are sheep found in the final occupation unit?	No

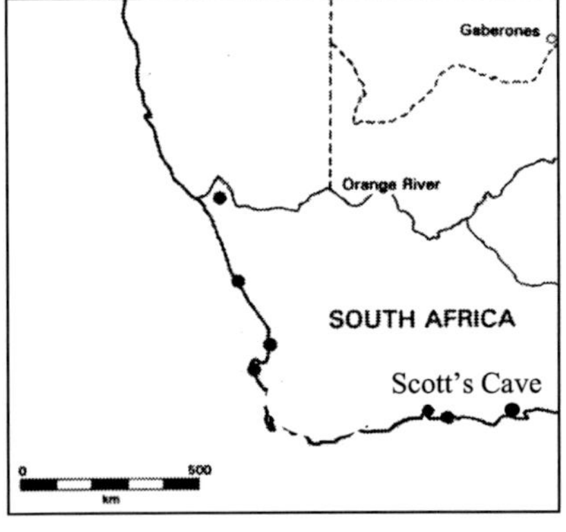

Figure 2.16 The location of Scott's Cave.

LAYER	LAB NUMBER	YEARS BP	MATERIAL
Top 8 cm	Y-1425	360 ± 80	Charcoal
60 cm depth	SR-82	1190 ± 100	Charcoal

Table 2.26 Conventional radiocarbon determination.

References: Beaumont & Vogel (1972), Deacon (1967), Klein & Scott (1974).

2.19 Site number 18

Site name:	Dikbosch 1 and 2
Co-ordinates:	28°39'08"S, 23°54'09"E
Site type:	2
AMS determinations:	None
Date grading:	3
Context and properties of site:	Two rock shelters 1 km apart.
Total sheep at site:	Dikbosch 1 (area A): 1 MNI, Dikbosch 2: 1 MNI
Date at which sheep first occur and MNI:	Dikbosch 1 (area A): Level III, before 3090 ± 60 bp (this date is disputed). Dikbosch 2, Level I, post-dating 570 ± 45 bp, 1 MNI
Date at which sheep peak and MNI:	N/A
Sheep found in the final occupation unit?	No

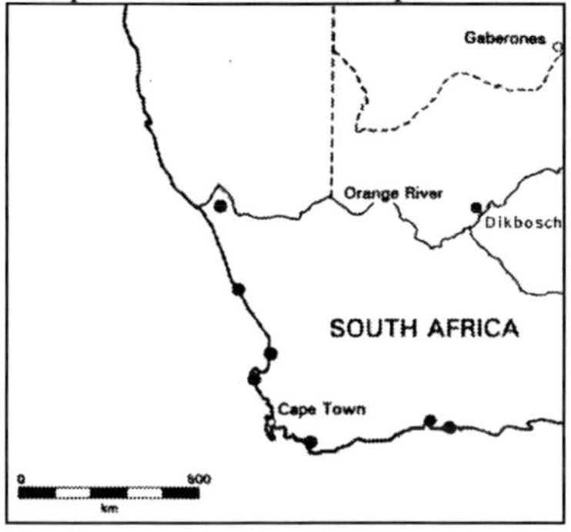

Figure 2.17 The location of Dikbosch

SITE	LAYER	LAB NUMBER	YEARS BP	MATERIAL
1 (area A) (10 to 15 cm)	II	Pta-1065	3110 ± 60	Charcoal
1 (area B) (25 cm)	II	Pta-3413	1720 ± 40	Charcoal
1 (area B) (35 cm)	III	Pta-3412	1570 ± 40	Charcoal
1 (area B) (55 cm)	V	Pta-3411	8010 ± 60	Charcoal
2	III (0.45m depth)	Pta-1507	570 ± 45	Charcoal

Table 2.27 Conventional radiocarbon determinations.

References: Humphreys (1974), Thackeray & Humphreys (1983), Vogel *et al.* (1986).

2.20 Site number 19

Site name:	Tortoise Cave
Co-ordinates:	32°18'02"S, 18°23'10"E
Site type:	2
AMS determinations:	None
Determination grading:	3
Context and properties of site:	Small rock shelter proximity to coast.
Total sheep at site:	The excavator of this site used the system of allocating X to denote the presence of fauna rather than actual numbers. 'X' denotes the presence of sheep. 'XX' means noteworthy and 'XXX' means significant.
Date at which sheep first occur and MNI:	4330 ± 50 bp, Layers 10-13, 'X'
Date at which sheep peak and MNI:	Layers 1-3 (see Table 2.28 for date range, 'XX'.
Sheep found in the final occupation unit?	XX

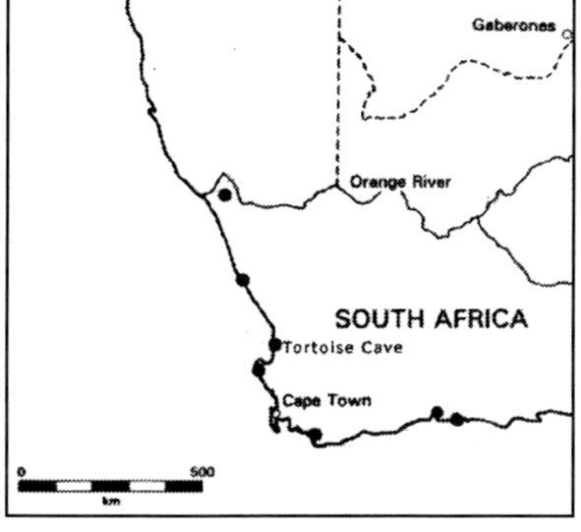

Figure 2.18 The location of Tortoise cave.

LAYER	LAB NUMBER	YEARS BP	MATERIAL
1a, Fold-up bedding/J3	Pta-3600	760 ± 50	Restio grass
2b,Fran/K2	Pta-3309	1580 ± 50	Charcoal
3, ABDII/J2	Pta-3310	1620 ± 50	Charcoal
3, Turner/Y2	Pta-3311	1610 ± 50	Charcoal
3, Aluin/X3	Pta-3312	1680 ± 50	Charcoal
6,SM2/AA2	Pta-3604	3520 ± 60	Charcoal
10, Melanie/S1	Pta-3608	4190 ± 60	Charcoal

Table 2.28 Conventional radiocarbon determination.

References: Robey (1984, 1987).

2.21 Site number 20

Site name: Limerock 1 and 2
Co-ordinates: 28°33'01"S, 24°00'01"E
Site type: 2?
AMS determinations: None
Determination grading: 3
Context and properties of site: Two inland rock shelters, 15m apart.
Total sheep at site: LR1: 1 MNI (sheep/goat)
LR2: 2 MNI (sheep/goat)
Date at which sheep first occur: LR1: 1620 ± 50 bp, 1 MNI
LR2: 1720 ± 40 bp, 1 MNI
Date at which sheep peak and MNI: N/A
Sheep in the final occupation unit? LR 2, MNI 1

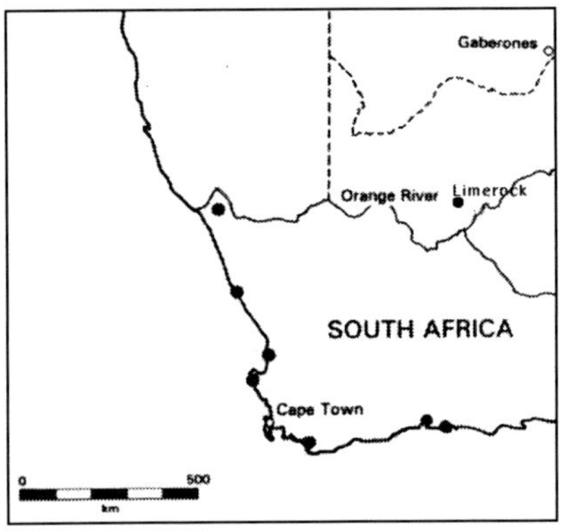

Figure 2.19 The location of Limerock 1/2.

SITE	LAYER	LAB NUMBER	YEARS BP	MATERIAL
1	III C (45 cm)	Pta-1621	1620 ± 50	Charcoal
2	IV A (47cm)	Pta-2095	1720 ± 40	Charcoal
2	V A (55 – 60 cm)	Pta-1759	1430 ± 50	Charcoal

Table 2.29 Conventional radiocarbon determinations.

References: Thackeray & Humphreys (1983), Vogel *et al.* (1986)

2.22 Site number 21

Site name: Drie Susters Main
Co-ordinates: 33°06'03"S, 18°00'02"E
Site type: 2
AMS determinations: none
Determination grading: 3
Context and properties of site: Open coastal shell midden site
Total sheep at site: 1 MNI/3 NISP
Date at which sheep first occur: 1050 ± 60 bp
Date at which sheep peak and MNI: Faunal categories are given as totals for the site as a whole.
Are sheep found in the final occupation unit? N/A

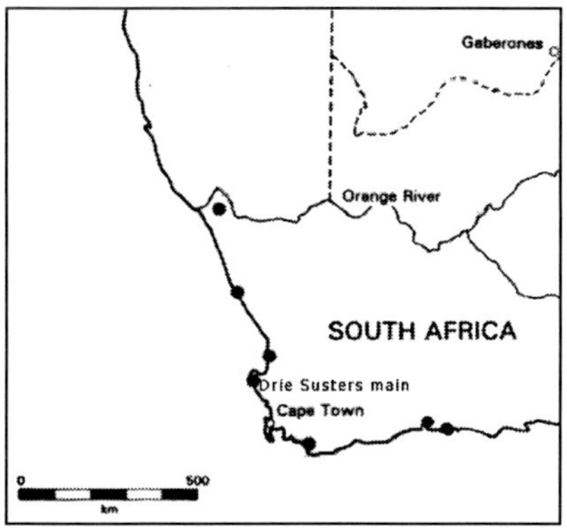

Figure 2.20 The location of Drie Susters Main.

LAYER	LAB NUMBER	YEARS BP	MATERIAL
Bottom of sq. C6 (on bedrock)	Pta-5478	1050 ± 60	Charcoal

Table 2.30 Conventional radiocarbon determinations.

References: Smith *et al.* (1991)

2.23 Site number 22

Site name:	Pearly Beach Area
Co-ordinates:	34°40'S, 19°30'E
Site type:	1?
AMS determinations:	none
Determination grading:	3
Context and properties of site:	Area of open-air coastal shell middens.
Total sheep at site:	Unknown
Date at which sheep first occur:	1680 ± 50 bp
Date at which sheep peak and MNI:	Unknown
Are sheep found in the final occupation unit?	Unknown

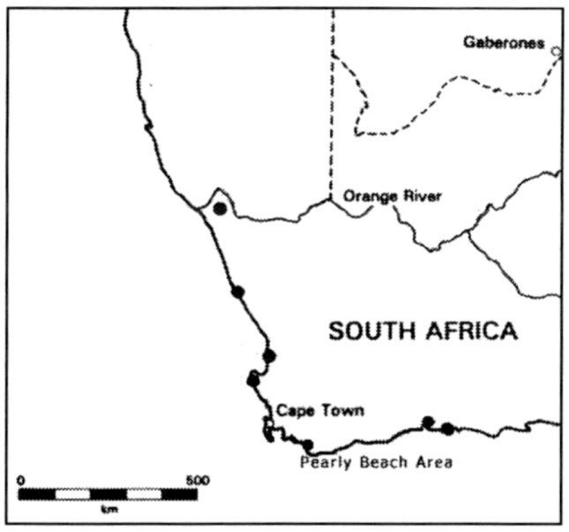

Figure 2.21 The location of Pearly Beach Area.

LAYER	LAB NUMBER	YEARS BP	MATERIAL
	Pta-1069	1680 ± 50	Charcoal

Table 2.31 Conventional radiocarbon determination.

References: Avery (1974)

2.24 Site number 23

Site name:	Smitswinkelbaai
Co-ordinates:	34°16'S, 18°28'E
Site type:	1
AMS determinations:	none
Date grading:	3
Context and properties of site:	Coastal cave site with difficult access from steep slope above
Total sheep at site:	9 MNI
Date at which sheep first occur:	Layer 6 (undated), 1 MNI

Date at which sheep peak and MNI:	Layer 3 (undated), 3 MNI
Sheep found in the final occupation unit?	Yes

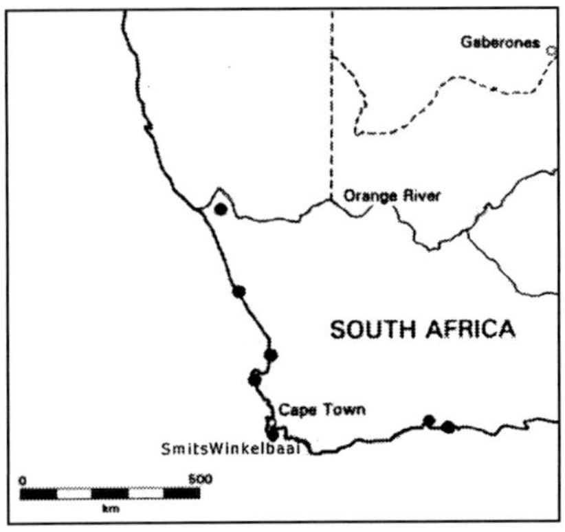

Figure 2.22 The location of Smitswinkelbaai.

LAYER	LAB NUMBER	YEARS BP	MATERIAL
4	Pta-2200	1175 ± 35	Charcoal
9	Pta-2198	1420 ± 35	Charcoal

Table 2.32 Conventional radiocarbon determination.

References: Marean (1985), Poggenpoel & Robertshaw (1981).

2.25 Site number 24

Site name:	Steenbokfontein
Co-ordinates:	32°09'42"S, 18°20'00"E
Site type:	2
AMS determinations:	none
Determination grading:	3
Context and properties of site:	Large cave close to the coast
Total sheep at site:	3 MNI
Date at which sheep first occur:	2360 ± 45 bp, 1 MNI
Date at which sheep peak and MNI:	N/A, one sheep bone occurs in each layer from layer 2 to the surface layer.
Sheep in the final occupation unit?	Yes

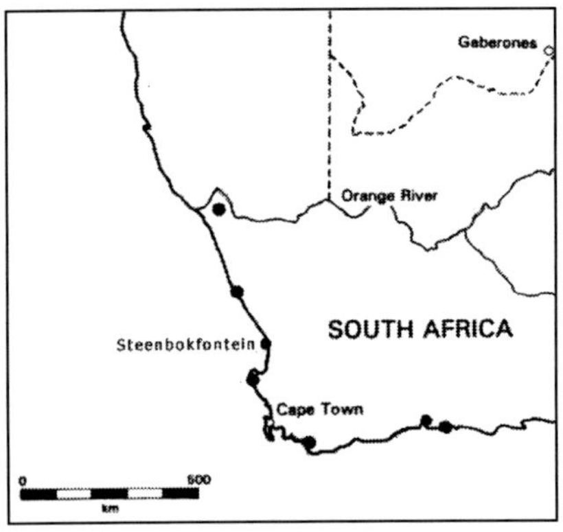

Figure 2.23 The location of Steenbokfontein.

LAYER	LAB NUMBER	YEARS BP	MATERIAL
1 Top	Pta-6136	2200 ± 60	charcoal
1 Bottom	Pta-6424	2200 ± 50	charcoal
2	Pta-6136	2360 ± 45	Microfaunal bone

Table 2.33 Conventional radiocarbon determination.

References: Jerardino & Yates (1996).

2.26 Site number 25

Site name:	Lame Sheep Shelter
Co-ordinates:	31°25S, 24°40E
Site type:	Unsure
AMS determinations:	none

Determination grading:	3
Context and properties of site:	Inland rock shelter
Total sheep at site:	Unknown
Date at which sheep first occur and MNI:	1350 ± 50 bp, unknown MNI
Date at which sheep peak and MNI:	Unknown
Sheep found in the final occupation unit?	Yes

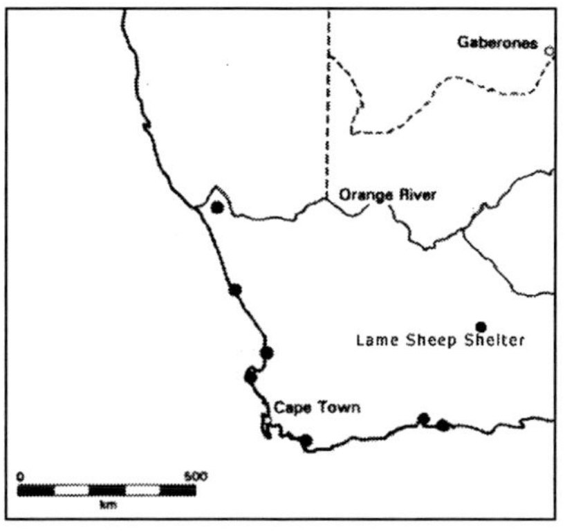

Figure 2.24 The location of Lame Sheep Shelter.

LAYER	LAB NUMBER	YEARS BP	MATERIAL
Unknown	Pta-6303	1350 ± 50	Charcoal

Table 2.34 Conventional radiocarbon determination.

References: Plug *et al.* (1994), Sampson & Vogel (1995).

2.27 Site number 26

Site name:	Blydefontein
Co-ordinates:	30°50S, 25°20E
Site type:	Unknown
AMS determinations:	none
Determination grading:	3
Context and properties of site:	Inland rock shelter
Total sheep at site:	There are no sheep bones at this site. In the words of the excavator Bousman (1998:142), "Phosphate analysis of sediments at Blydefontein indicates that layer CPB is composed primarily of dung and sheep are the most likely source".
Date at which sheep first occur and MNI:	Unknown
Date at which sheep peak and MNI:	Unknown
Are sheep found in the final occupation unit?	Unknown

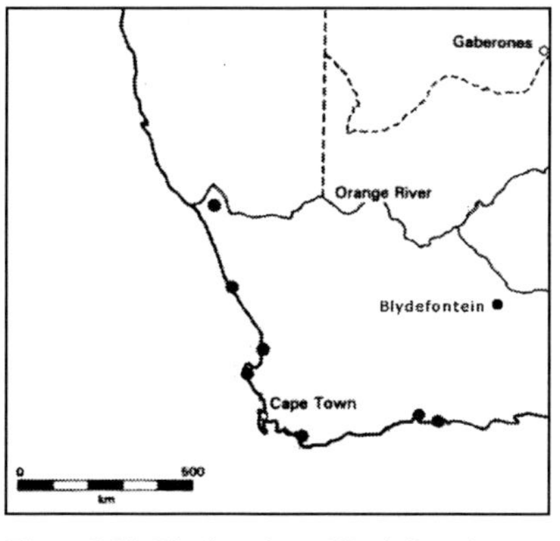

Figure 2.25 The location of Bydefontein.

LAYER	LAB NUMBER	YEARS BP	MATERIAL
Layer CPB	Smu-1925	1255 ± 109	Charcoal
Immediately below layer CPB	Smu-1850	1305 ± 31	Charcoal

Table 2.35 Conventional radiocarbon determinations.

References: Bousman (1998)

2.28 Site number 27

Site name:	Klein kliphuis
Co-ordinates:	32°07'S, 18°51'E
Site type:	2
AMS determinations:	none
Date grading:	3
Context and properties of site:	Inland rock shelter at the foothills of the Cedarberg mountains and near the Klein kliphuis river.
Total sheep at site:	1 MNI/3 NISP
Date at which sheep first occur:	Upper Unit A
Date at which sheep peak and MNI:	N/A
Sheep in the final occupation unit?	No

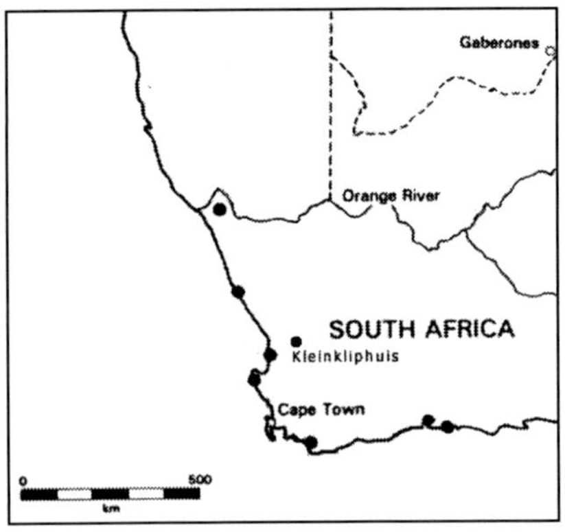

Figure 2.26 The location of KleinKliphuis.

LAYER	LAB NUMBER	YEARS BP	MATERIAL
70 mm below surface, Bedding unit	Pta-4672	1230 ± 50	Charcoal
180 mm below surface, Unit C	Pta-4671	1990 ± 50	Charcoal

Table 2.36 Conventional radiocarbon determination.

References: Avery (1994), Van Rijssen (1994).

2.29 Sites for which excavation details were not available at the time of data collection but which are known to contain sheep bones:

Bethelsklip (Webley 1992b)
De Hangen (Parkington & Poggenpoel 1987)
Diepkloof (Parkington 1976)
Heuningklip (Smith *et al.* 1992)
Haaskraal shelter (Plug *et al.* 1994)

Appendix 3: The Gkiasta European database: Late Mesolithic site list (107 sites) used to construct the isochron map (Chapter 4).

Site Name	Country	latitude (decimal degrees)	Longitude (decimal degrees)	Sample ID	Radiocarbon determination (BP)	SD
Serovo	Lithuania	54	22.41666667	Le-1181	4020	80
Serteja 2B	Russian Federation	56.33333333	31.5	TA-632	4080	60
Usviaty 4B	Russian Federation	56	31.83333333	TA-203	4100	70
Serteja 2B	Russian Federation	56.33333333	31.5	TA-633	4120	60
Serteja 2B	Russian Federation	56.33333333	31.5	TA-817	4150	80
Nainiekste	Latvia	57.91666667	27.53333333	Le-648	4170	130
Serovo	Lithuania	54	22.41666667	Le-848	4180	50
Sventoji Latvia	Latvia	56.08333333	21.66666667	Vib-1	4190	80
Usviaty 4B	Russian Federation	56	31.83333333	TA-202	4210	70
Serovo	Lithuania	54	22.41666667	Le-1176	4240	90
Piestinia	Latvia	57.91666667	27.5	Le-867	4250	50
Sarnate	Latvia	57.41666667	21.55	Vs-318	4260	90
Krivina 1B	Belarus	55.16666667	32	GrN-5125	4270	40
Usviaty 4B	Russian Federation	56	31.83333333	TA-243	4310	80
Zvidze	Latvia	57.83333333	27.25	TA-675	4370	80
Sventoji Latvia	Latvia	56.08333333	21.66666667	Vib-9	4410	70
Sarnate	Latvia	57.41666667	21.55	Le-814	4510	110
Usviaty 4B	Russian Federation	56	31.83333333	TA-244	4510	70
Piestinia	Latvia	57.91666667	27.5	Le-748	4520	120
Zvidze	Latvia	57.83333333	27.25	TA-674	4540	60
Usviaty 4B	Russian Federation	56	31.83333333	TA-105	4570	70
Sarnate	Latvia	57.41666667	21.55	TA-265	4630	70
Kepa	Estonia	58.08333333	27.58333333	TA-815	4640	100
Kepa	Estonia	58.08333333	27.58333333	TA-724	4740	60
Kepa	Estonia	58.08333333	27.58333333	TA-478	4760	80
Serovo	Lithuania	54	22.41666667	Le-1386	4870	80
Serovo	Lithuania	54	22.41666667	Le-1387	4900	80
Zvidze	Latvia	57.83333333	27.25	TA-395	4905	70
Hein	Norway	60.36666667	7.733333333	T-1076	4920	90
Serovo	Lithuania	54	22.41666667	Le-1388	4920	80
FAVERDINES	France	46.65	2.433333333	Ly-3837	5010	120
Hein	Norway	60.36666667	7.733333333	T-1001	5020	100
Serovo	Lithuania	54	22.41666667	LE-1389	5100	60
Serovo	Lithuania	54	22.41666667	Bln-2165	5120	50
Hein	Norway	60.36666667	7.733333333	T-1005	5150	130
Hein	Norway	60.36666667	7.733333333	GrN-7171	5255	80
Ferriters Cove	Ireland	52.16666667	-10.5	BM-2229A	5270	90
Serovo	Lithuania	54	22.41666667	Bln-2162	5280	50
Zatsene	Belarus	54.16666667	27.5	LE-960	5450	75
Windmill Farm	Great Britain	50.2	-5.416666667	HAR-5668	5470	130
Ferriters Cove	Ireland	52.16666667	-10.5	OxA-3869	5510	70
Zemaitiske 3	Belarus	55.16666667	27.36666667	Bln-2594	5510	60
Ferriters Cove	Ireland	52.16666667	-10.5	BM-2228A	5620	80
Dabki 9	Poland	54.4	16.4	GD-1698	5630	70
Dabki 9	Poland	54.4	16.4	GD-2162	5700	80
Dabki 9	Poland	54.4	16.4	GD-1279	5730	60
Osa	Latvia	57.91666667	27.5	LE-850	5730	50
Osa	Latvia	57.91666667	27.5	LE-962	5780	70
Asavets 4	Belarus			Ki-6213	5860	50
Dabki 9	Poland	54.4	16.4	GD-3125	5860	60
Osa	Latvia	57.91666667	27.5	LE-963	5880	80
Dabki 9	Poland	54.4	16.4	GD-1278	5890	60
Stratford's Yard	Great Britain	51.66666667	-0.666666667	BM-2404	5890	100
Deby 29	Poland	52.61666667	18.36666667	GD-1928	6090	70
Dabki 9	Poland	54.4	16.4	GD-2315	6160	70
Nab Head 2	Great Britain	51.71666667	-5.166666667	OxA-861	6210	90
Zatsene	Belarus	54.16666667	27.5	Ki-6214	6214	40
Dabki 9	Poland	54.4	16.4	GD-1703	6230	60
Dabki 9	Poland	54.4	16.4	GD-3127	6250	40
Moordenaarsven	Belgium	51.33333333	4.666666667	Lv-1294	6270	120
Moordenaarsven	Belgium	51.33333333	4.666666667	Lv-1295	6320	120
Pointe de la Torche	France	48.83333333	-4.283333333	Gif-6857	6370	70
Grotte de Rouffignac	France	45.05	0.966666667	GrN-5512	6400	40
Cabeco do Pez	Portugal	38.25	-8.3	Q-2496	6430	65
Thorpe Common Shelter	Great Britain	53.3	-1.2	Q-1116	6433	115
Felsdach Lauterach	Germany	48.33333333	9.566666667	GrN-4667	6440	45
Poldwrian	Great Britain	50.2	-5.416666667	HAR-4568	6450	110
Osa	Latvia	57.91666667	27.5	TA-862	6535	60
Tomaszow	Poland	51.36666667	21.26666667	GrN-7051	6555	45
Sutton	Ireland	53.38333333	6.083333333	OxA-3960	6560	75
Padina	Yugoslavia	44.5	22.41666667	GrN-7980	6565	40
Pointe de la Torche	France	48.83333333	-4.283333333	Gif-6859	6590	110
Abri Gigot	France	47.21666667	6.683333333	Lv-1227	6640	80
Trou al'Wesse	Belgium	50.46666667	5.25	Lv-1751	6650	70
Sutton	Ireland	53.38333333	6.083333333	OxA-3691	6660	80
Cabeco do Pez	Portugal	38.25	-8.3	Q-2497	6730	75
Osa	Latvia	57.91666667	27.5	TA-856	6770	60
Osa	Latvia	57.91666667	27.5	TA-861	6780	60
Pocas de Sao Bento	Portugal	38.21666667	-8.366666667	Q-2494	6780	65
Pocas de Sao Bento	Portugal	38.21666667	-8.366666667	Q-2495	6850	70
Baume de Montclus	France	44.26666667	4.433333333	MC-728	6880	100
Paardsdrank	Belgium	51.4	5.016666667	Lv-959	6990	135
Pocas de Sao Bento	Portugal	38.21666667	-8.366666667	Q-2493	7040	70
Chez-Jugie	France	45.11666667	1.6	Ly-1396	7060	140
Moita do Sebastiao	Portugal	39.1	-9	H-2119/1546	7080	130
Deby 29	Poland	52.61666667	18.36666667	GD-2555	7140	120
Baume de Montclus	France	44.26666667	4.433333333	MC-729	7200	100

Appendix 3: The Gkiasta European database: Late Mesolithic site list (107 sites) used to construct the isochron map (Chapter 4).

Site Name	Country	latitude (decimal degrees)	Longitude (decimal degrees)	Sample ID	Radiocarbon determination (BP)	SD
Chateauneuf Les Martigues	France	43.38333333	5.166666667	MC-2516	7220	100
Deby 29	Poland	52.61666667	18.36666667	GD-2278	7250	100
Chateauneuf Les Martigues	France	43.38333333	5.166666667	Ly-2830	7260	120
Pointe de la Torche	France	48.83333333	-4.283333333	Gif-6858	7280	80
Chateauneuf Les Martigues	France	43.38333333	5.166666667	Ly-2832	7290	130
Abri de Chauvaux	Belgium	50.33333333	5.85	Lv-1615	7350	75
Chateauneuf Les Martigues	France	43.38333333	5.166666667	MC-532	7360	100
Nab Head 2	Great Britain	51.71666667	-5.166666667	OxA-860	7360	90
Arapouco	Portugal	38.41666667	-8.583333333	Q-2492	7420	65
Falkensteinhole	Germany	48.08333333	9	B-767	7540	120
Riparo di Vatte di Zambana	Italy	46.16666667	11.08333333	R-488	7540	75
Riparo di Vatte di Zambana	Italy	46.16666667	11.08333333	R-488a	7585	75
Grotte Coufin	France	45.1	5.4	Ly-3648	7810	140
Falkensteinhole	Germany	48.08333333	9	B-768	7820	120
Grotte de l'Abbe Prialet	France	43.88333333	3.733333333	MC-1291	7820	100
Jagerhaus Hohle	Germany	48.96666667	8.75	B-939	7880	120
Franchthi	Greece	37.43333333	23.13333333	P-1527	7897	88
Franchthi	Greece	37.43333333	23.13333333	P-1526	8022	76
Mannlefelsen	France	47.45	7.233333333	Lv-858	8140	80
Curacchiaghiu	France	41.7	9.133333333	Gif-1963	8300	130

Appendix 4

```
% Script code that calculates the correlation coefficients
% for each lat, long point across a grid
%
% input: files containing dates and probabilities
% output: variable 'coeff' contain the correlation coefficients
%          surface plot of coefficients
%          contour and feather plot
%%%%%%%%%%%%%%%%%%%%%%%%%%%%%%%%%%%%%%%%%%%%%%%%%%%%%%%%%%%
% First part: loads individual date/prob files of format
%
% lat lon
% scale 0
% date1 prob1
% date2 prob2...
%
% sample, combine and convert them into format
% lat lon date weight
% ...
allSites='';                                        % initialise data storage matrix

for n=1:8                                           % number of sites
   sitefile=['sa',num2str(n),'.m'];              % filename 'sa'+<n>
   sitefile
   site=load(sitefile);
   sitecoord=site(1,:);                          % lat long in first row of file
   scale=site(2,1)*4;                            % scaling factor (area=1)
   n=size(site,1)-2;                               % number of data points (ignoring first 2 rows)
   interval=5;                                   % 5*4=20 year intervals

   intSite='';
   b=0;
   for a=3:interval:2+n
      a2=a+interval-1;                           % check for end of file
      if a2>(n+2)
         a2=n+2;
      end
      intProb=(sum(site(a:a2,2))*scale)^2;       % weight=(sum of interval prob)^2
      if site(a,1)>0                             % date at BC origin if negative
         intDate=abs(site(a,1)-1950+interval*2);% mid interval date
      else
         intDate=1950-site(a,1)-interval*2;  % converted to BP
      end
      if intProb>(0.01^2)
         b=b+1;
         intSite(b,:)=[sitecoord,intDate,intProb];
      end
   end
   bar(-intSite(:,3),intSite(:,4))
   allSites=[allSites;intSite];
end
%%%%%%%%%%%%%%%%%%%%%%%%%%%%%%%%%%%%%%%%%%%%%%%%%%%%%%%%%%%
% Calculate correlation coefficient
%
latmin=25;                                       % latitude grid range
latmax=35;
lonmin=15;                                       % longitude grid range
lonmax=33;
interval=1;
x=0;
y=0;
                                                    % initialise coeff
coeff=zeros((latmax-latmin)/interval+1,(lonmax-lonmin)/interval+1);
num_data=size(geo_data,1);
d_lat=allSites(:,1)/180*pi;                      % convert to rad
d_lon=allSites(:,2)/180*pi;
date=allSites(:,3);
w=allSites(:,4);
sum_w=sum(w);

for lat=latmin:interval:latmax
   x=x+1
   for lon=lonmin:interval:lonmax
      y=y+1;
      o_lat=lat/180*pi;
      o_lon=lon/180*pi;
                                                    % calculate geo-distance
      dist=acos(sin(o_lat).*sin(d_lat)+cos(o_lat).*cos(d_lat).*cos(o_lon-d_lon))*6378;

      wxy=w.*dist.*date;                         % calculate correlation
      wx=w.*dist;
      wy=w.*date;
      wx2=w.*dist.*dist;
```

```
        wy2=w.*date.*date;

        coeff(x,y)=...
           (sum_w*sum(wxy)-sum(wx)*sum(wy))/...
           sqrt((sum_w*sum(wx2)-sum(wx)^2)*(sum_w*sum(wy2)-sum(wy)^2));
    end
    y=0;
end                                                    % Surface plot of coeff
surf(lonmin:interval:lonmax,latmin:interval:latmax,coeff)
xlabel 'Longitude (East)'
ylabel 'Latitude (South)'
zlabel 'Correlation coefficient'
axis ij

figure(2)
                                                       % Contour plot (20 lines shown)
[C,h]=contour(lonmin:interval:lonmax,latmin:interval:latmax,coeff,20);
clabel(C,h)                                            % Label contour lines
axis ij
hold on
[U,V]=gradient(coeff,.2);
quiver(lonmin:interval:lonmax,latmin:interval:latmax,U,V)
r=min(min(coeff))

%%%%%%%%%% Weighted Model 1b regression; date on Distance %%%%%%%%%%%
o_lat=30/180*pi;
o_lon=21/180*pi;
d_lat=geo_data(:,1)/180*pi;
d_lon=geo_data(:,2)/180*pi;
dist=acos(sin(o_lat).*sin(d_lat)+cos(o_lat).*cos(d_lat).*cos(o_lon-d_lon))*6378;
dat=geo_data(:,3);
sDat=1./sqrt(geo_data(:,4));

[m,b,r,sm,sb,xm,ym]=lsqfityz(dist,dat,sDat);

plot(dist,dat,'k.')
xlabel 'Distance from an origin at 21E, 30S (km)'
ylabel 'Calibrated date (years BP)'
hold on
v=axis;
x1=v(1);
x2=v(2);
y1=m*x1+b;
y2=m*x2+b;
line([x1 x2],[y1 y2])
axis ij
r
m
sm
b
sb
```

[Authored by Tien Tran]

Appendix 5:
Correlation coefficients at 504 potential origins for linear regression to Dataset A sites.

	Longitude W																								
Latitude	10ºW	8ºW	6ºW	4ºW	2ºW	0ºW	2ºE	4ºE	6ºE	8ºE	10ºE	12ºE	14ºE	16ºE	18ºE	20ºE	22ºE	24ºE	26ºE	28ºE	30ºE	32ºE	34ºE	36ºE	Latitude
30ºN	0.1796	0.1107	0.037	-0.0402	-0.1196	-0.1996	-0.2784	-0.3538	-0.4239	-0.4871	-0.5421	-0.5884	-0.6261	-0.6558	-0.6786	-0.6956	-0.708	-0.717	-0.7234	-0.728	-0.7311	-0.7333	-0.7346	-0.7352	30ºN
32ºN	0.2258	0.1571	0.0827	0.0037	-0.0789	-0.1632	-0.2473	-0.329	-0.4059	-0.4757	-0.5365	-0.5874	-0.6284	-0.6601	-0.6837	-0.7007	-0.7125	-0.7206	-0.7261	-0.7298	-0.7322	-0.7337	-0.7344	-0.7344	32ºN
34ºN	0.2759	0.2082	0.1337	0.0534	-0.0317	-0.12	-0.2096	-0.2981	-0.3827	-0.4604	-0.5284	-0.5851	-0.6302	-0.6643	-0.689	-0.7059	-0.7169	-0.7239	-0.7282	-0.731	-0.7327	-0.7334	-0.7334	-0.7329	34ºN
36ºN	0.3303	0.2645	0.1907	0.11	0.0229	-0.0689	-0.1635	-0.259	-0.3523	-0.4395	-0.5164	-0.5805	-0.6307	-0.668	-0.6941	-0.7111	-0.7211	-0.7265	-0.7295	-0.7314	-0.7323	-0.7322	-0.7315	-0.7302	36ºN
38ºN	0.3897	0.3274	0.2561	0.1759	0.0868	-0.0084	-0.1069	-0.2091	-0.312	-0.4104	-0.4985	-0.5719	-0.6288	-0.6699	-0.698	-0.7157	-0.7249	-0.7282	-0.73	-0.7308	-0.7306	-0.7296	-0.7281	-0.7261	38ºN
40ºN	0.4538	0.3977	0.3312	0.2536	0.1649	0.0675	-0.036	-0.1451	-0.2582	-0.3698	-0.4716	-0.5573	-0.6227	-0.6683	-0.6987	-0.7179	-0.7273	-0.7292	-0.729	-0.7281	-0.7267	-0.7248	-0.7225	-0.7198	40ºN
42ºN	0.5206	0.4733	0.4147	0.3427	0.2564	0.157	0.0499	-0.0644	-0.1871	-0.3124	-0.4304	-0.5322	-0.6092	-0.6618	-0.6954	-0.7143	-0.7225	-0.7242	-0.7226	-0.7211	-0.7191	-0.7166	-0.7138	-0.7105	42ºN
44ºN	0.5862	0.5497	0.5024	0.441	0.3627	0.2663	0.1547	0.0335	-0.0949	-0.2334	-0.3696	-0.4901	-0.5825	-0.6449	-0.6816	-0.7	-0.7076	-0.7107	-0.7104	-0.7084	-0.7062	-0.7037	-0.7006	-0.6969	44ºN
46ºN	0.6462	0.6215	0.5879	0.542	0.4799	0.3971	0.2926	0.169	0.0336	-0.1155	-0.2739	-0.419	-0.5302	-0.6038	-0.6461	-0.6672	-0.6787	-0.6855	-0.6882	-0.688	-0.6865	-0.6843	-0.6813	-0.6776	46ºN
48ºN	0.6959	0.6824	0.6626	0.6334	0.5927	0.5322	0.4404	0.3247	0.1876	0.0309	-0.1405	-0.3064	-0.4383	-0.5266	-0.5801	-0.6108	-0.6308	-0.6443	-0.652	-0.6557	-0.6565	-0.6557	-0.6535	-0.6504	48ºN
50ºN	0.7314	0.727	0.7197	0.7075	0.6857	0.6421	0.5669	0.4606	0.3306	0.1842	0.0171	-0.151	-0.2961	-0.4052	-0.4791	-0.5272	-0.5594	-0.5818	-0.5968	-0.6062	-0.6115	-0.614	-0.6145	-0.6134	50ºN
52ºN	0.7499	0.7503	0.7493	0.7442	0.7308	0.6989	0.6432	0.5588	0.4487	0.3244	0.179	0.0261	-0.1179	-0.2406	-0.3357	-0.4051	-0.455	-0.491	-0.5167	-0.5348	-0.5472	-0.5554	-0.5605	-0.5633	52ºN
54ºN	0.7537	0.7534	0.7522	0.747	0.7342	0.7099	0.6693	0.6084	0.5257	0.4226	0.3037	0.1779	0.0523	-0.0648	-0.1671	-0.2514	-0.318	-0.3695	-0.4087	-0.4382	-0.4602	-0.4765	-0.4883	-0.4967	54ºN
56ºN	0.7468	0.7436	0.7386	0.7308	0.7185	0.6979	0.6655	0.6181	0.5536	0.4718	0.3769	0.2791	0.1804	0.0842	-0.0083	-0.0928	-0.1665	-0.2283	-0.2789	-0.3196	-0.3518	-0.3772	-0.397	-0.4123	56ºN
58ºN	0.7339	0.7282	0.7202	0.7092	0.6954	0.6766	0.6492	0.6111	0.5612	0.4996	0.4294	0.3545	0.2773	0.1995	0.1218	0.0466	-0.0235	-0.0868	-0.1423	-0.1898	-0.2297	-0.2629	-0.2901	-0.3124	58ºN
60ºN	0.7186	0.7112	0.7016	0.6896	0.6749	0.6565	0.6323	0.6012	0.5624	0.5165	0.4646	0.4081	0.3481	0.2858	0.2223	0.159	0.0976	0.0394	-0.0142	-0.0627	-0.1056	-0.1431	-0.1753	-0.2028	60ºN
62ºN	0.7032	0.6949	0.6849	0.6728	0.6582	0.6405	0.6189	0.5928	0.5618	0.5262	0.4867	0.4434	0.3969	0.3479	0.2971	0.2456	0.1943	0.1442	0.0962	0.0512	0.0097	-0.028	-0.0618	-0.0918	62ºN
64ºN	0.6882	0.6796	0.6695	0.6576	0.6437	0.6274	0.6084	0.5864	0.5613	0.533	0.5018	0.4677	0.4312	0.3926	0.3523	0.3109	0.2691	0.2275	0.1868	0.1475	0.1101	0.0751	0.0427	0.0131	64ºN
66ºN	0.6738	0.6652	0.6554	0.644	0.6311	0.6165	0.5998	0.5811	0.5603	0.5373	0.5121	0.485	0.4559	0.4253	0.3933	0.3603	0.3266	0.2927	0.2591	0.226	0.194	0.1632	0.1341	0.1068	66ºN
68ºN	0.6601	0.6518	0.6424	0.6318	0.6201	0.607	0.5925	0.5765	0.5591	0.5401	0.5195	0.4976	0.4743	0.4498	0.4243	0.3979	0.371	0.3437	0.3163	0.2892	0.2625	0.2365	0.2114	0.1874	68ºN
70ºN	0.6471	0.6391	0.6304	0.6208	0.6102	0.5986	0.5861	0.5724	0.5577	0.5419	0.5251	0.5072	0.4884	0.4687	0.4483	0.4272	0.4057	0.3838	0.3618	0.3397	0.3179	0.2965	0.2755	0.2553	70ºN
Max values	0.7537	0.7534	0.7522	0.747	0.7342	0.7099	0.6693	0.6181	0.5624	0.5419	0.5251	0.5072	0.4884	0.4687	0.4483	0.4272	0.4057	0.3838	0.3618	0.3397	0.3179	0.2965	0.2755	0.2553	
Min values	0.1796	0.1107	0.037	-0.0402	-0.1196	-0.1996	-0.2784	-0.3538	-0.4239	-0.4871	-0.5421	-0.5884	-0.6307	-0.6699	-0.6987	-0.7179	-0.7273	-0.7292	-0.73	-0.7314	-0.7327	-0.7337	-0.7346	-0.7352	

Overall max: 0.754 (10W,54N)

Overall min: -0.74 (36 E, 30N)

Dataset A (Chi result +oldest dates)

Appendix 5:
Correlation coefficients at 504 potential origins for linear regression to Dataset B sites.

Latitude	Longitude W 10°W	8°W	6°W	4°W	2°W	0°W	2°E	4°E	6°E	8°E	10°E
30°N	0.2124	0.145	0.0728	-0.0032	-0.0817	-0.1612	-0.2398	-0.3154	-0.3862	-0.4503	-0.5063
32°N	0.2561	0.1891	0.1164	0.0388	-0.0426	-0.1263	-0.2102	-0.2922	-0.3698	-0.4407	-0.5028
34°N	0.3033	0.2375	0.1649	0.0863	0.0026	-0.0849	-0.1742	-0.263	-0.3485	-0.4276	-0.4973
36°N	0.3545	0.2907	0.219	0.1405	0.0551	-0.0358	-0.13	-0.2259	-0.3204	-0.4094	-0.4886
38°N	0.41	0.3499	0.281	0.2035	0.1167	0.0224	-0.0755	-0.1781	-0.2825	-0.3836	-0.4749
40°N	0.4697	0.4158	0.352	0.2773	0.1914	0.0962	-0.0062	-0.1159	-0.2315	-0.3469	-0.4536
42°N	0.5312	0.4859	0.4297	0.3608	0.2781	0.1825	0.0779	-0.0375	-0.1642	-0.2953	-0.4199
44°N	0.5909	0.5557	0.51	0.451	0.376	0.2837	0.1759	0.054	-0.0796	-0.2242	-0.3685
46°N	0.6445	0.6201	0.5867	0.5415	0.4808	0.4001	0.2986	0.1753	0.0355	-0.1207	-0.2864
48°N	0.6883	0.6741	0.653	0.6221	0.5802	0.519	0.4265	0.3088	0.1675	0.0052	-0.1734
50°N	0.7188	0.7132	0.7042	0.6896	0.6645	0.6167	0.5367	0.4247	0.2887	0.1366	-0.0361
52°N	0.7334	0.7324	0.7293	0.7213	0.7038	0.6673	0.6052	0.513	0.3957	0.2644	0.1139
54°N	0.7338	0.7322	0.7285	0.7199	0.7028	0.6735	0.6266	0.5583	0.468	0.3583	0.2347
56°N	0.7232	0.7186	0.7114	0.7002	0.6837	0.6582	0.6201	0.5665	0.4957	0.4084	0.3095
58°N	0.7066	0.6993	0.6892	0.6754	0.6582	0.6352	0.603	0.5598	0.5049	0.4389	0.3651
60°N	0.688	0.679	0.6676	0.6534	0.6359	0.6139	0.5858	0.5506	0.5079	0.4582	0.4031
62°N	0.6696	0.6598	0.648	0.6339	0.6168	0.5964	0.5718	0.5426	0.5085	0.4699	0.4276
64°N	0.6519	0.6419	0.6302	0.6165	0.6006	0.5821	0.5607	0.5363	0.5088	0.4781	0.4446
66°N	0.6353	0.6254	0.6141	0.6012	0.5866	0.5702	0.5517	0.5311	0.5083	0.4834	0.4564
68°N	0.6196	0.6101	0.5994	0.5876	0.5745	0.5599	0.5439	0.5264	0.5074	0.4869	0.4649
70°N	0.6049	0.5959	0.5861	0.5754	0.5637	0.551	0.5372	0.5223	0.5064	0.4893	0.4713
Max values by column	0.7338	0.7324	0.7293	0.7213	0.7038	0.6735	0.6266	0.5665	0.5088	0.4893	0.4713
Min values	0.2124	0.145	0.0728	-0.0032	-0.0817	-0.1612	-0.2398	-0.3154	-0.3862	-0.4503	-0.5063

Latitude	12°E	14°E	16°E	18°E	20°E	22°E	24°E	26°E	28°E	30°E	32°E	34°E	36°E	Latitude
30°N	-0.5538	-0.5926	-0.6234	-0.6471	-0.665	-0.6784	-0.6885	-0.6963	-0.7027	-0.7079	-0.7122	-0.7156	-0.7183	30°N
32°N	-0.5551	-0.5973	-0.63	-0.6545	-0.6721	-0.6845	-0.6935	-0.7004	-0.7061	-0.7109	-0.7147	-0.7177	-0.7199	32°N
34°N	-0.5556	-0.6022	-0.6375	-0.6628	-0.6801	-0.6912	-0.6984	-0.7041	-0.7095	-0.7139	-0.7173	-0.7197	-0.7213	34°N
36°N	-0.5548	-0.6069	-0.6456	-0.6723	-0.6893	-0.6988	-0.7038	-0.7084	-0.7134	-0.7172	-0.7198	-0.7215	-0.7224	36°N
38°N	-0.5515	-0.6108	-0.6538	-0.6826	-0.7	-0.7081	-0.7108	-0.715	-0.7186	-0.7209	-0.7223	-0.7228	-0.7227	38°N
40°N	-0.5439	-0.613	-0.6609	-0.6921	-0.7111	-0.7196	-0.7218	-0.723	-0.7238	-0.724	-0.7237	-0.7229	-0.7216	40°N
42°N	-0.5281	-0.61	-0.6651	-0.6995	-0.7189	-0.7269	-0.7287	-0.7276	-0.7262	-0.7246	-0.7227	-0.7206	-0.7183	42°N
44°N	-0.4971	-0.5952	-0.6605	-0.6985	-0.7174	-0.725	-0.7269	-0.7256	-0.7228	-0.7201	-0.7174	-0.7144	-0.7112	44°N
46°N	-0.4393	-0.5572	-0.6338	-0.677	-0.6977	-0.7077	-0.7124	-0.7131	-0.7114	-0.7088	-0.7059	-0.7025	-0.6986	46°N
48°N	-0.3455	-0.4819	-0.572	-0.6249	-0.6535	-0.6704	-0.6811	-0.6863	-0.6876	-0.687	-0.6851	-0.6822	-0.6785	48°N
50°N	-0.2082	-0.3551	-0.4638	-0.5358	-0.5807	-0.6094	-0.6284	-0.6404	-0.6473	-0.6506	-0.6516	-0.6508	-0.6487	50°N
52°N	-0.042	-0.1867	-0.3083	-0.4011	-0.4673	-0.5136	-0.5465	-0.5692	-0.5845	-0.5946	-0.6008	-0.6043	-0.6057	52°N
54°N	0.1065	-0.0197	-0.1365	-0.2375	-0.3195	-0.3834	-0.4321	-0.4684	-0.4953	-0.5149	-0.529	-0.539	-0.5458	54°N
56°N	0.2095	0.1097	0.0126	-0.0799	-0.1635	-0.2355	-0.2953	-0.3437	-0.3821	-0.4122	-0.4356	-0.4535	-0.4672	56°N
58°N	0.2875	0.2084	0.1292	0.051	-0.0241	-0.0936	-0.1557	-0.2097	-0.2556	-0.2939	-0.3254	-0.351	-0.3717	58°N
60°N	0.344	0.2818	0.2179	0.1534	0.0896	0.0282	-0.0295	-0.0825	-0.13	-0.1718	-0.208	-0.239	-0.2652	60°N
62°N	0.382	0.3334	0.2827	0.2307	0.1783	0.1265	0.0763	0.0286	-0.016	-0.0569	-0.0939	-0.1268	-0.1559	62°N
64°N	0.4085	0.3702	0.3299	0.2884	0.246	0.2035	0.1614	0.1205	0.0812	0.044	0.0094	-0.0225	-0.0516	64°N
66°N	0.4275	0.3969	0.3649	0.3316	0.2976	0.2631	0.2286	0.1945	0.1612	0.1291	0.0984	0.0694	0.0424	66°N
68°N	0.4416	0.417	0.3912	0.3646	0.3373	0.3095	0.2815	0.2536	0.2261	0.1991	0.1729	0.1478	0.1239	68°N
70°N	0.4523	0.4324	0.4117	0.3903	0.3684	0.346	0.3235	0.3009	0.2784	0.2562	0.2344	0.2133	0.1929	70°N
Max values by column	0.4523	0.4324	0.4117	0.3903	0.3684	0.346	0.3235	0.3009	0.2784	0.2562	0.2344	0.2133	0.1929	
Min values	-0.5556	-0.613	-0.6651	-0.6995	-0.7189	-0.7269	-0.7287	-0.7276	-0.7262	-0.7246	-0.7237	-0.7229	-0.7227	

Overall max: 0.734 (10W,54N)

Overall min: -0.73 (24E, 42N)

Dataset B (Chi result + most recent date)

Appendix 5:
Correlation coefficients at 504 potential origins for linear regression to Dataset C sites.

Latitude	Longitude W 10°W	8°W	6°W	4°W	2°W	0°W	2°E	4°E	6°E	8°E	10°E	12°E	14°E	16°E	18°E	20°E	22°E	24°E	26°E	28°E	30°E	32°E	34°E	36°E	Latitude
30°N	0.1735	0.1065	0.0349	-0.0404	-0.1182	-0.1968	-0.2745	-0.3493	-0.4192	-0.4821	-0.537	-0.583	-0.6204	-0.6496	-0.6719	-0.6883	-0.7001	-0.7084	-0.7142	-0.7183	-0.7209	-0.7226	-0.7235	-0.7236	30°N
32°N	0.2187	0.1521	0.0799	0.003	-0.0775	-0.1602	-0.2432	-0.3242	-0.4007	-0.4704	-0.5311	-0.5819	-0.6225	-0.6537	-0.6768	-0.6931	-0.7042	-0.7117	-0.7165	-0.7197	-0.7217	-0.7227	-0.7229	-0.7225	32°N
34°N	0.2677	0.2021	0.1299	0.0521	-0.0307	-0.117	-0.2051	-0.2927	-0.377	-0.4546	-0.5226	-0.5792	-0.624	-0.6576	-0.6817	-0.698	-0.7083	-0.7145	-0.7183	-0.7205	-0.7217	-0.722	-0.7216	-0.7205	34°N
36°N	0.321	0.2572	0.1857	0.1078	0.0234	-0.066	-0.1588	-0.2531	-0.3459	-0.433	-0.51	-0.574	-0.624	-0.6608	-0.6863	-0.7028	-0.7121	-0.7167	-0.7191	-0.7205	-0.7208	-0.7203	-0.7191	-0.7174	36°N
38°N	0.3791	0.3187	0.2497	0.1723	0.0864	-0.0058	-0.102	-0.2027	-0.3048	-0.4031	-0.4914	-0.5648	-0.6213	-0.6619	-0.6894	-0.7067	-0.7154	-0.7179	-0.719	-0.7193	-0.7186	-0.7172	-0.7151	-0.7127	38°N
40°N	0.4421	0.3876	0.3234	0.2486	0.1632	0.0692	-0.0313	-0.1382	-0.2502	-0.3615	-0.4634	-0.5491	-0.6141	-0.6591	-0.6889	-0.7079	-0.7171	-0.7183	-0.7175	-0.7161	-0.7141	-0.7116	-0.7088	-0.7056	40°N
42°N	0.5079	0.4619	0.4051	0.3358	0.2528	0.1572	0.054	-0.0575	-0.1786	-0.3032	-0.421	-0.5226	-0.5991	-0.6509	-0.684	-0.7026	-0.7107	-0.7121	-0.7103	-0.7081	-0.7054	-0.7024	-0.699	-0.6953	42°N
44°N	0.5728	0.5372	0.4911	0.4318	0.3564	0.2639	0.1565	0.0389	-0.0867	-0.2237	-0.3589	-0.4783	-0.5696	-0.6312	-0.6675	-0.6856	-0.6933	-0.6963	-0.6958	-0.6934	-0.6908	-0.6879	-0.6845	-0.6805	44°N
46°N	0.6323	0.608	0.5751	0.5304	0.4702	0.3906	0.2905	0.1714	0.0399	-0.1064	-0.2616	-0.404	-0.5133	-0.5861	-0.6281	-0.6494	-0.6609	-0.668	-0.6709	-0.6707	-0.6691	-0.6668	-0.6636	-0.6597	46°N
48°N	0.6818	0.6684	0.6488	0.6199	0.5797	0.5207	0.4326	0.3229	0.1914	0.0402	-0.126	-0.2878	-0.4176	-0.5052	-0.5586	-0.5893	-0.6097	-0.6238	-0.6321	-0.6361	-0.6372	-0.6365	-0.6343	-0.6311	48°N
50°N	0.7172	0.7128	0.7053	0.6927	0.6704	0.6276	0.5556	0.4554	0.3325	0.1925	0.0319	-0.1314	-0.274	-0.3823	-0.456	-0.504	-0.5365	-0.5594	-0.575	-0.5849	-0.5907	-0.5935	-0.5941	-0.593	50°N
52°N	0.736	0.7363	0.7351	0.7296	0.716	0.6852	0.6319	0.5521	0.4479	0.33	0.191	0.0427	-0.0987	-0.2196	-0.3137	-0.3826	-0.4323	-0.4685	-0.4945	-0.513	-0.5257	-0.5342	-0.5396	-0.5426	52°N
54°N	0.7405	0.7402	0.7389	0.7339	0.7216	0.6984	0.6598	0.6023	0.5245	0.4272	0.3133	0.1904	0.0669	-0.0484	-0.1493	-0.2322	-0.2981	-0.349	-0.388	-0.4174	-0.4395	-0.4558	-0.4678	-0.4764	54°N
56°N	0.7346	0.7316	0.7269	0.7196	0.708	0.6885	0.658	0.6137	0.5536	0.4768	0.3862	0.2898	0.192	0.0969	0.0056	-0.0777	-0.1502	-0.2111	-0.261	-0.3011	-0.333	-0.3582	-0.3779	-0.3932	56°N
58°N	0.7229	0.7176	0.7101	0.6998	0.6869	0.6692	0.6436	0.6081	0.5614	0.5035	0.4363	0.3631	0.287	0.21	0.1331	0.0589	-0.0104	-0.0728	-0.1274	-0.1741	-0.2135	-0.2462	-0.2731	-0.2951	58°N
60°N	0.7088	0.7018	0.6928	0.6815	0.6678	0.6505	0.6278	0.5986	0.562	0.5183	0.4687	0.414	0.3553	0.294	0.2314	0.169	0.1083	0.0509	-0.002	-0.0498	-0.0921	-0.129	-0.1607	-0.1878	60°N
62°N	0.6944	0.6867	0.6772	0.6658	0.6521	0.6354	0.615	0.5902	0.5608	0.5268	0.4889	0.4471	0.4019	0.354	0.3042	0.2535	0.2029	0.1535	0.1062	0.0618	0.0209	-0.0163	-0.0496	-0.0791	62°N
64°N	0.6804	0.6723	0.6628	0.6516	0.6384	0.6229	0.6049	0.584	0.56	0.5329	0.5029	0.47	0.4346	0.3969	0.3575	0.317	0.2759	0.2349	0.1948	0.1561	0.1192	0.0847	0.0528	0.0236	64°N
66°N	0.6669	0.6587	0.6494	0.6387	0.6264	0.6124	0.5966	0.5788	0.5588	0.5367	0.5125	0.4862	0.4581	0.4283	0.3971	0.3648	0.3318	0.2985	0.2654	0.2329	0.2013	0.171	0.1423	0.1154	66°N
68°N	0.6539	0.646	0.637	0.627	0.6158	0.6033	0.5895	0.5742	0.5574	0.5391	0.5194	0.4981	0.4756	0.4518	0.4269	0.4011	0.3748	0.348	0.3212	0.2945	0.2682	0.2426	0.218	0.1944	68°N
70°N	0.6415	0.634	0.6256	0.6164	0.6063	0.5953	0.5832	0.5701	0.556	0.5407	0.5245	0.5072	0.4889	0.4698	0.4499	0.4294	0.4083	0.3869	0.3653	0.3438	0.3223	0.3012	0.2807	0.2607	70°N
Max values by column	0.7405	0.7402	0.7389	0.7339	0.7216	0.6984	0.6598	0.6137	0.562	0.5407	0.5245	0.5072	0.4889	0.4698	0.4499	0.4294	0.4083	0.3869	0.3653	0.3438	0.3223	0.3012	0.2807	0.2607	
Min values	0.1735	0.1065	0.0349	-0.0404	-0.1182	-0.1968	-0.2745	-0.3493	-0.4192	-0.4821	-0.537	-0.583	-0.624	-0.6619	-0.6894	-0.7079	-0.7171	-0.7183	-0.7191	-0.7205	-0.7217	-0.7227	-0.7235	-0.7236	

Overall max: 0.741 (10W,54N)
Overall min: -0.72 (36E, 30N)
Dataset C (oldest date from each site)

Appendix 5:
Correlation coefficients at 504 potential origins for linear regression to Dataset D sites.

	Longitude W											
Latitude	10ºW	8ºW	6ºW	4ºW	2ºW	0ºW	2ºE	4ºE	6ºE	8ºE	10ºE	12ºE
30ºN	0.1978	0.1331	0.0638	-0.0093	-0.085	-0.1619	-0.2381	-0.3118	-0.3808	-0.4432	-0.4978	-0.5437
32ºN	0.2397	0.1755	0.1057	0.0312	-0.047	-0.1277	-0.209	-0.2888	-0.3646	-0.4338	-0.4944	-0.5451
34ºN	0.285	0.2218	0.1523	0.0771	-0.0031	-0.0872	-0.1736	-0.2599	-0.3435	-0.4208	-0.489	-0.5458
36ºN	0.3341	0.2728	0.2041	0.1292	0.0477	-0.0393	-0.1302	-0.2231	-0.3154	-0.4026	-0.4804	-0.5452
38ºN	0.3876	0.3296	0.2635	0.1897	0.1072	0.0174	-0.0766	-0.1758	-0.2776	-0.3767	-0.4667	-0.5421
40ºN	0.4454	0.3931	0.3318	0.2606	0.179	0.0886	-0.009	-0.1144	-0.2267	-0.3398	-0.4451	-0.5345
42ºN	0.5054	0.4612	0.4068	0.3407	0.2619	0.1715	0.0727	-0.0373	-0.1597	-0.2881	-0.4113	-0.5186
44ºN	0.5641	0.5294	0.4845	0.4273	0.3553	0.2677	0.1661	0.0513	-0.0762	-0.217	-0.3598	-0.4873
46ºN	0.6177	0.5931	0.5598	0.515	0.4555	0.3777	0.2818	0.1662	0.0339	-0.1169	-0.279	-0.4291
48ºN	0.6622	0.6477	0.6263	0.595	0.5525	0.4919	0.4025	0.2912	0.1583	0.0034	-0.169	-0.3358
50ºN	0.6939	0.6883	0.6792	0.6642	0.6382	0.5901	0.5118	0.4056	0.2771	0.1309	-0.0352	-0.2009
52ºN	0.71	0.7091	0.7063	0.6986	0.6814	0.6452	0.5845	0.4962	0.3842	0.2576	0.1123	-0.0382
54ºN	0.7121	0.7108	0.7075	0.6993	0.683	0.6549	0.6096	0.5442	0.4582	0.353	0.2337	0.1085
56ºN	0.703	0.6989	0.6923	0.6818	0.6663	0.6422	0.6061	0.5555	0.4888	0.4062	0.3114	0.2125
58ºN	0.688	0.6814	0.672	0.6592	0.6431	0.6215	0.5912	0.5508	0.4992	0.4368	0.3658	0.2897
60ºN	0.671	0.6626	0.652	0.6387	0.6223	0.6017	0.5753	0.5421	0.5016	0.4543	0.4013	0.3438
62ºN	0.6539	0.6448	0.6338	0.6205	0.6045	0.5852	0.562	0.5342	0.5017	0.4648	0.4241	0.3799
64ºN	0.6374	0.628	0.617	0.6041	0.5891	0.5716	0.5513	0.5281	0.5018	0.4724	0.4401	0.4052
66ºN	0.6217	0.6124	0.6017	0.5896	0.5758	0.5602	0.5426	0.5229	0.5011	0.4772	0.4512	0.4233
68ºN	0.6069	0.5979	0.5879	0.5766	0.5642	0.5503	0.5351	0.5184	0.5001	0.4804	0.4592	0.4367
70ºN	0.5929	0.5845	0.5752	0.565	0.5538	0.5417	0.5285	0.5143	0.499	0.4826	0.4652	0.4468
Max values by column	0.7121	0.7108	0.7075	0.6993	0.683	0.6549	0.6096	0.5555	0.5018	0.4826	0.4652	0.4468
Min values by column	0.1978	0.1331	0.0638	-0.0093	-0.085	-0.1619	-0.2381	-0.3118	-0.3808	-0.4432	-0.4978	-0.5458

Latitude	14ºE	16ºE	18ºE	20ºE	22ºE	24ºE	26ºE	28ºE	30ºE	32ºE	34ºE	36ºE	Latitude
30ºN	-0.581	-0.6103	-0.6325	-0.649	-0.661	-0.67	-0.677	-0.6826	-0.6873	-0.6911	-0.6942	-0.6965	30ºN
32ºN	-0.5857	-0.6168	-0.6396	-0.6557	-0.6667	-0.6744	-0.6804	-0.6855	-0.6898	-0.6933	-0.6959	-0.6979	32ºN
34ºN	-0.5907	-0.6243	-0.6479	-0.6633	-0.6728	-0.6787	-0.6835	-0.6884	-0.6924	-0.6955	-0.6977	-0.699	34ºN
36ºN	-0.5957	-0.6325	-0.6574	-0.6723	-0.6798	-0.6834	-0.6872	-0.6919	-0.6955	-0.6979	-0.6993	-0.6999	36ºN
38ºN	-0.5998	-0.6411	-0.6679	-0.6832	-0.6891	-0.6901	-0.6938	-0.6971	-0.6992	-0.7003	-0.7005	-0.7001	38ºN
40ºN	-0.6024	-0.6486	-0.6779	-0.6951	-0.7017	-0.7022	-0.7024	-0.7027	-0.7025	-0.7018	-0.7006	-0.699	40ºN
42ºN	-0.5996	-0.6532	-0.6857	-0.7031	-0.7094	-0.7099	-0.7078	-0.7056	-0.7033	-0.7009	-0.6983	-0.6955	42ºN
44ºN	-0.5843	-0.6481	-0.6845	-0.7017	-0.7077	-0.7083	-0.7059	-0.7022	-0.6988	-0.6956	-0.692	-0.6883	44ºN
46ºN	-0.545	-0.6201	-0.6621	-0.6816	-0.6901	-0.6935	-0.6932	-0.6906	-0.6873	-0.6838	-0.6799	-0.6756	46ºN
48ºN	-0.4687	-0.557	-0.6089	-0.6364	-0.652	-0.6617	-0.6659	-0.6665	-0.6653	-0.6629	-0.6595	-0.6553	48ºN
50ºN	-0.3431	-0.4491	-0.5195	-0.5631	-0.5906	-0.6087	-0.6199	-0.6261	-0.6289	-0.6294	-0.6282	-0.6257	50ºN
52ºN	-0.1785	-0.2967	-0.387	-0.4512	-0.496	-0.5277	-0.5495	-0.5641	-0.5736	-0.5793	-0.5824	-0.5835	52ºN
54ºN	-0.0152	-0.1293	-0.2277	-0.3073	-0.3692	-0.4162	-0.4512	-0.477	-0.4958	-0.5092	-0.5186	-0.525	54ºN
56ºN	0.1133	0.0174	-0.0735	-0.1552	-0.2253	-0.2833	-0.33	-0.3671	-0.3961	-0.4185	-0.4356	-0.4486	56ºN
58ºN	0.2114	0.1329	0.0555	-0.0184	-0.0865	-0.1472	-0.1997	-0.2442	-0.2812	-0.3115	-0.3362	-0.3561	58ºN
60ºN	0.2829	0.2199	0.1563	0.0935	0.0331	-0.0236	-0.0753	-0.1216	-0.1623	-0.1974	-0.2274	-0.2527	60ºN
62ºN	0.3326	0.283	0.232	0.1805	0.1296	0.0803	0.0335	-0.0102	-0.0501	-0.0861	-0.1182	-0.1464	62ºN
64ºN	0.368	0.3288	0.2881	0.2466	0.2049	0.1637	0.1235	0.085	0.0487	0.0148	-0.0164	-0.0447	64ºN
66ºN	0.3936	0.3625	0.3301	0.2968	0.2631	0.2294	0.196	0.1634	0.1319	0.1019	0.0736	0.0471	66ºN
68ºN	0.4128	0.3879	0.362	0.3354	0.3083	0.281	0.2537	0.2267	0.2003	0.1747	0.1502	0.1268	68ºN
70ºN	0.4276	0.4075	0.3868	0.3654	0.3437	0.3217	0.2997	0.2777	0.256	0.2348	0.2141	0.1942	70ºN
Max values by column	0.4276	0.4075	0.3868	0.3654	0.3437	0.3217	0.2997	0.2777	0.256	0.2348	0.2141	0.1942	
Min values by column	-0.6024	-0.6532	-0.6857	-0.7031	-0.7094	-0.7099	-0.7078	-0.7056	-0.7033	-0.7018	-0.7006	-0.7001	

Overall max: 0.71 (10W,54N)
Overall min: -0.71 (24E, 42N)
Dataset D (most recent date, all sites)

Appendix 5:
Correlation coefficients at 504 potential origins for linear regression to Dataset E sites.

	Longitude W																								
Latitude	10ºW	8ºW	6ºW	4ºW	2ºW	0ºW	2ºE	4ºE	6ºE	8ºE	10ºE	12ºE	14ºE	16ºE	18ºE	20ºE	22ºE	24ºE	26ºE	28ºE	30ºE	32ºE	34ºE	36ºE	Latitude
30ºN	0.2398	0.1673	0.0896	0.008	-0.0757	-0.1598	-0.2421	-0.3207	-0.3936	-0.4591	-0.5159	-0.5637	-0.6023	-0.6325	-0.6554	-0.6722	-0.6846	-0.694	-0.7016	-0.7083	-0.7143	-0.7196	-0.7243	-0.7284	30ºN
32ºN	0.2876	0.2162	0.1383	0.0552	-0.0315	-0.1199	-0.208	-0.2935	-0.374	-0.4471	-0.5109	-0.5643	-0.607	-0.6397	-0.6635	-0.68	-0.6912	-0.6993	-0.706	-0.7123	-0.7182	-0.7234	-0.7279	-0.7317	32ºN
34ºN	0.3384	0.269	0.1919	0.108	0.0191	-0.0731	-0.1667	-0.2594	-0.3484	-0.4306	-0.503	-0.5637	-0.6119	-0.648	-0.6731	-0.6893	-0.6988	-0.7047	-0.7103	-0.7167	-0.7227	-0.7278	-0.732	-0.7354	34ºN
36ºN	0.3921	0.3259	0.2507	0.1671	0.0767	-0.0185	-0.1167	-0.2163	-0.3146	-0.4075	-0.4907	-0.5609	-0.6164	-0.6574	-0.6848	-0.7008	-0.7082	-0.7113	-0.7159	-0.7228	-0.7286	-0.7332	-0.7368	-0.7395	36ºN
38ºN	0.4485	0.3871	0.3153	0.2334	0.1425	0.0447	-0.0559	-0.1615	-0.2696	-0.375	-0.4717	-0.5542	-0.6194	-0.6673	-0.6983	-0.7154	-0.7212	-0.7213	-0.7267	-0.7322	-0.7365	-0.7397	-0.742	-0.7436	38ºN
40ºN	0.5067	0.4521	0.3862	0.3084	0.2199	0.1228	0.0192	-0.0921	-0.2104	-0.3297	-0.4426	-0.5416	-0.6202	-0.6758	-0.7116	-0.7317	-0.74	-0.7409	-0.742	-0.7438	-0.7453	-0.7464	-0.7471	-0.7472	40ºN
42ºN	0.5649	0.5192	0.4619	0.3913	0.307	0.2105	0.107	-0.0072	-0.134	-0.2673	-0.3977	-0.5173	-0.6129	-0.6792	-0.7206	-0.7439	-0.7533	-0.7554	-0.7543	-0.7534	-0.7525	-0.7515	-0.7504	-0.7491	42ºN
44ºN	0.6201	0.5852	0.5393	0.4799	0.4046	0.3137	0.2094	0.0909	-0.0408	-0.1852	-0.3333	-0.4741	-0.5899	-0.6709	-0.7198	-0.7451	-0.7561	-0.7593	-0.7586	-0.7567	-0.7545	-0.7522	-0.7498	-0.7474	44ºN
46ºN	0.6686	0.6447	0.612	0.5673	0.5071	0.4281	0.3298	0.2113	0.077	-0.0731	-0.238	-0.4007	-0.5394	-0.637	-0.6952	-0.7249	-0.7403	-0.7476	-0.7498	-0.7493	-0.7475	-0.7452	-0.7426	-0.7398	46ºN
48ºN	0.7063	0.6925	0.6722	0.6424	0.6005	0.5387	0.4492	0.3387	0.2077	0.0569	-0.1151	-0.2932	-0.4482	-0.5618	-0.6338	-0.6751	-0.699	-0.7147	-0.7228	-0.7258	-0.727	-0.7268	-0.7254	-0.7232	48ºN
50ºN	0.7307	0.7243	0.7142	0.6981	0.6711	0.6238	0.5497	0.4494	0.3287	0.1868	0.0212	-0.1499	-0.3066	-0.4345	-0.527	-0.5882	-0.628	-0.6548	-0.6722	-0.6829	-0.6892	-0.6927	-0.6941	-0.6941	50ºN
52ºN	0.7406	0.738	0.7332	0.7234	0.7041	0.6678	0.6104	0.5274	0.4214	0.3013	0.1593	0.0109	-0.1333	-0.2641	-0.3725	-0.4548	-0.5149	-0.5591	-0.5904	-0.6122	-0.6271	-0.6373	-0.6439	-0.6479	52ºN
54ºN	0.7374	0.7347	0.7305	0.7216	0.7042	0.6751	0.6302	0.5665	0.4835	0.3828	0.2688	0.15	0.0296	-0.0878	-0.1961	-0.2898	-0.3664	-0.427	-0.4738	-0.5092	-0.5357	-0.5554	-0.5698	-0.5802	54ºN
56ºN	0.7257	0.7208	0.7137	0.7035	0.6874	0.6625	0.626	0.5756	0.5098	0.4292	0.3379	0.2454	0.1518	0.058	-0.0351	-0.1237	-0.2037	-0.273	-0.3311	-0.3785	-0.4167	-0.4469	-0.4706	-0.4891	56ºN
58ºN	0.7097	0.7026	0.6932	0.6805	0.6643	0.6423	0.6119	0.5717	0.5208	0.46	0.3917	0.3193	0.2452	0.1703	0.0942	0.0185	-0.0542	-0.1217	-0.1824	-0.2356	-0.2812	-0.3196	-0.3515	-0.3777	58ºN
60ºN	0.692	0.6835	0.6729	0.6597	0.6435	0.6231	0.5973	0.565	0.5259	0.4804	0.4295	0.3743	0.3161	0.2558	0.194	0.1315	0.0698	0.01	-0.0464	-0.0985	-0.1455	-0.1872	-0.2237	-0.2551	60ºN
62ºN	0.6746	0.6655	0.6546	0.6416	0.626	0.6073	0.5849	0.5584	0.5276	0.4925	0.4534	0.4109	0.3655	0.3179	0.2685	0.218	0.1672	0.117	0.0681	0.0215	-0.0223	-0.0628	-0.0996	-0.1327	62ºN
64ºN	0.658	0.6488	0.638	0.6255	0.611	0.5941	0.5748	0.5526	0.5276	0.4996	0.4689	0.4356	0.4	0.3624	0.3232	0.283	0.242	0.2009	0.1602	0.1205	0.0822	0.0459	0.0118	-0.0198	64ºN
66ºN	0.6424	0.6333	0.623	0.6113	0.598	0.583	0.5662	0.5475	0.5268	0.5041	0.4794	0.4529	0.4246	0.3948	0.3638	0.3317	0.299	0.2658	0.2327	0.1998	0.1677	0.1366	0.1068	0.0786	66ºN
68ºN	0.6278	0.6191	0.6094	0.5986	0.5866	0.5734	0.5589	0.543	0.5257	0.5069	0.4869	0.4654	0.4428	0.419	0.3943	0.3688	0.3426	0.3161	0.2894	0.2627	0.2364	0.2105	0.1855	0.1613	68ºN
70ºN	0.6141	0.6059	0.597	0.5872	0.5766	0.565	0.5524	0.5389	0.5244	0.5089	0.4924	0.475	0.4567	0.4376	0.4179	0.3975	0.3766	0.3554	0.3341	0.3126	0.2913	0.2703	0.2496	0.2295	70ºN

Max values by column	0.7406	0.738	0.7332	0.7234	0.7042	0.6751	0.6302	0.5756	0.5276	0.5089	0.4924	0.475	0.4567	0.4376	0.4179	0.3975	0.3766	0.3554	0.3341	0.3126	0.2913	0.2703	0.2496	0.2295
Min values	0.2398	0.1673	0.0896	0.008	-0.0757	-0.1598	-0.2421	-0.3207	-0.3936	-0.4591	-0.5159	-0.5643	-0.6202	-0.6792	-0.7206	-0.7451	-0.7561	-0.7593	-0.7586	-0.7567	-0.7545	-0.7522	-0.7504	-0.7491

Overall max: 0.741 (10W,52N)
Overall min: -0.76 (24E, 44N)
Dataset E (most recent date, only sites with multiple dates)

Appendix 5:
Correlation coefficients at 504 potential origins for linear regression to Dataset F sites.

	Longitude W																								
Latitude	10°W	8°W	6°W	4°W	2°W	0°W	2°E	4°E	6°E	8°E	10°E	12°E	14°E	16°E	18°E	20°E	22°E	24°E	26°E	28°E	30°E	32°E	34°E	36°E	Latitude
30°N	0.1485	0.073	-0.0068	-0.0893	-0.1729	-0.2555	-0.3354	-0.4105	-0.4791	-0.5398	-0.5917	-0.6344	-0.6682	-0.694	-0.7129	-0.7263	-0.7355	-0.7418	-0.7461	-0.7492	-0.7515	-0.753	-0.7539	-0.7544	30°N
32°N	0.2011	0.1261	0.0455	-0.0392	-0.1262	-0.2136	-0.2994	-0.3814	-0.4574	-0.5253	-0.5835	-0.6314	-0.669	-0.697	-0.7168	-0.7301	-0.7386	-0.7439	-0.7474	-0.7499	-0.7517	-0.7528	-0.7534	-0.7534	32°N
34°N	0.2576	0.184	0.1035	0.0173	-0.0725	-0.1642	-0.2557	-0.3448	-0.429	-0.5054	-0.5716	-0.6262	-0.6686	-0.6998	-0.7209	-0.7341	-0.7415	-0.7455	-0.7482	-0.7501	-0.7515	-0.7522	-0.7523	-0.7519	34°N
36°N	0.3182	0.2472	0.1677	0.0809	-0.0111	-0.1062	-0.2026	-0.2988	-0.3918	-0.4781	-0.554	-0.6171	-0.6661	-0.7016	-0.7249	-0.7382	-0.7441	-0.7464	-0.7481	-0.7499	-0.7509	-0.7511	-0.7506	-0.7496	36°N
38°N	0.3828	0.3161	0.2393	0.153	0.0592	-0.0386	-0.1384	-0.2407	-0.3428	-0.4401	-0.5278	-0.6018	-0.6594	-0.701	-0.7277	-0.7423	-0.7474	-0.7474	-0.7486	-0.7496	-0.7497	-0.7491	-0.7479	-0.7461	38°N
40°N	0.4509	0.3909	0.3193	0.2363	0.1433	0.0437	-0.0601	-0.1679	-0.2786	-0.3877	-0.4889	-0.5771	-0.6461	-0.6949	-0.7263	-0.7437	-0.7508	-0.7508	-0.7496	-0.7485	-0.7471	-0.7453	-0.7432	-0.7406	40°N
42°N	0.5208	0.47	0.4071	0.3306	0.2403	0.1385	0.0321	-0.0786	-0.1958	-0.3153	-0.4313	-0.5375	-0.6212	-0.6791	-0.7161	-0.7366	-0.7445	-0.7461	-0.7444	-0.7427	-0.7406	-0.738	-0.7351	-0.7319	42°N
44°N	0.5893	0.5502	0.4998	0.435	0.3535	0.2558	0.1459	0.0293	-0.091	-0.2203	-0.3514	-0.4759	-0.5786	-0.65	-0.693	-0.7154	-0.7255	-0.73	-0.7306	-0.7296	-0.7276	-0.7249	-0.7217	-0.7182	44°N
46°N	0.6519	0.6257	0.5905	0.5428	0.4788	0.3956	0.2929	0.1757	0.0516	-0.0828	-0.2309	-0.3774	-0.5022	-0.5902	-0.6429	-0.6709	-0.6881	-0.6983	-0.7033	-0.7049	-0.7044	-0.7028	-0.7003	-0.6972	46°N
48°N	0.7039	0.6895	0.6692	0.6399	0.5987	0.5378	0.4508	0.3466	0.2246	0.0876	-0.0687	-0.2331	-0.3777	-0.485	-0.5546	-0.5968	-0.6251	-0.6449	-0.6569	-0.6635	-0.6668	-0.668	-0.6677	-0.6661	48°N
50°N	0.7415	0.736	0.7274	0.7141	0.6916	0.6508	0.5847	0.4945	0.3853	0.2589	0.1093	-0.051	-0.2025	-0.3289	-0.4229	-0.4878	-0.5323	-0.5642	-0.5863	-0.601	-0.6106	-0.6166	-0.62	-0.6215	50°N
52°N	0.7632	0.7625	0.7605	0.7547	0.7416	0.7133	0.6662	0.5959	0.5039	0.4005	0.2739	0.1346	-0.006	-0.1369	-0.2478	-0.3344	-0.3995	-0.4487	-0.4849	-0.5112	-0.5301	-0.5435	-0.5529	-0.5592	52°N
54°N	0.7709	0.7706	0.7702	0.7663	0.7558	0.735	0.7007	0.6499	0.5812	0.4951	0.3933	0.2818	0.1653	0.0497	-0.059	-0.1551	-0.2354	-0.3002	-0.3514	-0.3912	-0.4218	-0.4451	-0.4627	-0.4759	54°N
56°N	0.7694	0.7676	0.7648	0.76	0.7506	0.7335	0.7063	0.6667	0.6126	0.5432	0.4611	0.3748	0.2858	0.1963	0.1055	0.0173	-0.064	-0.1357	-0.1968	-0.2476	-0.2892	-0.3228	-0.3497	-0.371	56°N
58°N	0.7621	0.7586	0.7535	0.746	0.7354	0.7201	0.6974	0.6657	0.6238	0.5716	0.5109	0.445	0.3764	0.3061	0.2335	0.1599	0.088	0.0201	-0.042	-0.0971	-0.145	-0.1859	-0.2203	-0.249	58°N
60°N	0.7514	0.7463	0.7394	0.7305	0.7193	0.7048	0.6854	0.66	0.6281	0.5896	0.5452	0.4961	0.4433	0.3876	0.3296	0.2701	0.2103	0.1515	0.0953	0.0428	-0.0052	-0.0483	-0.0863	-0.1195	60°N
62°N	0.7398	0.7338	0.7264	0.7172	0.706	0.6921	0.675	0.6541	0.6289	0.5995	0.566	0.5288	0.4883	0.445	0.3994	0.3522	0.304	0.2556	0.208	0.1619	0.1182	0.0775	0.0401	0.0062	62°N
64°N	0.7283	0.7219	0.7142	0.7051	0.6943	0.6815	0.6665	0.6488	0.6284	0.6051	0.579	0.5501	0.5187	0.4851	0.4495	0.4124	0.3741	0.3353	0.2964	0.258	0.2206	0.1848	0.1509	0.1192	64°N
66°N	0.717	0.7104	0.7029	0.6941	0.684	0.6725	0.6592	0.6442	0.6273	0.6083	0.5875	0.5646	0.54	0.5136	0.4858	0.4566	0.4265	0.3956	0.3645	0.3333	0.3025	0.2724	0.2433	0.2156	66°N
68°N	0.7061	0.6996	0.6924	0.6841	0.6749	0.6645	0.6529	0.6401	0.6259	0.6103	0.5933	0.575	0.5554	0.5345	0.5126	0.4896	0.4659	0.4415	0.4168	0.3919	0.367	0.3424	0.3184	0.295	68°N
70°N	0.6956	0.6894	0.6826	0.675	0.6667	0.6575	0.6474	0.6364	0.6244	0.6115	0.5976	0.5827	0.567	0.5503	0.5329	0.5148	0.4961	0.4769	0.4573	0.4375	0.4177	0.398	0.3785	0.3595	70°N

Max values by column	0.7709	0.7706	0.7702	0.7663	0.7558	0.735	0.7063	0.6667	0.6289	0.6115	0.5976	0.5827	0.567	0.5503	0.5329	0.5148	0.4961	0.4769	0.4573	0.4375	0.4177	0.398	0.3785	0.3595
Min values	0.1485	0.073	-0.0068	-0.0893	-0.1729	-0.2555	-0.3354	-0.4105	-0.4791	-0.5398	-0.5917	-0.6344	-0.669	-0.7016	-0.7277	-0.7437	-0.7508	-0.7508	-0.7496	-0.7501	-0.7517	-0.753	-0.7539	-0.7544

Overall max: 0.771 (10W, 54N)
Overall min: -0.75 (36E, 30N)
Dataset F (oldest date, sites with mulitple dates only)

Appendix 5:
Correlation coefficients at 504 potential origins for linear regression to Dataset G sites.

	Longitude W																								
Latitude	10ºW	8ºW	6ºW	4ºW	2ºW	0ºW	2ºE	4ºE	6ºE	8ºE	10ºE	12ºE	14ºE	16ºE	18ºE	20ºE	22ºE	24ºE	26ºE	28ºE	30ºE	32ºE	34ºE	36ºE	Latitude
30ºN	0.2266	0.1438	0.0554	-0.0364	-0.1292	-0.2206	-0.308	-0.3895	-0.4634	-0.5286	-0.5845	-0.631	-0.6687	-0.6984	-0.7213	-0.7385	-0.7513	-0.7609	-0.7683	-0.7742	-0.7789	-0.7828	-0.7859	-0.7884	30ºN
32ºN	0.2818	0.1999	0.1108	0.0165	-0.0806	-0.1777	-0.272	-0.3611	-0.4428	-0.5153	-0.5777	-0.6294	-0.6709	-0.703	-0.727	-0.7443	-0.7566	-0.7654	-0.772	-0.7772	-0.7815	-0.7849	-0.7877	-0.7898	32ºN
34ºN	0.3407	0.261	0.1723	0.0763	-0.0244	-0.1268	-0.228	-0.3253	-0.4159	-0.4973	-0.5676	-0.626	-0.6724	-0.7077	-0.7332	-0.7506	-0.7621	-0.7698	-0.7755	-0.7801	-0.7839	-0.7869	-0.7892	-0.7908	34ºN
36ºN	0.403	0.3272	0.2404	0.144	0.0404	-0.0669	-0.1744	-0.28	-0.3805	-0.4724	-0.5527	-0.6195	-0.6724	-0.7121	-0.7398	-0.7576	-0.7678	-0.7738	-0.7786	-0.7828	-0.7862	-0.7886	-0.7903	-0.7914	36ºN
38ºN	0.4682	0.3982	0.3155	0.2207	0.1155	0.0033	-0.1089	-0.2223	-0.3335	-0.4378	-0.5304	-0.6081	-0.6694	-0.715	-0.7461	-0.7652	-0.7744	-0.7778	-0.7822	-0.7858	-0.7883	-0.7899	-0.7908	-0.7912	38ºN
40ºN	0.535	0.4733	0.3977	0.3075	0.2038	0.0907	-0.0273	-0.1487	-0.2713	-0.3896	-0.4971	-0.5891	-0.6616	-0.7142	-0.7497	-0.7714	-0.7823	-0.7856	-0.7872	-0.7886	-0.7896	-0.7901	-0.7901	-0.7897	40ºN
42ºN	0.6015	0.5506	0.4855	0.404	0.305	0.191	0.0698	-0.0571	-0.1896	-0.322	-0.4467	-0.5572	-0.6444	-0.7063	-0.7472	-0.7715	-0.783	-0.7871	-0.7874	-0.788	-0.7882	-0.7878	-0.787	-0.7858	42ºN
44ºN	0.6642	0.626	0.5751	0.5078	0.4206	0.3131	0.1891	0.0543	-0.0865	-0.2328	-0.376	-0.5068	-0.6131	-0.6882	-0.735	-0.7608	-0.7734	-0.7796	-0.7817	-0.7823	-0.7821	-0.7812	-0.7799	-0.7782	44ºN
46ºN	0.7189	0.6938	0.6589	0.6102	0.5432	0.4533	0.3376	0.2001	0.051	-0.1069	-0.2721	-0.4275	-0.5565	-0.6467	-0.7017	-0.7315	-0.7494	-0.7598	-0.7651	-0.7675	-0.7681	-0.7677	-0.7665	-0.7647	46ºN
48ºN	0.7618	0.7481	0.7279	0.6976	0.6541	0.5883	0.488	0.3603	0.2092	0.0438	-0.1355	-0.3131	-0.4613	-0.5675	-0.6354	-0.6769	-0.7035	-0.7211	-0.7321	-0.7385	-0.7419	-0.7433	-0.7434	-0.7425	48ºN
50ºN	0.7901	0.7849	0.7763	0.7619	0.7369	0.6899	0.6102	0.4954	0.3562	0.2007	0.0232	-0.156	-0.3144	-0.4384	-0.5267	-0.5866	-0.6269	-0.6557	-0.6757	-0.6891	-0.6979	-0.7033	-0.7063	-0.7077	50ºN
52ºN	0.8027	0.8018	0.7999	0.7927	0.7756	0.7409	0.6822	0.5922	0.4754	0.3439	0.1893	0.0273	-0.1266	-0.2614	-0.3695	-0.4506	-0.5103	-0.5556	-0.5888	-0.6127	-0.6298	-0.6419	-0.6502	-0.6557	52ºN
54ºN	0.8031	0.8021	0.8005	0.7943	0.7792	0.7518	0.7079	0.6418	0.5524	0.4422	0.3168	0.1868	0.056	-0.0695	-0.1828	-0.279	-0.3571	-0.419	-0.4673	-0.5044	-0.5328	-0.5542	-0.5703	-0.5822	54ºN
56ºN	0.7948	0.7913	0.7861	0.7778	0.7638	0.7407	0.7048	0.6526	0.5819	0.4929	0.3921	0.293	0.1928	0.0922	-0.0073	-0.1009	-0.1848	-0.257	-0.3175	-0.3672	-0.4073	-0.4394	-0.4648	-0.4848	56ºN
58ºN	0.7809	0.775	0.7667	0.755	0.7402	0.7195	0.6894	0.6478	0.5934	0.5271	0.4531	0.3759	0.2968	0.2168	0.1354	0.0545	-0.0229	-0.0946	-0.1589	-0.2152	-0.2634	-0.3042	-0.3381	-0.3662	58ºN
60ºN	0.7646	0.7569	0.747	0.7345	0.7191	0.6994	0.6736	0.6405	0.5996	0.5513	0.497	0.4382	0.3762	0.3118	0.2457	0.1789	0.1127	0.0487	-0.0116	-0.0673	-0.1175	-0.1621	-0.2011	-0.2348	60ºN
62ºN	0.7482	0.7397	0.7294	0.7169	0.7018	0.6834	0.661	0.634	0.6021	0.5655	0.5244	0.4795	0.4314	0.3807	0.3281	0.2741	0.2197	0.1659	0.1134	0.0633	0.0162	-0.0273	-0.0669	-0.1024	62ºN
64ºN	0.7325	0.7237	0.7134	0.7012	0.687	0.6703	0.6509	0.6284	0.6028	0.5739	0.5419	0.507	0.4695	0.4298	0.3883	0.3454	0.3017	0.2577	0.2141	0.1714	0.1303	0.0912	0.0544	0.0204	64ºN
66ºN	0.7175	0.7088	0.6988	0.6874	0.6743	0.6594	0.6426	0.6236	0.6025	0.5792	0.5537	0.5261	0.4966	0.4653	0.4326	0.3987	0.3639	0.3286	0.2932	0.258	0.2236	0.1901	0.1581	0.1276	66ºN
68ºN	0.7035	0.6951	0.6856	0.675	0.6632	0.65	0.6354	0.6194	0.6018	0.5827	0.562	0.5399	0.5164	0.4916	0.4657	0.4389	0.4113	0.3832	0.3548	0.3264	0.2982	0.2706	0.2437	0.2177	68ºN
70ºN	0.6902	0.6823	0.6735	0.6639	0.6533	0.6418	0.6292	0.6156	0.6009	0.5851	0.5682	0.5503	0.5314	0.5116	0.491	0.4697	0.4479	0.4256	0.403	0.3803	0.3577	0.3352	0.3132	0.2917	70ºN
Max values by column	0.8031	0.8021	0.8005	0.7943	0.7792	0.7518	0.7079	0.6526	0.6028	0.5851	0.5682	0.5503	0.5314	0.5116	0.491	0.4697	0.4479	0.4256	0.403	0.3803	0.3577	0.3352	0.3132	0.2917	
Min values by column	0.2266	0.1438	0.0554	-0.0364	-0.1292	-0.2206	-0.308	-0.3895	-0.4634	-0.5286	-0.5845	-0.631	-0.6724	-0.715	-0.7497	-0.7715	-0.783	-0.7871	-0.7874	-0.7886	-0.7896	-0.7901	-0.7908	-0.7914	

Overall max: 0.8 (10W,54N)

Overall min: -0.79 36E, 36 N)

Dataset G (Chi + Median)

Appendix 6:
Dataset A: Correlation coefficients at 504 potential origin points in an exercise to find the most likely origin for sites in Zvelebil and Lillie's demic diffusion region.

	Longitude W																								
Latitude	10ºW	8ºW	6ºW	4ºW	2ºW	0ºW	2ºE	4ºE	6ºE	8ºE	10ºE	12ºE	14ºE	16ºE	18ºE	20ºE	22ºE	24ºE	26ºE	28ºE	30ºE	32ºE	34ºE	36ºE	Latitude
30ºN	0.218	0.179	0.13	0.074	0.009	-0.06	-0.14	-0.21	-0.27	-0.33	-0.38	-0.42	-0.45	-0.48	-0.5	-0.51	-0.52	-0.53	-0.53	-0.53	-0.52	-0.52	-0.52	-0.51	30ºN
32ºN	0.262	0.229	0.19	0.137	0.077	0.006	-0.07	-0.15	-0.23	-0.29	-0.35	-0.4	-0.44	-0.47	-0.49	-0.51	-0.52	-0.53	-0.53	-0.53	-0.52	-0.52	-0.51	-0.51	32ºN
34ºN	0.304	0.278	0.24	0.202	0.149	0.084	0.008	-0.08	-0.16	-0.24	-0.31	-0.37	-0.42	-0.46	-0.49	-0.51	-0.53	-0.53	-0.53	-0.53	-0.52	-0.51	-0.51	-0.5	34ºN
36ºN	0.342	0.322	0.3	0.265	0.223	0.168	0.099	0.015	-0.08	-0.17	-0.25	-0.33	-0.39	-0.44	-0.48	-0.51	-0.53	-0.53	-0.53	-0.52	-0.51	-0.51	-0.5	-0.49	36ºN
38ºN	0.375	0.362	0.35	0.323	0.293	0.253	0.197	0.124	0.032	-0.07	-0.17	-0.26	-0.34	-0.41	-0.46	-0.51	-0.53	-0.53	-0.52	-0.51	-0.5	-0.49	-0.48	-0.47	38ºN
40ºN	0.403	0.395	0.39	0.372	0.355	0.33	0.293	0.24	0.164	0.065	-0.05	-0.16	-0.26	-0.35	-0.42	-0.48	-0.52	-0.52	-0.5	-0.49	-0.48	-0.47	-0.46	-0.45	40ºN
42ºN	0.426	0.422	0.42	0.411	0.403	0.392	0.374	0.346	0.298	0.224	0.12	-0	-0.13	-0.26	-0.36	-0.43	-0.46	-0.47	-0.46	-0.45	-0.44	-0.44	-0.43	-0.42	42ºN
44ºN	0.443	0.442	0.44	0.439	0.438	0.436	0.433	0.425	0.408	0.372	0.304	0.2	0.061	-0.1	-0.23	-0.31	-0.36	-0.38	-0.39	-0.39	-0.39	-0.38	-0.38	-0.38	44ºN
46ºN	0.457	0.457	0.46	0.458	0.46	0.462	0.466	0.471	0.475	0.471	0.449	0.396	0.298	0.15	-0.01	-0.12	-0.2	-0.25	-0.28	-0.3	-0.31	-0.32	-0.32	-0.32	46ºN
48ºN	0.467	0.467	0.47	0.469	0.471	0.474	0.479	0.487	0.499	0.512	0.518	0.509	0.464	0.365	0.229	0.097	-0.01	-0.09	-0.15	-0.19	-0.22	-0.23	-0.24	-0.25	48ºN
50ºN	0.474	0.475	0.48	0.476	0.477	0.478	0.48	0.483	0.493	0.512	0.528	0.534	0.517	0.463	0.373	0.267	0.165	0.076	7E-04	-0.06	-0.1	-0.13	-0.15	-0.16	50ºN
52ºN	0.48	0.481	0.48	0.481	0.482	0.482	0.483	0.484	0.491	0.508	0.525	0.533	0.528	0.501	0.451	0.381	0.301	0.221	0.147	0.083	0.03	-0.01	-0.04	-0.07	52ºN
54ºN	0.486	0.487	0.49	0.488	0.488	0.49	0.492	0.495	0.502	0.512	0.523	0.529	0.528	0.514	0.487	0.445	0.391	0.331	0.269	0.209	0.155	0.108	0.068	0.035	54ºN
56ºN	0.492	0.492	0.49	0.494	0.495	0.497	0.5	0.504	0.51	0.516	0.522	0.526	0.525	0.517	0.501	0.476	0.442	0.401	0.355	0.307	0.26	0.215	0.175	0.139	56ºN
58ºN	0.497	0.498	0.5	0.5	0.502	0.504	0.507	0.51	0.514	0.518	0.521	0.523	0.522	0.517	0.507	0.491	0.47	0.442	0.41	0.375	0.338	0.301	0.265	0.231	58ºN
60ºN	0.502	0.503	0.5	0.505	0.507	0.509	0.511	0.514	0.516	0.519	0.521	0.521	0.52	0.516	0.509	0.499	0.484	0.466	0.444	0.419	0.392	0.364	0.335	0.306	60ºN
62ºN	0.506	0.507	0.51	0.51	0.511	0.513	0.514	0.516	0.518	0.519	0.52	0.52	0.518	0.515	0.51	0.503	0.493	0.48	0.465	0.448	0.428	0.407	0.385	0.363	62ºN
64ºN	0.51	0.511	0.51	0.513	0.514	0.515	0.516	0.517	0.518	0.519	0.519	0.518	0.517	0.514	0.51	0.505	0.498	0.489	0.478	0.466	0.452	0.437	0.421	0.404	64ºN
66ºN	0.513	0.513	0.51	0.515	0.516	0.517	0.517	0.518	0.518	0.518	0.518	0.517	0.516	0.513	0.51	0.506	0.501	0.494	0.486	0.478	0.468	0.457	0.445	0.433	66ºN
68ºN	0.515	0.515	0.52	0.516	0.517	0.517	0.518	0.518	0.518	0.518	0.517	0.516	0.515	0.513	0.51	0.507	0.503	0.498	0.492	0.486	0.478	0.471	0.462	0.453	68ºN
70ºN	0.516	0.517	0.52	0.517	0.517	0.518	0.518	0.518	0.517	0.517	0.516	0.515	0.514	0.512	0.51	0.507	0.504	0.5	0.496	0.491	0.486	0.48	0.474	0.467	70ºN
Max values	0.516	0.517	0.52	0.517	0.517	0.518	0.518	0.518	0.518	0.519	0.528	0.534	0.528	0.517	0.51	0.507	0.504	0.5	0.496	0.491	0.486	0.48	0.474	0.467	
Min values	0.218	0.179	0.13	0.074	0.009	-0.06	-0.14	-0.21	-0.27	-0.33	-0.38	-0.42	-0.45	-0.48	-0.5	-0.51	-0.53	-0.53	-0.53	-0.53	-0.52	-0.52	-0.52	-0.51	

Overall max	0.534
Overall min	-0.53

Appendix 6:

Dataset A: Correlation coefficients at 504 potential origin points in an exercise to find the most likely origin to sites in Zvelebil & Lillie's demic region with the exclusion of Italian sites.

	Longitude W																								
Latitude	10ºW	8ºW	6ºW	4ºW	2ºW	0ºW	2ºE	4ºE	6ºE	8ºE	10ºE	12ºE	14ºE	16ºE	18ºE	20ºE	22ºE	24ºE	26ºE	28ºE	30ºE	32ºE	34ºE	36ºE	Latitude
30ºN	0.274	0.229	0.17	0.084	-0.03	-0.18	-0.33	-0.46	-0.54	-0.59	-0.61	-0.61	-0.61	-0.61	-0.61	-0.6	-0.6	-0.59	-0.59	-0.58	-0.57	-0.57	-0.56	-0.55	30ºN
32ºN	0.322	0.289	0.24	0.181	0.09	-0.04	-0.2	-0.37	-0.5	-0.57	-0.6	-0.61	-0.61	-0.61	-0.61	-0.6	-0.6	-0.59	-0.58	-0.58	-0.57	-0.56	-0.55	-0.55	32ºN
34ºN	0.363	0.34	0.31	0.265	0.202	0.106	-0.04	-0.23	-0.41	-0.52	-0.58	-0.6	-0.6	-0.61	-0.6	-0.6	-0.6	-0.59	-0.58	-0.57	-0.56	-0.55	-0.54	-0.54	34ºN
36ºN	0.397	0.382	0.36	0.335	0.295	0.234	0.135	-0.03	-0.24	-0.43	-0.54	-0.58	-0.59	-0.6	-0.6	-0.6	-0.59	-0.59	-0.57	-0.56	-0.55	-0.54	-0.53	-0.52	36ºN
38ºN	0.426	0.416	0.4	0.389	0.367	0.334	0.28	0.183	0.009	-0.24	-0.43	-0.52	-0.56	-0.58	-0.59	-0.59	-0.59	-0.58	-0.56	-0.55	-0.54	-0.52	-0.51	-0.5	38ºN
40ºN	0.45	0.444	0.44	0.43	0.42	0.405	0.382	0.34	0.256	0.08	-0.19	-0.39	-0.48	-0.52	-0.55	-0.56	-0.57	-0.56	-0.54	-0.52	-0.51	-0.5	-0.49	-0.48	40ºN
42ºN	0.469	0.466	0.46	0.461	0.458	0.454	0.448	0.437	0.412	0.349	0.188	-0.08	-0.29	-0.39	-0.45	-0.48	-0.5	-0.49	-0.48	-0.47	-0.46	-0.46	-0.45	-0.45	42ºN
44ºN	0.485	0.484	0.48	0.483	0.484	0.485	0.488	0.491	0.493	0.485	0.442	0.301	0.061	-0.14	-0.26	-0.33	-0.37	-0.39	-0.4	-0.4	-0.4	-0.4	-0.4	-0.4	44ºN
46ºN	0.498	0.498	0.5	0.499	0.501	0.505	0.511	0.52	0.532	0.545	0.548	0.512	0.388	0.193	0.017	-0.1	-0.18	-0.24	-0.28	-0.3	-0.32	-0.33	-0.34	-0.34	46ºN
48ºN	0.51	0.51	0.51	0.512	0.514	0.517	0.522	0.532	0.547	0.568	0.587	0.591	0.552	0.444	0.294	0.151	0.032	-0.06	-0.13	-0.18	-0.21	-0.24	-0.25	-0.27	48ºN
50ºN	0.52	0.52	0.52	0.522	0.524	0.526	0.528	0.533	0.545	0.572	0.599	0.616	0.609	0.557	0.463	0.347	0.234	0.132	0.044	-0.03	-0.08	-0.12	-0.15	-0.17	50ºN
52ºN	0.53	0.531	0.53	0.533	0.535	0.537	0.54	0.545	0.556	0.58	0.605	0.623	0.626	0.606	0.558	0.482	0.394	0.303	0.216	0.139	0.074	0.021	-0.02	-0.06	52ºN
54ºN	0.54	0.541	0.54	0.545	0.548	0.551	0.557	0.565	0.577	0.593	0.611	0.625	0.631	0.624	0.601	0.56	0.503	0.435	0.364	0.293	0.227	0.168	0.117	0.074	54ºN
56ºN	0.549	0.551	0.55	0.557	0.56	0.565	0.572	0.58	0.59	0.602	0.614	0.624	0.63	0.63	0.619	0.598	0.565	0.522	0.471	0.416	0.36	0.306	0.255	0.209	56ºN
58ºN	0.559	0.561	0.56	0.568	0.572	0.577	0.583	0.59	0.599	0.608	0.616	0.624	0.629	0.63	0.626	0.615	0.597	0.572	0.54	0.502	0.461	0.418	0.375	0.333	58ºN
60ºN	0.568	0.571	0.57	0.577	0.581	0.586	0.592	0.598	0.605	0.611	0.618	0.624	0.628	0.63	0.628	0.623	0.613	0.599	0.579	0.556	0.528	0.498	0.465	0.432	60ºN
62ºN	0.576	0.579	0.58	0.585	0.589	0.594	0.598	0.604	0.609	0.614	0.619	0.624	0.627	0.629	0.629	0.626	0.621	0.613	0.602	0.587	0.57	0.55	0.528	0.504	62ºN
64ºN	0.583	0.586	0.59	0.592	0.596	0.599	0.603	0.608	0.612	0.616	0.62	0.623	0.626	0.628	0.629	0.628	0.625	0.62	0.614	0.605	0.595	0.582	0.568	0.552	64ºN
66ºN	0.59	0.592	0.59	0.598	0.601	0.604	0.607	0.611	0.614	0.617	0.621	0.623	0.626	0.627	0.628	0.628	0.627	0.624	0.621	0.615	0.609	0.601	0.592	0.582	66ºN
68ºN	0.595	0.598	0.6	0.602	0.605	0.608	0.61	0.613	0.616	0.618	0.621	0.623	0.625	0.626	0.627	0.628	0.627	0.626	0.624	0.621	0.617	0.613	0.607	0.601	68ºN
70ºN	0.6	0.602	0.6	0.606	0.608	0.611	0.613	0.615	0.617	0.619	0.621	0.623	0.625	0.626	0.627	0.627	0.627	0.627	0.626	0.624	0.622	0.619	0.616	0.612	70ºN
Max values	0.6	0.602	0.6	0.606	0.608	0.611	0.613	0.615	0.617	0.619	0.621	0.625	0.631	0.63	0.629	0.628	0.627	0.627	0.626	0.624	0.622	0.619	0.616	0.612	
Min values	0.274	0.229	0.17	0.084	-0.03	-0.18	-0.33	-0.46	-0.54	-0.59	-0.61	-0.61	-0.61	-0.61	-0.61	-0.6	-0.6	-0.59	-0.59	-0.58	-0.57	-0.57	-0.56	-0.55	

Overall max 0.631

overall min -0.61

Appendix 6:
Dataset A: Correlation coefficient at 504 potential origin points in an exercise to find the most likely origin to sites in Italy.

	Longitude W																								
Latitude	10ºW	8ºW	6ºW	4ºW	2ºW	0ºW	2ºE	4ºE	6ºE	8ºE	10ºE	12ºE	14ºE	16ºE	18ºE	20ºE	22ºE	24ºE	26ºE	28ºE	30ºE	32ºE	34ºE	36ºE	Latitude
30ºN	-0.0745	-0.1058	-0.138	-0.1695	-0.198	-0.2219	-0.2405	-0.2535	-0.2616	-0.2653	-0.2655	-0.2626	-0.2571	-0.2498	-0.2416	-0.2334	-0.2257	-0.2188	-0.2128	-0.2075	-0.2027	-0.1983	-0.1941	-0.19	30ºN
32ºN	-0.0321	-0.0624	-0.096	-0.1318	-0.1673	-0.1999	-0.227	-0.2472	-0.2606	-0.2679	-0.2699	-0.2672	-0.2606	-0.2511	-0.2403	-0.2298	-0.2206	-0.2129	-0.2065	-0.201	-0.1961	-0.1916	-0.1873	-0.1832	32ºN
34ºN	0.0128	-0.0132	-0.044	-0.08	-0.1199	-0.1612	-0.1999	-0.2321	-0.2553	-0.2693	-0.2753	-0.2741	-0.266	-0.2526	-0.2373	-0.2236	-0.2128	-0.2046	-0.1982	-0.1928	-0.188	-0.1835	-0.1791	-0.1748	34ºN
36ºN	0.0564	0.0369	0.0128	-0.0171	-0.054	-0.0979	-0.1469	-0.1956	-0.2368	-0.2655	-0.281	-0.2845	-0.2751	-0.2541	-0.2306	-0.2124	-0.2007	-0.1931	-0.1874	-0.1824	-0.1778	-0.1733	-0.1688	-0.1643	36ºN
38ºN	0.0953	0.0827	0.0671	0.0473	0.0216	-0.0124	-0.0573	-0.1142	-0.1788	-0.2383	-0.2793	-0.2985	-0.2898	-0.2539	-0.2152	-0.1924	-0.1836	-0.1786	-0.174	-0.1695	-0.1649	-0.1602	-0.1555	-0.1507	38ºN
40ºN	0.1277	0.1209	0.1128	0.1028	0.09	0.0729	0.0491	0.0142	-0.0387	-0.1166	-0.2113	-0.2822	-0.2932	-0.2627	-0.1919	-0.1693	-0.167	-0.1627	-0.1577	-0.1526	-0.1475	-0.1425	-0.1375	-0.1325	40ºN
42ºN	0.1534	0.1507	0.1477	0.1445	0.1407	0.1363	0.1308	0.1234	0.1121	0.0918	0.0485	-0.0512	-0.1986	-0.2846	-0.2329	-0.1775	-0.1535	-0.1413	-0.1333	-0.127	-0.1215	-0.1165	-0.1116	-0.1069	42ºN
44ºN	0.1731	0.1729	0.1729	0.1732	0.1741	0.1756	0.178	0.1815	0.1864	0.1927	0.1992	0.1915	0.1175	-0.0142	-0.0904	-0.1002	-0.0941	-0.0884	-0.0845	-0.0814	-0.0787	-0.0759	-0.0731	-0.07	44ºN
46ºN	0.1879	0.189	0.1904	0.1923	0.1949	0.1982	0.2023	0.2073	0.2132	0.2195	0.2241	0.2247	0.2085	0.171	0.1253	0.0797	0.0443	0.0205	0.0052	-0.0048	-0.0112	-0.0153	-0.0177	-0.0188	46ºN
48ºN	0.199	0.2005	0.2025	0.2048	0.2076	0.2109	0.2146	0.2186	0.2226	0.2261	0.2284	0.2281	0.2221	0.2098	0.1929	0.1725	0.1495	0.126	0.1042	0.0857	0.0707	0.0591	0.0503	0.0437	48ºN
50ºN	0.2072	0.2089	0.2108	0.2131	0.2156	0.2183	0.2211	0.224	0.2266	0.2287	0.2298	0.2294	0.2268	0.2218	0.2146	0.2054	0.1942	0.1813	0.1671	0.1526	0.1386	0.1258	0.1147	0.1053	50ºN
52ºN	0.2133	0.2149	0.2167	0.2186	0.2207	0.2228	0.225	0.227	0.2288	0.2301	0.2308	0.2307	0.2295	0.2272	0.2238	0.2192	0.2135	0.2067	0.1988	0.19	0.1806	0.1711	0.1619	0.1534	52ºN
54ºN	0.2179	0.2193	0.2209	0.2225	0.2242	0.2259	0.2275	0.229	0.2303	0.2312	0.2318	0.2319	0.2314	0.2303	0.2286	0.2262	0.2231	0.2194	0.2149	0.2098	0.2041	0.198	0.1917	0.1854	54ºN
56ºN	0.2215	0.2227	0.224	0.2253	0.2267	0.228	0.2293	0.2304	0.2314	0.2322	0.2327	0.2329	0.2328	0.2323	0.2314	0.2302	0.2285	0.2263	0.2237	0.2207	0.2173	0.2135	0.2095	0.2053	56ºN
58ºN	0.2243	0.2253	0.2264	0.2275	0.2286	0.2297	0.2307	0.2316	0.2324	0.2331	0.2335	0.2338	0.2339	0.2337	0.2333	0.2327	0.2317	0.2305	0.229	0.2272	0.2251	0.2228	0.2202	0.2175	58ºN
60ºN	0.2265	0.2274	0.2283	0.2292	0.2301	0.231	0.2318	0.2326	0.2333	0.2338	0.2343	0.2346	0.2348	0.2348	0.2347	0.2343	0.2339	0.2332	0.2323	0.2312	0.23	0.2285	0.2269	0.2252	60ºN
62ºN	0.2284	0.2292	0.2299	0.2307	0.2314	0.2321	0.2328	0.2335	0.234	0.2345	0.235	0.2353	0.2355	0.2356	0.2357	0.2356	0.2353	0.235	0.2345	0.2338	0.2331	0.2322	0.2312	0.2302	62ºN
64ºN	0.23	0.2306	0.2312	0.2319	0.2325	0.2331	0.2337	0.2342	0.2347	0.2352	0.2356	0.2359	0.2361	0.2363	0.2364	0.2364	0.2364	0.2362	0.236	0.2356	0.2352	0.2347	0.2341	0.2334	64ºN
66ºN	0.2313	0.2319	0.2324	0.2329	0.2334	0.234	0.2344	0.2349	0.2353	0.2357	0.2361	0.2364	0.2367	0.2369	0.237	0.2371	0.2371	0.2371	0.237	0.2369	0.2367	0.2364	0.236	0.2357	66ºN
68ºN	0.2325	0.2329	0.2334	0.2338	0.2343	0.2347	0.2351	0.2355	0.2359	0.2363	0.2366	0.2369	0.2371	0.2373	0.2375	0.2376	0.2377	0.2378	0.2378	0.2378	0.2377	0.2375	0.2374	0.2372	68ºN
70ºN	0.2335	0.2339	0.2343	0.2347	0.235	0.2354	0.2358	0.2361	0.2364	0.2367	0.237	0.2373	0.2375	0.2377	0.2379	0.2381	0.2382	0.2383	0.2384	0.2384	0.2384	0.2384	0.2383	0.2382	70ºN
Max values	0.2335	0.2339	0.2343	0.2347	0.235	0.2354	0.2358	0.2361	0.2364	0.2367	0.237	0.2373	0.2375	0.2377	0.2379	0.2381	0.2382	0.2383	0.2384	0.2384	0.2384	0.2384	0.2383	0.2382	
Min values	-0.0745	-0.1058	-0.138	-0.1695	-0.198	-0.2219	-0.2405	-0.2535	-0.2616	-0.2693	-0.281	-0.2985	-0.2932	-0.2846	-0.2416	-0.2334	-0.2257	-0.2188	-0.2128	-0.2075	-0.2027	-0.1983	-0.1941	-0.19	

Overall max	0.2384
overall min	-0.2985

Appendix 6:

Dataset A: Correlation coefficients at 504 potential origin points in an exercise to find the most likely origin to sites in Italy and the enclaves.

	Longitude W																								
Latitude	10ºW	8ºW	6ºW	4ºW	2ºW	0ºW	2ºE	4ºE	6ºE	8ºE	10ºE	12ºE	14ºE	16ºE	18ºE	20ºE	22ºE	24ºE	26ºE	28ºE	30ºE	32ºE	34ºE	36ºE	Latitude
30ºN	0.1993	0.186	0.1745	0.1605	0.1423	0.1165	0.0781	0.0206	-0.0593	-0.1508	-0.2287	-0.2776	-0.3024	-0.313	-0.3164	-0.3164	-0.315	-0.3129	-0.3105	-0.3081	-0.3056	-0.3032	-0.3007	-0.2983	30ºN
32ºN	0.2084	0.1993	0.1895	0.1784	0.164	0.1428	0.1088	0.0528	-0.0328	-0.1368	-0.2246	-0.2764	-0.3004	-0.3096	-0.3117	-0.3109	-0.309	-0.3067	-0.3044	-0.3021	-0.2997	-0.2974	-0.295	-0.2927	32ºN
34ºN	0.2218	0.2135	0.2052	0.1972	0.1878	0.1731	0.1461	0.0947	0.0052	-0.1138	-0.2155	-0.2726	-0.2971	-0.305	-0.3057	-0.3041	-0.3019	-0.2997	-0.2975	-0.2954	-0.2932	-0.291	-0.2887	-0.2865	34ºN
36ºN	0.2376	0.2296	0.2217	0.2181	0.2144	0.2082	0.1916	0.1489	0.0593	-0.076	-0.198	-0.2659	-0.2929	-0.2994	-0.2982	-0.2956	-0.2935	-0.2916	-0.2898	-0.2879	-0.2859	-0.2839	-0.2818	-0.2797	36ºN
38ºN	0.258	0.2521	0.2485	0.2486	0.2462	0.2517	0.248	0.2164	0.1314	-0.0174	-0.1651	-0.2544	-0.2875	-0.2923	-0.2882	-0.2849	-0.2836	-0.2825	-0.2811	-0.2794	-0.2777	-0.2758	-0.2739	-0.272	38ºN
40ºN	0.2845	0.2841	0.2881	0.2967	0.3075	0.3205	0.3251	0.3014	0.2182	0.0579	-0.1048	-0.2181	-0.2712	-0.2847	-0.2758	-0.273	-0.2731	-0.2723	-0.2711	-0.2696	-0.2681	-0.2665	-0.2649	-0.2633	40ºN
42ºN	0.3146	0.3209	0.3315	0.3469	0.3669	0.3894	0.4072	0.3871	0.2988	0.1548	-0.0017	-0.1487	-0.2329	-0.2699	-0.2689	-0.2632	-0.2609	-0.2596	-0.2586	-0.2577	-0.2566	-0.2556	-0.2544	-0.2532	42ºN
44ºN	0.3426	0.3526	0.3661	0.383	0.4022	0.4202	0.4293	0.403	0.3124	0.169	0.0472	-0.0758	-0.1618	-0.21	-0.2302	-0.2373	-0.24	-0.2414	-0.2422	-0.2425	-0.2426	-0.2425	-0.2421	-0.2416	44ºN
46ºN	0.3649	0.3754	0.3878	0.4009	0.4122	0.4167	0.406	0.3668	0.2885	0.1796	0.0671	-0.0274	-0.0973	-0.1483	-0.1804	-0.1991	-0.21	-0.2166	-0.2209	-0.2237	-0.2256	-0.2269	-0.2277	-0.2282	46ºN
48ºN	0.3799	0.3886	0.3973	0.4042	0.4065	0.3996	0.3776	0.334	0.2661	0.18	0.0908	0.0126	-0.0502	-0.0984	-0.1338	-0.1587	-0.1757	-0.1875	-0.1957	-0.2016	-0.2058	-0.209	-0.2112	-0.2129	48ºN
50ºN	0.3879	0.3934	0.3974	0.398	0.3927	0.378	0.3505	0.3072	0.2486	0.1796	0.1087	0.0434	-0.0122	-0.0575	-0.0933	-0.1207	-0.1413	-0.1568	-0.1684	-0.1771	-0.1838	-0.1889	-0.1928	-0.1958	50ºN
52ºN	0.3896	0.3916	0.391	0.3862	0.3753	0.356	0.3265	0.2861	0.2358	0.1794	0.1216	0.0669	0.0181	-0.0236	-0.0581	-0.0862	-0.1086	-0.1264	-0.1404	-0.1515	-0.1603	-0.1672	-0.1728	-0.1772	52ºN
54ºN	0.3863	0.3848	0.3802	0.3713	0.3568	0.3354	0.3062	0.2694	0.2263	0.1793	0.1313	0.085	0.0425	0.0048	-0.0277	-0.0553	-0.0782	-0.0972	-0.1128	-0.1255	-0.136	-0.1446	-0.1516	-0.1574	54ºN
56ºN	0.3791	0.3746	0.3669	0.3552	0.3387	0.3168	0.2891	0.2562	0.219	0.1793	0.1388	0.0994	0.0623	0.0286	-0.0015	-0.0277	-0.0504	-0.0697	-0.0861	-0.0999	-0.1116	-0.1214	-0.1296	-0.1366	56ºN
58ºN	0.3691	0.3621	0.3523	0.3389	0.3217	0.3002	0.2746	0.2454	0.2133	0.1793	0.1449	0.111	0.0787	0.0487	0.0213	-0.0033	-0.025	-0.0441	-0.0607	-0.0751	-0.0875	-0.0981	-0.1073	-0.1152	58ºN
60ºN	0.3571	0.3484	0.3372	0.3231	0.3059	0.2856	0.2623	0.2364	0.2085	0.1794	0.1499	0.1207	0.0925	0.0659	0.0411	0.0185	-0.002	-0.0204	-0.0367	-0.0512	-0.0639	-0.0751	-0.0849	-0.0934	60ºN
62ºN	0.3439	0.3341	0.3223	0.3081	0.2916	0.2727	0.2517	0.2288	0.2046	0.1795	0.154	0.1288	0.1042	0.0807	0.0585	0.0379	0.0189	0.0015	-0.0142	-0.0284	-0.0411	-0.0525	-0.0626	-0.0715	62ºN
64ºN	0.33	0.3198	0.3078	0.2941	0.2785	0.2612	0.2424	0.2224	0.2013	0.1796	0.1576	0.1357	0.1143	0.0936	0.0739	0.0553	0.0378	0.0217	0.0068	-0.0068	-0.0192	-0.0305	-0.0406	-0.0498	64ºN
66ºN	0.3159	0.3056	0.294	0.281	0.2666	0.251	0.2343	0.2167	0.1984	0.1796	0.1607	0.1418	0.1231	0.105	0.0875	0.0709	0.0551	0.0403	0.0264	0.0136	0.0017	-0.0092	-0.0191	-0.0282	66ºN
68ºN	0.3019	0.2919	0.2809	0.2688	0.2558	0.2419	0.2271	0.2118	0.1959	0.1797	0.1634	0.147	0.1309	0.1151	0.0997	0.0849	0.0708	0.0574	0.0447	0.0328	0.0217	0.0113	0.0018	-0.0071	68ºN
70ºN	0.2882	0.2788	0.2686	0.2576	0.2459	0.2336	0.2207	0.2074	0.1937	0.1798	0.1657	0.1517	0.1377	0.124	0.1107	0.0977	0.0852	0.0732	0.0617	0.0509	0.0406	0.031	0.022	0.0135	70ºN

Max values	0.3896	0.3934	0.3974	0.4042	0.4122	0.4202	0.4293	0.403	0.3124	0.18	0.1657	0.1517	0.1377	0.124	0.1107	0.0977	0.0852	0.0732	0.0617	0.0509	0.0406	0.031	0.022	0.0135
Min values	0.1993	0.186	0.1745	0.1605	0.1423	0.1165	0.0781	0.0206	-0.0593	-0.1508	-0.2287	-0.2776	-0.3024	-0.313	-0.3164	-0.3164	-0.315	-0.3129	-0.3105	-0.3081	-0.3056	-0.3032	-0.3007	-0.2983

Overall max	0.4293
overall min	-0.3164

Appendix 6:
Dataset A: Correlation coefficients at 504 potential origin points to sites in Zvelebil and Lillie's southern acculturation zone.

	Longitude W																								
Latitude	10ºW	8ºW	6ºW	4ºW	2ºW	0ºW	2ºE	4ºE	6ºE	8ºE	10ºE	12ºE	14ºE	16ºE	18ºE	20ºE	22ºE	24ºE	26ºE	28ºE	30ºE	32ºE	34ºE	36ºE	Latitude
30ºN	-0.499	-0.5202	-0.538	-0.5525	-0.5642	-0.5734	-0.5803	-0.5852	-0.5884	-0.5902	-0.5908	-0.5904	-0.5892	-0.5872	-0.5848	-0.5819	-0.5786	-0.5751	-0.5712	-0.5672	-0.5628	-0.5583	-0.5535	-0.5484	30ºN
32ºN	-0.485	-0.5101	-0.531	-0.5481	-0.5616	-0.572	-0.5797	-0.585	-0.5883	-0.5899	-0.5901	-0.5892	-0.5874	-0.5848	-0.5818	-0.5783	-0.5744	-0.5703	-0.5659	-0.5613	-0.5565	-0.5514	-0.546	-0.5404	32ºN
34ºN	-0.466	-0.4962	-0.522	-0.542	-0.5581	-0.5701	-0.5787	-0.5844	-0.5878	-0.5891	-0.5888	-0.5873	-0.5847	-0.5814	-0.5776	-0.5734	-0.5689	-0.5641	-0.5591	-0.5539	-0.5485	-0.5428	-0.5368	-0.5306	34ºN
36ºN	-0.438	-0.4765	-0.508	-0.5335	-0.553	-0.5672	-0.577	-0.5833	-0.5867	-0.5877	-0.5867	-0.5843	-0.5807	-0.5764	-0.5717	-0.5666	-0.5614	-0.5559	-0.5502	-0.5444	-0.5383	-0.5319	-0.5253	-0.5184	36ºN
38ºN	-0.397	-0.4466	-0.488	-0.5205	-0.5451	-0.5621	-0.5737	-0.5809	-0.5846	-0.5852	-0.5832	-0.5794	-0.5746	-0.5691	-0.5632	-0.5572	-0.5511	-0.5449	-0.5385	-0.5319	-0.5251	-0.5181	-0.5107	-0.503	38ºN
40ºN	-0.33	-0.3971	-0.453	-0.497	-0.5297	-0.5521	-0.5668	-0.5762	-0.5809	-0.5809	-0.5772	-0.5715	-0.5648	-0.5578	-0.5508	-0.5439	-0.5369	-0.5299	-0.5228	-0.5155	-0.5079	-0.5001	-0.4919	-0.4833	40ºN
42ºN	-0.218	-0.3077	-0.387	-0.4512	-0.4998	-0.5332	-0.5529	-0.5677	-0.5748	-0.5733	-0.5662	-0.5573	-0.5485	-0.5401	-0.5321	-0.5244	-0.5168	-0.5091	-0.5013	-0.4933	-0.485	-0.4764	-0.4673	-0.4578	42ºN
44ºN	-0.036	-0.1457	-0.254	-0.3532	-0.4343	-0.4931	-0.5327	-0.5554	-0.5595	-0.5555	-0.542	-0.5302	-0.5204	-0.5117	-0.5036	-0.4958	-0.4879	-0.4798	-0.4715	-0.4629	-0.454	-0.4446	-0.4347	-0.4243	44ºN
46ºN	0.196	0.1004	-0.015	-0.1425	-0.2715	-0.381	-0.46	-0.4981	-0.5005	-0.4922	-0.4841	-0.4781	-0.4723	-0.4662	-0.4597	-0.4528	-0.4454	-0.4376	-0.4293	-0.4204	-0.4111	-0.4011	-0.3906	-0.3795	46ºN
48ºN	0.4014	0.3519	0.2804	0.1828	0.0588	-0.0933	-0.241	-0.3344	-0.3721	-0.3785	-0.3839	-0.3911	-0.3939	-0.3936	-0.3912	-0.3871	-0.3818	-0.3754	-0.3681	-0.3599	-0.3509	-0.3411	-0.3306	-0.3193	48ºN
50ºN	0.5286	0.5185	0.5005	0.4715	0.4228	0.3244	0.1684	0.0109	-0.104	-0.1754	-0.2182	-0.2457	-0.2639	-0.2754	-0.2819	-0.2848	-0.2849	-0.2828	-0.2788	-0.2733	-0.2664	-0.2584	-0.2492	-0.239	50ºN
52ºN	0.5867	0.5936	0.6001	0.5989	0.5845	0.544	0.4772	0.3825	0.2693	0.1586	0.0671	-0.0018	-0.0516	-0.0872	-0.1124	-0.1299	-0.1416	-0.1488	-0.1524	-0.1531	-0.1515	-0.1478	-0.1424	-0.1355	52ºN
54ºN	0.6071	0.6132	0.6215	0.6261	0.622	0.6076	0.5835	0.5452	0.4906	0.4219	0.3461	0.272	0.2059	0.1505	0.1059	0.0709	0.0439	0.0235	0.0086	-0.0018	-0.0085	-0.0121	-0.013	-0.0117	54ºN
56ºN	0.6088	0.6105	0.6117	0.6128	0.614	0.6118	0.6033	0.5873	0.5628	0.5296	0.4885	0.442	0.3933	0.3457	0.3015	0.2622	0.2283	0.1999	0.1767	0.158	0.1435	0.1325	0.1248	0.1198	56ºN
58ºN	0.6039	0.6029	0.6008	0.5973	0.5968	0.598	0.596	0.5897	0.5785	0.5622	0.5411	0.5156	0.4868	0.456	0.4246	0.3939	0.365	0.3385	0.3148	0.2942	0.2766	0.2619	0.25	0.2407	58ºN
60ºN	0.5985	0.5965	0.5938	0.5909	0.5895	0.5894	0.5882	0.5851	0.5793	0.5707	0.5591	0.5449	0.5282	0.5096	0.4896	0.469	0.4483	0.4282	0.4091	0.3914	0.3754	0.3611	0.3486	0.338	60ºN
62ºN	0.5951	0.5935	0.5916	0.5897	0.5881	0.5866	0.5848	0.5822	0.5783	0.573	0.566	0.5574	0.5473	0.5357	0.5231	0.5097	0.4958	0.4817	0.4678	0.4544	0.4417	0.4298	0.419	0.4093	62ºN
64ºN	0.5932	0.5918	0.5904	0.5888	0.5872	0.5855	0.5835	0.5811	0.578	0.5742	0.5694	0.5638	0.5571	0.5496	0.5414	0.5325	0.5231	0.5135	0.5037	0.494	0.4846	0.4755	0.4669	0.4589	64ºN
66ºN	0.5918	0.5907	0.5894	0.588	0.5865	0.5849	0.583	0.5808	0.5782	0.5752	0.5717	0.5677	0.563	0.5579	0.5522	0.5462	0.5397	0.533	0.5262	0.5193	0.5124	0.5057	0.4991	0.4929	66ºN
68ºN	0.5907	0.5897	0.5885	0.5873	0.5859	0.5844	0.5827	0.5808	0.5786	0.5762	0.5735	0.5704	0.567	0.5633	0.5593	0.555	0.5505	0.5457	0.5408	0.5359	0.5309	0.526	0.5211	0.5164	68ºN
70ºN	0.5897	0.5888	0.5877	0.5866	0.5854	0.584	0.5825	0.5809	0.5791	0.5771	0.575	0.5726	0.57	0.5672	0.5643	0.5611	0.5578	0.5544	0.5509	0.5473	0.5437	0.54	0.5365	0.5329	70ºN
Max values by column	0.6088	0.6132	0.6215	0.6261	0.622	0.6118	0.6033	0.5897	0.5793	0.5771	0.575	0.5726	0.57	0.5672	0.5643	0.5611	0.5578	0.5544	0.5509	0.5473	0.5437	0.54	0.5365	0.5329	
Min values	-0.499	-0.5202	-0.538	-0.5525	-0.5642	-0.5734	-0.5803	-0.5852	-0.5884	-0.5902	-0.5908	-0.5904	-0.5892	-0.5872	-0.5848	-0.5819	-0.5786	-0.5751	-0.5712	-0.5672	-0.5628	-0.5583	-0.5535	-0.5484	

Overall max	0.6261
Overall min	-0.591

Appendix 6:
Correlation coefficients at 504 potential origin points to sites in Zvelebil and Lillie's southern acculturation zone with the exclusion of enclave sites.

	Longitude W																								
Latitude	10ºW	8ºW	6ºW	4ºW	2ºW	0ºW	2ºE	4ºE	6ºE	8ºE	10ºE	12ºE	14ºE	16ºE	18ºE	20ºE	22ºE	24ºE	26ºE	28ºE	30ºE	32ºE	34ºE	36ºE	Latitude
30ºN	-0.485	-0.5045	-0.521	-0.5338	-0.5443	-0.5524	-0.5585	-0.5629	-0.5657	-0.5673	-0.5679	-0.5675	-0.5664	-0.5647	-0.5626	-0.5601	-0.5573	-0.5542	-0.5509	-0.5474	-0.5438	-0.5399	-0.5358	-0.5315	30ºN
32ºN	-0.472	-0.4958	-0.515	-0.5304	-0.5425	-0.5517	-0.5586	-0.5633	-0.5663	-0.5677	-0.5679	-0.5671	-0.5655	-0.5632	-0.5605	-0.5575	-0.5542	-0.5506	-0.5469	-0.5429	-0.5388	-0.5344	-0.5299	-0.5251	32ºN
34ºN	-0.455	-0.4837	-0.507	-0.5256	-0.54	-0.5508	-0.5587	-0.5639	-0.5669	-0.5681	-0.5678	-0.5663	-0.564	-0.561	-0.5576	-0.5539	-0.55	-0.5459	-0.5416	-0.5371	-0.5325	-0.5276	-0.5226	-0.5173	34ºN
36ºN	-0.43	-0.4663	-0.496	-0.5187	-0.5364	-0.5495	-0.5587	-0.5646	-0.5676	-0.5683	-0.5672	-0.5648	-0.5615	-0.5576	-0.5533	-0.5489	-0.5443	-0.5395	-0.5347	-0.5296	-0.5244	-0.519	-0.5134	-0.5075	36ºN
38ºN	-0.392	-0.4397	-0.478	-0.5082	-0.5309	-0.5474	-0.5586	-0.5653	-0.5683	-0.5682	-0.5658	-0.562	-0.5573	-0.5522	-0.547	-0.5417	-0.5364	-0.531	-0.5255	-0.5198	-0.514	-0.508	-0.5017	-0.4952	38ºN
40ºN	-0.332	-0.396	-0.449	-0.4904	-0.5217	-0.5437	-0.558	-0.5662	-0.5689	-0.5671	-0.5625	-0.5565	-0.5501	-0.5437	-0.5375	-0.5313	-0.5252	-0.5192	-0.513	-0.5068	-0.5003	-0.4936	-0.4866	-0.4793	40ºN
42ºN	-0.233	-0.3196	-0.395	-0.4564	-0.5035	-0.5366	-0.5557	-0.567	-0.5687	-0.5631	-0.5547	-0.5458	-0.5376	-0.5299	-0.5228	-0.516	-0.5094	-0.5027	-0.496	-0.4891	-0.482	-0.4746	-0.4668	-0.4586	42ºN
44ºN	-0.075	-0.1821	-0.287	-0.3808	-0.4582	-0.514	-0.5505	-0.5673	-0.5619	-0.5491	-0.5356	-0.5245	-0.5154	-0.5076	-0.5003	-0.4934	-0.4864	-0.4794	-0.4722	-0.4648	-0.457	-0.4488	-0.4402	-0.4311	44ºN
46ºN	0.137	0.0369	-0.08	-0.2062	-0.3305	-0.4327	-0.5028	-0.5304	-0.5215	-0.5038	-0.4912	-0.4838	-0.4777	-0.4717	-0.4656	-0.4593	-0.4526	-0.4456	-0.4382	-0.4304	-0.422	-0.4132	-0.4039	-0.394	46ºN
48ºN	0.3432	0.2861	0.2062	0.1001	-0.0328	-0.1885	-0.3275	-0.4052	-0.4259	-0.418	-0.4143	-0.4165	-0.4165	-0.4145	-0.4112	-0.4067	-0.4013	-0.3951	-0.3882	-0.3805	-0.3722	-0.3631	-0.3534	-0.3429	48ºN
50ºN	0.4831	0.4699	0.4482	0.4146	0.3585	0.2452	0.0713	-0.0923	-0.198	-0.2539	-0.2829	-0.3006	-0.3119	-0.3187	-0.3219	-0.3225	-0.3209	-0.3177	-0.313	-0.307	-0.2999	-0.2918	-0.2827	-0.2727	50ºN
52ºN	0.5511	0.5584	0.5656	0.5647	0.5489	0.5026	0.4254	0.3164	0.1893	0.0721	-0.0176	-0.0804	-0.1234	-0.1528	-0.1728	-0.1861	-0.1945	-0.199	-0.2005	-0.1996	-0.1966	-0.1919	-0.1856	-0.178	52ºN
54ºN	0.5764	0.5834	0.5929	0.5985	0.5944	0.579	0.553	0.5114	0.4513	0.3749	0.2911	0.211	0.142	0.0862	0.0428	0.0097	-0.015	-0.0331	-0.0459	-0.0544	-0.0594	-0.0614	-0.0609	-0.0584	54ºN
56ºN	0.5799	0.5821	0.5837	0.5852	0.587	0.5851	0.5768	0.5607	0.5355	0.5006	0.4568	0.4066	0.354	0.3029	0.256	0.2149	0.1802	0.1516	0.1285	0.1104	0.0966	0.0866	0.0798	0.0757	56ºN
58ºN	0.5756	0.5748	0.5727	0.5692	0.5688	0.5706	0.5691	0.5632	0.5523	0.5362	0.5148	0.4886	0.4587	0.4264	0.3933	0.361	0.3306	0.3029	0.2784	0.2572	0.2393	0.2246	0.2128	0.2037	58ºN
60ºN	0.5704	0.5685	0.5658	0.5628	0.5614	0.5616	0.5608	0.5581	0.5528	0.5446	0.5334	0.5194	0.5027	0.4839	0.4636	0.4424	0.4211	0.4004	0.3807	0.3625	0.346	0.3314	0.3187	0.308	60ºN
62ºN	0.5672	0.5655	0.5637	0.5619	0.5603	0.559	0.5575	0.5552	0.5517	0.5467	0.5401	0.5319	0.5221	0.5108	0.4983	0.485	0.471	0.4569	0.4429	0.4293	0.4164	0.4044	0.3934	0.3836	62ºN
64ºN	0.5654	0.5641	0.5626	0.5612	0.5597	0.5581	0.5563	0.5542	0.5514	0.5478	0.5434	0.5381	0.5318	0.5246	0.5167	0.508	0.4989	0.4894	0.4798	0.4702	0.4608	0.4518	0.4432	0.4352	64ºN
66ºN	0.5642	0.5631	0.5619	0.5606	0.5592	0.5577	0.556	0.554	0.5516	0.5489	0.5456	0.5419	0.5376	0.5327	0.5274	0.5216	0.5154	0.5089	0.5023	0.4956	0.4889	0.4823	0.476	0.4699	66ºN
68ºN	0.5633	0.5623	0.5612	0.5601	0.5588	0.5574	0.5559	0.5541	0.5521	0.5499	0.5474	0.5446	0.5414	0.538	0.5342	0.5302	0.5259	0.5214	0.5167	0.512	0.5072	0.5025	0.4978	0.4933	68ºN
70ºN	0.5625	0.5616	0.5606	0.5596	0.5585	0.5572	0.5559	0.5544	0.5527	0.5509	0.5489	0.5467	0.5443	0.5418	0.539	0.5361	0.533	0.5298	0.5265	0.5231	0.5197	0.5162	0.5128	0.5095	70ºN
Max values by column	0.5799	0.5834	0.5929	0.5985	0.5944	0.5851	0.5768	0.5632	0.5528	0.5509	0.5489	0.5467	0.5443	0.5418	0.539	0.5361	0.533	0.5298	0.5265	0.5231	0.5197	0.5162	0.5128	0.5095	
Min values	-0.485	-0.5045	-0.521	-0.5338	-0.5443	-0.5524	-0.5587	-0.5673	-0.5689	-0.5683	-0.5679	-0.5675	-0.5664	-0.5647	-0.5626	-0.5601	-0.5573	-0.5542	-0.5509	-0.5474	-0.5438	-0.5399	-0.5358	-0.5315	

Overall max	0.5985
Overall min	-0.569

Appendix 6:
Dataset A: Correlation coefficients at 504 potential origin points in an exercise to find the most likely origin to sites in Zvelebil and Lillie's northern acculturation zone.

	Longitude W																								
Latitude	10°W	8°W	6°W	4°W	2°W	0°W	2°E	4°E	6°E	8°E	10°E	12°E	14°E	16°E	18°E	20°E	22°E	24°E	26°E	28°E	30°E	32°E	34°E	36°E	Latitude
30°N	0.3233	0.2975	0.2703	0.2421	0.213	0.1832	0.1531	0.1228	0.0926	0.0628	0.0335	0.005	-0.0226	-0.0492	-0.0746	-0.0989	-0.1218	-0.1436	-0.1642	-0.1836	-0.2019	-0.2193	-0.2357	-0.2512	30°N
32°N	0.3375	0.3115	0.2838	0.2548	0.2247	0.1937	0.1621	0.1302	0.0984	0.0668	0.0358	0.0055	-0.0237	-0.0517	-0.0785	-0.1039	-0.1278	-0.1504	-0.1716	-0.1915	-0.2102	-0.2278	-0.2444	-0.2601	32°N
34°N	0.353	0.3269	0.2989	0.2692	0.238	0.2057	0.1726	0.1389	0.1051	0.0715	0.0384	0.0062	-0.0249	-0.0547	-0.083	-0.1097	-0.1347	-0.1581	-0.18	-0.2004	-0.2194	-0.2373	-0.254	-0.2698	34°N
36°N	0.37	0.3441	0.3159	0.2855	0.2534	0.2197	0.1848	0.1492	0.1132	0.0772	0.0416	0.007	-0.0264	-0.0582	-0.0883	-0.1165	-0.1427	-0.167	-0.1896	-0.2104	-0.2298	-0.2479	-0.2647	-0.2804	36°N
38°N	0.3886	0.3632	0.335	0.3043	0.2712	0.2361	0.1994	0.1615	0.1229	0.084	0.0456	0.008	-0.0282	-0.0625	-0.0947	-0.1245	-0.1521	-0.1774	-0.2006	-0.2219	-0.2416	-0.2597	-0.2765	-0.2922	38°N
40°N	0.4087	0.3843	0.3566	0.3258	0.2921	0.2557	0.217	0.1766	0.1349	0.0926	0.0505	0.0093	-0.0303	-0.0677	-0.1025	-0.1343	-0.1633	-0.1896	-0.2135	-0.2351	-0.2549	-0.273	-0.2897	-0.3053	40°N
42°N	0.4301	0.4074	0.3809	0.3506	0.3166	0.2791	0.2385	0.1953	0.1501	0.1037	0.057	0.011	-0.0331	-0.0743	-0.1122	-0.1464	-0.177	-0.2042	-0.2285	-0.2503	-0.2701	-0.288	-0.3045	-0.3197	42°N
44°N	0.4525	0.4324	0.408	0.3791	0.3455	0.3075	0.2652	0.2191	0.1698	0.1183	0.0656	0.0135	-0.0366	-0.083	-0.1248	-0.1616	-0.1937	-0.2217	-0.2462	-0.268	-0.2874	-0.305	-0.321	-0.3357	44°N
46°N	0.4748	0.4585	0.4375	0.4113	0.3796	0.342	0.2987	0.2499	0.1961	0.1383	0.078	0.0172	-0.0413	-0.0948	-0.1416	-0.1813	-0.2147	-0.2429	-0.2672	-0.2884	-0.3072	-0.3241	-0.3394	-0.3534	46°N
48°N	0.4951	0.4841	0.4683	0.4469	0.419	0.3839	0.3411	0.2905	0.2323	0.1672	0.0965	0.0232	-0.048	-0.1118	-0.1648	-0.2071	-0.2409	-0.2685	-0.2917	-0.3119	-0.3297	-0.3455	-0.3599	-0.3729	48°N
50°N	0.5104	0.5059	0.4972	0.4833	0.4627	0.4338	0.395	0.3452	0.2836	0.2106	0.1267	0.0345	-0.0576	-0.1377	-0.1981	-0.2412	-0.2732	-0.2989	-0.3204	-0.3389	-0.3551	-0.3695	-0.3824	-0.394	50°N
52°N	0.5161	0.5183	0.5182	0.5144	0.5054	0.4888	0.4616	0.42	0.36	0.2805	0.1814	0.0586	-0.0715	-0.18	-0.2458	-0.284	-0.3119	-0.3345	-0.3534	-0.3696	-0.3836	-0.3959	-0.4067	-0.4164	52°N
54°N	0.5068	0.5141	0.521	0.5271	0.5315	0.5327	0.5284	0.5141	0.4828	0.4246	0.3374	0.1461	-0.0932	-0.2421	-0.3064	-0.3364	-0.3587	-0.3765	-0.3911	-0.4034	-0.414	-0.4232	-0.4312	-0.4383	54°N
56°N	0.4788	0.487	0.4958	0.5055	0.5161	0.5277	0.54	0.552	0.5611	0.5595	0.5013	0.209	-0.1327	-0.2807	-0.3546	-0.3882	-0.4057	-0.4175	-0.4266	-0.4343	-0.4407	-0.4464	-0.4512	-0.4555	56°N
58°N	0.4336	0.4378	0.442	0.446	0.4497	0.4528	0.4542	0.4518	0.4402	0.4067	0.3212	0.1402	-0.0652	-0.2081	-0.3062	-0.368	-0.4035	-0.4238	-0.4361	-0.4442	-0.4501	-0.4545	-0.458	-0.4608	58°N
60°N	0.3782	0.3761	0.3725	0.3667	0.358	0.3449	0.3256	0.2976	0.2579	0.2082	0.1517	0.0737	-0.0232	-0.1219	-0.2098	-0.2808	-0.3337	-0.371	-0.3965	-0.414	-0.4262	-0.4348	-0.441	-0.4456	60°N
62°N	0.3215	0.3141	0.3043	0.2916	0.275	0.2536	0.2263	0.1919	0.1508	0.1078	0.0703	0.028	-0.0239	-0.0822	-0.1419	-0.1983	-0.2482	-0.2901	-0.3238	-0.3503	-0.3707	-0.3863	-0.3983	-0.4075	62°N
64°N	0.27	0.2598	0.2473	0.2323	0.2144	0.1934	0.1692	0.142	0.1126	0.0818	0.0497	0.0156	-0.0214	-0.0611	-0.1023	-0.1435	-0.183	-0.2193	-0.2518	-0.2799	-0.3038	-0.3238	-0.3403	-0.3538	64°N
66°N	0.2263	0.2153	0.2026	0.1882	0.1718	0.1536	0.1335	0.1118	0.0887	0.0642	0.0385	0.0114	-0.0171	-0.0467	-0.0772	-0.1079	-0.1382	-0.1673	-0.1948	-0.22	-0.2428	-0.2631	-0.2809	-0.2963	66°N
68°N	0.19	0.1794	0.1676	0.1545	0.1402	0.1247	0.108	0.0902	0.0714	0.0516	0.0308	0.0092	-0.0132	-0.0362	-0.0597	-0.0833	-0.1068	-0.1299	-0.1521	-0.1732	-0.1931	-0.2114	-0.2282	-0.2433	68°N
70°N	0.1597	0.1499	0.1393	0.1279	0.1155	0.1024	0.0884	0.0736	0.0582	0.042	0.0252	0.0078	-0.01	-0.0282	-0.0467	-0.0653	-0.0838	-0.1021	-0.1199	-0.1372	-0.1538	-0.1695	-0.1842	-0.1979	70°N

Max values by column	0.5161	0.5183	0.521	0.5271	0.5315	0.5327	0.54	0.552	0.5611	0.5595	0.5013	0.209	-0.01	-0.0282	-0.0467	-0.0653	-0.0838	-0.1021	-0.1199	-0.1372	-0.1538	-0.1695	-0.1842	-0.1979
Min values	0.1597	0.1499	0.1393	0.1279	0.1155	0.1024	0.0884	0.0736	0.0582	0.042	0.0252	0.005	-0.1327	-0.2807	-0.3546	-0.3882	-0.4057	-0.4238	-0.4361	-0.4442	-0.4501	-0.4545	-0.458	-0.4608

Overall max	0.5611
Overall min	-0.461

Appendix 7:

Dataset G: Correlation coefficients at 504 origin points in an exercise to find the most likely origin to sites in Zvelebil and Lillie's demic diffusion region.

	Longitude W																							
Latitude	10ºW	8ºW	6ºW	4ºW	2ºW	0ºW	2ºE	4ºE	6ºE	8ºE	10ºE	12ºE	14ºE	16ºE	18ºE	20ºE	22ºE	24ºE	26ºE	28ºE	30ºE	32ºE	34ºE	36ºE	Latitude
30ºN	0.152	0.1031	0.0488	-0.0103	-0.0731	-0.1378	-0.2024	-0.2644	-0.3221	-0.3739	-0.419	-0.4569	-0.4877	-0.5112	-0.5281	-0.5393	-0.546	-0.5499	-0.552	-0.5531	-0.5535	-0.5535	-0.5528	-0.5516	30ºN
32ºN	0.211	0.1674	0.1175	0.0612	-0.001	-0.068	-0.1378	-0.2078	-0.2752	-0.3377	-0.3934	-0.4413	-0.4805	-0.5105	-0.5313	-0.5438	-0.5502	-0.5529	-0.5539	-0.5542	-0.5542	-0.5536	-0.5524	-0.5505	32ºN
34ºN	0.2684	0.2316	0.1885	0.1383	0.0804	0.0151	-0.0566	-0.1326	-0.2096	-0.2842	-0.3534	-0.4151	-0.4676	-0.5085	-0.536	-0.5505	-0.5556	-0.5562	-0.5555	-0.5548	-0.5543	-0.5532	-0.5512	-0.5483	34ºN
36ºN	0.322	0.2929	0.2582	0.2167	0.1672	0.1087	0.0407	-0.036	-0.1193	-0.2054	-0.2899	-0.3692	-0.4417	-0.5025	-0.5431	-0.5617	-0.5632	-0.5571	-0.5539	-0.5544	-0.554	-0.5521	-0.5487	-0.5443	36ºN
38ºN	0.37	0.3485	0.3227	0.2914	0.2531	0.2061	0.1487	0.0795	-0.0017	-0.0934	-0.1915	-0.2902	-0.3888	-0.4849	-0.5511	-0.5817	-0.5802	-0.5571	-0.5551	-0.5561	-0.5535	-0.5489	-0.543	-0.5365	38ºN
40ºN	0.4113	0.3966	0.3789	0.3576	0.3312	0.2982	0.2565	0.2033	0.136	0.0524	-0.0483	-0.1655	-0.2963	-0.4317	-0.5434	-0.6017	-0.6162	-0.5927	-0.5711	-0.5581	-0.5481	-0.5389	-0.5299	-0.5209	40ºN
42ºN	0.4454	0.4361	0.4252	0.4122	0.3964	0.3767	0.3514	0.3182	0.2738	0.2139	0.1337	0.0262	-0.1202	-0.3036	-0.4727	-0.5629	-0.5857	-0.5759	-0.5487	-0.5344	-0.5225	-0.5113	-0.5007	-0.4905	42ºN
44ºN	0.4725	0.4672	0.4612	0.4544	0.4464	0.4368	0.4248	0.4092	0.3879	0.3574	0.3124	0.2449	0.133	-0.0423	-0.2354	-0.3643	-0.4301	-0.461	-0.4645	-0.4612	-0.4561	-0.4496	-0.4424	-0.4347	44ºN
46ºN	0.4933	0.4907	0.4879	0.4849	0.4817	0.4783	0.4744	0.4698	0.4639	0.4557	0.4425	0.4182	0.3676	0.2674	0.1172	-0.0225	-0.1376	-0.2213	-0.2741	-0.3055	-0.3239	-0.3343	-0.3396	-0.3416	46ºN
48ºN	0.5088	0.5077	0.5066	0.5057	0.5049	0.5042	0.5037	0.5034	0.5032	0.5028	0.5014	0.4961	0.4782	0.4337	0.3554	0.2551	0.1485	0.0601	-0.0179	-0.0809	-0.1289	-0.1642	-0.1897	-0.2078	48ºN
50ºN	0.5199	0.5196	0.5193	0.5192	0.5191	0.5192	0.5194	0.5195	0.5194	0.5191	0.5185	0.5199	0.5145	0.4909	0.4517	0.3976	0.3314	0.2605	0.1908	0.1259	0.0685	0.0199	-0.02	-0.052	50ºN
52ºN	0.5277	0.5277	0.5276	0.5277	0.5277	0.5277	0.5276	0.5272	0.5264	0.5251	0.5236	0.5253	0.5228	0.5105	0.4889	0.4579	0.4172	0.3684	0.3171	0.2658	0.2165	0.1707	0.1294	0.0933	52ºN
54ºN	0.5331	0.533	0.533	0.533	0.5329	0.5327	0.5323	0.5317	0.5309	0.5299	0.5289	0.5274	0.5235	0.515	0.5007	0.48	0.4532	0.4216	0.3866	0.3497	0.3124	0.2757	0.2406	0.2079	54ºN
56ºN	0.5366	0.5365	0.5364	0.5362	0.5359	0.5355	0.535	0.5343	0.5333	0.532	0.5303	0.5275	0.5229	0.5156	0.5048	0.4903	0.472	0.4503	0.4261	0.3999	0.3726	0.3448	0.3173	0.2906	56ºN
58ºN	0.5387	0.5385	0.5383	0.5379	0.5375	0.5369	0.5362	0.5352	0.5339	0.5322	0.5298	0.5265	0.5218	0.5154	0.5069	0.496	0.4827	0.4673	0.4499	0.431	0.411	0.3902	0.3692	0.3483	58ºN
60ºN	0.5398	0.5395	0.5391	0.5386	0.538	0.5372	0.5362	0.5349	0.5333	0.5313	0.5286	0.5252	0.5207	0.5151	0.5081	0.4996	0.4895	0.478	0.4652	0.4512	0.4363	0.4207	0.4047	0.3886	60ºN
62ºN	0.5401	0.5396	0.5391	0.5384	0.5376	0.5366	0.5354	0.5339	0.5321	0.5299	0.5272	0.5238	0.5197	0.5148	0.5089	0.5021	0.4942	0.4854	0.4756	0.465	0.4537	0.4419	0.4297	0.4172	62ºN
64ºN	0.5397	0.5391	0.5384	0.5376	0.5366	0.5355	0.5341	0.5325	0.5306	0.5284	0.5257	0.5225	0.5188	0.5145	0.5096	0.5039	0.4976	0.4907	0.4831	0.4749	0.4662	0.4571	0.4477	0.438	64ºN
66ºN	0.5388	0.5381	0.5373	0.5363	0.5353	0.534	0.5326	0.5309	0.529	0.5268	0.5242	0.5213	0.518	0.5143	0.5101	0.5054	0.5003	0.4947	0.4887	0.4823	0.4755	0.4684	0.461	0.4535	66ºN
68ºN	0.5375	0.5367	0.5358	0.5348	0.5336	0.5323	0.5308	0.5292	0.5273	0.5252	0.5228	0.5202	0.5173	0.514	0.5105	0.5066	0.5024	0.4979	0.493	0.4879	0.4826	0.477	0.4712	0.4653	68ºN
70ºN	0.5358	0.5349	0.534	0.533	0.5318	0.5305	0.529	0.5274	0.5257	0.5237	0.5216	0.5192	0.5166	0.5139	0.5108	0.5076	0.5041	0.5004	0.4965	0.4925	0.4882	0.4838	0.4792	0.4746	70ºN

Max values	0.5401	0.5396	0.5391	0.5386	0.538	0.5372	0.5362	0.5352	0.5339	0.5322	0.5303	0.5275	0.5235	0.5156	0.5108	0.5076	0.5041	0.5004	0.4965	0.4925	0.4882	0.4838	0.4792	0.4746
Min values	0.152	0.1031	0.0488	-0.0103	-0.0731	-0.1378	-0.2024	-0.2644	-0.3221	-0.3739	-0.419	-0.4569	-0.4877	-0.5112	-0.5511	-0.6017	-0.6162	-0.5927	-0.5711	-0.5581	-0.5543	-0.5536	-0.5528	-0.5516

Overall max	0.5401
Overall min	-0.6162

Appendix 7:
Dataset G: Correlation coefficients at 504 potential origin points in an exercise to find the most likely origin to sites in Zvelebil and Lillie's demic region when Italian sites are excluded.

	Longitude W																								
Latitude	10ºW	8ºW	6ºW	4ºW	2ºW	0ºW	2ºE	4ºE	6ºE	8ºE	10ºE	12ºE	14ºE	16ºE	18ºE	20ºE	22ºE	24ºE	26ºE	28ºE	30ºE	32ºE	34ºE	36ºE	Latitude
30ºN	0.1051	0.0176	-0.0872	-0.203	-0.3164	-0.4124	-0.4831	-0.5293	-0.557	-0.572	-0.579	-0.581	-0.58	-0.577	-0.5729	-0.5682	-0.5637	-0.5595	-0.556	-0.5532	-0.5508	-0.5486	-0.5465	-0.5441	30ºN
32ºN	0.1995	0.1315	0.0438	-0.0654	-0.191	-0.317	-0.4234	-0.4989	-0.5452	-0.5705	-0.5825	-0.5866	-0.5859	-0.5823	-0.577	-0.5708	-0.5647	-0.5593	-0.5551	-0.5519	-0.5494	-0.5472	-0.5448	-0.5422	32ºN
34ºN	0.278	0.2303	0.1668	0.0818	-0.0294	-0.1649	-0.3078	-0.4299	-0.5136	-0.5612	-0.5843	-0.5931	-0.5939	-0.5902	-0.5835	-0.5751	-0.5662	-0.5588	-0.5536	-0.55	-0.5477	-0.5454	-0.5427	-0.5395	34ºN
36ºN	0.3398	0.3084	0.267	0.2109	0.1333	0.0255	-0.1168	-0.2802	-0.4268	-0.525	-0.5763	-0.5979	-0.6038	-0.6016	-0.5944	-0.5833	-0.5691	-0.5556	-0.5488	-0.5473	-0.5459	-0.5433	-0.5397	-0.5355	36ºN
38ºN	0.3869	0.3671	0.3418	0.3083	0.2625	0.1973	0.1018	-0.0368	-0.2179	-0.4	-0.5253	-0.5866	-0.6094	-0.6143	-0.6105	-0.6005	-0.5812	-0.5515	-0.5472	-0.5474	-0.5445	-0.5399	-0.5344	-0.5284	38ºN
40ºN	0.4227	0.4105	0.3955	0.3766	0.3518	0.3178	0.2689	0.1949	0.0783	-0.0986	-0.3153	-0.4872	-0.5738	-0.6071	-0.6175	-0.6183	-0.6133	-0.5848	-0.5618	-0.5489	-0.5393	-0.5308	-0.5226	-0.5146	40ºN
42ºN	0.4499	0.4426	0.434	0.4237	0.4108	0.3943	0.3719	0.3398	0.2906	0.2098	0.0735	-0.1318	-0.3445	-0.4793	-0.5431	-0.57	-0.5763	-0.5652	-0.5389	-0.5258	-0.5152	-0.5053	-0.496	-0.487	42ºN
44ºN	0.4709	0.4667	0.4619	0.4565	0.4502	0.4427	0.4332	0.4205	0.4024	0.3742	0.3262	0.2398	0.0941	-0.0924	-0.2514	-0.3519	-0.4126	-0.446	-0.4528	-0.4522	-0.4494	-0.4448	-0.4391	-0.4328	44ºN
46ºN	0.4874	0.4851	0.4827	0.4801	0.4774	0.4744	0.471	0.467	0.4619	0.4542	0.4411	0.4152	0.3621	0.2652	0.13	0.0052	-0.1043	-0.1898	-0.2475	-0.2843	-0.3071	-0.3211	-0.3292	-0.3332	46ºN
48ºN	0.5006	0.4995	0.4985	0.4975	0.4967	0.496	0.4956	0.4953	0.4954	0.4957	0.4958	0.4938	0.4817	0.4479	0.3846	0.2989	0.2021	0.1201	0.0426	-0.0244	-0.0783	-0.1199	-0.151	-0.1737	48ºN
50ºN	0.5112	0.5109	0.5107	0.5107	0.5109	0.5114	0.5122	0.5134	0.515	0.5175	0.5213	0.5285	0.5309	0.5177	0.491	0.4503	0.3964	0.3362	0.2744	0.2137	0.1567	0.1055	0.0612	0.0242	50ºN
52ºN	0.52	0.5202	0.5205	0.5211	0.5218	0.5229	0.5242	0.5258	0.5279	0.5304	0.534	0.542	0.5469	0.5435	0.5323	0.5126	0.4831	0.4456	0.405	0.363	0.3207	0.2792	0.2399	0.2035	52ºN
54ºN	0.5273	0.5278	0.5285	0.5293	0.5304	0.5317	0.5333	0.5353	0.5377	0.5407	0.5445	0.5487	0.5515	0.5507	0.5451	0.5339	0.5171	0.4957	0.4711	0.4444	0.4164	0.3877	0.359	0.3311	54ºN
56ºN	0.5334	0.5341	0.535	0.536	0.5372	0.5387	0.5404	0.5423	0.5446	0.5472	0.5498	0.5522	0.5535	0.5528	0.5495	0.5429	0.5331	0.5204	0.5054	0.4885	0.4703	0.4512	0.4316	0.412	56ºN
58ºN	0.5385	0.5393	0.5402	0.5413	0.5425	0.5439	0.5455	0.5472	0.549	0.5509	0.5526	0.5539	0.5544	0.5537	0.5514	0.5473	0.5413	0.5334	0.524	0.5132	0.5013	0.4887	0.4754	0.4619	58ºN
60ºN	0.5427	0.5435	0.5444	0.5454	0.5466	0.5478	0.5491	0.5505	0.5518	0.5531	0.5541	0.5548	0.5549	0.5542	0.5525	0.5497	0.5458	0.5408	0.5347	0.5276	0.5198	0.5113	0.5023	0.493	60ºN
62ºN	0.5461	0.5469	0.5477	0.5486	0.5496	0.5506	0.5516	0.5526	0.5535	0.5544	0.555	0.5553	0.5551	0.5545	0.5532	0.5513	0.5486	0.5453	0.5412	0.5365	0.5312	0.5254	0.5192	0.5128	62ºN
64ºN	0.5489	0.5495	0.5502	0.551	0.5517	0.5525	0.5533	0.554	0.5546	0.5551	0.5554	0.5555	0.5553	0.5547	0.5538	0.5524	0.5505	0.5481	0.5453	0.5421	0.5385	0.5345	0.5303	0.5258	64ºN
66ºN	0.551	0.5516	0.5522	0.5527	0.5533	0.5539	0.5544	0.5549	0.5553	0.5556	0.5557	0.5557	0.5554	0.555	0.5542	0.5531	0.5518	0.5501	0.5482	0.5459	0.5434	0.5406	0.5376	0.5345	66ºN
68ºN	0.5527	0.5531	0.5536	0.554	0.5544	0.5548	0.5552	0.5555	0.5557	0.5559	0.5559	0.5558	0.5556	0.5551	0.5545	0.5537	0.5527	0.5515	0.5501	0.5485	0.5468	0.5448	0.5427	0.5405	68ºN
70ºN	0.554	0.5543	0.5546	0.5549	0.5552	0.5555	0.5557	0.5559	0.556	0.5561	0.556	0.5559	0.5557	0.5553	0.5548	0.5542	0.5535	0.5526	0.5516	0.5504	0.5492	0.5478	0.5463	0.5447	70ºN

Max values	0.554	0.5543	0.5546	0.5549	0.5552	0.5555	0.5557	0.5559	0.556	0.5561	0.556	0.5559	0.5557	0.5553	0.5548	0.5542	0.5535	0.5526	0.5516	0.5504	0.5492	0.5478	0.5463	0.5447
Min values	0.1051	0.0176	-0.0872	-0.203	-0.3164	-0.4124	-0.4831	-0.5293	-0.557	-0.572	-0.5843	-0.5979	-0.6094	-0.6143	-0.6175	-0.6183	-0.6133	-0.5848	-0.5618	-0.5532	-0.5508	-0.5486	-0.5465	-0.5441

Overall max	0.5561
Overall min	-0.6183

Appendix 7:
Dataset G: Correlation coefficient at 504 potential origin points in an exercise to find the most likely origin to sites in Italy.

	Longitude W											
Latitude	10°W	8°W	6°W	4°W	2°W	0°W	2°E	4°E	6°E	8°E	10°E	12°E
30°N	0.386	0.2572	0.1119	-0.0371	-0.1762	-0.2959	-0.3936	-0.471	-0.5318	-0.5802	-0.6196	-0.6528
32°N	0.5299	0.4262	0.2932	0.1365	-0.0287	-0.1834	-0.3146	-0.4184	-0.498	-0.5588	-0.6065	-0.646
34°N	0.6466	0.579	0.4823	0.3494	0.1814	-0.0052	-0.1833	-0.3318	-0.4447	-0.5271	-0.5877	-0.6358
36°N	0.7259	0.6903	0.6367	0.555	0.432	0.2584	0.045	-0.1699	-0.3474	-0.4747	-0.56	-0.6198
38°N	0.7717	0.756	0.7326	0.6964	0.6381	0.5413	0.3823	0.1469	-0.1276	-0.362	-0.5177	-0.605
40°N	0.7938	0.7877	0.779	0.7661	0.7462	0.7134	0.6555	0.5461	0.3365	-0.0042	-0.3664	-0.6096
42°N	0.8016	0.7992	0.7959	0.7913	0.7848	0.775	0.7591	0.7309	0.6747	0.5486	0.2646	-0.2091
44°N	0.8016	0.8002	0.7984	0.796	0.7928	0.7885	0.7824	0.7736	0.7597	0.7356	0.6816	0.5567
46°N	0.7978	0.7964	0.7946	0.7924	0.7896	0.7862	0.7819	0.7766	0.7706	0.7641	0.7568	0.7529
48°N	0.7921	0.7904	0.7883	0.7858	0.7828	0.7791	0.7746	0.7692	0.7626	0.754	0.7425	0.7266
50°N	0.7857	0.7837	0.7813	0.7785	0.7751	0.7711	0.7663	0.7605	0.7534	0.7446	0.7333	0.7175
52°N	0.7791	0.7769	0.7742	0.7711	0.7675	0.7633	0.7583	0.7524	0.7454	0.7367	0.7258	0.7116
54°N	0.7726	0.7702	0.7674	0.7641	0.7604	0.7561	0.7511	0.7452	0.7383	0.73	0.7199	0.7074
56°N	0.7663	0.7638	0.7608	0.7575	0.7537	0.7494	0.7445	0.7387	0.7321	0.7243	0.7152	0.7042
58°N	0.7602	0.7576	0.7546	0.7513	0.7475	0.7433	0.7385	0.733	0.7267	0.7195	0.7113	0.7018
60°N	0.7544	0.7517	0.7488	0.7454	0.7418	0.7376	0.733	0.7278	0.722	0.7155	0.7081	0.6998
62°N	0.7487	0.7461	0.7432	0.7399	0.7364	0.7324	0.7281	0.7232	0.7179	0.712	0.7054	0.6981
64°N	0.7433	0.7407	0.7379	0.7348	0.7314	0.7277	0.7236	0.7191	0.7142	0.7089	0.7031	0.6968
66°N	0.7381	0.7356	0.7329	0.7299	0.7267	0.7232	0.7195	0.7154	0.711	0.7062	0.7011	0.6956
68°N	0.7331	0.7307	0.7281	0.7254	0.7224	0.7192	0.7157	0.712	0.7081	0.7039	0.6994	0.6946
70°N	0.7283	0.726	0.7236	0.7211	0.7183	0.7154	0.7123	0.709	0.7055	0.7018	0.6978	0.6937
Max values	0.8016	0.8002	0.7984	0.796	0.7928	0.7885	0.7824	0.7766	0.7706	0.7641	0.7568	0.7529
Min values	0.386	0.2572	0.1119	-0.0371	-0.1762	-0.2959	-0.3936	-0.471	-0.5318	-0.5802	-0.6196	-0.6528

Latitude	14°E	16°E	18°E	20°E	22°E	24°E	26°E	28°E	30°E	32°E	34°E	36°E	Latitude
30°N	-0.6817	-0.7069	-0.7286	-0.7466	-0.7613	-0.773	-0.7823	-0.7898	-0.7959	-0.8009	-0.805	-0.8083	30°N
32°N	-0.6804	-0.7105	-0.7357	-0.7557	-0.771	-0.7826	-0.7914	-0.7981	-0.8033	-0.8075	-0.8108	-0.8135	32°N
34°N	-0.6783	-0.7163	-0.7467	-0.7686	-0.7838	-0.7942	-0.8016	-0.807	-0.8111	-0.8141	-0.8165	-0.8182	34°N
36°N	-0.675	-0.7284	-0.7654	-0.787	-0.7996	-0.8074	-0.8125	-0.8159	-0.8183	-0.82	-0.8211	-0.8218	36°N
38°N	-0.677	-0.7608	-0.7961	-0.8101	-0.8165	-0.82	-0.8219	-0.8231	-0.8236	-0.8236	-0.8233	-0.8226	38°N
40°N	-0.7444	-0.8156	-0.827	-0.8277	-0.8274	-0.8269	-0.826	-0.8249	-0.8235	-0.822	-0.8201	-0.818	40°N
42°N	-0.6153	-0.7956	-0.8169	-0.8185	-0.8182	-0.817	-0.8153	-0.8132	-0.8108	-0.8082	-0.8052	-0.802	42°N
44°N	0.2261	-0.3694	-0.6318	-0.7099	-0.7402	-0.7545	-0.7617	-0.7651	-0.7662	-0.7658	-0.7643	-0.7618	44°N
46°N	0.6762	0.4837	0.1491	-0.1843	-0.3959	-0.5128	-0.5787	-0.6179	-0.6421	-0.6573	-0.6666	-0.6719	46°N
48°N	0.6941	0.6269	0.5117	0.3459	0.1515	-0.0335	-0.1835	-0.2946	-0.3737	-0.4295	-0.4687	-0.4963	48°N
50°N	0.6933	0.6556	0.5993	0.5216	0.423	0.3091	0.1898	0.0757	-0.0257	-0.111	-0.1801	-0.2348	50°N
52°N	0.6924	0.6664	0.6315	0.5865	0.5309	0.4654	0.3923	0.3149	0.237	0.1623	0.0937	0.0327	52°N
54°N	0.6917	0.6721	0.6476	0.6176	0.582	0.5407	0.4944	0.4441	0.3913	0.3377	0.2849	0.2345	54°N
56°N	0.6912	0.6756	0.6572	0.6355	0.6105	0.5822	0.5508	0.5167	0.4805	0.443	0.405	0.3672	56°N
58°N	0.6908	0.6781	0.6636	0.6471	0.6285	0.6079	0.5853	0.561	0.5352	0.5083	0.4808	0.453	58°N
60°N	0.6904	0.6799	0.6682	0.6551	0.6408	0.6251	0.6082	0.5901	0.571	0.5512	0.5308	0.5101	60°N
62°N	0.6901	0.6813	0.6717	0.6611	0.6497	0.6374	0.6243	0.6105	0.596	0.5809	0.5654	0.5497	62°N
64°N	0.6899	0.6824	0.6744	0.6658	0.6565	0.6467	0.6363	0.6255	0.6142	0.6025	0.5905	0.5783	64°N
66°N	0.6897	0.6834	0.6766	0.6695	0.6619	0.6539	0.6456	0.6369	0.628	0.6187	0.6093	0.5998	66°N
68°N	0.6895	0.6841	0.6785	0.6725	0.6663	0.6597	0.653	0.646	0.6388	0.6314	0.624	0.6164	68°N
70°N	0.6894	0.6848	0.68	0.675	0.6699	0.6645	0.659	0.6533	0.6475	0.6416	0.6356	0.6296	70°N
Max values	0.6941	0.6848	0.68	0.675	0.6699	0.6645	0.659	0.6533	0.6475	0.6416	0.6356	0.6296	
Min values	-0.7444	-0.8156	-0.827	-0.8277	-0.8274	-0.8269	-0.826	-0.8249	-0.8236	-0.8236	-0.8233	-0.8226	

Overall max 0.8016
Overall min -0.8277

Appendix 7:
Dataset G: Correlation coefficients at 504 potential origin points in an exercise to find the most likely origin to sites in Italy and the enclave regions.

	Longitude W																								
Latitude	10ºW	8ºW	6ºW	4ºW	2ºW	0ºW	2ºE	4ºE	6ºE	8ºE	10ºE	12ºE	14ºE	16ºE	18ºE	20ºE	22ºE	24ºE	26ºE	28ºE	30ºE	32ºE	34ºE	36ºE	Latitude
30ºN	0.2852	0.2518	0.2116	0.1629	0.1037	0.0325	-0.051	-0.1443	-0.2415	-0.3338	-0.413	-0.4745	-0.5187	-0.5482	-0.5668	-0.5776	-0.5829	-0.5845	-0.5835	-0.581	-0.5773	-0.5729	-0.568	-0.5627	30ºN
32ºN	0.3233	0.2943	0.2588	0.2149	0.1599	0.091	0.0061	-0.0943	-0.2043	-0.3124	-0.4055	-0.4759	-0.524	-0.5542	-0.5717	-0.5804	-0.5835	-0.583	-0.5803	-0.5762	-0.5713	-0.5658	-0.5601	-0.5541	32ºN
34ºN	0.3617	0.3377	0.308	0.2706	0.2224	0.1593	0.0767	-0.0283	-0.152	-0.2808	-0.3938	-0.4766	-0.5294	-0.5602	-0.5759	-0.5819	-0.5821	-0.5791	-0.5744	-0.5687	-0.5625	-0.5561	-0.5496	-0.5431	34ºN
36ºN	0.3996	0.3809	0.3578	0.3284	0.2897	0.2366	0.1624	0.059	-0.0764	-0.2309	-0.374	-0.4752	-0.5339	-0.5661	-0.579	-0.5807	-0.5771	-0.5713	-0.5644	-0.5572	-0.5499	-0.5427	-0.5357	-0.5288	36ºN
38ºN	0.436	0.4227	0.4065	0.3859	0.3585	0.32	0.2622	0.1716	0.0338	-0.1472	-0.3354	-0.4724	-0.537	-0.5731	-0.5783	-0.5733	-0.5654	-0.5568	-0.5482	-0.54	-0.5321	-0.5245	-0.5174	-0.5105	38ºN
40ºN	0.47	0.4617	0.4518	0.4398	0.4242	0.4026	0.3687	0.3083	0.1939	-0.0028	-0.2342	-0.4558	-0.5501	-0.5711	-0.5638	-0.5522	-0.5414	-0.5315	-0.5226	-0.5145	-0.507	-0.5	-0.4933	-0.487	40ºN
42ºN	0.5008	0.4966	0.4921	0.4871	0.4814	0.4744	0.4645	0.4424	0.3939	0.3154	0.1429	-0.2367	-0.4402	-0.5088	-0.5105	-0.503	-0.4958	-0.4892	-0.4832	-0.4776	-0.4723	-0.4672	-0.4622	-0.4574	42ºN
44ºN	0.5276	0.5267	0.5261	0.526	0.5265	0.5279	0.5305	0.5266	0.5006	0.4679	0.4195	0.1416	-0.139	-0.3063	-0.3762	-0.4042	-0.4167	-0.4226	-0.4253	-0.4261	-0.4257	-0.4245	-0.4227	-0.4204	44ºN
46ºN	0.5499	0.5512	0.553	0.5556	0.5591	0.5635	0.5681	0.5695	0.5607	0.5275	0.4315	0.2974	0.1209	-0.0571	-0.1828	-0.2584	-0.303	-0.3302	-0.3475	-0.3588	-0.3662	-0.371	-0.374	-0.3756	46ºN
48ºN	0.5673	0.5696	0.5723	0.5756	0.5792	0.5826	0.5841	0.5794	0.5601	0.5133	0.4338	0.3353	0.2216	0.0966	-0.0188	-0.1099	-0.176	-0.2225	-0.2552	-0.2785	-0.2954	-0.3076	-0.3165	-0.323	48ºN
50ºN	0.5795	0.5816	0.5839	0.586	0.5875	0.5871	0.5826	0.5703	0.5448	0.5016	0.4401	0.3644	0.278	0.185	0.0927	0.0094	-0.0603	-0.1158	-0.1588	-0.1919	-0.2173	-0.2369	-0.252	-0.2637	50ºN
52ºN	0.5865	0.5877	0.5885	0.5885	0.5868	0.5823	0.573	0.5566	0.5305	0.4927	0.4433	0.3836	0.316	0.2433	0.1694	0.0986	0.0344	-0.0212	-0.0679	-0.1064	-0.1377	-0.163	-0.1835	-0.1999	52ºN
54ºN	0.589	0.5888	0.5877	0.5853	0.5808	0.5731	0.5611	0.5433	0.5184	0.4856	0.4448	0.3968	0.3427	0.2846	0.2248	0.1656	0.1095	0.0581	0.0125	-0.0272	-0.0612	-0.0899	-0.1141	-0.1342	54ºN
56ºN	0.5876	0.5859	0.583	0.5786	0.5719	0.5624	0.5491	0.5313	0.5083	0.4797	0.4455	0.4061	0.3622	0.3151	0.2661	0.2169	0.1688	0.1233	0.0811	0.043	0.009	-0.0208	-0.0468	-0.0692	56ºN
58ºN	0.5834	0.5803	0.5759	0.5699	0.5619	0.5514	0.5378	0.5207	0.4998	0.4748	0.4458	0.4129	0.3768	0.3382	0.2978	0.2568	0.216	0.1765	0.1389	0.1038	0.0716	0.0424	0.0163	-0.007	58ºN
60ºN	0.577	0.5728	0.5673	0.5603	0.5516	0.5407	0.5274	0.5114	0.4925	0.4706	0.4458	0.4182	0.3882	0.3561	0.3227	0.2884	0.254	0.2201	0.1872	0.1559	0.1264	0.099	0.0739	0.051	60ºN
62ºN	0.5692	0.5642	0.558	0.5504	0.5414	0.5306	0.5179	0.5031	0.4862	0.467	0.4457	0.4223	0.3971	0.3704	0.3425	0.3138	0.2848	0.256	0.2276	0.2001	0.1738	0.1489	0.1256	0.1039	62ºN
64ºN	0.5605	0.5549	0.5483	0.5406	0.5316	0.5212	0.5093	0.4958	0.4807	0.4639	0.4455	0.4256	0.4044	0.382	0.3586	0.3346	0.3102	0.2857	0.2615	0.2377	0.2146	0.1925	0.1714	0.1515	64ºN
66ºN	0.5512	0.5453	0.5386	0.5309	0.5222	0.5124	0.5014	0.4892	0.4758	0.4612	0.4453	0.4284	0.4104	0.3915	0.3719	0.3518	0.3313	0.3106	0.29	0.2697	0.2497	0.2304	0.2117	0.1939	66ºN
68ºN	0.5416	0.5357	0.529	0.5216	0.5134	0.5043	0.4943	0.4834	0.4715	0.4587	0.4451	0.4306	0.4154	0.3995	0.3831	0.3662	0.349	0.3317	0.3143	0.297	0.2799	0.2633	0.247	0.2314	68ºN
70ºN	0.5318	0.5261	0.5197	0.5127	0.5051	0.4967	0.4877	0.478	0.4676	0.4566	0.4449	0.4326	0.4197	0.4063	0.3926	0.3785	0.3641	0.3496	0.335	0.3204	0.306	0.2918	0.2779	0.2644	70ºN

Max values	0.589	0.5888	0.5885	0.5885	0.5875	0.5871	0.5841	0.5794	0.5607	0.5275	0.4458	0.4326	0.4197	0.4063	0.3926	0.3785	0.3641	0.3496	0.335	0.3204	0.306	0.2918	0.2779	0.2644
Min values	0.2852	0.2518	0.2116	0.1629	0.1037	0.0325	-0.051	-0.1443	-0.2415	-0.3338	-0.413	-0.4766	-0.5501	-0.5731	-0.579	-0.5819	-0.5835	-0.5845	-0.5835	-0.581	-0.5773	-0.5729	-0.568	-0.5627

Overall max	0.589
Overall min	-0.5845

Appendix 7:

Dataset G: Correlation coefficients at 504 potential origin points in an exercise to find the most likely origin to sites in Zvelebil and Lillie's western acculturation region.

	Longitude W																								
Latitude	10ºW	8ºW	6ºW	4ºW	2ºW	0ºW	2ºE	4ºE	6ºE	8ºE	10ºE	12ºE	14ºE	16ºE	18ºE	20ºE	22ºE	24ºE	26ºE	28ºE	30ºE	32ºE	34ºE	36ºE	Latitude
30ºN	-0.3564	-0.3935	-0.4254	-0.4528	-0.4763	-0.4964	-0.5136	-0.5284	-0.541	-0.5518	-0.5612	-0.5692	-0.5761	-0.5821	-0.5874	-0.592	-0.596	-0.5995	-0.6026	-0.6053	-0.6077	-0.6097	-0.6114	-0.6127	30ºN
32ºN	-0.336	-0.3786	-0.4151	-0.4461	-0.4725	-0.4947	-0.5136	-0.5295	-0.5429	-0.5542	-0.5638	-0.572	-0.5789	-0.5848	-0.5899	-0.5943	-0.5981	-0.6015	-0.6043	-0.6068	-0.6089	-0.6106	-0.612	-0.6131	32ºN
34ºN	-0.3089	-0.3589	-0.4015	-0.4374	-0.4674	-0.4925	-0.5134	-0.5307	-0.5451	-0.557	-0.5669	-0.575	-0.5819	-0.5876	-0.5925	-0.5967	-0.6002	-0.6033	-0.6059	-0.608	-0.6098	-0.6112	-0.6123	-0.613	34ºN
36ºN	-0.2717	-0.3317	-0.3827	-0.4253	-0.4605	-0.4894	-0.513	-0.5322	-0.5478	-0.5603	-0.5704	-0.5785	-0.5851	-0.5905	-0.595	-0.5989	-0.6021	-0.6048	-0.6071	-0.6089	-0.6103	-0.6113	-0.6119	-0.6122	36ºN
38ºN	-0.2183	-0.2922	-0.3553	-0.4078	-0.4505	-0.4848	-0.5122	-0.5339	-0.551	-0.5642	-0.5743	-0.5821	-0.5883	-0.5932	-0.5973	-0.6007	-0.6036	-0.6059	-0.6077	-0.609	-0.61	-0.6105	-0.6106	-0.6103	38ºN
40ºN	-0.139	-0.2316	-0.3126	-0.3803	-0.4348	-0.4775	-0.5103	-0.5357	-0.5549	-0.5688	-0.5786	-0.5857	-0.5911	-0.5954	-0.5989	-0.6018	-0.6041	-0.6059	-0.6072	-0.6081	-0.6084	-0.6084	-0.6078	-0.6069	40ºN
42ºN	-0.0198	-0.1339	-0.2402	-0.3324	-0.4071	-0.4644	-0.5052	-0.5365	-0.5595	-0.5741	-0.5826	-0.588	-0.5924	-0.596	-0.599	-0.6014	-0.6032	-0.6044	-0.6051	-0.6052	-0.6049	-0.6041	-0.6027	-0.601	42ºN
44ºN	0.1481	0.023	-0.1102	-0.2382	-0.3487	-0.4331	-0.4926	-0.5362	-0.5628	-0.5779	-0.5821	-0.5859	-0.5899	-0.5934	-0.5961	-0.5981	-0.5994	-0.5999	-0.5999	-0.5992	-0.598	-0.5963	-0.5941	-0.5913	44ºN
46ºN	0.3398	0.2375	0.1063	-0.0479	-0.21	-0.3545	-0.4589	-0.5203	-0.5463	-0.5563	-0.565	-0.5738	-0.5805	-0.5851	-0.5881	-0.5898	-0.5905	-0.5903	-0.5894	-0.5879	-0.5857	-0.5829	-0.5796	-0.5757	46ºN
48ºN	0.4949	0.4402	0.358	0.2386	0.0769	-0.1317	-0.333	-0.4598	-0.5148	-0.5254	-0.5367	-0.5518	-0.5613	-0.5669	-0.5702	-0.5717	-0.5719	-0.5712	-0.5695	-0.5671	-0.564	-0.5603	-0.5559	-0.5509	48ºN
50ºN	0.5828	0.5644	0.533	0.4808	0.3929	0.2264	-0.0301	-0.2756	-0.4152	-0.4682	-0.4916	-0.507	-0.5182	-0.5259	-0.5308	-0.5335	-0.5346	-0.5342	-0.5327	-0.5302	-0.5268	-0.5226	-0.5176	-0.5119	50ºN
52ºN	0.6156	0.6129	0.6085	0.5875	0.5378	0.4381	0.2697	0.0432	-0.1521	-0.2651	-0.3353	-0.381	-0.4116	-0.4324	-0.4467	-0.4562	-0.4623	-0.4658	-0.4671	-0.4667	-0.4649	-0.4618	-0.4575	-0.4523	52ºN
54ºN	0.621	0.6198	0.6189	0.6077	0.5751	0.5157	0.4246	0.2978	0.1579	0.0332	-0.0694	-0.1504	-0.2125	-0.259	-0.2936	-0.319	-0.3376	-0.3508	-0.3599	-0.3657	-0.3688	-0.3698	-0.3688	-0.3662	54ºN
56ºN	0.6129	0.6081	0.6009	0.5892	0.5692	0.536	0.4864	0.4199	0.3405	0.2553	0.1708	0.0917	0.021	-0.0398	-0.0905	-0.1319	-0.165	-0.1911	-0.2114	-0.2267	-0.238	-0.2458	-0.2508	-0.2533	56ºN
58ºN	0.5989	0.5917	0.5815	0.5674	0.5514	0.5311	0.5023	0.4646	0.4189	0.367	0.3114	0.2544	0.1982	0.1448	0.0957	0.0517	0.0131	-0.0199	-0.0477	-0.0707	-0.0893	-0.1041	-0.1155	-0.1239	58ºN
60ºN	0.5844	0.5764	0.5663	0.5542	0.5407	0.5249	0.5052	0.481	0.4522	0.4194	0.3833	0.345	0.3055	0.266	0.2273	0.1905	0.1561	0.1247	0.0966	0.0717	0.0502	0.0319	0.0166	0.0042	60ºN
62ºN	0.5726	0.565	0.5562	0.5461	0.5346	0.5215	0.5064	0.4889	0.4689	0.4465	0.4219	0.3957	0.3681	0.3399	0.3115	0.2834	0.2563	0.2304	0.2062	0.1839	0.1636	0.1455	0.1296	0.1158	62ºN
64ºN	0.5631	0.5563	0.5486	0.5399	0.5302	0.5194	0.5073	0.4939	0.479	0.4626	0.445	0.4262	0.4064	0.386	0.3652	0.3444	0.3237	0.3036	0.2843	0.2659	0.2487	0.2328	0.2184	0.2053	64ºN
66ºN	0.5553	0.5493	0.5425	0.5351	0.5269	0.5179	0.5081	0.4974	0.4858	0.4734	0.4602	0.4462	0.4315	0.4164	0.401	0.3854	0.3698	0.3544	0.3393	0.3248	0.3109	0.2978	0.2855	0.2742	66ºN
68ºN	0.5488	0.5434	0.5375	0.5311	0.5242	0.5167	0.5087	0.5001	0.4909	0.4812	0.4709	0.4602	0.4491	0.4376	0.426	0.4141	0.4022	0.3904	0.3788	0.3674	0.3564	0.3459	0.3359	0.3264	68ºN
70ºN	0.5431	0.5384	0.5333	0.5279	0.522	0.5158	0.5092	0.5022	0.4948	0.4871	0.479	0.4707	0.4621	0.4532	0.4442	0.4351	0.426	0.4169	0.4079	0.399	0.3903	0.382	0.374	0.3663	70ºN

Max values	0.621	0.6198	0.6189	0.6077	0.5751	0.536	0.5092	0.5022	0.4948	0.4871	0.479	0.4707	0.4621	0.4532	0.4442	0.4351	0.426	0.4169	0.4079	0.399	0.3903	0.382	0.374	0.3663
Min values	-0.3564	-0.3935	-0.4254	-0.4528	-0.4763	-0.4964	-0.5136	-0.5365	-0.5628	-0.5779	-0.5826	-0.588	-0.5924	-0.596	-0.599	-0.6018	-0.6041	-0.6059	-0.6077	-0.609	-0.6103	-0.6113	-0.6123	-0.6131

Overall max	0.621
Overall min	-0.6131

Appendix 7:

Dataset G: Correlation coefficients at 504 potential origin points in an exercise to find the most likely origin to sites in the western acculturation region when enclave sites are excluded.

	Longitude W																								
Latitude	10ºW	8ºW	6ºW	4ºW	2ºW	0ºW	2ºE	4ºE	6ºE	8ºE	10ºE	12ºE	14ºE	16ºE	18ºE	20ºE	22ºE	24ºE	26ºE	28ºE	30ºE	32ºE	34ºE	36ºE	Latitude
30ºN	-0.34	-0.3762	-0.4075	-0.4344	-0.4576	-0.4774	-0.4944	-0.509	-0.5215	-0.5323	-0.5416	-0.5497	-0.5567	-0.5628	-0.5682	-0.5729	-0.5771	-0.5808	-0.5841	-0.587	-0.5895	-0.5916	-0.5934	-0.5949	30ºN
32ºN	-0.321	-0.3624	-0.398	-0.4284	-0.4542	-0.4761	-0.4946	-0.5103	-0.5235	-0.5347	-0.5443	-0.5524	-0.5594	-0.5655	-0.5707	-0.5753	-0.5793	-0.5828	-0.5859	-0.5885	-0.5908	-0.5927	-0.5942	-0.5954	32ºN
34ºN	-0.2958	-0.3442	-0.3856	-0.4206	-0.4499	-0.4744	-0.4948	-0.5117	-0.5258	-0.5375	-0.5473	-0.5554	-0.5624	-0.5683	-0.5733	-0.5777	-0.5815	-0.5847	-0.5875	-0.5899	-0.5919	-0.5934	-0.5947	-0.5955	34ºN
36ºN	-0.2613	-0.3191	-0.3684	-0.4098	-0.4441	-0.4721	-0.495	-0.5135	-0.5286	-0.5407	-0.5506	-0.5588	-0.5655	-0.5711	-0.5759	-0.58	-0.5835	-0.5864	-0.5889	-0.5909	-0.5925	-0.5937	-0.5945	-0.5949	36ºN
38ºN	-0.212	-0.2828	-0.3436	-0.3943	-0.4356	-0.4687	-0.495	-0.5156	-0.5318	-0.5445	-0.5544	-0.5623	-0.5686	-0.5739	-0.5783	-0.582	-0.5851	-0.5877	-0.5898	-0.5914	-0.5925	-0.5932	-0.5935	-0.5933	38ºN
40ºN	-0.139	-0.2273	-0.3048	-0.3699	-0.4225	-0.4634	-0.4945	-0.5181	-0.5358	-0.5488	-0.5584	-0.5657	-0.5714	-0.5762	-0.5801	-0.5834	-0.5861	-0.5882	-0.5897	-0.5908	-0.5914	-0.5915	-0.5911	-0.5903	40ºN
42ºN	-0.029	-0.1377	-0.2391	-0.3272	-0.3989	-0.4538	-0.4922	-0.5204	-0.5402	-0.5534	-0.5619	-0.5681	-0.5731	-0.5773	-0.5807	-0.5835	-0.5857	-0.5872	-0.5882	-0.5885	-0.5884	-0.5878	-0.5866	-0.585	42ºN
44ºN	0.1279	0.0071	-0.1203	-0.2423	-0.3477	-0.428	-0.484	-0.5234	-0.5447	-0.5569	-0.5623	-0.5671	-0.5718	-0.5759	-0.5791	-0.5814	-0.5829	-0.5838	-0.5839	-0.5835	-0.5824	-0.5809	-0.5788	-0.5761	44ºN
46ºN	0.3126	0.2105	0.0814	-0.0681	-0.2234	-0.3606	-0.4586	-0.5145	-0.5359	-0.5428	-0.5501	-0.5586	-0.5654	-0.5701	-0.5732	-0.5751	-0.5758	-0.5758	-0.575	-0.5735	-0.5715	-0.5688	-0.5656	-0.5618	46ºN
48ºN	0.4691	0.4121	0.3273	0.206	0.0439	-0.1609	-0.3537	-0.4713	-0.519	-0.5241	-0.5314	-0.544	-0.5519	-0.5566	-0.5592	-0.5603	-0.5603	-0.5593	-0.5576	-0.5551	-0.552	-0.5483	-0.544	-0.539	48ºN
50ºN	0.5615	0.5418	0.5085	0.4534	0.3611	0.1867	-0.0755	-0.3156	-0.444	-0.4863	-0.5016	-0.5113	-0.5185	-0.5235	-0.5265	-0.5279	-0.5279	-0.5269	-0.5248	-0.522	-0.5183	-0.514	-0.5089	-0.5032	50ºN
52ºN	0.5971	0.5942	0.5893	0.567	0.5141	0.407	0.2268	-0.009	-0.2024	-0.3066	-0.3668	-0.4036	-0.4273	-0.4429	-0.4533	-0.46	-0.464	-0.4658	-0.4659	-0.4646	-0.4621	-0.4585	-0.454	-0.4485	52ºN
54ºN	0.6036	0.6024	0.6013	0.589	0.5539	0.4898	0.3915	0.2558	0.1094	-0.0164	-0.1155	-0.1905	-0.2458	-0.2861	-0.3153	-0.3363	-0.3513	-0.3618	-0.3686	-0.3727	-0.3745	-0.3744	-0.3726	-0.3694	54ºN
56ºN	0.5955	0.5905	0.5828	0.5703	0.5486	0.5128	0.4592	0.3878	0.3034	0.2142	0.1276	0.0487	-0.0199	-0.0774	-0.1242	-0.1617	-0.1912	-0.214	-0.2315	-0.2444	-0.2537	-0.26	-0.2636	-0.2651	56ºN
58ºN	0.5811	0.5736	0.5629	0.548	0.5308	0.5088	0.4779	0.4374	0.3887	0.3339	0.2757	0.217	0.1601	0.107	0.059	0.0166	-0.0198	-0.0506	-0.0762	-0.0971	-0.1138	-0.1269	-0.1368	-0.1439	58ºN
60ºN	0.5661	0.5577	0.5472	0.5344	0.52	0.503	0.482	0.4561	0.4254	0.3907	0.3528	0.3129	0.2722	0.2318	0.1929	0.1562	0.1224	0.0919	0.0648	0.0412	0.0209	0.0039	-0.0102	-0.0216	60ºN
62ºN	0.5537	0.5458	0.5366	0.5259	0.5137	0.4998	0.4837	0.4651	0.4439	0.4202	0.3944	0.3669	0.3383	0.3092	0.2801	0.2517	0.2244	0.1986	0.1746	0.1528	0.1331	0.1156	0.1004	0.0873	62ºN
64ºN	0.5438	0.5367	0.5286	0.5194	0.5092	0.4978	0.485	0.4707	0.455	0.4378	0.4192	0.3996	0.379	0.3578	0.3364	0.315	0.294	0.2737	0.2542	0.2359	0.2188	0.2031	0.1889	0.1762	64ºN
66ºN	0.5357	0.5293	0.5222	0.5144	0.5057	0.4963	0.4859	0.4747	0.4625	0.4495	0.4356	0.4209	0.4057	0.39	0.374	0.3579	0.3419	0.3262	0.3109	0.2962	0.2822	0.2691	0.2569	0.2456	66ºN
68ºN	0.5288	0.5231	0.517	0.5103	0.503	0.4951	0.4867	0.4777	0.468	0.4578	0.4471	0.4359	0.4243	0.4124	0.4003	0.3881	0.3759	0.3637	0.3518	0.3402	0.3291	0.3184	0.3083	0.2989	68ºN
70ºN	0.5229	0.5179	0.5126	0.5069	0.5008	0.4942	0.4873	0.48	0.4723	0.4642	0.4558	0.4471	0.4381	0.4289	0.4196	0.4102	0.4007	0.3914	0.3821	0.373	0.3642	0.3557	0.3475	0.3398	70ºN
Max values	0.6036	0.6024	0.6013	0.589	0.5539	0.5128	0.4873	0.48	0.4723	0.4642	0.4558	0.4471	0.4381	0.4289	0.4196	0.4102	0.4007	0.3914	0.3821	0.373	0.3642	0.3557	0.3475	0.3398	
Min values	-0.34	-0.3762	-0.4075	-0.4344	-0.4576	-0.4774	-0.495	-0.5234	-0.5447	-0.5569	-0.5623	-0.5681	-0.5731	-0.5773	-0.5807	-0.5835	-0.5861	-0.5882	-0.5898	-0.5914	-0.5925	-0.5937	-0.5947	-0.5955	

Overall max 0.6036

overall min -0.5955

Appendix 7:
Dataset G: Correlation coefficients at 504 potential origin points in an exercise to find the most likely origin to sites in Zvelebil and Lillie's northern acculturation region.

	Longitude W																								
Latitude	10ºW	8ºW	6ºW	4ºW	2ºW	0ºW	2ºE	4ºE	6ºE	8ºE	10ºE	12ºE	14ºE	16ºE	18ºE	20ºE	22ºE	24ºE	26ºE	28ºE	30ºE	32ºE	34ºE	36ºE	Latitude
30ºN	0.2685	0.2394	0.2099	0.1801	0.15	0.1197	0.0893	0.0588	0.0282	-0.0024	-0.033	-0.0636	-0.0942	-0.1248	-0.1552	-0.1856	-0.2158	-0.2457	-0.2754	-0.3046	-0.3333	-0.3614	-0.3888	-0.4153	30ºN
32ºN	0.2843	0.2541	0.2234	0.1923	0.1608	0.129	0.097	0.0649	0.0327	0.0005	-0.0317	-0.064	-0.0962	-0.1284	-0.1604	-0.1923	-0.224	-0.2554	-0.2864	-0.3169	-0.3467	-0.3757	-0.4039	-0.431	32ºN
34ºN	0.3023	0.271	0.239	0.2064	0.1733	0.1398	0.106	0.0721	0.038	0.0039	-0.0303	-0.0644	-0.0985	-0.1325	-0.1664	-0.2001	-0.2336	-0.2666	-0.2991	-0.3309	-0.3619	-0.3919	-0.4208	-0.4484	34ºN
36ºN	0.3231	0.2906	0.2571	0.2228	0.1879	0.1525	0.1166	0.0805	0.0443	0.0079	-0.0285	-0.0649	-0.1012	-0.1374	-0.1735	-0.2093	-0.2448	-0.2797	-0.3139	-0.3472	-0.3795	-0.4104	-0.4399	-0.4678	36ºN
38ºN	0.3473	0.3136	0.2786	0.2424	0.2054	0.1677	0.1294	0.0907	0.0518	0.0127	-0.0264	-0.0654	-0.1045	-0.1433	-0.182	-0.2204	-0.2582	-0.2953	-0.3314	-0.3664	-0.3999	-0.4317	-0.4616	-0.4895	38ºN
40ºN	0.3756	0.3408	0.3042	0.2661	0.2268	0.1863	0.145	0.1032	0.061	0.0187	-0.0238	-0.0661	-0.1085	-0.1506	-0.1925	-0.2339	-0.2745	-0.3142	-0.3525	-0.3891	-0.4237	-0.4562	-0.4862	-0.5137	40ºN
42ºN	0.409	0.3734	0.3354	0.2953	0.2533	0.2097	0.1648	0.1191	0.0728	0.0262	-0.0204	-0.067	-0.1135	-0.1598	-0.2056	-0.2508	-0.2949	-0.3375	-0.3781	-0.4163	-0.4518	-0.4844	-0.5139	-0.5403	42ºN
44ºN	0.4484	0.4128	0.3739	0.332	0.2872	0.2399	0.1907	0.1399	0.0882	0.0361	-0.0161	-0.0682	-0.1201	-0.1717	-0.2228	-0.2727	-0.321	-0.3669	-0.4098	-0.4492	-0.4849	-0.5167	-0.5446	-0.5688	44ºN
46ºN	0.4947	0.4603	0.4217	0.3788	0.3316	0.2804	0.2258	0.1686	0.1096	0.0497	-0.0102	-0.0698	-0.1291	-0.188	-0.246	-0.3022	-0.3555	-0.4048	-0.4494	-0.4889	-0.5233	-0.5526	-0.5773	-0.5977	46ºN
48ºN	0.5473	0.5167	0.4809	0.4391	0.3912	0.3368	0.2763	0.2107	0.1413	0.07	-0.0015	-0.0722	-0.1422	-0.2117	-0.2796	-0.3439	-0.4026	-0.4543	-0.4985	-0.5354	-0.5656	-0.5899	-0.6092	-0.6241	48ºN
50ºN	0.6028	0.5798	0.551	0.5152	0.4712	0.4175	0.3532	0.2783	0.1942	0.1041	0.0128	-0.0759	-0.1628	-0.2493	-0.332	-0.4057	-0.4675	-0.517	-0.5555	-0.5849	-0.6069	-0.6231	-0.6348	-0.6429	50ºN
52ºN	0.6524	0.6403	0.624	0.6017	0.5713	0.5297	0.4728	0.3963	0.2976	0.178	0.0437	-0.0821	-0.1995	-0.319	-0.4206	-0.4963	-0.5491	-0.5849	-0.609	-0.6251	-0.6356	-0.6422	-0.6458	-0.6473	52ºN
54ºN	0.6809	0.6797	0.677	0.6724	0.6646	0.6516	0.6293	0.5891	0.5143	0.3816	0.215	-0.0103	-0.2831	-0.4537	-0.5413	-0.5861	-0.6101	-0.6234	-0.6307	-0.6343	-0.6356	-0.6353	-0.6338	-0.6313	54ºN
56ºN	0.6745	0.6775	0.6806	0.6838	0.6872	0.691	0.6952	0.7002	0.7061	0.7087	0.6009	0.1187	-0.3748	-0.5009	-0.5471	-0.5704	-0.5834	-0.5909	-0.5952	-0.5972	-0.5977	-0.597	-0.5955	-0.5933	56ºN
58ºN	0.6325	0.632	0.6302	0.6266	0.6203	0.6097	0.592	0.5627	0.5136	0.4317	0.2822	-0.0002	-0.2519	-0.3609	-0.4182	-0.4558	-0.482	-0.5005	-0.5137	-0.523	-0.5294	-0.5336	-0.5361	-0.5373	58ºN
60ºN	0.5682	0.5611	0.5512	0.5377	0.5193	0.4944	0.4608	0.4154	0.3541	0.2714	0.1617	0.0306	-0.097	-0.1985	-0.2719	-0.3249	-0.3641	-0.3938	-0.4165	-0.4339	-0.4474	-0.4577	-0.4655	-0.4712	60ºN
62ºN	0.497	0.4845	0.4688	0.4493	0.425	0.3949	0.358	0.3129	0.2584	0.1938	0.1201	0.0411	-0.0372	-0.1087	-0.1702	-0.2212	-0.263	-0.297	-0.3246	-0.3472	-0.3656	-0.3805	-0.3926	-0.4023	62ºN
64ºN	0.4286	0.4132	0.395	0.3734	0.3481	0.3185	0.2843	0.2452	0.2011	0.1525	0.1002	0.0461	-0.0077	-0.0592	-0.1067	-0.1493	-0.1867	-0.2193	-0.2473	-0.2712	-0.2917	-0.309	-0.3237	-0.336	64ºN
66ºN	0.3674	0.3512	0.3327	0.3117	0.288	0.2614	0.232	0.1996	0.1647	0.1275	0.0887	0.0491	0.0096	-0.0289	-0.0656	-0.0998	-0.1313	-0.1598	-0.1855	-0.2083	-0.2285	-0.2463	-0.2618	-0.2753	66ºN
68ºN	0.3146	0.2989	0.2816	0.2624	0.2414	0.2185	0.1939	0.1676	0.1398	0.1109	0.0811	0.051	0.0209	-0.0087	-0.0373	-0.0647	-0.0906	-0.1147	-0.137	-0.1574	-0.1761	-0.1929	-0.208	-0.2216	68ºN
70ºN	0.2697	0.2553	0.2398	0.2229	0.2049	0.1857	0.1654	0.1441	0.1219	0.0991	0.0758	0.0524	0.0289	0.0057	-0.0169	-0.0388	-0.0599	-0.0799	-0.0988	-0.1165	-0.1329	-0.1481	-0.1621	-0.1749	70ºN

Max values	0.6809	0.6797	0.6806	0.6838	0.6872	0.691	0.6952	0.7002	0.7061	0.7087	0.6009	0.1187	0.0289	0.0057	-0.0169	-0.0388	-0.0599	-0.0799	-0.0988	-0.1165	-0.1329	-0.1481	-0.1621	-0.1749
Min values	0.2685	0.2394	0.2099	0.1801	0.15	0.1197	0.0893	0.0588	0.0282	-0.0024	-0.033	-0.0821	-0.3748	-0.5009	-0.5471	-0.5861	-0.6101	-0.6234	-0.6307	-0.6343	-0.6356	-0.6422	-0.6458	-0.6473

Overall max	0.7087
Overall min	-0.6473

Appendix 8: The First Event tool in OxCal ver.3.5 (Bronk Ramsey 2000)

The following description of the last and first events is from the OxCal ver. 3.5 manual pages. They can be accessed at http://www.rlaha.ox.ac.uk/oxcal/oxcal.htm

First and Last Dated Events in a group (Bronk Ramsey 2000)

For a group of dates which may be constrained in some way by stratigraphic information it is useful to be able to obtain a probability distribution for the first and last members of the group.
It should be stressed that to use this to estimate the start and end of phases relies on the fact that the distribution of dated samples within the group are representative of the archaeological phase in question. If no objects have been recovered from the first century of a period no amount of statistical analysis can determine when that period began! Furthermore if there are no dated events prior to the period and a large number of dated events within it statistical analysis is liable to indicate that the period started earlier than it actually did simply because of the inevitable scatter in the measurements. These caveats are no more or less relevant to non-mathematical methods of analysis and simply imply good archaeological practice in bracketing periods.

The probability of being after a single event is given by:

$$r(t) = \int_{t'=-\infty}^{t'=t} p(t')dt'$$

and so if a group of events are independent the probability of being after all of them is given by:

$$r(t) = \prod_{i} \int_{t'=-\infty}^{t'=t} p_i(t')dt'$$

This is the distribution (normalised to a maximum of 1) returned by After. From this a distribution *r'(t)* can be calculated which gives a probability distribution for the last of the group of events:

$r'(t) = dr(t)/dt$

The probabilities of being before a group of events and a distribution for the first event of a phase can be similarly defined.

WARNING: These methods assume that the events are entirely independent.

Appendix 9: Sites used in the analysis of the first sheep spread in southern Africa.
Key:
Site Type 1 = Herder
Site Type 2 = Hunter-gatherer with stock
Site Type 3 = unsure

Date Grading:
Grade 1 = AMS
Grade 2 = Conventional (good)
Grade 3 = Convenional (poor)

Site name	No.	Latitude (decimal degrees)	Longitude (decimal degrees)	Date BP	SD	Sample ID	Sample material	Dating method	Date grading	Site Type
Die Kelders	1	34.53556	19.37583	1325	60	OxA-3860	Sheep bone	AMS	1	1
Die Kelders	1	34.53556	19.37583	1290	60	OxA-3961	Sheep bone	AMS	1	1
Elands Bay	2	32.275	18.27222	1120	85	Gak-4335	charcoal	Conv.	2	2
Boomplaas	3	33.38333	22.18333	1630	50	UW-337	charcoal	Conv.	3	1
Boomplaas	3	33.38333	22.18333	1510	75	UW-307	charcoal	Conv.	3	1
Boomplaas	3	33.38333	22.18333	1700	55	UW-338	charcoal	Conv.	3	1
Blombos	4	34.41667	21.21667	1960	50	OxA-4543	Sheep bone	Ams	1	2
Blombos	4	34.41667	21.21667	1880	55	OxA-4544	Sheep bone	Ams	1	2
Blombos	4	34.41667	21.21667	1840	50	Pta-6185	Charcoal	Conv.	1	2
Blombos	4	34.41667	21.21667	1900	40	Pta-6175	Marine shell	Conv.	1	2
Blombos	4	34.41667	21.21667	1880	50	Pta-6246	Marine shell	Conv.	1	2
Blombos	4	34.41667	21.21667	1940	50	Pta-6247	Marine shell	Conv.	1	2
Spoegrivier	5	30.49444	17.36667	1390	50	Pta-4753	Charcoal	Conv.	1	1
Spoegrivier	5	30.49444	17.36667	1920	40	Pta-4745	Charcoal	Conv.	1	1
Spoegrivier	5	30.49444	17.36667	2105	65	OxA-3862	Sheep bone	AMS	1	1
Jakkalsberg A/B	6	28.175	16.88889	1330	60	Pta-5958	Charcoal	Conv.	2	1
Jakkalsberg A/B	6	28.175	16.88889	1300	25	Pta-6100	Charcoal	Conv.	2	1
Jakkalsberg A/B	6	28.175	16.88889	1420	25	Pta-6122	Charcoal	Conv.	2	1
Jakkalsberg A/B	6	28.175	16.88889	1380	50	Pta-6101	Charcoal	Conv.	2	1
Kasteelberg B	7	32.80222	17.93556	190	40	Pta-3737	Charcoal	Conv.	2	1
Kasteelberg B	7	32.80222	17.93556	920	40	Pta-3742	Charcoal	Conv.	2	1
Kasteelberg B	7	32.80222	17.93556	880	50	Pta-3747	Charcoal	Conv.	2	1
Kasteelberg B	7	32.80222	17.93556	910	50	Pta-3787	Charcoal	Conv.	2	1
Kasteelberg B	7	32.80222	17.93556	1200	45	Pta-3994	Charcoal	Conv.	2	1
Kasteelberg B	7	32.80222	17.93556	1220	45	Pta-3998	Charcoal	Conv.	2	1
Kasteelberg B	7	32.80222	17.93556	1300	60	Pta-3995	Charcoal	Conv.	2	1
Kasteelberg A	8	32.80222	17.93556	1790	40	Pta-3461	Charcoal	Conv.	2	1
Kasteelberg A	8	32.80222	17.93556	1630	60	OxA-3864	Sheep bone	AMS	1	1
Kasteelberg A	8	32.80222	17.93556	1430	55	OxA-3865	Sheep bone	AMS	1	1
Witklip	9	32.91694	17.98333	500	50	Pta-5467	Charcoal	Conv.	2	2
Witklip	9	32.91694	17.98333	330	45	Pta-5467	Charcoal	Conv.	2	2
Witklip	9	32.91694	17.98333	1380	50	Pta-4608	Charcoal	Conv.	2	2
Witklip	9	32.91694	17.98333	1860	50	Pta-4609	Charcoal	Conv.	2	2
Byneskranskop	10	34.58333	19.46667	255	50	Pta-1864	Charcoal	Conv.	2	1
Byneskranskop	10	34.58333	19.46667	535	50	Pta-1866	Charcoal	Conv.	2	1
Byneskranskop	10	34.58333	19.46667	1880	50	Pta-1865	Charcoal	Conv.	3	1
Byneskranskop	10	34.58333	19.46667	1370	60	OxA-3863	Sheep bone	AMS	1	1
Nelson Bay Cave	11	34.10556	23.38056	1100	80	OxA-873	Sheep bone	AMS	1	2
Wonderwerk	12	27.84583	23.55528	1210	50	Pta-2779	Charcoal	Conv.	3	2
Jubilee Shelter	13	25.7	27.91667	310	50	OxA-6039	Sheep bone	AMS	1	2
Jubilee Shelter	13	25.7	27.91667	1840	50	WITS-1398	Charcoal	Conv.	3	2
Blinklipkop	14	28.30278	23.11667	80	35	Pta-2841	Grass	Conv.	2	1
Blinklipkop	14	28.30278	23.11667	260	45	Pta-2833	Charcoal	Conv.	2	1
Blinklipkop	14	28.30278	23.11667	280	50	Pta-2839	Charcoal	Conv.	2	1
Blinklipkop	14	28.30278	23.11667	830	45	Pta-3419	Charcoal	Conv.	2	1
Blinklipkop	14	28.30278	23.11667	1160	50	Pta-2840	Charcoal	Conv.	2	1
Blinklipkop	14	28.30278	23.11667	1150	40	Pta-2835	Charcoal	Conv.	2	1
/Ai Tomas	15	28.58333	17.08333	420	50	Pta-5458	Charcoal	Conv.	3	3
/Ai Tomas	15	28.58333	17.08333	330	45	Pta-5452	Charcoal	Conv.	3	3

Appendix 9: Sites used in the analysis of the first sheep spread in southern Africa.
Key:
Site Type 1 = Herder
Site Type 2 = Hunter-gatherer with stock
Site Type 3 = unsure

Date Grading:
Grade 1 = AMS
Grade 2 = Conventional (good)
Grade 3 = Convenional (poor)

Site name	No.	Latitude (decimal degrees)	Longitude (decimal degrees)	Date BP	SD	Sample ID	Sample material	Dating method	Date grading	Site Type
Hawston	16	34.41667	19.13333	1860	60	Pta-834	Charcoal	Conv.	3	3
Scott's Cave	17	33.73333	25.71667	360	80	Y-1425	Charcoal	Conv.	3	2
Dikbosch 2	18	28.65222	23.9025	570	45	Pta-1507	Charcoal	Conv.	3	2
Tortoise Cave	19	32.30056	18.38611	1580	50	Pta-3309	Charcoal	Conv.	3	2
Tortoise Cave	19	32.30056	18.38611	1620	50	Pta-3310	Charcoal	Conv.	3	2
Tortoise Cave	19	32.30056	18.38611	1610	50	Pta-3311	Charcoal	Conv.	3	2
Tortoise Cave	19	32.30056	18.38611	1680	50	Pta-3312	Charcoal	Conv.	3	2
Limerock 1/2	20	28.55028	24.00028	1620	50	Pta-1621	Charcoal	Conv.	3	2?
Limerock 1/2	20	28.55028	24.00028	1720	40	Pta-2095	Charcoal	Conv.	3	2?
Drie Susters Main	21	33.10083	18.00056	1050	60	Pta-5478	Charcoal	Conv.	3	2
Pearly Beach area	22	34.66667	19.5	1680	50	Pta-1069	Charcoal	Conv.	3	1?
Smitswinkelbaai	23	34.2667	18.46667	1175	35	Pta-2200	Charcoal	Conv.	3	1
Steenbokfontein	24	32.16167	18.33333	2200	60	Pta-6136	charcoal	Conv.	3	2
Steenbokfontein	24	32.16167	18.33333	2200	50	Pta-6424	charcoal	Conv.	3	2
Lame Sheep Shelter	25	31.41667	24.66667	1350	50	Pta-6303	Charcoal	Conv.	3	3
Blydefontein	27	30.83333	25.33333	1255	109	Smu-1925	Charcoal	Conv.	3	3
Blydefontein	27	30.83333	25.33333	1305	31	Smu-1850	Charcoal	Conv.	3	3